VISUALIZING
ENVIRONMENTAL SCIENCE

VISUALIZING
ENVIRONMENTAL
SCIENCE

THIRD EDITION

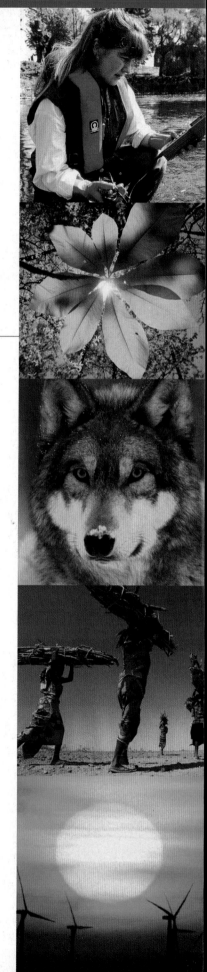

LINDA R. BERG

MARY CATHERINE HAGER

DAVID M. HASSENZAHL

WILEY

In collaboration with
The National Geographic Society

Credits

EXECUTIVE VP AND PUBLISHER Kaye Pace
SENIOR EDITOR Rachel Falk
PROJECT EDITOR Merillat Staat
WILEY VISUALIZING PROJECT EDITOR Brian B. Baker
ASSISTANT EDITOR Jenna Paleski
DIRECTOR OF DEVELOPMENT Barbara Heaney
MANAGER, PRODUCT DEVELOPMENT Nancy Perry
ASSOCIATE DIRECTOR, MARKETING Jeffrey Rucker
MARKETING MANAGER Kristine Ruff
SENIOR PRODUCTION MANAGER Micheline Frederick
SENIOR MEDIA EDITOR Linda Muriello
MEDIA SPECIALIST Daniela DiMaggio
CREATIVE DIRECTOR Harry Nolan
COVER DESIGN Harry Nolan
INTERIOR DESIGN Jim O'Shea
SENIOR PHOTO EDITOR Elle Wagner
PHOTO RESEARCHER Stacy Gold/National Geographic Society
ART DEVELOPER Elizabeth Morales
SENIOR ILLUSTRATION EDITOR Sandra Rigby
PRODUCTION SERVICES Furino Production

COVER CREDITS Main: Tyrone Turner/NG Image Collection, (left): MICHAEL MELFORD/ NG Image Collection, (center left): HIROYA MINAKUCHI/MINDEN PICTURES/ NG Image Collection, (center): JOHAN ELZENGA/FOTO NATURA/MINDEN PICTURES/ NG Image Collection, (center right): BRIAN J. SKERRY/NG Image Collection, (right): JUSTIN GUARIGLIA/NG Image Collection

This book was set in New Baskerville by PreMediaGlobal, printed and bound by Quad/Graphics. The cover was printed by Quad/Graphics.

ISBN-13: 978-0470-56918-4
BRV ISBN: 978-0470-91744-2

Printed in the United States of America
10 9 8 7 6 5 4

How is Wiley Visualizing different?

Wiley Visualizing differs from competing textbooks by uniquely combining three powerful elements: a visual pedagogy, integrated with comprehensive text, the use of authentic situations and issues from the National Geographic Society collections, and the inclusion of interactive multimedia in the Wiley Resource Kit. Together these elements deliver a level of rigor in ways that maximize student learning and involvement. Each key concept and its supporting details have been analyzed and carefully crafted to maximize student learning and engagement.

(1) Visual Pedagogy. Wiley Visualizing is based on decades of research on the use of visuals in learning (Mayer, 2005).[1] Using the Cognitive Theory of Multimedia Learning, which is backed up by hundreds of empirical research studies, Wiley's authors select visualizations for their texts that specifically support students' thinking and learning—for example, the selection of relevant materials, the organization of the new information, or the integration of the new knowledge with prior knowledge. Visuals and text are conceived and planned together in ways that clarify and reinforce major concepts while allowing students to understand the details. This commitment to distinctive and consistent visual pedagogy sets Wiley Visualizing apart from other textbooks.

(2) Authentic Situations and Problems. Through Wiley's exclusive publishing partnership with National Geographic, *Visualizing Environmental Science 3e* has benefited from National Geographic's more than century-long recording of the world and offers an array of remarkable photographs, maps, media, and film from the National Geographic Society collections. These authentic materials immerse the student in real-life issues and problems in environmental science, thereby enhancing motivation, learning, and retention (Donovan & Bransford, 2005).[2] These authentic situations, using high-quality materials from the National Geographic Society collections, are unique to Wiley Visualizing.

(3) Interactive Multimedia. Wiley Visualizing is based on the understanding that learning is an active process of knowledge construction. *Visualizing Environmental Science 3e* is therefore tightly integrated with interactive multimedia activities provided in the Wiley Resource Kit.

Wiley Visualizing is designed as a natural extension of how we learn

Visuals, comprehensive text, and learning aids, are integrated to display facts, concepts, processes, and principles more effectively than words alone can. To understand why the visualizing approach is effective, it is first helpful to understand how we learn.

1. Our brain processes information using two channels: visual and verbal. Our *working memory*, also called *short-term memory*, holds information that our minds process as we learn. In working memory we begin to make sense of words and pictures, and build verbal and visual models of the information.

2. When the verbal and visual models of corresponding information are connected in working memory, we form more comprehensive, or integrated, mental models.

3. After we link these integrated mental models to our prior knowledge, which is stored in our *long-term memory*, we build even stronger mental models. When an integrated mental model is formed and stored in long-term memory, real learning begins.

The effort our brains put forth to make sense of instructional information is called *cognitive load*. There are two kinds of cognitive load: productive cognitive load, such as when we're engaged in learning or exert positive effort to create mental models; and unproductive cognitive load, which occurs when the brain is trying to make sense of needlessly complex content or when information is not presented well. The learning process can be impaired when the amount of information to be processed exceeds the capacity of working memory. Well-designed visuals and text with effective pedagogical guidance can reduce the unproductive cognitive load in our working memory.

---

[1]Mayer, R. E. (Ed.) (2005.) *The Cambridge Handbook of Multimedia Learning*. New York: Cambridge University Press.

[2]Donovan, M. S., & Bransford, J. (Eds.) (2005). *How Students Learn: Science in the Classroom*. Washington, DC: The National Academy Press. Available online at http://www.nap.edu/openbook.php?record_id=11102&page=1

Wiley Visualizing is designed for engaging and effective learning

The visuals and text in *Visualizing Environmental Science 3e* are specially integrated to present complex processes in clear steps and with clear representations, organize related pieces of information, and integrate related information with one another. This approach, along with the use of interactive multimedia, minimizes unproductive cognitive load and helps students engage with the content. When students are engaged, they're reading and learning, which can lead to greater knowledge and academic success.

Research shows that well-designed visuals, integrated with comprehensive text, can improve the efficiency with which a learner processes information. In this regard, SEG Research, an independent research firm, conducted a national, multisite study evaluating the effectiveness of Wiley Visualizing. Its findings indicate that students using Wiley Visualizing products (both print and multimedia) were more engaged in the course, exhibited greater retention throughout the course, and made significantly greater gains in content area knowledge and skills, as compared to students in similar classes that did not use Wiley Visualizing.[3]

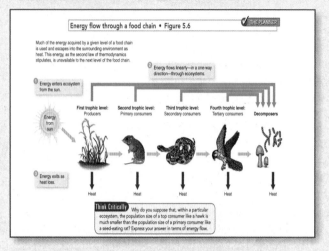

Figure 2: Health Effects of Several Major Air Pollutants (Table 8.1) This matrix visually organizes abstract information to reduce cognitive load.

Figure 3: Energy Flow through a Food Chain (Fig. 5.6) Textual elements have been physically integrated with the visual elements. This eliminates split attention (when we must divide our attention between several sources of different information). The arrows visually display processes, easing the way we recognize relationships.

Figure 1: Timeline of Selected Environment Events (Fig. 3.8) This timeline visually organizes information to integrate related events, dates, and figures. It links new information to prior knowledge.

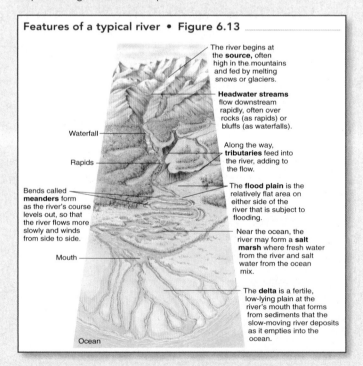

Figure 4: Features of a Typical River (Fig. 6.13) This illustration, by bolding or italicizing words, directs learners' attention to important features, characteristics, steps, or relationships.

[3]SEG Research (2009). Improving Student-Learning with Graphically-Enhanced Textbooks: A Study of the Effectiveness of the Wiley Visualizing Series. Available online at www.segmeasurement.com/

What is the Wiley Visualizing chapter organization?

Student engagement requires more than just providing visuals, text, and interactivity—it entails motivating students to learn. Student engagement can be behavioral, cognitive, social, and/or emotional. It is easy to get bored or lose focus when presented with large amounts of information, and it is easy to lose motivation when the relevance of the information is unclear. Wiley Visualizing reorganizes course content into manageable learning objectives and relate it to everyday life.

The content in Wiley Visualizing is organized in learning modules. Each module has a clear instructional objective, one or more examples, and an opportunity for assessment. These modules are the building blocks of Wiley Visualizing.

Each Wiley Visualizing chapter engages students from the start

Chapter opening text and visuals introduce the subject and connect the student with the material that follows.

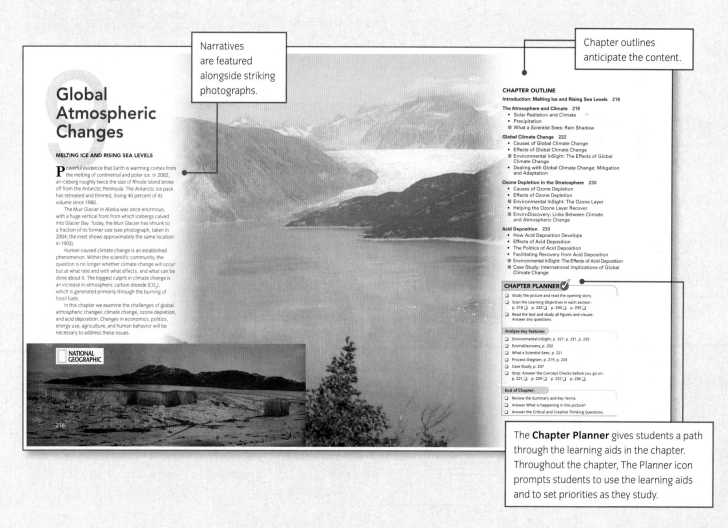

Narratives are featured alongside striking photographs.

Chapter outlines anticipate the content.

The **Chapter Planner** gives students a path through the learning aids in the chapter. Throughout the chapter, The Planner icon prompts students to use the learning aids and to set priorities as they study.

Wiley Visualizing media engages students in the chapter

Learning Objectives at the start of each section indicate in behavioral terms the concepts that students are expected to master while reading the section.

Human Impacts on the Envi

LEARNING OBJECTIVES

1. **Distinguish** among highly developed countries, moderately developed countries, and less developed countries.
2. **Relate** human population size to natural resources and resource consumption.
3. **Distinguish** between people overpopulation and consumption overpopulation.
4. **Describe** the three factors that are most important in determining human impact on the environment.

The satellite photograph in **Figure 1.1a** is a portrait of about 450 million people. The tiny specks of light represent cities, and the great metropolitan areas, such as New York City along the northeastern seacoast, are ablaze with light.

Earth's central environmental problem, which links all others together, is that there are many people, and

purch
of the
povert
illitera
water
T
of the
that a
Popul
in the
2.6 chi
to co
expert
popula
7.7 bil
fertility
No
people
one of

Process Diagrams provide in-depth coverage of processes correlated with clear, step-by-step narrative, enabling students to grasp important topics with less effort.

WILEY RESOURCE KIT

Interactive Process Diagrams provide additional visual examples and descriptive narration of a difficult concept, process, or theory, allowing the students to interact and engage with the content. Many of these diagrams are built around a specific feature such as a Process Diagram. Look for them in the WIley Resource Kit when you see this icon.

PROCESS DIAGRAM

The carbon cycle • Figure 5.8

✓ THE PLANNER

The movement of carbon between the abiotic environment (the atmosphere and ocean) and living organisms is known as the carbon cycle. Because proteins, carbohydrates, and other living molecules contain carbon, the process is essential to life.

Air (CO₂)

Animal and plant respiration
Soil microorganism respiration
Decomposition (involves respiration)
Photosynthesis by land plants
Combustion of coal, oil, natural gas, and wood
Chemical compounds in living organisms
Erosion of limestone
Dissolved CO₂ in water
Carbon incorporated into shells of marine organisms
Soil
Partly decomposed plant remains (ancient trees)
Remains of ancient unicellular marine organisms
Coal
Natural gas
Coal
Oil

Think Critically How would removing a large forest affect the carbon cycle? What would be the additional effects if the wood from those trees were used as fuel?

WILEY RESOURCE KIT

Think Critically questions let students analyze the material and develop insights into essential concepts.

Environmental InSight features are multipart visual sections that focus on a key concept or topic in the chapter, exploring it in detail or in broader context using a combination of photos, figures, and data.

WHAT A SCIENTIST SEES

Ocean Warming and Coral Bleaching

a. Bleached coral off the coast of Panama. Scientists have linked coral bleaching to ocean warming. Warmer than usual temperatures stress the coral animals, causing them to lose their zooxanthellae. Without their algae, the corals can't get enough food, and they die.

b. This 1920–2009 time series of annual global mean temperature changes of the ocean surface indicates that the ocean has warmed, particularly during the past 25 years. Most warming has occurred in shallow waters where corals live. Mean temperature anomaly data relative to 1901–2000.

Interpreting Data If this warming trend continues, what will the mean temperature change be by 2050?

Human Impacts on the Ocean 287

Interpreting Data questions help students evaluate graphs, figures, and data sets.

Environmental InSight The effects of global climate change • Figure 9.11

✓ THE PLANNER

a. Warmer temperatures in Antarctica threaten the Adélie penguin's food supply and reduce its reproductive success.

b. A coqui tree frog in Puerto Rico. These once-ubiquitous little frogs have become rarer, an indirect casualty of climate change.

1995

Not corrosive

2100

Acid levels

c. Ocean warming and acidification stress corals, causing them to become bleached. Photographed near the Maldives in the Indian Ocean.

How might plants and animals where you live be affected by a 2°C increase in the lowest winter temperature each year? The same increase in the highest summer temperature?

Global-Local questions help students apply environmental conditions in distant locations to the places where they live.

What a Scientist Sees highlights a concept or phenomenon that would stand out to a professional in the field. Photos and figures are used to compare how a nonscientist and a scientist see the issues, and students apply their observational skills to answer questions.

EnviroDiscovery is an essay that explores an area or topic of relevance. Students synthesize the material for greater understanding.

population: A group of individuals of the same species in a particular area at a specific time.

Driven by instructor feedback on the most important topics for students to understand about environmental science, **Environmental Science Basics** provides a suite of animated concepts and tutorials to give students a solid grounding in the key basic environmental concepts. Concepts ranging from global climate change to sustainable agriculture are presented across 21 modules in easy-to-understand language.

Global Locator Maps, prepared specifically for this book by the National Geographic Society cartographers, help students visualize where the area depicted in the photo is situated on Earth.

Students analyze and engage in global and local issues with **National Geographic maps**. They master the location of the issues through critical thinking questions. There are 46 National Geographic maps in *Visualizing Environmental Science 3e* and its ancillaries.

what we
n and its
kinds of
npact on
e respon-
on? How
mentally
ess such
ion mak-
policies
environ-
l should
humann-
ne ability
riches of

onsump-
s impact
Success

in achieving sustainability in population size and consumption will require the cooperation of all the world's peoples.

CONCEPT CHECK 🛑 STOP

1. **How** do highly developed countries, moderately developed countries, and less developed countries differ regarding population growth and per person incomes?

2. **How** is human population growth related to natural resource depletion and environmental degradation?

3. **What** is the difference between people overpopulation and consumption overpopulation?

4. **According** to the IPAT equation, what three factors are most important in determining environmental impact?

Concept Check questions at the end of each section allow students to test their comprehension of the learning objectives.

RESOURCE KIT

Students think critically and solve the problems of real-life situations with a rich collection of videos from a variety of sources, including 50 **National Geographic videos** from their award-winning collection. Each video is linked to the text, and questions allow students to solve problems online. Videos are also available as lecture launcher PowerPoint presentations designed for in-class viewing and can be easily integrated into existing presentations.

Student understanding is assessed at different levels

Wiley Visualizing offers students lots of practice material for assessing their understanding of each study objective. Students know exactly what they are getting out of each study session through immediate feedback and coaching.

Case Studies use a critical-thinking approach to walk students through the particular characteristics defining a real-life environmental challenge. The case studies present a problem and in some cases describe attempted solutions. Students then explore the information necessary to appreciate the significance of the featured example.

Critical and Creative Thinking Questions

1. What two climate factors are most important in determining an area's characteristic biome?

2. In which biome do you live? Where would you place your biome in the figure below? How would that compare with your placement of the biome in northern Siberia or the biome dominating northern Africa and Saudi Arabia?

3. What environmental factors are most important in determining the kinds of organisms found in aquatic environments?

4. Distinguish between freshwater wetlands and estuaries and between flowing-water and standing-water ecosystems.

5. Name and compare temperate and tropical estuaries. What types of plants are characteristic of each?

6. During the mating season, male giraffes slam their necks together in fighting bouts to determine which male is stronger and can therefore mate with females. Explain how the long necks of giraffes may have evolved, using Darwin's theory of evolution by natural selection.

7. Explain why evolution, by definition, cannot take place within one individual and during that individual's life span.

8. Describe the process and stages of ecological succession.

9. Which type of ecological succession might occur in a forest where a stand of trees has been cleared? Explain your choice of primary and secondary succession.

10. Although most salamanders have four legs, the aquatic salamander shown below resembles an eel. It lacks hind limbs and has very tiny forelimbs. Propose a hypothesis to explain how these salamanders evolved according to Darwin's theory of natural selection.

11. How could you test the hypothesis you proposed in question 10? What type of evidence might you produce?

12. Which biome discussed in this chapter is depicted by the information in the graph below? Explain your answer.

CASE **STUDY**

The New Orleans Disaster

Hurricane Katrina, which hit the north-central Gulf Coast in August 2005, was one of the most devastating storms in U.S. history. It produced a storm surge that caused severe damage to New Orleans as well as to other coastal cities and towns in the region. The high waters caused levees and canals to fail, flooding 80 percent of New Orleans and many nearby neighborhoods.

Most people are aware of the catastrophic loss of life and property caused by Katrina. Here we focus on how humans altered the geography and geology of the New Orleans area in ways that exacerbated the storm damage.

Over the years, engineers constructed a system of canals to aid navigation and a system of levees to control flooding because the city is at or below sea level. The canals allowed salt water to intrude and kill the freshwater marsh vegetation. The levees prevented the deposition of sediments that remain behind after floodwaters subside (The sediments are now deposited in the Gulf of Mexico.) Under natural conditions, these sediments replenish and maintain the delta, building up coastal wetlands.

As the city has grown, new development has [encroached] on wetlands—bayous, waterways, and marshes—[that were] drained and filled in. Before their destruction, these wetlands provided some protection against flood [storm] surges. We are not implying that had Louisiana's [wetlands been] intact, New Orleans would not have suffered any [damage from a] hurricane of Katrina's magnitude. However, had the [wetlands] been largely unaltered, they would have moderate[d the] damage by absorbing much of the water from the [storm surge.]

Another reason that Katrina devastated New [Orleans is that] the city has been subsiding (sinking) for many yea[rs, in part] because New Orleans is built on unconsolidated se[diments (with] bedrock underneath). Many wetlands scientists a[ttribute] this subsidence to the extraction of the area's rich [supply of] underground natural resources—groundwater, oi[l, and natural] gas. As these resources are removed, the land co[mpacts,] lowering the city. New Orleans and nearby coasta[l areas are] subsiding an average of 6 mm each year (see imag[e). At the] same time, the sea level has been rising an averag[e of] 2.5 mm per year due to human-induced changes in climate.

Satellite image of flooding in New Orleans following Hurricane Katrina

Along the left (west) side is a levee from Lake Ponchartrain (top) that failed so that water inundated the New Orleans area east of the levee. Areas on the far left top remained dry. Part of the Mississippi River is shown at lower center.

Global Locator

New Orleans

NATIONAL GEOGRAPHIC

Critical and Creative Thinking Questions challenge students to think more broadly about chapter concepts. The level of these questions ranges from simple to advanced; they encourage students to think critically and develop an analytical understanding of the ideas discussed in the chapter.

The **Summary** revisits each learning objective, with relevant accompanying images taken from the chapter; these visual clues reinforce important elements.

4 **Environmental Economics 62**

1. **Economics** is the study of how people use their limited resources to try to satisfy their unlimited wants. Economies depend on the natural environment as sources for raw materials and sinks for waste products. Both sources and sinks contribute to **natural capital**, which is Earth's resources and processes that sustain living organisms, including humans. Natural capital includes minerals, forests, soils, water, clean air, wildlife, and fisheries.

2. **National income accounts** are measures of the total income of a nation's goods and services for a given year. An **external cost** is a harmful environmental or social cost that is borne by people not directly involved in buying or selling a product. National income accounts are incomplete estimates of national economic performance because they do not include both natural resource depletion and the environmental costs of economic activities. Many economists, government planners, and scientists support more comprehensive income accounting that includes these estimates.

3. From an economic point of view, the appropriate amount of pollution is a trade-off between harm to the environment and inhibition of development. The **marginal cost of pollution** is the added cost of an additional unit of pollution. The **marginal cost of pollution abatement** is the added cost of reducing one unit of a given type of pollution. Economists think the use of resources for pollution abatement should increase only until the cost of abatement equals the cost of the pollution damage. This results in the **optimum amount of pollution**—the amount of pollution that is economically most desirable.

4. **Incentive-based regulations** take advantage of economic markets to reduce environmental damage. **Environmental taxes** require polluters to pay an amount equal to the harm they cause. **Tradable permit** systems limit the total amount of a pollutant that can be released, allowing people to buy and sell rights to emit and reduce emissions as inexpensively as possible.

Key Terms

- biocentric preservationist 53
- command and control regulation 67
- cost–benefit diagram 66
- external c[ost]
- incentive-based regulation 67
- marginal cost of pollution 65
- marginal cost of pollution abatement 66
- natur[al]
- optim[um]
- utilita[rian]

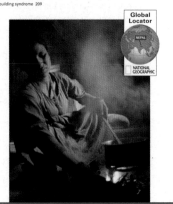

Global Locator

NEPAL

NATIONAL GEOGRAPHIC

What is happening in this picture?

- This Nepalese woman is preparing a meal inside a poorly ventilated room. Cooking meals can take up many hours each day. In this picture, where is the smoke most dense? What does this imply for the health of women, who do much of the cooking in developing countries?

- Young children in developing countries tend to spend much of their time with their mothers; in fact, an infant may be strapped to the mother while she cooks. Explain what sorts of health effects you might expect these children to suffer as a result.

214 CHAPTER 8 Air and Air Pollution

WILEY RESOURCE KIT

Students go to 9 different places around the world on all new **Virtual Field Trips** and gain a better understanding of the environment and our impact on it. Through these video-based field trips students gain virtual on-the-ground experience using their Wiley Resource Kit.

What is happening in this picture? presents an uncaptioned photograph that is relevant to a chapter topic and illustrates a situation students are not likely to have encountered previously. The photograph is paired with questions that ask the students to describe and explain what they can observe in the photo based on what they have learned.

What is the organization of this book?

We begin *Visualizing Environmental Science 3e* with an introduction of the environmental dilemmas we face in our world today, emphasizing particularly how unchecked population growth and economic inequity complicate our ability to solve these problems. We stress that solutions rest in creativity and diligence at all levels, from individual commitment to international cooperation. Indeed, a key theme integrated throughout the third edition is the local to global scales of environmental science. We offer concrete suggestions that students can adopt to make their own difference in solving environmental problems, and we explain the complications that arise when solutions are tackled on a local, regional, national, or global scale.

Yet *Visualizing Environmental Science 3e* is not simply a checklist of "to do" items to save the planet. In the context of an engaging visual presentation, we offer solid discussions of such critical environmental concepts as sustainability, conservation and preservation, and risk analysis. We weave the threads of these concepts throughout our treatment of ecological principles and their application to various ecosystems, the impacts of human population change, and the problems associated with our use of the world's resources. We particularly instruct students in the importance of ecosystem services to a functioning world, and the threats that restrict our planet's ability to provide such services.

This text is intended to provide introductory content primarily for nonscience undergraduate students. The accessible format of *Visualizing Environmental Science 3e*, coupled with our assumption that students have little prior knowledge of ecosystem ecology, allows students to easily make the transition from jumping-off points in the early chapters to the more complex concepts they encounter later. With its interdisciplinary presentation, which mirrors the nature of environmental science itself, this book is appropriate for use in one-semester and one-quarter environmental science courses offered by a variety of departments, including environmental studies and sciences, biology, ecology, agriculture, earth sciences, and geography.

Visualizing Environmental Science 3e is organized around the premise that humans are inextricably linked to the world's environmental dilemmas. We must address these issues as we use Earth's resources and seek to avoid future disasters so often predicted in the media.

- Chapters 1 through 4 establish the groundwork for understanding the environmental issues we face, how environmental sustainability and human values play a critical role in addressing these issues, how the environmental movement developed over time and was shaped by economics, and how environmental threats from many sources create health hazards that must be evaluated.

- Chapters 5, 6, and 7 present the intricacies of ecological concepts in a human-dominated world, including energy flow and the cycling of matter through ecosystems, and the various ways that species interact and divide resources. Gaining familiarity with these concepts allows students to better appreciate the variety of terrestrial and aquatic ecosystems that we then introduce, and to develop a richer understanding of the implications of human population change for the environment.

- The remaining 11 chapters deal with the world's resources as we use them today and as we assess their availability and impacts for the future. These issues cover a broad spectrum, including the sources and effects of air pollution, climate and global atmospheric change, freshwater resources, causes and effects of water pollution, the ocean and fisheries, mineral and soil resources, land resources, agriculture and food resources, biological resources, solid and hazardous waste, and nonrenewable and renewable energy resources. Recognizing the importance of the global ocean to environmental issues, we are particularly pleased to dedicate an entire chapter to a discussion of ocean processes and resources.

New to this edition

This edition of *Visualizing Environmental Science* welcomes a new author, Dr. David M. Hassenzahl. Dr. Hassenzahl is an internationally recognized expert on sustainability and risk analysis whose research focuses on how science, experts, and uncertainty impact environmental decision making. His research and practice in teaching and learning for higher education include numerous publications, grants, and awards, and he has spoken on risk and climate change education on three continents. His extensive knowledge of energy, climate change, policy, and human health brings additional depth and insight to an already strong author team.

Several features and topics are new to this third edition. To actively engage students with the content, three types of questions have been paired with graphics and images throughout the book. "Think Critically" questions encourage students to develop insights into essential ideas. "Global-Local" questions relate environmental conditions in distant locations to the places where students live. "Interpreting Data" questions develop students' skills as they negotiate graphs, figures, and data sets. A few examples of new material in this edition include:

- A new EnviroDiscovery on an emerging urban feature: Green Roofs

- A chapter opener on drought in Australia

- Case studies on the environmental impacts of China's Three Gorges Dam and Loess Plateau

- A Process Diagram explaining and providing examples of Environmental Impact Analyses

Finally, recognizing the educational value of integrating text with graphics and imagery, we have greatly increased the number and quality of process diagrams and have carefully revised our art program, layout, and design to provide students with a visually stunning image-based learning experience.

Also available

Environmental Science: Active Learning Laboratories and Applied Problem Sets, 2e by Travis Wagner and Robert Sanford, both of the University of Southern Maine, is designed to introduce environmental science students to the broad, interdisciplinary field of environmental science. It presents specific labs that use natural and social science concepts and encourages a "hands-on" approach to evaluating the impacts from the environmental/human interface. The laboratory and homework activities are designed to be low cost and to reflect a sustainable approach in both practice and theory. *Environmental Science: Active Learning Laboratories and Applied Problem Sets, 2e* is available as a stand-alone or in a package with *Visualizing Environmental Science 3e*. Contact your Wiley representative for more information.

EarthPulse. Utilizing full-color imagery and National Geographic photographs, *EarthPulse* takes you on a journey of discovery covering topics such as *The Human Condition, Our Relationship with Nature, and Our Connected World*. Illustrated by specific examples, each section focuses on trends affecting our world today. Included are extensive full-color world and regional maps for reference. *EarthPulse* is available only in a package with *Visualizing Environmental Science 3e*. Contact your Wiley representative for more information or visit www.wiley.com/college/earthpulse.

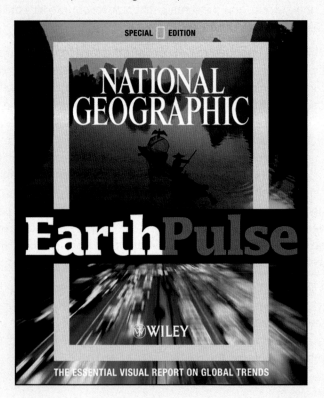

How does Wiley Visualizing support instructors?

Wiley Visualizing Site WILEY VISUALIZING™ | NATIONAL GEOGRAPHIC

The Wiley Visualizing site hosts a wealth of information for instructors using Wiley Visualizing, including ways to maximize the visual approach in the classroom and a white paper titled "How Visuals Can Help Students Learn," by Matt Leavitt, instructional design consultant. You can also find information about our relationship with the National Geographic Society and other texts published in our program. Visit Wiley Visualizing at www.wiley.com/college/visualizing.

Wiley Custom Select WILEY Custom LEARNING SOLUTIONS

Wiley Custom Select gives you the freedom to build your course materials exactly the way you want them. Offer your students a cost-efficient alternative to traditional texts. In a simple three-step process create a solution containing the content you want, in the sequence you want, delivered how you want. Visit Wiley Custom Select at http://customselect.wiley.com.

PowerPoint Presentations (available in the Wiley Resource Kit and on the book companion site)

A complete set of highly visual PowerPoint presentations—one per chapter—by Barbara Nemeth Rhode, Woodland Community College, is available online and in the Wiley Resource Kit to enhance classroom presentations. Tailored to the text's topical coverage and learning objectives, these presentations are designed to convey key text concepts, illustrated by embedded text art. Lecture Launcher PowerPoints also offer embedded links to videos to help introduce classroom discussions with short, engaging video clips.

Test Bank (available in the Wiley Resource Kit and on the book companion site)

The visuals from the textbook are also included in the Test Bank by Shamili Sandiford, College of DuPage. The Test Bank has approximately 1,100 test items, with at least 25 percent of them incorporating visuals from the book. The test items include multiple-choice and essay questions testing a variety of comprehension levels. The test bank is available online in MS Word files, as a Computerized Test Bank, and within the Wiley Resource Kit. The easy-to-use test-generation program fully supports graphics, print tests, student answer sheets, and answer keys. The software's advanced features allow you to produce an exam to your exact specifications.

Instructor's Manual (available in the Wiley Resource Kit and on the book companion site)

The Instructor's Manual includes over 50 creative ideas for in-class activities by co-author David M. Hassenzahl of the Chatham University and Jody Terrell of Texas Woman's University. It also includes answers to Critical and Creative Thinking questions and Concept Check questions.

Guidance is also provided on how to maximize the effectiveness of visuals in the classroom.

1. **Use visuals during class discussions or presentations.** Point out important information as the students look at the visuals, to help them integrate separate visual and verbal mental models.

2. **Use visuals for assignments and to assess learning.** For example, learners could be asked to identify samples of concepts portrayed in visuals.

3. **Use visuals to encourage group activities.** Students can study together, make sense of, discuss, hypothesize, or make decisions about the content. Students can work together to interpret and describe the diagram, or use the diagram to solve problems, conduct related research, or work through a case study activity.

4. **Use visuals during reviews.** Students can review key vocabulary, concepts, principles, processes, and relationships displayed visually. This recall helps link prior knowledge to new information in working memory, building integrated mental models.

5. **Use visuals for assignments and to assess learning.** For example, learners could be asked to identify samples of concepts portrayed in visuals.

6. **Use visuals to apply facts or concepts to realistic situations or examples.** For example, a familiar photograph, such as Earth from space, can illustrate key information about the thinness of the atmosphere and the distribution of land and water across the planet.

Image Gallery

All photographs, figures, maps, and other visuals from the text are online and in the WIiley Resource Kit and can be used as you wish in the classroom. These online electronic files allow you to easily incorporate images into your PowerPoint presentations as you choose, or to create your own handouts.

Wiley Faculty Network

The Wiley Faculty Network (WFN) is a global community of faculty, connected by a passion for teaching and a drive to learn, share, and collaborate. Their mission is to promote the effective use of technology and enrich the teaching experience. Connect with the Wiley Faculty Network to collaborate with your colleagues, find a mentor, attend virtual and live events, and view a wealth of resources all designed to help you grow as an educator. Visit the Wiley Faculty Network at www.wherefacultyconnect.com.

How has Wiley Visualizing been shaped by contributors?

Wiley Visualizing would not have come about without lots of people, each of whom played a part in sharing their research and contributing to this new approach.

National Geographic Society

Visualizing Environmental Science 3e offers an array of remarkable photographs, maps, illustrations, multimedia, and film from the National Geographic Society collections. Students using the book benefit from the rich, fascinating resources of National Geographic.

National Geographic School Publishing performed an invaluable service in fact-checking *Visualizing Environmental Science 3e*. They have verified every fact in the book with two outside sources, to ensure that the text is accurate and up-to-date. This kind of fact-checking is rare in textbooks and unheard-of in most online media.

National Geographic Image Collection provided access to National Geographic's award-winning image and illustrations collection to identify the most appropriate and effective images and illustrations to accompany the content. Each image and illustration has been chosen to be instructive, supporting the processes of selecting, organizing, and integrating information, rather than being merely decorative.

National Geographic Digital Media TV enabled the use of National Geographic videos to accompany *Visualizing Environmental Science 3e* and enrich the text. Available for each chapter are video clips that illustrate and expand on a concept or topic to aid student understanding.

National Geographic Maps Group provided access to National Geographic's extensive map collection, or their team of cartographers designed new maps for the text.

Academic Research Consultants

Richard Mayer, Professor of Psychology, UC Santa Barbara. Mayer's *Cognitive Theory of Multimedia Learning* provided the basis on which we designed our program. He continues to provide guidance to our author and editorial teams on how to develop and implement strong, pedagogically effective visuals and use them in the classroom.

Jan L. Plass, Professor of Educational Communication and Technology in the Steinhardt School of Culture, Education, and Human Development at New York University. Plass co-directs the NYU Games for Learning Institute and is the founding director of the CREATE Consortium for Research and Evaluation of Advanced Technology in Education.

Matthew Leavitt, Instructional Design Consultant, advises the Visualizing team on the effective design and use of visuals in instruction and has made virtual and live presentations to university faculty around the country regarding effective design and use of instructional visuals.

Independent Research Studies

SEG Research, an independent research and assessment firm, conducted a national, multi-site effectiveness study of students enrolled in entry-level college Psychology and Geology courses. The study was designed to evaluate the effectiveness of Wiley Visualizing. You can view the full research paper at www.wiley.com/college/visualizing/huffman/efficacy.html

Instructor and Student Contributions

Throughout the process of developing the concept of guided visual pedagogy for Wiley's Visualizing, we benefited from the comments and constructive criticism provided by the instructors and colleagues listed below. We offer our sincere appreciation to these individuals for their helpful reviews and general feedback:

Reviewers, Focus Group Participants, and Survey Respondents

James Abbott, Temple University
Melissa Acevedo, Westchester Community College
Shiva Achet, Roosevelt University
Denise Addorisio, Westchester Community College
Dave Alan, University of Phoenix
Sue Allen-Long, Indiana University Purdue
Robert Amey, Bridgewater State College
Nancy Bain, Ohio University
Corinne Balducci, Westchester Community College
Steve Barnhart, Middlesex County Community College
Stefan Becker, University of Washington—Oshkosh
Callan Bentley, Northern Virginia Community College
Valerie Bergeron, Delaware Technical & Community College
Andrew Berns, Milwaukee Area Technical College
Gregory Bishop, Orange Coast College
Rebecca Boger, Brooklyn College
Scott Brame, Clemson University
Joan Brandt, Central Piedmont Community College
Richard Brinn, Florida International University
Jim Bruno, University of Phoenix
William Chamberlin, Fullerton College
Oiyin Pauline Chow, Harrisburg Area Community College
Laurie Corey, Westchester Community College
Ozeas Costas, Ohio State University at Mansfield
Christopher Di Leonardo, Foothill College
Dani Ducharme, Waubonsee Community College
Mark Eastman, Diablo Valley College
Ben Elman, Baruch College
Staussa Ervin, Tarrant County College
Michael Farabee, Estrella Mountain Community College
Laurie Flaherty, Eastern Washington University
Susan Fuhr, Maryville College
Peter Galvin, Indiana University at Southeast
Andrew Getzfeld, New Jersey City University
Janet Gingold, Prince George's Community College
Donald Glassman, Des Moines Area Community College

Richard Goode, Porterville College
Peggy Green, Broward Community College
Stelian Grigoras, Northwood University
Paul Grogger, University of Colorado
Michael Hackett, Westchester Community College
Duane Hampton, Western Michigan University
Thomas Hancock, Eastern Washington University
Gregory Harris, Polk State College
John Haworth, Chattanooga State Technical Community College
James Hayes-Bohanan, Bridgewater State College
Peter Ingmire, San Francisco State University
Mark Jackson, Central Connecticut State University
Heather Jennings, Mercer County Community College
Eric Jerde, Morehead State University
Jennifer Johnson, Ferris State University
Richard Kandus, Mt. San Jacinto College District
Christopher Kent, Spokane Community College
Gerald Ketterling, North Dakota State University
Lynnel Kiely, Harold Washington College
Eryn Klosko, Westchester Community College
Cary T. Komoto, University of Wisconsin—Barron County
John Kupfer, University of South Carolina
Nicole Lafleur, University of Phoenix
Arthur Lee, Roane State Community College
Mary Lynam, Margrove College
Heidi Marcum, Baylor University
Beth Marshall, Washington State University
Dr. Theresa Martin, Eastern Washington University
Charles Mason, Morehead State University
Susan Massey, Art Institute of Philadelphia
Linda McCollum, Eastern Washington University
Mary L. Meiners, San Diego Miramar College
Shawn Mikulay, Elgin Community College
Cassandra Moe, Century Community College
Lynn Hanson Mooney, Art Institute of Charlotte
Kristy Moreno, University of Phoenix

Jacob Napieralski, University of Michigan—Dearborn
Gisele Nasar, Brevard Community College, Cocoa Campus
Daria Nikitina, West Chester University
Robin O'Quinn, Eastern Washington University
Richard Orndorff, Eastern Washington University
Sharen Orndorff, Eastern Washington University
Clair Ossian, Tarrant County College
Debra Parish, North Harris Montgomery Community College District
Linda Peters, Holyoke Community College
Robin Popp, Chattanooga State Technical Community College
Michael Priano, Westchester Community College
Alan "Paul" Price, University of Wisconsin—Washington County
Max Reams, Olivet Nazarene University
Mary Celeste Reese, Mississippi State University
Bruce Rengers, Metropolitan State College of Denver
Guillermo Rocha, Brooklyn College
Penny Sadler, College of William and Mary
Shamili Sandiford, College of DuPage
Thomas Sasek, University of Louisiana at Monroe
Donna Seagle, Chattanooga State Technical Community College
Diane Shakes, College of William and Mary
Jennie Silva, Louisiana State University
Michael Siola, Chicago State University
Morgan Slusher, Community College of Baltimore County

Julia Smith, Eastern Washington University
Darlene Smucny, University of Maryland University College
Jeff Snyder, Bowling Green State University
Alice Stefaniak, St. Xavier University
Alicia Steinhardt, Hartnell Community College
Kurt Stellwagen, Eastern Washington University
Charlotte Stromfors, University of Phoenix
Shane Strup, University of Phoenix
Donald Thieme, Georgia Perimeter College
Pamela Thinesen, Century Community College
Chad Thompson, SUNY Westchester Community College
Lensyl Urbano, University of Memphis
Gopal Venugopal, Roosevelt University
Daniel Vogt, University of Washington—College of Forest Resources
Dr. Laura J. Vosejpka, Northwood University
Brenda L. Walker, Kirkwood Community College
Stephen Wareham, Cal State Fullerton
Fred William Whitford, Montana State University
Katie Wiedman, University of St. Francis
Harry Williams, University of North Texas
Emily Williamson, Mississippi State University
Bridget Wyatt, San Francisco State University
Van Youngman, Art Institute of Philadelphia
Alexander Zemcov, Westchester Community College

Student Participants

Lucy DiAroscia, Westchester Community College
Estelle Rizzin, Westchester Community College
Eric Torres, Westchester Community College
Pia Chawla, Westchester Community College
Michael Maczuga, Westchester Community College
Joshua Watson, Eastern Washington University
Karl Beall, Eastern Washington University
Patty Hosner, Eastern Washington University
Brenden Hayden, Eastern Washington University

Tonya Karunartue, Eastern Washington University
Lindsey Harris, Eastern Washington University
Jessica Bryant, Eastern Washington University
Melissa Michael, Eastern Washington University
Channel DeWitt, Eastern Washington University
Andrew Rowley, Eastern Washington University
Sydney Lindgren, Eastern Washington University
Heather Gregg, Eastern Washington University

Reviewers of Previous Editions

Jennifer Andersen, Johnson County Community College
Mark Anderson, University of Maine
Nancy Bain, Ohio University
Raymond Beiersdorfer, Youngstown State University
Cheryl Berg, Gateway Community College, Phoenix
Richard Bowden, Allegheny College
Scott Brame, Clemson University
James A. Brenneman, University of Evansville
Huntting W. Brown, Wright State University
Stephan Bullard, University of Hartford
Oiyin Pauline Chow, Harrisburg Area Community College
Michael S. Dann, Penn State University
R. Laurence Davis, Northeastern Cave Conservancy, Inc.
JodyLee Estrada Duek, Pima Community College
Catherine M. Etter, Cape Cod Community College

Brad C. Fiero, Pima Community College
Michael Freake, Lee University
Jennifer Frick-Ruppert, Brevard College
Todd G. Fritch, Northeastern University
Marcia L. Gillette, Indiana University, Kokomo
Arthur Goldsmith, Hallandale High
Cliff Gottlieb, Shasta College
Peggy Green, Broward Community College
Stelian Grigoras, Northwood University
Syed E. Hasan, University of Missouri—Kansas City
Carol Hoban, Kennesaw State University
Guang Jin, Illinois State University
Dawn Keller, Hawkeye Community College
Martin Kelly, Genesee Community College
David Kitchen, University of Richmond

Paul Kramer, Farmingdale State College

Meredith Gooding Lassiter, Wiona State University

Ernesto Lasso de la Vega, Edison College

Madelyn E. Logan, North Shore Community College

Linda Lyon, Frostburg State University

Timothy F. Lyon, Ball State University

Robert S. Mahoney, Johnson & Wales University at Florida

Heidi Marcum, Baylor University

Matthew H. McConeghy, Johnson & Wales University

Rick McDaniel, Henderson State University

Brian Mooney, Johnson & Wales University at North Carolina

Jacob Napieralski, University of Michigan, Dearborn

Renee Nerish, Mercer County Community College

Leslie Nesbitt, Niagara University

Ken Nolte, Shasta College

Natalie Osterhoudt, Broward Community College

Barry Perlmutter, Community College of Southern Nevada

Neal Phillip, Bronx Community College

Thomas E. Pliske, Florida International University

Katherine Prater, Texas Wesleyan University

Uma Ramakrishnan, Juniata College

Sabine Rech, San Jose State University

Shamili A. Sandiford, College of DuPage

Thomas Sasek, University of Louisiana at Monroe

Howie Scher, University of Rochester

Nan Schmidt, Pima Community College

Richard B. Schultz, Elmhurst College

Richard Shaker, University of Wisconsin, Milwaukee

Charles Shorten, West Chester University

Jerry Skinner, Keystone College

Roy Sofield, Chattanooga State Technical Community College

Bo Sosnicki, Florida Community College at Jacksonville

Ravi Srinivas, University of St. Thomas

David Steffy, Jacksonville State University

Andrew Suarez, University of Illinois

Charles Venuto, Brevard Community College, Cocoa Campus

Margaret E. Vorndam, Colorado State University Pueblo

Laura J. Vosejpka, Northwood University

Maud M. Walsh, Louisiana State University

John F. Weishampel, University of Central Florida

Karen Wellner, Arizona State University

Arlene Westhoven, Ferris State University

Susan M. Whitehead, Becker College

John Wielichowski, Milwaukee Area Technical College

Reviewers of the Third Edition

Marc Albrecht, University of Nebraska—Kearney

Frederick Bein, Indiana University-Purdue University

Bart Bergquist, University of Northern Iowa

Ralph L. Bonati, Pima Community College

David L. Bowman, De Moines Area Community College

Ravi Chandran, DeVry University

Reggie Cobb, Nash Community College

Michael L Denniston, Georgia Perimeter College

Robert Dill, Bergen Community College

Danielle M. DuCharme, Waubonsee Community College

Karen L Duston, San Jacinto College South

Brad Fiero, Pima Community College

Catherine Folio, Brookdale Community College

Stelian Grigoras, Northwood University

Donald Hunsaker, California State University Fresno

Catherine Hurlbut, Florida State College at Jacksonville

Catherine Hurlbut, Florida State College at Jacksonville

Matthew Isles, SUNY-Environmental Science and Forestry

Erica Kosal, North Carolina Wesleyan College

John Krolak, Georgia Perimeter College

Dale Lambert, Tarrant County College

Jill Malmstad, Florida State University

Diana Merkel, Passaic County Community College

Sharon Moran, SUNY-Environmental Science and Forestry

Kathleen Nuckolls, University of Kansas

Barry Perlmutter, College of Southern Nevada

Shamili Ajgaonkar Sandiford, College of DuPage

Michelle Pulich Stewart, Mesa Community College

Keith S. Summerville, Drake University

Mike Tveten, Pima Community College Northwest

Karen Wellner, Arizona State University

David Wyatt, Sacramento City College

Students and Class Testers

To make certain that *Visualizing Environmental Science 3e* met the needs of current students, we asked several instructors to class-test a chapter. The feedback that we received from students and instructors confirmed our belief that the visualizing approach taken in this book is highly effective in helping students to learn. We wish to thank the following instructors and their students who provided us with helpful feedback and suggestions:

Hernan Aubert, Pima Community College; Keith Hench, Kirkwood Community College; Dawn Keller, Hawkeye Community College; Dale Lambert, Tarrant Community College; Janice Padula, Clinton College; and Ashok Malik, Evergreen Valley College.

Special Thanks

We are extremely grateful to the many members of the editorial and production staff at John Wiley and Sons who guided us through the challenging steps of developing this book. Their tireless enthusiasm, professional assistance, and endless patience smoothed the path as we found our way. We thank in particular Senior Editor Rachel Falk, who expertly launched and directed the revision; Merillat Staat, Project Editor, for coordinating the development and revision process; Jeffrey Rucker, Associate Director of Marketing, and Kristine Ruff, Marketing Manager, for a superior marketing effort; and Jenna Paleski, Editorial Program Assistant, for her constant attention to detail. Thanks also to Linda Muriello, Senior Media Editor, and Daniela DiMaggio, Interactive Project Manager, for their insightful work in developing our Wiley Resource Kit as well as the other media components. We also thank Micheline Frederick, Production Manager, and Jeanine Furino of Furino Production for expertly helping us through the production process. Thanks to Sandra Rigby, Senior Illustration Editor, who managed our illustration program, much of which was masterfully developed by Elizabeth Morales. We thank Elle Wagner for her unflagging, always swift work in researching and obtaining many of our text images, and Stacy Gold of the National Geographic Image Collection for her valuable expertise in selecting NGS photos. We thank James O'Shea for the beautiful new interior design and for his constant attention to page layout, as well as Harry Nolan and Wendy Lai for our stunning new cover. Thank you to Kaye Pace, Vice President and Executive Publisher; Anne Smith, Vice President and Executive Publisher; Brian Baker, Project Editor; and Beth Tripmacher, Project Editor, for providing guidance and support to the rest of the team throughout the revision.

Many other individuals at National Geographic offered their expertise and assistance in developing this book: Francis Downey, Vice President and Publisher, and Richard Easby, Supervising Editor, National Geographic School Division; Mimi Dornack, Sales Manager, National Geographic Image Collection; Dierdre Bevington-Attardi, Project Manager, and Kevin Allen, Director of Map Services, National Geographic Maps; and Jim Burch and Michael Garrity of the National Geographic Film Library. We appreciate their contributions and support.

About the Authors

Linda R. Berg is an award-winning teacher and textbook author. She received a B.S. in science education, an M.S. in botany, and a Ph.D. in plant physiology from the University of Maryland. Dr. Berg taught at the University of Maryland—College Park for 17 years and at St. Petersburg College in Florida for 8 years. She has taught introductory courses in environmental science, biology, and botany to thousands of students and has received numerous teaching and service awards. Dr. Berg is also the recipient of many national and regional awards, including the National Science Teachers Association Award for Innovations in College Science Teaching, the Nation's Capital Area Disabled Student Services Award, and the Washington Academy of Sciences Award in University Science Teaching. During her career as a professional science writer, Dr. Berg has authored or co-authored numerous editions of several leading college science textbooks. Her writing reflects her teaching style and love of science.

Mary Catherine Hager is a professional science writer and editor specializing in educational materials for life and earth sciences. She received a double-major B.A. in environmental science and biology from the University of Virginia and an M.S. in zoology from the University of Georgia. Ms. Hager worked as an editor for an environmental consulting firm and as a senior editor for a scientific reference publisher. For the past 15 years, she has published articles in environmental trade magazines, edited federal and state reports addressing wetlands conservation issues, and written and edited components of environmental science and biology textbooks for target audiences ranging from middle school to college. Her writing and editing pursuits are a natural outcome of her scientific training and curiosity, coupled with her love of reading and effective communication.

David M. Hassenzahl is the Founding Dean of the School of Sustainability and the Environment at Chatham University. An internationally recognized scholar of sustainability and risk analysis, his research focuses on incorporating scientific information and expertise into public decision. He holds a B.A. in Environmental Science and Paleontology from the University of California at Berkeley, and a Ph.D. in Princeton University's Woodrow Wilson School. His efforts in climate change education have been supported by the National Science Foundation, and recognition of his work includes the Society for Risk Analysis Outstanding Educator Award and the UNLV Foundation Distinguished Teaching Award. Dr. Hassenzahl is a Senior Fellow of the National Council for Science and the Environment, and serves on the Council of the Association of Environmental Studies and Sciences. Prior to his academic career, Dr. Hassenzahl worked in the private sector as an environmental manager, and as an inspector for the (San Francisco) Bay Area Air Quality Management District.

Environmental InSight Features

Multi-part visual presentations that focus on a key concept or topic in the chapter.

Process Diagrams

A series or combination of figures and photos that describe and depict a complex process.

THIRD EDITION

VISUALIZING
ENVIRONMENTAL
SCIENCE

1 The Environmental Challenges We Face

A WORLD IN CRISIS

Over three billion years ago, just before the first life forms arose, Earth's surface and climate were inhospitable by modern standards but contained abundant raw materials and energy. As early life forms evolved, their growth shaped the landscape, altered the global climate, and modified the chemical makeup of the ocean and soils. These changes, along with the shifts in Earth's orbit and the sun, caused conditions in which the millions of species, including humans, that now inhabit our planet arose and adapted.

Today the human species is the most significant agent of environmental change on our planet. Our intellectual capacity has even made it possible for us to venture into space, allowing us a view of the uniqueness of our planet in the solar system (see photograph). However, we are overpowering the planet with our burgeoning population; transforming forests, prairies, and deserts to meet our needs and desires; and consuming ever-increasing amounts of Earth's abundant but finite resources—rich topsoil, clean water, and breathable air. We are eradicating thousands upon thousands of unique species as we destroy or alter their habitats. Evidence continues to accumulate that human-induced climate change is putting the natural environment at risk.

This book introduces the major environmental problems that humans have created and considers ways to address these issues. Most importantly, it explains why we must minimize human impact on our planet. We can't afford to ignore the environment because our lives, as well as those of future generations, depend on it. As a wise proverb says, "We have not inherited the world from our ancestors; we have only borrowed it from our children."

NATIONAL GEOGRAPHIC

CHAPTER OUTLINE

CHAPTER PLANNER ✓

- ☐ Study the picture and read the opening story
- ☐ Scan the Learning Objectives in each section:
  p. 4 ☐   p. 12 ☐   p. 16 ☐   p. 20 ☐
- ☐ Read the text and study all figures and visuals. Answer any questions.

Analyze key features

- ☐ Environmental InSight, p. 5 ☐   p.13 ☐
- ☐ EnviroDiscovery, p. 8 ☐   p. 22 ☐
- ☐ National Geographic Map, pp. 14–15 ☐
- ☐ Process Diagram, p. 19 ☐   p. 21 ☐
- ☐ Case Study, p. 23 ☐
- ☐ Stop: Answer the Concept Checks before you go on:
  p. 11 ☐   p. 12 ☐   p. 20 ☐   p. 21 ☐

End of Chapter

- ☐ Review the Summary and Key Terms.
- ☐ Answer What is happening in this picture?
- ☐ Answer the Critical and Creative Thinking Questions.

Human Impacts on the Environment

LEARNING OBJECTIVES

1. **Distinguish** among highly developed countries, moderately developed countries, and less developed countries.
2. **Relate** human population size to natural resources and resource consumption.
3. **Distinguish** between people overpopulation and consumption overpopulation.
4. **Describe** the three factors that are most important in determining human impact on the environment.

The satellite photograph in **Figure 1.1a** is a portrait of about 450 million people. The tiny specks of light represent cities, and the great metropolitan areas, such as New York City along the northeastern seacoast, are ablaze with light.

Earth's central environmental problem, which links all others together, is that there are many people, and the number, both in North America and worldwide, continues to grow. In 2009 the human population as a whole passed 6.8 billion individuals. Not only is this number incomprehensibly large, but our population has grown this large in a very brief span of time. In 1960 the human population was only 3 billion (**Figure 1.1b**). By 1975 there were 4 billion people, and by 1987 there were 5 billion. The 6.8 billion people who currently inhabit our planet consume vast quantities of food and water, use a great deal of energy and raw materials, and produce much waste.

Despite most countries' involvement with family planning, population growth rates won't change overnight. Several billion people will be added to the world in the 21st century, so even if we remain concerned about overpopulation and even if our solutions are very effective, the coming decades may very well see many problems.

On a global level, nearly one in four people lives in extreme **poverty** (**Figure 1.1c**). By one measure, living in poverty is defined as having a per person income of less than $2 per day, expressed in U.S. dollars adjusted for purchasing power. About 3.3 billion people—nearly half of the world's population—currently live at this level of poverty. Poverty is associated with a short life expectancy, illiteracy, and inadequate access to health services, safe water, and balanced nutrition.

The world population may stabilize toward the end of the 21st century, given the family planning efforts that are currently under way. Population experts at the Population Reference Bureau have noticed a decrease in the fertility rate worldwide to a current average of 2.6 children per woman, and the fertility rate is projected to continue to decline in coming decades. Population experts have made various projections for the world population at the end of the 21st century, from about 7.7 billion to 10.6 billion, depending on how fast the fertility rate decreases.

No one knows whether Earth can support so many people indefinitely. Finding ways for it to do so represents one of the greatest challenges of our times. Among the tasks to be accomplished is feeding a world population considerably larger than today's without destroying the biological communities that support life on our planet. The quality of life available to our children and grandchildren will depend to a large extent on our ability to develop a sustainable system of agriculture to feed the world's people.

A factor as important as population size is a population's level of **consumption**, which is the human use of material and energy. Consumption is intimately connected to a country's **economic growth**, the expansion in output of a nation's goods and services. The world's economy is growing at an enormous rate, yet this growth is unevenly distributed across the nations of the world.

> **poverty** A condition in which people are unable to meet their basic needs for food, clothing, shelter, education, or health.

> **highly developed countries** Countries with complex industrialized bases, low rates of population growth, and high per person incomes.

The Gap Between Rich and Poor Countries

Generally speaking, countries are divided into rich (the "haves") and poor (the "have-nots"). Rich countries are known as **highly developed countries**. The United States, Canada, most of

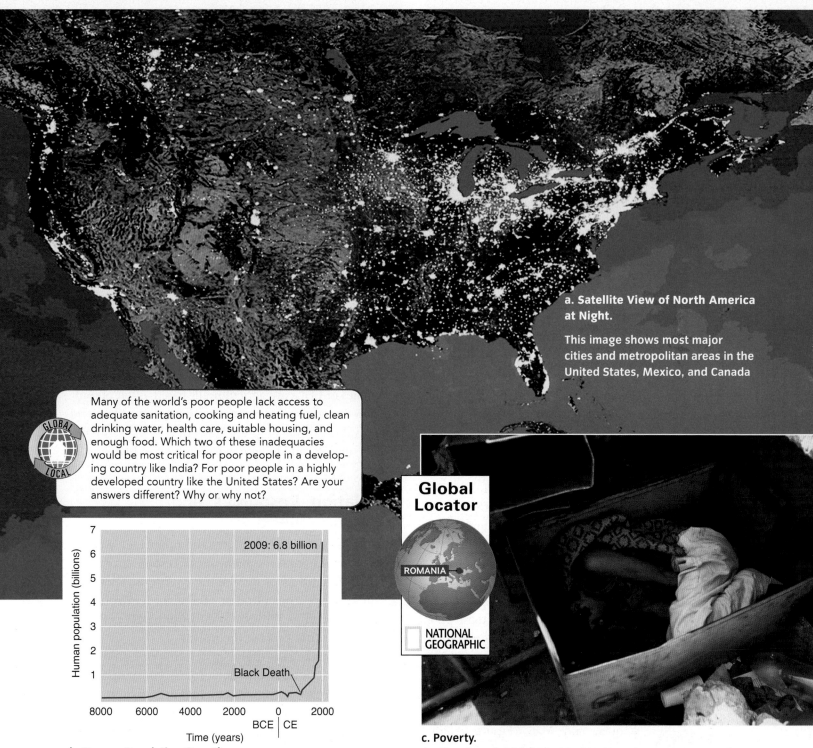

a. Satellite View of North America at Night.

This image shows most major cities and metropolitan areas in the United States, Mexico, and Canada

Many of the world's poor people lack access to adequate sanitation, cooking and heating fuel, clean drinking water, health care, suitable housing, and enough food. Which two of these inadequacies would be most critical for poor people in a developing country like India? For poor people in a highly developed country like the United States? Are your answers different? Why or why not?

Global Locator

ROMANIA

NATIONAL GEOGRAPHIC

b. Human Population Growth.
It took thousands of years for the human population to reach 1 billion (in 1800). By mid-2009, Earth's human population had reached 6.8 billion. (Black Death refers to a devastating disease, probably bubonic plague, that decimated Europe and Asia in the 14th century.)

Graph: Human population (billions) vs. Time (years), 8000 BCE to 2000 CE. 2009: 6.8 billion. Black Death labeled.

c. Poverty.
An abandoned child sleeps in a box in Bucharest, Romania. The human population problem requires not only a stabilization of population numbers but also an improvement of the economic conditions of people living in extreme poverty.

WILEY
RESOURCE KIT

a. A typical Japanese family, from Tokyo, with their possessions. People in highly developed countries consume a disproportionate share of natural resources.

b. A typical Mexican family, from Guadalajara, with their possessions. Economic development in this moderately developed country has allowed many people to enjoy a middle-class lifestyle. Other Mexicans live in poverty, however.

Europe, and Japan, which represent about 18 percent of the world's population, are highly developed countries (**Figure 1.2a**).

Poor countries, in which about 82 percent of the world's population live, fall into two subcategories: moderately developed and less developed. Turkey, South Africa, Thailand and Mexico are examples of **moderately developed countries** (MDCs) (**Figure 1.2b**). People living in MDCs have fewer opportunities for income, education, and health care than people living in highly developed countries.

Examples of **less developed countries** (LDCs) include Haiti, Bangladesh, Rwanda, Laos, Ethiopia, and Mali (**Figure 1.2c**). Cheap, unskilled labor is abundant in LDCs, but capital for investment is scarce. To improve their economic conditions, many LDCs must borrow money from banks in highly developed countries. Most economies of LDCs are agriculturally based, often on only one or a few crops. As a result, crop failure or a low world market value for that crop is catastrophic to the economy. Hunger, disease, and illiteracy are common in LDCs.

Population, Resources, and the Environment

Inhabitants of the United States and other highly developed countries consume many more resources per person than do citizens of developing countries. This high rate of resource consumption affects the environment at least as much as the explosion in population that is occurring in other parts of the world.

We can make two useful generalizations about the relationships among population growth, consumption of natural resources, and environmental degradation. First, the amount of resources essential to an individual's survival is small, but rapid population growth (often found in developing countries) tends to overwhelm and deplete a country's soils, forests, and other natural resources. Second, in highly developed nations, individual demands on natural resources are far greater than the requirements for mere survival. To satisfy their desires as well as their basic needs, many people in more affluent nations

moderately developed countries Countries with medium levels of industrialization and per person incomes lower than those of highly developed countries.

less developed countries Countries with low levels of industrialization, very high rates of population growth, very high infant mortality rates, and very low per person incomes relative to highly developed countries.

c. A typical family from Kouakourou, Mali, with all their possessions. The rapidly increasing number of people in less developed countries overwhelms their natural resources, even though individual resource requirements may be low.

Some examples of **renewable resources** are trees, fishes, fertile agricultural soil, and fresh water. Nature replaces these resources fairly rapidly, on a scale of days to decades. Forests, fisheries, and agricultural land are particularly important renewable resources in developing countries because they provide food. Indeed, many people in developing countries are subsistence farmers who harvest just enough food for their families to survive.

> **renewable resources** Resources that are replaced by natural processes and that can be used forever, provided they are not overexploited in the short term.

Rapid population growth can cause renewable resources to be overexploited. For example, large numbers of poor people must grow crops on land—such as mountain slopes or tropical rain forests—that is poorly suited for farming. Although this practice may provide a short-term solution to the need for food, it does not work in the long run because when these lands are cleared for farming, their agricultural productivity declines rapidly and severe environmental deterioration occurs. Renewable resources, then, are *potentially* renewable. They must be used in a manner that allows natural processes time to replace or replenish them.

The effects of population growth on natural resources are particularly critical in developing countries. The economic growth of developing countries is frequently tied to the exploitation of their natural resources, often for export to highly developed countries. Developing countries are faced with the difficult choice of exploiting natural resources to provide for their expanding populations in the short term (that is, to pay for food or to cover debts) or conserving those resources for future generations. It is instructive to note that the economic growth and development of the United States and of other highly developed nations came about through the exploitation—and in some cases the destruction—of their resources. Continued growth and development in highly developed countries now relies significantly on the importation of these resources from less developed countries.

Poverty is tied to the effects of population pressures on natural resources and the environment. Poor people in developing countries find themselves trapped in a vicious cycle of poverty. They use environmental resources unwisely for short-term gain (that is, to survive), but this exploitation degrades the resources and diminishes long-term prospects of economic development.

deplete resources and degrade the global environment through increased consumption of nonessential items such as televisions, jet skis, and cellular phones.

Types of Resources When examining the effects of population on the environment, it is important to distinguish between nonrenewable and renewable natural resources. **Nonrenewable resources** include minerals (such as aluminum, tin, and copper) and fossil fuels (coal, oil, and natural gas). Natural processes do not replenish nonrenewable resources within a reasonable duration on the human timescale. Fossil fuels, for example, take millions of years to form.

> **nonrenewable resources** Natural resources that are present in limited supplies and are depleted as they are used.

In addition to a nation's population and its level of resource use, several other factors affect the way nonrenewable resources are used—including how efficiently the resource is extracted and processed and how much of it is required or consumed. Nonetheless, the inescapable fact is that Earth has a finite supply of nonrenewable resources that sooner or later will be exhausted. In time, technological advances may help find or develop substitutes for nonrenewable resources. Slowing the rate of population growth and resource consumption will help us buy time to develop such alternatives.

EnviroDiscovery

Green Roofs

Green roof
City Hall is one of many buildings in Chicago with a living green roof.

A roof that is completely or partially covered with vegetation and soil is known as a *green roof*. Also called *eco-roofs*, green roofs can provide several environmental benefits. For one thing, the plants and soil are effective insulators, reducing heating costs in winter and cooling costs in summer. The rooftop mini-ecosystem filters pollutants out of rainwater and reduces the amount of stormwater draining into sewers. In urban areas, green roofs provide wildlife habitat, even on the tops of tall buildings. A city with multiple green roofs provides "stepping stones" of habitat that enable migrating birds and insects to pass unharmed through the city. Green roofs can also be used to grow vegetable and fruit crops and to provide an outdoor refuge for people living or working in the building. Green roofs allow urban systems to more closely resemble the natural systems they have replaced.

Green roofs may be added to existing buildings, but it is often easier and less expensive to install them in new buildings. Modern green roofs, which are designed to support the additional weight of soil and plants, consist of several layers that hold the soil in place, stop plant roots from growing through the rooftop, and drain excess water, thereby preventing leaks. Currently, Chicago, Illinois, is the U.S. city with the largest total area of green roofs (see photograph). One of the largest individual green roofs in the United States is on the Ford Motor Company's Plant in Dearborn, Michigan.

Population Size and Resource Consumption

Resource issues are clearly related to population size: More people use more resources. An equally important factor is a population's resource consumption. Consumption is both an economic and a social act. Consumption provides the consumer with a sense of identity as well as status among peers. The media, including the advertising industry, promote consumption as a way to achieve happiness. We are encouraged to spend, to consume.

People in highly developed countries are extravagant and wasteful consumers; their use of resources is greatly out of proportion to their numbers. A single child born in a highly developed country such as the United States causes a greater impact on the environment and on resource depletion than perhaps 20 children born in a developing country. Many natural resources are needed to provide the automobiles, air conditioners, disposable diapers, cell phones, DVD players, computers, clothes, newspapers, athletic shoes, furniture, books, and other "comforts" of life in highly developed nations. Thus, the disproportionately large consumption of resources by the United States and other highly developed countries affects natural resources and the environment as much as or more than the population explosion in the developing world.

People overpopulation and consumption overpopulation
A country is overpopulated if the level of demand on its resource base results in damage to the environment. In comparing human impact on the environment in developing and highly developed countries, we see that a country can be overpopulated in two ways. **People overpopulation** occurs when the environment is worsening because there are too many people, even if those people consume few resources per person. People overpopulation is the current problem in many developing nations. In contrast, **consumption overpopulation** results from the consumption-oriented lifestyles popular in highly developed countries. The effects of consumption overpopulation on the environment are the same as those of people overpopulation—pollution, resource depletion, and degradation of the environment.

Many affluent, highly developed nations, including the United States, suffer from consumption overpopulation. Highly developed nations represent less than 20 percent of the world's population, yet they consume significantly more than half of its resources. According to the Worldwatch Institute, highly developed nations account for the lion's share of total resources consumed:

- 86 percent of aluminum used.
- 76 percent of timber harvested.
- 68 percent of energy produced.
- 61 percent of meat eaten.
- 42 percent of the fresh water consumed.

These nations also generate 75 percent of the world's pollution and waste (**Figure 1.3**).

> **people overpopulation** A situation in which there are too many people in a given geographic area.
>
> **consumption overpopulation** A situation in which each individual in a population consumes too large a share of resources.

Consumption overpopulation • Figure 1.3

American consumption is actively promoted in Times Square advertisements.
Highly developed nations, such as the United States, consume more than 50 percent of the world's resources, produce 75 percent of its pollution and waste, and represent only 18 percent of its total population.

Ecological footprints Each person has an **ecological footprint**, an amount of productive land, fresh water, and ocean required on a continuous basis to supply that person with food, wood, energy, water, housing, clothing, transportation, and waste disposal. In the *Living Planet Report 2008*, scientists calculated that Earth has about 11.4 billion hectares (28.2 billion acres) of productive land and water. If we divide this area by the global human population, it indicates that each person is allotted about 1.8 hectares (4.3 acres). However, the average global ecological footprint is currently about 2.7 hectares (6.7 acres) per person, which means we have an *ecological overshoot*—we have depleted our allotment. We can see the short-term results around us—forest destruction, degradation of croplands, loss of biological diversity, declining ocean fisheries, local water shortages, and increasing pollution. The long-term outlook, if we do not seriously address our consumption of natural resources, is potentially disastrous.

The developing nation of India is the world's second largest country in terms of population, so even though its per capita footprint is low, the country's total footprint is high (**Figure 1.4**). In France, the per capita ecological footprint is high at 4.9 hectares (12.1 acres), but its footprint as a country is relatively low, at 298.1 million hectares (736.6 million acres). In the United States, the world's third largest country, the per capita ecological footprint is 9.4 hectares (23.3 acres); the U.S. footprint as a country is a whopping 2809.7 million hectares (6943 million acres). If all people in the world had the same lifestyle and level of consumption as the average American, and assuming no changes in technology, we would need four additional planets the size of Earth.

As developing nations increase their economic growth and improve their standard of living, more and more people in those countries purchase consumer goods. By the early 2000s, more new cars were sold annually in Asia than in North America and western Europe combined. These new consumers may not consume at the high level of the average consumer in a highly developed nation, but their consumption has increasingly adverse effects on the environment. For example, air pollution from traffic in urban centers in developing countries is bad and getting worse every year.

Millions of dollars are lost to health problems caused by air pollution in these cities.

Population, consumption, and environmental impact When you turn on the tap to brush your teeth in the morning, you probably do not think about where the water comes from or about the environmental consequences of removing it from a river or the ground. All the materials that make up the products we use every day come from Earth, and these materials eventually are returned to Earth, mainly in sanitary landfills.

Such human impacts on the environment are difficult to assess. One way to estimate them is to use the three factors most important in determining environmental impact (I):

- The number of people (P).
- The affluence per person, which is a measure of the consumption, or amount of resources used per person (A).
- The environmental effects (resources needed and wastes produced) of the technologies used to obtain and consume the resources (T).

This method of assessment is usually referred to as the IPAT equation: $I = P \times A \times T$.

Biologist **Paul R. Ehrlich** and physicist **John P. Holdren** first proposed the IPAT model in the 1970s. It shows the mathematical relationship between environmental impacts and the forces that drive them. To determine the environmental impact of carbon dioxide (CO_2) emissions from motor vehicles, for example, multiply the population by the number of cars per person (affluence or consumption per person) by the average annual CO_2 emissions per year (technological impact). This model demonstrates that although improving motor vehicle efficiency and developing cleaner technologies will reduce pollution and environmental degradation, a larger reduction will result if population and per person consumption are also controlled.

The three factors in the IPAT equation are always changing in relation to each other. For example, consumption of a particular resource may increase, but technological advances may decrease the environmental impact of the increased consumption.

Ecological footprints • Figure 1.4

a. The average ecological footprint of a person living in India, France, or the United States. For example, each Indian requires 0.9 hectare (2.2 acres) of productive land and ocean to meet his or her resource requirements.

b. The total ecological footprint for India, France, and the United States. Notice that India, although having a low per capita ecological footprint, has a relatively large *total* footprint as a country because of its large population. If everyone in the world had the same level of consumption as the average American, it would take the resources and area of 5 Earths.

c. Earth's ecological footprint has been increasing over time. By 2005, humans were using the equivalent of 1.3 Earths, a situation that is not sustainable.

 Calculate your individual ecological footprint online. (Search for "ecological footprint.") Are you living sustainably? Suggest two things that you could do to lower your ecological footprint.

The IPAT equation helps to identify what we don't know or understand about consumption and its environmental impact. For example, which kinds of consumption have the greatest destructive impact on the environment? Which groups in society are responsible for the greatest environmental disruption? How can we alter the activities of these environmentally disruptive groups? It will take years to address such questions, but the answers should help decision makers in business and government formulate policies that will alter consumption patterns in an environmentally responsible way. The ultimate goal should be to make consumption sustainable so that humanity's current practices do not compromise the ability of future generations to use and enjoy the riches of our planet.

To summarize, as human numbers and consumption increase worldwide, so does humanity's impact on Earth, posing new challenges to us all. Success in achieving sustainability in population size and consumption will require the cooperation of all the world's peoples.

CONCEPT CHECK

1. **How** do highly developed countries, moderately developed countries, and less developed countries differ regarding population growth and per person incomes?

2. **How** is human population growth related to natural resource depletion and environmental degradation?

3. **What** is the difference between people overpopulation and consumption overpopulation?

4. **According** to the IPAT equation, what three factors are most important in determining environmental impact?

Sustainability and the Environment

LEARNING OBJECTIVES

1. **Define** *environmental sustainability.*
2. **Identify** human behaviors that threaten environmental sustainability.

Environmental sustainability is an organizing principle for this text. *Sustainability* implies that the environment will function indefinitely without going into a decline from the stresses that human society imposes on natural systems (such as fertile soil, water, and air) (**Figure 1.5**). Environmental sustainability applies to many levels, including the individual, communal, regional, national, and global levels.

Environmental sustainability is based in part on the following ideas:

- We must consider the effects of our actions on the health and well being of the natural environment, including all living things.
- Earth's resources are not present in infinite supply. We must live within limits that let renewable resources such as fresh water regenerate for future needs.
- We must understand all the costs to the environment and to society of products we consume.
- We must each share responsibility for environmental sustainability.

environmental sustainability The ability to meet humanity's current needs without compromising the ability of future generations to meet their needs.

Focus on sustainability • Figure 1.5

Environmental sustainability requires a long-term perspective to protect human welfare and natural resource assets, such as the goals shown here.

Many environmental experts think that human society is not operating sustainably because of the following human behaviors (**Figures 1.6** and **1.7** [the National Geographic map]):

- We are using nonrenewable resources such as fossil fuels as if they were present in unlimited supplies.
- We are using renewable resources such as fresh water and forests faster than they are replenished naturally.
- We are polluting the environment—the land, rivers, ocean, and atmosphere—with toxins as if the capacity of the environment to absorb them were limitless.
- Our numbers continue to grow, despite Earth's finite ability to feed us and to absorb our wastes.

If left unchecked, these activities may threaten the life-support systems of Earth to the extent that recovery is impossible. Our first goal should be to critically evaluate which changes our society is willing to make.

At first glance, the issues may seem simple. The solutions are more complex and challenging, in part because of various interacting ecological, societal, and economic factors. Our inadequate scientific understanding of how the environment works and how human choices affect the environment is a major reason that environmental sustainability is difficult to achieve. Even for established environmental problems, political and social controversy often prevents widespread acceptance that an environmental threat is real.

CONCEPT CHECK STOP

1. **What** is environmental sustainability?
2. **Which** human behaviors threaten environmental sustainability?

a. Frozen Giant Bluefin Tuna.
These tuna are laid out at Tsukiji Market in Tokyo, Japan, so buyers can pick the ones they want. Bluefin tuna are overexploited in certain parts of the ocean.

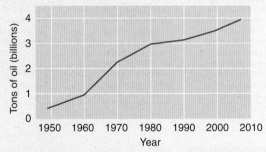

Annual World Oil Consumption, 1950 to 2006.
This increase is unsustainable.

Interpreting Data
By approximately how much did annual world oil consumption increase in the past 20 years? Why is the increase unsustainable?

b. Clear Cutting In.
Forest harvest is unsustainable in many parts of the world.

c. Oil Refinery at Grangemouth, United Kingdom.
Both highly developed and developing countries depend largely on oil for economic development.

WILEY
RESOURCE KIT

These issues occur locally at so many places around the planet that they are global in scope.

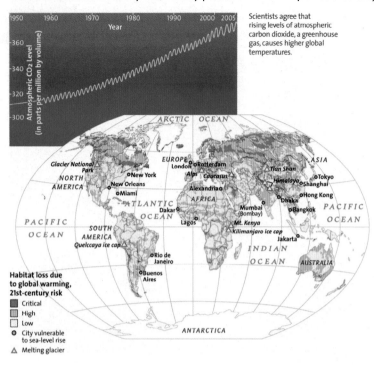

Scientists agree that rising levels of atmospheric carbon dioxide, a greenhouse gas, causes higher global temperatures.

Habitat loss due to global warming, 21st-century risk
- ■ Critical
- ■ High
- □ Low
- ⊙ City vulnerable to sea-level rise
- △ Melting glacier

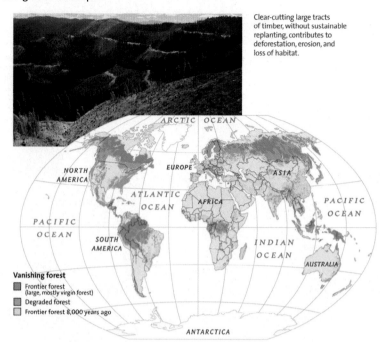

Clear-cutting large tracts of timber, without sustainable replanting, contributes to deforestation, erosion, and loss of habitat.

Vanishing forest
- ■ Frontier forest (large, mostly virgin forest)
- ■ Degraded forest
- □ Frontier forest 8,000 years ago

▲ GLOBAL WARMING

Temperatures across the world are increasing at a rate not seen at any other time in the past 10,000 years. Although climate variation is a natural phenomenon, human activities that release carbon dioxide and other greenhouse gases into the atmosphere—industrial processes, fossil fuel consumption, deforestation, and land use change—are contributing to this warming trend. Scientists predict that if this trend continues, one-third of plant and animal habitats will be dramatically altered and more than 1 million species will be threatened with extinction in the next 50 years. And even small increases in global temperatures can melt glaciers and polar ice sheets, raising sea levels and flooding coastal cities and towns.

▲ DEFORESTATION

Of the 13 million hectares (32 million acres) of forest lost each year, mostly to make room for agriculture, more than half are in South America and Africa, where many of the world's tropical rain forests and terrestrial plant and animal species can be found. Loss of habitat in such species-rich areas takes a toll on the world's biodiversity. Deforested areas also release, instead of absorb, carbon dioxide into the atmosphere, contributing to global climate change. Deforestation can also affect local climates by reducing evaporative cooling, leading to decreased rainfall and higher temperatures.

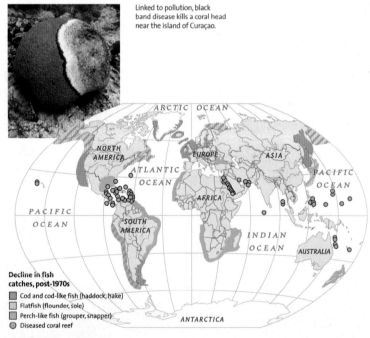

Linked to pollution, black band disease kills a coral head near the island of Curaçao.

Decline in fish catches, post-1970s
- ■ Cod and cod-like fish (haddock, hake)
- ■ Flatfish (flounder, sole)
- ■ Perch-like fish (grouper, snapper)
- ⊙ Diseased coral reef

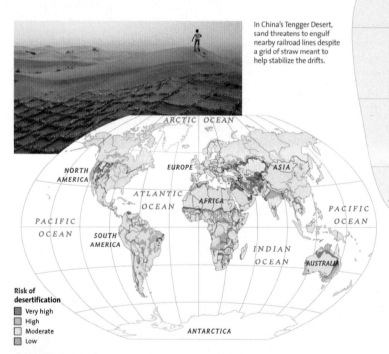

In China's Tengger Desert, sand threatens to engulf nearby railroad lines despite a grid of straw meant to help stabilize the drifts.

Risk of desertification
- ■ Very high
- ■ High
- □ Moderate
- ■ Low

▲ THREATENED OCEANS

Oceans cover more than two-thirds of Earth's surface and are home to at least half of the world's biodiversity, yet they are the least understood ecosystems. The combined stresses of overfishing, pollution, increased carbon dioxide emissions, global climate change, and coastal development are having a serious impact on the health of oceans and ocean species. Over 70 percent of the world's commercial fish species are depleted or nearing depletion, and 50 percent of coral reefs worldwide are threatened by human activities.

▲ DESERTIFICATION

Climate variability and human activities, such as grazing and conversion of natural areas to agricultural use, are leading causes of desertification, the degradation of land in arid, semiarid, and dry subhumid areas. The environmental consequences of desertification are great—loss of topsoil, increased soil salinity, damaged vegetation, regional climate change, and a decline in biodiversity. Equally critical are the social consequences—more than 2 billion people live in and make a living off these dryland areas, covering about 41 percent of Earth's surface.

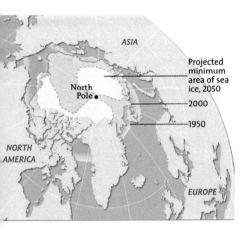

POLAR ICE CAP

ver the past 50 years, the xtent of polar sea ice has oticeably decreased. Since 970 alone, an area larger an Norway, Sweden, and Denmark combined has melted. This trend is predicted to accelerate as temperatures rise in the Arctic and across the globe.

GeoBytes

ACIDIFYING OCEANS
Oceans are absorbing an unprecedented 20 to 25 million tonnes (22 to 28 millions tons) of carbon dioxide each day, increasing the water's acidity.

ENDANGERED REEFS
Some 95 percent of coral reefs in Southeast Asia have been destroyed or are threatened.

RECORD TEMPERATURES
The 1990s was the warmest decade on record in the past century.

WARMING ARCTIC
While the world as a whole has warmed nearly 0.6°C (1°F) over the past 100 years, parts of the Arctic have warmed 4 to 5 times as much in only the past 50 years.

DISAPPEARING RAIN FORESTS
Scientists predict that the world's rain forests will disappear within the next 100 years if the current rate of deforestation continues.

OIL POLLUTION
Nearly 1.3 million tonnes (1.4 million tons) of oil seep into the world's oceans each year from the combined sources of natural seepage, extraction, transportation, and consumption.

ACCIDENTAL DROWNINGS
Entanglement in fishing gear is one of the greatest threats to marine mammals.

A FAREWELL TO FROGS?
Worldwide, almost half of the 5700 named amphibian species are in decline.

With the growth of scientific record keeping, observation, modeling, and analysis, our understanding of Earth's environment is improving. Yet even as we deepen our insight into environmental processes, we are changing what we are studying. At no other time in history have humans altered their environment with such speed and force. Nothing occurs in isolation, and stress in one area has impacts elsewhere. Our agricultural and fishing practices, industrial processes, extraction of resources, and transportation methods are leading to extinctions, destroying habitats, devastating fish stocks, disturbing the soil, and polluting the oceans and the air. As a result, biodiversity is declining, global temperatures are rising, polar ice is shrinking, and the ozone layer continues to thin.

▼ **POLLUTION**
No corner of Earth is immune to pollution, be it in the air, soil, or water. Concentrations of pollution can be found in the industrial centers of North America, Europe, and, increasingly, Asia—and areas downwind or downstream from them. Shipping routes are sources of pollution, from oil spills to garbage dumpings.

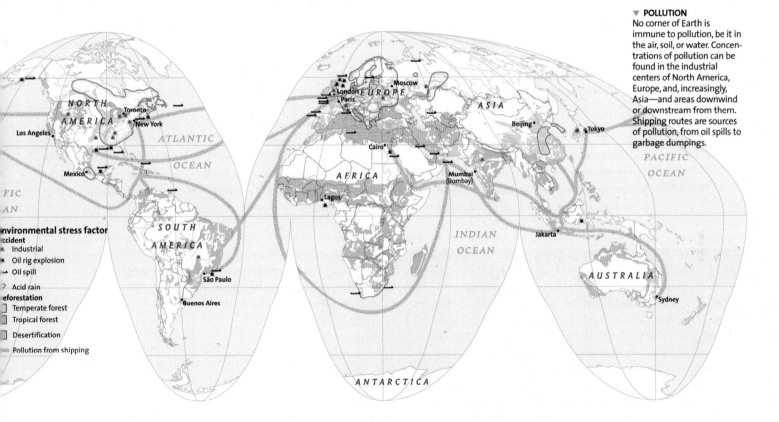

Environmental stress factor
ccident
* Industrial
* Oil rig explosion
- Oil spill
? Acid rain
eforestation
- Temperate forest
- Tropical forest
- Desertification
- Pollution from shipping

NATIONAL GEOGRAPHIC

OZONE DEPLETION

irst noted in the mid-1980s, he springtime "ozone hole" ver the Antarctic continues o thin. With sustained fforts to restrict chlorofluo- ocarbons (CFCs) and other zone-depleting chemicals, cientists have begun to ee what they hope is a eveling off in the rate of depletion. Stratospheric ozone shields the Earth from the sun's ultraviolet radiation. Thinning of this protective layer puts people at risk for skin cancer and cataracts. It can also have devastating effects on the Earth's biological functions.

1980

2000

2004

Atmospheric ozone
(Dobson Unit - DU)
Low   High

Environmental Science

LEARNING OBJECTIVES

1. **Define** *environmental science.*
2. **Outline** the steps of the scientific method.

Environmental science is an interdisciplinary field that combines information from many disciplines, such as biology, geography, chemistry, geology, physics, economics, sociology (particularly *demography*, the study of populations), cultural anthropology, natural resource management, agriculture, engineering, law, politics, and ethics.

Ecology, the discipline of biology that studies the interrelationships between organisms and their environment, is a basic tool of environmental science. **Atmospheric science** is a branch of environmental science that includes the study of weather and climate, greenhouse gases, and other airborne pollutants. **Environmental chemistry** examines chemicals in the environment, including air, soil, and water pollution (**Figure 1.8**). **Geosciences**—for example, environmental geology and physical geography—study a wide range of environmental topics, such as soil erosion, groundwater use, ocean pollution, and climate. Scientists in these subdisciplines not only evaluate environmental quality but also develop ways to restore damaged environments.

> **environmental science** The interdisciplinary study of humanity's relationship with other organisms and the physical environment.

The Goals of Environmental Science

Environmental scientists try to establish general principles about how the natural world functions. They use these principles to develop viable solutions to environmental problems—solutions that are based as much as possible on scientific knowledge.

Environmental problems are generally complex, however, and scientific understanding of them is often less complete than we would like. Environmental scientists are often called on to reach scientific consensus before the data are complete. As a result, they often cannot give precise answers and so instead make recommendations based on what they think is most likely to occur.

Many of the environmental problems considered in this book require urgent attention. Yet environmental science is not simply a "doom and gloom" listing of problems, coupled with predictions of a bleak future. To the contrary, the focus of environmental science, and our

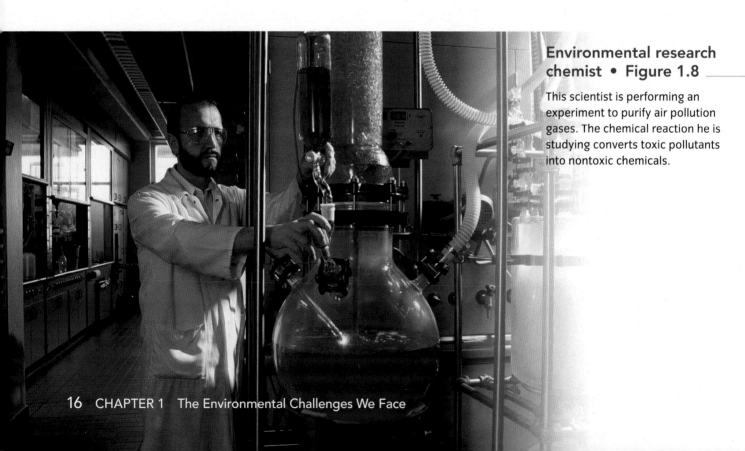

Environmental research chemist • Figure 1.8

This scientist is performing an experiment to purify air pollution gases. The chemical reaction he is studying converts toxic pollutants into nontoxic chemicals.

focus as individuals and as world citizens, is on identifying, understanding, and solving problems that we as a society have generated. It is encouraging that individuals, businesses, and governments are already doing a great deal, although more must be done to address the problems of today's world.

Science as a Process

The key to successfully solving any environmental problem is rigorous scientific evaluation. It is important to understand clearly just what science is, as well as what it is not. Most people think of **science** as a body of knowledge—a collection of facts about the natural world. However, science is also a dynamic *process*, a systematic way to investigate the natural world. Science seeks to reduce the apparent complexity of our world to general principles, which are then used to make predictions, solve problems, or provide new insights.

Scientists collect objective **data** (singular, *datum*), the information with which science works. Data are collected through observation and experimentation and then analyzed or interpreted (**Figure 1.9**). Scientific conclusions are inferred from the available data and are not based on faith, emotion, or intuition. Scientists publish their findings in scientific journals, and other scientists examine and critique their work. A requirement of science is repeatability—that is, observations and experiments must produce consistent data when they are repeated. The scrutiny by other scientists reveals any inconsistencies in results or interpretation. These errors are discussed openly, and ways to eliminate them are developed.

There is no absolute certainty or universal agreement about anything in science. Science is an ongoing

Data collection • Figure 1.9

A researcher observes genetically modified rice plants. Photographed at Cornell University, New York. (Inset) A scientist records his data on a computer in the laboratory.

enterprise, and scientific concepts must be reevaluated in light of newly discovered data. Thus, scientists never claim to know the final answer about anything because scientific understanding changes.

While science tells us what *is* and what *can be*, it cannot tell us what *should be*. Questions about what should be are in the realm of religion, ethics, policy, and philosophy. Once we have used these approaches to decide on our priorities and preferences, science is the most useful tool available to help us achieve them. Science aims to discover and better understand the general principles that govern the operation of the natural world.

The Scientific Method The established processes that scientists use to answer questions or solve problems are collectively called the scientific method. Although there are many variations of the scientific method, it basically involves five steps:

> **scientific method**
> The way a scientist approaches a problem, by formulating a hypothesis and then testing it.

1. Recognize a question or an unexplained phenomenon in the natural world.

2. Develop a *hypothesis*, or the expected answer to the question.

3. Design and perform an experiment to test the hypothesis.

4. Analyze and interpret the data to reach a conclusion.

5. Share new knowledge with the scientific community.

Although the scientific method is often portrayed as a linear sequence of events, science is rarely as straightforward or tidy as the scientific method implies (**Figure 1.10**). Good science involves creativity, not only in recognizing questions and developing hypotheses but also in designing experiments. Because scientists try to expand our current knowledge, their work is in the realm of the unknown. Many creative ideas end up as dead ends, and there are often temporary setbacks or reversals of direction as scientific knowledge progresses. Scientific knowledge often expands haphazardly, with the "big picture" emerging slowly from confusing and sometimes contradictory details.

Scientific discoveries are often incorrectly portrayed in the media as "new facts" that have just come to light. At a later time, additional "new facts" that question the validity of the original study are reported. If you were to read the scientific papers on which such media reports are based, however, you would find that all the scientists involved probably made very tentative conclusions based on their data. Science progresses from uncertainty to less uncertainty, not from certainty to greater certainty. Thus, science is self-correcting over time, despite the fact that it never "proves" anything.

The Importance of Prediction A scientific **hypothesis** needs to be useful—it needs to tell you something you want to know. A hypothesis is most useful when it makes predictions because the predictions provide a way to test the validity of the hypothesis. If your experiment refutes your prediction, then you must carefully recheck the entire experiment. If the prediction is still refuted, then you must reject the hypothesis. The more verifiable predictions a hypothesis makes, the more valid that hypothesis is.

Each of the many factors that influence a process is called a **variable**. To evaluate alternative hypotheses about a specific variable, it is necessary to hold all other variables constant so that they are not misleading or confusing. To test a hypothesis about a variable, we carry out two forms of the experiment in parallel. In the **experimental group**, the chosen variable is altered in a known way. In the **control group**, that variable isn't altered. In all other respects, the experimental group and the control group are the same. We then ask, "What is the difference, if any, between the outcomes for the two groups?" Any difference must be due to the influence of the variable we changed because all other variables remained the same. Much of the challenge of science lies in designing control groups and in successfully isolating a single variable from all other variables.

Theories A **theory** is an integrated explanation of numerous hypotheses, each of which is supported by a large body of observations and experiments. A theory condenses and simplifies many data that previously appeared to be unrelated. Because a theory demonstrates the relationships among different data, it simplifies and clarifies our understanding of the natural world. A good theory grows as additional information becomes known. It predicts new data and suggests new relationships among a range of natural phenomena.

Theories are the solid ground of science, the explanations of which we are most sure. This definition contrasts sharply with the general public's use of the word *theory*, which implies lack of knowledge or a guess. In this book, the word *theory* is always used in its scientific sense, to refer to a broadly conceived, logically coherent, and well-supported explanation.

The scientific method • Figure 1.10

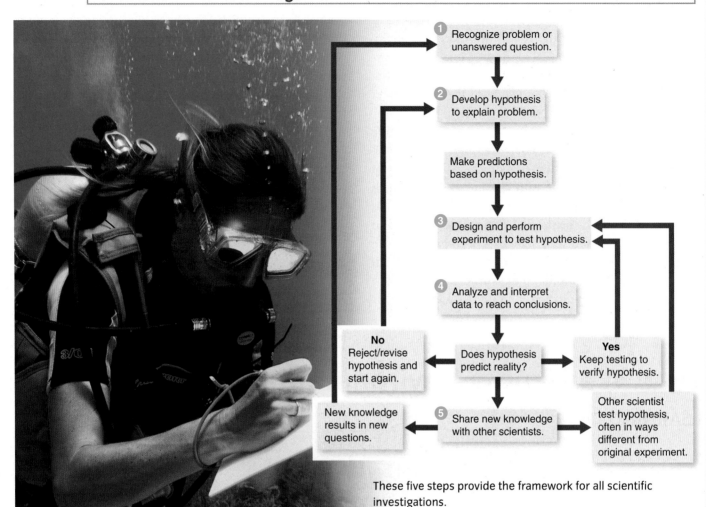

1. Recognize problem or unanswered question.

2. Develop hypothesis to explain problem.

Make predictions based on hypothesis.

3. Design and perform experiment to test hypothesis.

4. Analyze and interpret data to reach conclusions.

No
Reject/revise hypothesis and start again.

Does hypothesis predict reality?

Yes
Keep testing to verify hypothesis.

New knowledge results in new questions.

5. Share new knowledge with other scientists.

Other scientist test hypothesis, often in ways different from original experiment.

These five steps provide the framework for all scientific investigations.

3. A field scientist makes observations critical to understanding damage to coral reefs from global climate change. Photographed at Turneffe Atoll, Belize.

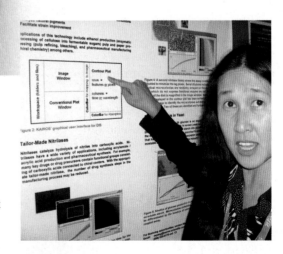

5. Many scientists present their research during poster sessions at scientific meetings. This allows their work to be critically assessed by others in the scientific community.

NATIONAL GEOGRAPHIC

WILEY
RESOURCE KIT

Think Critically What is the relationship between a hypothesis and an experiment?

Despite the fact that theories are generally accepted, there is no absolute truth in science, only varying degrees of uncertainty. Science is continually evolving as new evidence comes to light, and therefore its conclusions are always provisional or uncertain. It is always possible that the results of a future experiment will contradict a prevailing theory and show at least one aspect of it to be false.

Uncertainty, however, does not mean that scientific conclusions are invalid. For example, overwhelming evidence links cigarette smoking and incidence of lung cancer. We can't state with absolute certainty which smokers will be diagnosed with lung cancer, but this uncertainty does not mean that there is no correlation between smoking and lung cancer. On the basis of the available evidence, we say that people who smoke have an increased risk of developing lung cancer.

In conclusion, the aim of science is to increase human comprehension by explaining the processes and events of nature. Scientists work under the assumption that all phenomena in the natural world have natural causes, and they formulate theories to explain these phenomena. The process of science as a human endeavor has shaped the world we live in and transformed our views of the universe and how it works.

CONCEPT CHECK STOP

1. **What** is environmental science? What are some of the disciplines involved in environmental science?

2. **What** are the five steps of the scientific method? Why is each important?

How We Handle Environmental Problems

LEARNING OBJECTIVES

1. **List** and briefly describe the five stages of solving environmental problems.

Before examining the environmental problems discussed in the remaining chapters of this book, let's consider the elements that contribute to solving those problems. How, for example, can we handle water pollution in a river (**Figure 1.11**)? At what point are conclusions regarded as certain enough to warrant action? Who makes the decisions, and what are the trade-offs? Viewed simply,

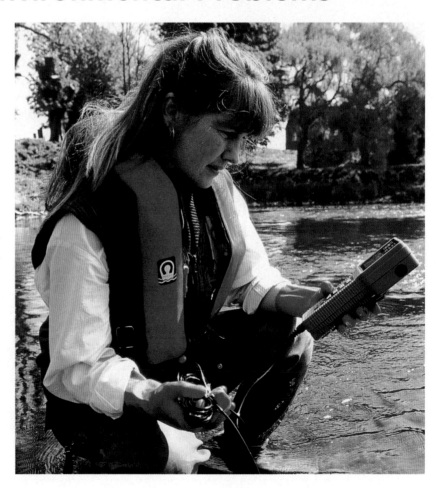

Monitoring water pollution • Figure 1.11

This pollution control officer is measuring the oxygen level in the Severn River near Shrewsbury, England. When dissolved oxygen levels are high, pollution levels (of sewage, fertilizer, and such) are low.

Addressing environmental problems • Figure 1.12

These five steps provide a framework for addressing environmental problems.

① Scientific assessment:
Problem is defined, hypotheses are tested, and models are constructed to show how present situation developed and to predict future course of events.

Example:
Scientists find higher-than-normal levels of bacteria are threatening a lake's native fish and determine the cause is human-produced pollution.

② Risk analysis:
Potential effects of various interventions—including doing nothing—are analyzed to determine risks associated with each particular course of action.

If no action is taken, fishing resources —a major source of income in the region—will be harmed. If pollution is reduced appreciably, fishery will recover.

③ Public engagement:
Changing public attitudes involves explaining the problem, presenting available alternatives for action, and revealing probable risks, results, and cost of each choice.

Public is informed of the ramifications—in this case, loss of income—if problem is not addressed.

⑤ Long-term environmental management:
Results of any action taken should be carefully monitored to see if environmental problem is being addressed.

Water quality in lake is tested frequently, and fish populations are monitored to ensure they do not decline.

④ Political considerations:
Elected officials, often at urging of their constituencies, implement a course of action based on scientific evidence as well as economic and social considerations.

Elected officials, supported by the public, pass legislation to protect lake and develop lake cleanup plan.

Think Critically Despite having a framework for addressing environmental problems, many problems are either incorrectly addressed or not addressed adequately. Offer at least one possible reason for such failures.

---

there are five stages in addressing an environmental problem (**Figure 1.12**):

1. Scientific assessment

2. Risk analysis

3. Public engagement

4. Political considerations

5. Long-term environmental management

These five stages represent an ideal approach to systematically addressing environmental problems. In real life, seeking solutions to environmental problems is rarely so neat and tidy, particularly when the problem is exceedingly complex, of regional or global scale, or has high costs and unclear benefits for the money invested. Quite often, the public becomes aware of a problem, which triggers discussion of remediation before the problem is clearly identified and scientifically assessed.

CONCEPT CHECK 🛑 STOP

1. **What** are the five steps used to solve an environmental problem?

EnviroDiscovery:

NIMBY and NIMTOO

The NIMBY response is an unavoidable complication in addressing many environmental issues. NIMBY stands for "not in my backyard." As soon as people hear that a power plant, an incinerator, or a hazardous waste disposal site may be situated nearby, the NIMBY response rears its head. Part of the reason NIMBYism is so prevalent is that, despite the assurances experts give that a site will be safe, no one can guarantee complete safety and no possibility of accident. In many cases, people who are labeled as NIMBYs simply wish to be involved in the decision-making process. Public planners and developers often do not engage people living in low-income urban areas, older suburban areas, or rural areas to help make decisions that affect their neighborhoods.

A sister response to NIMBY is NIMTOO, which stands for "not in my term of office." Politicians who wish to get reelected are sensitive to their constituents' concerns and are not likely to support the construction of power plants or waste disposal sites in their districts.

Exacerbating the NIMBY and NIMTOO responses is the failure by companies and government to develop processes for listening and responding to public concerns. When experts are not trusted, people don't believe their analyses, no matter how scientifically valid. Experts, who typically do not have training in effective communication, interpret this distrust as ignorance or emotion. Resentment and conflict follow.

Consider the disposal of radioactive waste from nuclear power plants. There is broad agreement that the best long-term solution is to safely isolate radioactive waste, preferably deep underground, for thousands of years. However, rather than explore a range of possible disposal sites, the US government, backed by the nuclear energy industry, committed in 1982 to explore only a single disposal site, Yucca Mountain in Nevada. It then spent the next three decades studying only that site. As Nevada became more politically powerful, its residents objected to the process, which was often interpreted as an "antiscientific" attitude. Only recently (in 2010) did a new process begin, one that incorporates broad perspectives and stakeholders in a national conversation.

Most people agree that our generation has the responsibility to dispose of hazardous waste. Only we want to put it in some other state, in someone else's backyard. Arguing against any disposal scheme that is proposed will simply result in letting the waste remain where it is now. Although this may be the only politically acceptable solution, it is unacceptable from a safety viewpoint.

Not in my backyard

Steam rises from two of the cooling towers of a nuclear power plant. All nuclear power plants store highly radioactive spent fuel on site because there is currently no place to safely dispose of it.

NATIONAL GEOGRAPHIC

The New Orleans Disaster

Hurricane Katrina, which hit the north-central Gulf Coast in August 2005, was one of the most devastating storms in U.S. history. It produced a storm surge that caused severe damage to New Orleans as well as to other coastal cities and towns in the region. The high waters caused levees and canals to fail, flooding 80 percent of New Orleans and many nearby neighborhoods.

Most people are aware of the catastrophic loss of life and property caused by Katrina. Here we focus on how humans altered the geography and geology of the New Orleans area in ways that exacerbated the storm damage.

Over the years, engineers constructed a system of canals to aid navigation and a system of levees to control flooding because the city is at or below sea level. The canals allowed salt water to intrude and kill the freshwater marsh vegetation. The levees prevented the deposition of sediments that remain behind after floodwaters subside (The sediments are now deposited in the Gulf of Mexico.) Under natural conditions, these sediments replenish and maintain the delta, building up coastal wetlands.

As the city has grown, new development has taken place on wetlands—bayous, waterways, and marshes—that were drained and filled in. Before their destruction, these coastal wetlands provided some protection against flooding from storm surges. We are not implying that had Louisiana's wetlands been intact, New Orleans would not have suffered any damage from a hurricane of Katrina's magnitude. However, had these wetlands been largely unaltered, they would have moderated the storm's damage by absorbing much of the water from the storm surge.

Another reason that Katrina devastated New Orleans is that the city has been subsiding (sinking) for many years, primarily because New Orleans is built on unconsolidated sediment (no bedrock underneath). Many wetlands scientists also attribute this subsidence to the extraction of the area's rich supply of underground natural resources—groundwater, oil, and natural gas. As these resources are removed, the land compacts, lowering the city. New Orleans and nearby coastal areas are subsiding an average of 6 mm each year (see image). At the same time, the sea level has been rising an average of 1 mm to 2.5 mm per year due to human-induced changes in climate.

Satellite image of flooding in New Orleans following Hurricane Katrina

Along the left (west) side is a levee from Lake Ponchartrain (top) that failed so that water inundated the New Orleans area east of the levee. Areas on the far left top remained dry. Part of the Mississippi River is shown at lower center.

WILEY
RESOURCE KIT

Summary

1 Human Impacts on the Environment 4

1. **Highly developed countries** are countries that have complex industrialized bases, low rates of population growth, and high per person incomes. **Moderately developed countries** are developing countries that have medium levels of industrialization and average per person incomes lower than those of highly developed countries. **Less developed countries (LDCs)** are developing countries with low levels of industrialization, very high rates of population growth, very high infant mortality rates, and very low per person incomes (relative to highly developed countries). **Poverty**, which is common in LDCs, is a condition in which people are unable to meet their basic needs for food, clothing, shelter, education, or health.

2. The increasing global population is placing stresses on the environment, as humans consume ever-increasing quantities of food and water, use more energy and raw materials, and produce enormous amounts of waste and pollution. **Nonrenewable resources** are natural resources that are present in limited supplies and are depleted as they are used. **Renewable resources** are resources that natural processes replace and that therefore can be used forever, provided that they are not exploited in the short term.

3. **People overpopulation** is a situation in which too many people live in a given geographic area. Developing countries have people overpopulation. **Consumption overpopulation** is a situation that occurs when each individual in a population consumes too large a share of resources. Highly developed countries have consumption overpopulation.

4. The forces that drive environmental impact can be modeled by the IPAT equation, $I = P \times A \times T$. Environmental impact (I) has three factors: the number of people (P); the affluence per person (A), which is a measure of the consumption, or amount of resources used per person; and the environmental effect of the technologies used to obtain and consume those resources (T).

2 Sustainability and the Environment 12

1. **Environmental sustainability** is the ability to meet humanity's current needs without compromising the ability of future generations to meet their needs. Sustainability implies that the environment can function indefinitely without going into a decline from the stresses that human society imposes on natural systems.

2. Human behaviors that threaten environmental sustainability include overuse of renewable and nonrenewable resources, pollution, and overpopulation.

Stabilize human population	Prevent pollution where possible	Restore degraded environments
Protect natural ecosystems	**Focus on Sustainability**	Use resources efficiently
Educate children and adults	Prevent and reduce waste	Eradicate hunger and poverty

3 Environmental Science 16

1. **Environmental science** is the interdisciplinary study of humanity's relationship with other organisms and the nonliving physical environment. Environmental science encompasses many problems involving human numbers, Earth's natural resources, and environmental pollution.

2. The **scientific method** is the way a scientist approaches a problem, by formulating a **hypothesis** and then testing it by means of an experiment. (1) A scientist recognizes and states the problem or unanswered question. (2) The scientist develops a hypothesis, or an educated guess, to explain the problem. (3) An experiment is designed and performed to test the hypothesis. (4) **Data**, the results obtained from the experiment, are analyzed and interpreted to reach a conclusion. (5) The conclusion is shared with the scientific community.

4 How We Handle Environmental Problems 20

1. Addressing environmental problems ideally requires five stages. (1) Scientific assessment involves identifying a potential environmental problem and collecting data to construct a model. (2) Risk analysis evaluates the potential effects of intervention. (3) Public engagement occurs when the results of scientific assessment and risk analysis are placed in the public arena. (4) In political considerations, elected or appointed officials implement a particular risk-management strategy. (5) Long-term environmental management monitors the effects of the action taken.

Key Terms

- consumption overpopulation 9
- environmental science 16
- environmental sustainability 12
- highly developed countries 4
- less developed countries 6
- moderately developed countries 6
- nonrenewable resources 7
- people overpopulation 9
- poverty 4
- renewable resources 7
- scientific method 18

What is happening in this picture?

- What are these people protesting?

- How is this opposition an example of NIMBY?

- What sorts of political processes and scientific information might make it easier to find a long-term solution to nuclear waste management?

Critical and Creative Thinking Questions

1. Criticize the following statement: "Population growth in developing countries is of much more concern than is population growth in highly developed countries."

2. Why is population growth often linked to excess resource extraction and consumption?

3. Explain why a country with the world's largest level of consumption may not have the largest population.

4. Explain how population, affluence, and technology interact in complex ways.

5. Do you think our current worldwide population growth and economic growth are sustainable? Why or why not?

6. Give at least two examples of things that you can do as an individual to promote environmental sustainability.

7. How does the field of environmental science involve science? economics? politics?

8. Your throat feels scratchy, and you think you're coming down with a cold. You take a couple of vitamin C pills and feel better. You conclude that vitamin C helps prevent colds. Is your conclusion valid from a scientific standpoint? Why or why not?

9. People want scientists to give them precise, definitive answers to environmental problems. Explain why this is not possible.

10. Examine the graph, which shows an estimate of the discrepancy between the wealth of the world's poorest countries and that of the richest countries.

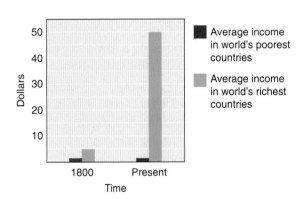

a. How has the distribution of wealth changed from the 1800s to the present? How would you explain this difference?

b. Based on the trend evident in this graph, predict what the graph might look like in 100 years.

c. Some economists think that our current path of economic growth is unsustainable. Do the data in this graph support or refute this idea? Explain your answer.

Environmental Sustainability and Human Values

THE GLOBAL COMMONS

Ecologist **Garrett Hardin** (1915–2003, see inset) is best known for his 1968 essay "The Tragedy of the Commons." In it he contended that our inability to solve complex environmental problems is the result of a struggle between short-term individual welfare and long-term *environmental sustainability*.

In today's world, Hardin's essay has particular relevance at the global level. The "commons" that he refers to in the title of his essay are those parts of our environment that are available to everyone but for which no single individual has responsibility: the atmosphere, water, wildlife, forests, and fisheries. For example, the photograph shows an entire shoal of tuna caught in a purse seine and surrounded by a towing cage, which will take them to a tuna farm for fattening and harvest.

The world needs effective legal and economic policies to prevent the degradation of our global commons. We must foster a strong sense of *stewardship*, or shared responsibility, for the sustainable care of our planet.

This chapter examines the role of ethics and values in environmental issues. As you read, keep in mind these words from the *Earth Charter*, formulated in 1992 by representatives from 178 countries: "Let ours be a time remembered for the awakening of a new reverence for life, the firm resolve to achieve sustainability, the quickening of the struggle for justice and peace, and the joyful celebration of life."

CHAPTER OUTLINE

CHAPTER PLANNER ✓

- ❏ Study the picture and read the opening story
- ❏ Scan the Learning Objectives in each section:
  p. 28 ❏   p. 31 ❏   p. 35 ❏   p. 36 ❏
- ❏ Read the text and study all figures and visuals. Answer any questions.

Analyze key features

- ❏ Environmental InSight, p. 37 ❏
- ❏ Process Diagram, p. 42 ❏
- ❏ Case Study, p. 44 ❏
- ❏ Stop: Answer the Concept Checks before you go on:
  p. 30 ❏   p. 34 ❏   p. 36 ❏   p. 43 ❏

End of Chapter

- ❏ Review the Summary and Key Terms.
- ❏ Answer What is happening in this picture?
- ❏ Answer the Critical and Creative Thinking Questions.

Human Use of the Earth

LEARNING OBJECTIVES

1. **Define** *sustainable development.*
2. **Outline** some of the complexities associated with the concept of sustainable consumption.
3. **Define** *voluntary simplicity.*

Environmental sustainability is a concept that people have discussed for many years. *Our Common Future*, the 1987 report of the U.N. World Commission on Environment and Development, presented the closely related concept of **sustainable development** (**Figure 2.1**). The authors of *Our Common Future* pointed out that sustainable development includes meeting the needs of the world's poor. The report also linked the environment's ability to meet present and future needs to the state of technology and social organization existing at a given time and in a given place. The number of people, their degree of affluence (that is, their level of consumption), and their choices of technology all interact to produce the total effect of a given society, or of society at large, on the sustainability of the environment.

> **sustainable development**
> Economic growth that meets the needs of the present without compromising the ability of future generations to meet their needs.

Even using the best technologies imaginable, Earth's productivity still has limits, and our use of it can't be expanded indefinitely. *Sustainable development can occur only within the limits of the environment.* To live within these limits, population growth must be held at a level that we can sustain, and the wealthy must first stabilize their use of natural resources and then reduce that use to a level that can be maintained. The world does not contain nearly enough resources to sustain everyone at the level of consumption that is enjoyed in the United States, Europe, and Japan. Suitable strategies, however, do exist to reduce these levels of consumption without concurrently reducing the real quality of life.

Sustainable Consumption

As you saw in Chapter 1, **consumption overpopulation** is pollution and degradation of the environment that occurs when each individual in a population consumes too large a share of resources. Consumption overpopulation stems from the lifestyles of people living in highly developed

Sustainable development • Figure 2.1

Three factors—environmentally sound decisions, economically viable decisions, and socially equitable decisions—interact to promote sustainable development.

Is sustainable development a reasonable goal at the local level? at the global level? Explain your answers.

The challenge of eradicating poverty • Figure 2.2

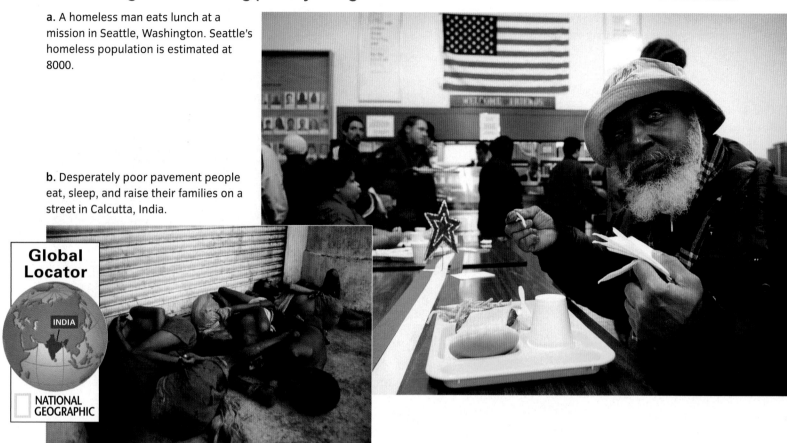

a. A homeless man eats lunch at a mission in Seattle, Washington. Seattle's homeless population is estimated at 8000.

b. Desperately poor pavement people eat, sleep, and raise their families on a street in Calcutta, India.

Global Locator

INDIA

NATIONAL GEOGRAPHIC

nations. *Lifestyle* is interpreted broadly to include goods and services bought for food, clothing, housing, travel, recreation, and entertainment. In evaluating consumption overpopulation, all aspects of the production, use, and disposal of these goods and services are taken into account, including environmental costs. Such an analysis provides a sense of what it means to consume sustainably versus unsustainably.

Sustainable consumption, like sustainable development, forces us to address whether our present actions undermine the long-term ability of the environment to meet the needs of future generations. Factors that affect sustainable consumption include population, economic activities, technology choices, social values, and government policies.

At the global level, sustainable consumption requires the eradication of poverty (**Figure 2.2**). This in turn requires that poor people *increase* their consumption of certain essential resources.

> **sustainable consumption**
> The use of goods and services that satisfy basic human needs and improve the quality of life but that also minimize resource use.

For their increased consumption to be sustainable, however, the consumption patterns of people in highly developed countries must change.

Widespread adoption of sustainable consumption will not be easy. It will require major changes in the consumption patterns and lifestyles of most people in highly developed countries. Some examples of promoting sustainable consumption include switching from motor vehicles to public transport and bicycles and developing durable, repairable, recyclable products.

An increasing number of people in the United States and other highly developed nations have embraced a type of sustainable consumption known as **voluntary simplicity**, which recognizes that individual happiness and quality of life are not necessarily linked to the accumulation of material goods. People who embrace voluntary simplicity recognize that a person's values and character define that individual more than how many things he or she owns.

More than 30,000 Philadelphians have signed up for PhillyCarShare, the world's largest community-based car-sharing organization.

This belief requires a change in behavior as people purchase and use fewer items than they might have formerly. It is a commitment at the individual level to saving the planet for future generations.

One example of voluntary simplicity is car sharing. Car-sharing programs, which are designed for people who use a car occasionally, offer an economical alternative to individual ownership (**Figure 2.3**). Car sharing may also reduce the numbers of cars manufactured. Studies show that most car sharers drive significantly fewer miles than they did before they joined the program.

Sustainable consumption and voluntary simplicity are not popular ideas with policymakers, politicians, and economists, or even with consumers such as you and me. They contain an inherent threat to "business as usual," but they also offer new and exciting opportunities. Many scientists and population experts increasingly advocate adoption of sustainable consumption before it is forced on us by an environmentally degraded, resource-depleted world.

As people adopt new lifestyles, they must be educated so that they understand the reasons for changing practices that may be highly ingrained or traditional. Formal education and informal education are both important in bringing about change and in contributing to sustainable consumption. If people understand the way the natural world functions, they can appreciate their own place in it and value sustainable actions.

Any long-term involvement in the condition of the world must start with individuals—our values, attitudes, and practices. Each of us makes a difference, and it is ultimately our collective activities that make the world what it is.

CONCEPT CHECK

1. **What** is sustainable development?

2. **What** is sustainable consumption? How is it linked to a reduction in world poverty?

3. **How** is voluntary simplicity an example of sustainable consumption?

Human Values and Environmental Problems

LEARNING OBJECTIVES

1. **Define** *environmental ethics.*
2. **Discuss** distinguishing features of the Western and deep ecology worldviews.

We now shift our attention to the views of different individuals and societies and how those views affect our ability to understand and solve sustainability problems. **Ethics** is the branch of philosophy that is derived through the logical application of human **values**. These values are the principles that an individual or a society considers important or worthwhile. Values are not static entities but change as societal, cultural, political, and economic priorities change. Ethics helps us determine which forms of conduct are morally acceptable and unacceptable, right and wrong. Ethics plays a role in any types of human activities that involve intelligent judgment and voluntary action. Whenever alternative, conflicting values occur, ethics helps us choose which value is better, or worthier, than other values.

Environmental ethics examines moral values to determine how humans should relate to the natural environment. Environmental ethicists consider such

> **environmental ethics** A field of applied ethics that considers the moral basis of environmental responsibility.

issues as what role we should play in determining the fate of Earth's resources, including other species, or how we might develop an environmental ethic that is acceptable in the short term for us as individuals and also in the long term for our species and the planet. These issues and others like them are difficult intellectual questions that involve political, economic, societal, and individual trade-offs.

Environmental ethics considers not only the rights of people living today, both individually and collectively, but also the rights of future generations (**Figure 2.4**). This aspect of environmental ethics is critical because the impacts of today's activities and technologies are changing the environment. In some cases these impacts may be felt for hundreds or even

Tomorrow's generation • Figure 2.4

The choices made today will determine whether future generations, such as these students from Bailey Elementary School in Falls Church, Virginia, will inherit a sustainable world.

thousands of years. Addressing issues of environmental ethics puts us in a better position to use science and technology for long-term environmental sustainability.

Worldviews

Each of us has a particular **worldview**—that is, a personal perspective based on a collection of our basic values that helps us make sense of the world, understand our place and purpose in it, and determine right and wrong behaviors. These worldviews lead to behaviors and lifestyles that may or may not be compatible with environmental sustainability.

Two extreme, competing environmental worldviews are the Western worldview and the deep ecology worldview. These two worldviews, admittedly broad generalizations, are at nearly opposite ends of a spectrum of worldviews relevant to global sustainability problems, and they approach environmental responsibility in radically different ways.

The traditional Western worldview, also known as the *expansionist worldview*, is human centered and utilitarian. It mirrors the beliefs of the 19th-century **frontier attitude**, a desire to conquer and exploit nature as quickly as possible (**Figure 2.5**). The Western worldview also advocates the inherent rights of individuals, accumulation of wealth, and unlimited consumption of

goods and services to provide material comforts. According to the Western worldview, humans have a primary obligation to humans and are therefore responsible for managing natural resources to benefit human society. Thus, any concerns about the environment are derived from human interests.

The **deep ecology worldview** is a diverse set of viewpoints that dates from the 1970s and is based on the work of **Arne Naess**, a Norwegian philosopher, and others, including ecologist **Bill Devall** and philosopher **George Sessions**. The principles of

> **environmental worldview** A worldview based on how the environment works, our place in the environment, and right and wrong environmental behaviors.
>
> **Western worldview** A worldview based on human superiority over nature, the unrestricted use of natural resources, and economic growth to manage an expanding industrial base.

> **deep ecology worldview** A worldview based on harmony with nature, a spiritual respect for life, and the belief that humans and all other species have an equal worth.

Western worldview • Figure 2.5

a. Logging operations in 1884. This huge logjam occurred on the St. Croix River near Taylors Falls, Minnesota.

b. The Western worldview in operation today. These logs were cut from plantations of non-native eucalyptus trees, which have replaced 30 million hectares (75 million acres) of tropical rain forest in Brazil's Atlantic forest.

deep ecology, as expressed by Naess in *Ecology, Community and Lifestyle* (1989), include:

1. Both human and nonhuman life have **intrinsic value** (**Figure 2.6**). The value of nonhuman life forms is independent of the usefulness they may have for narrow human purposes.

2. Richness and diversity of life forms contribute to the flourishing of human and nonhuman life on Earth.

3. Humans have no right to reduce this richness and diversity except to satisfy vital needs.

4. Present human interference with the nonhuman world is excessive, and the situation is rapidly worsening.

5. The flourishing of human life and cultures is compatible with a substantial decrease in the human population. The flourishing of nonhuman life requires such a decrease.

6. Significant change of life conditions for the better requires changes in economic, technological, and ideological structures.

7. The ideological change is mainly that of appreciating life quality rather than adhering to a high standard of living.

8. Those who subscribe to the foregoing points have an obligation to participate in the attempt to implement the necessary changes.

Philosophers recognize two kinds of value, instrumental and intrinsic • Figure 2.6

a. According to the deep ecology worldview (right side of triangle), organisms have intrinsic value—that is, they are valued for their own sake, not for the goods and services they provide.

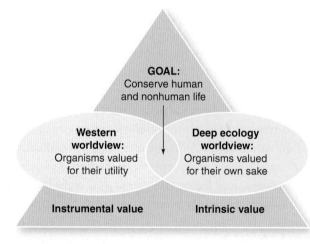

GOAL:
Conserve human and nonhuman life

Western worldview:
Organisms valued for their utility

Deep ecology worldview:
Organisms valued for their own sake

Instrumental value   **Intrinsic value**

Think Critically Why is there an overlapping goal between these two extreme worldviews?

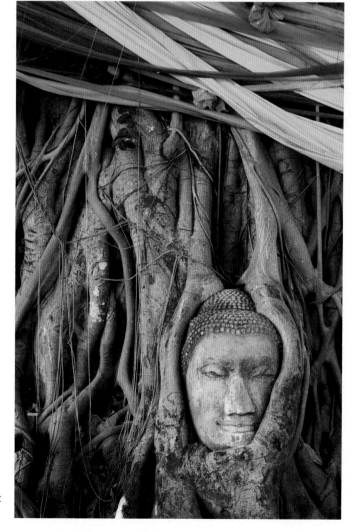

b. A tree trunk has grown around the head of Buddha at Wat Mahathat in Thailand, symbolizing the oneness of Buddha with nature. Buddhists practice the stilling of human desires, the reduction of consumption, and the contemplation of nature. Like Buddhism, many of the world's other religions espouse the intrinsic value of living things.

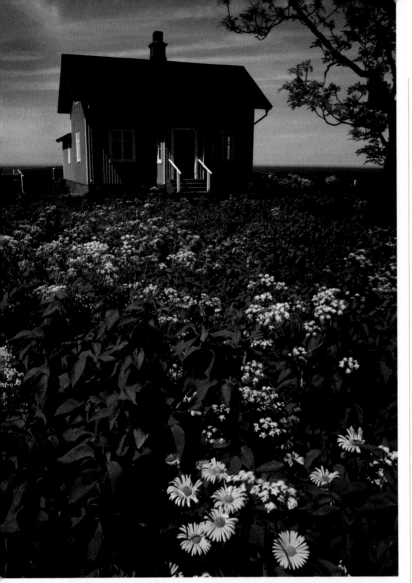

Embracing deep ecology • Figure 2.7

At one time or another, most of us yearn for the simpler life that the tenets of deep ecology advocate. However, there are far too many people and there is far too little land for us all to embrace this lifestyle. Photographed on Gotland Island, Sweden.

Compared to the Western worldview, the deep ecology worldview represents a radical shift in how humans relate to the environment. The deep ecology worldview stresses that all forms of life have the right to exist and that humans are not different or separate from other organisms. Humans have an obligation to themselves and to all other organisms living on Earth. The deep ecology worldview advocates sharply curbing human population growth. It does not advocate returning to a society free of today's technological advances but instead proposes a significant rethinking of our use of current technologies and alternatives. It asks individuals and societies to share an inner spirituality connected to the natural world.

Most people today do not fully embrace either the Western worldview or the deep ecology worldview. The Western worldview is **anthropocentric** and emphasizes the importance of humans as the overriding concern in the grand scheme of things. In contrast, the deep ecology worldview is **biocentric** and views humans as one species among others. The planet's natural resources could not support its more than 6.8 billion humans if each consumed the high level of goods and services sanctioned by the Western worldview. On the other hand, the world as envisioned by the deep ecology worldview could support only a fraction of the existing human population (**Figure 2.7**).

These worldviews, while not practical for widespread adoption, are useful to keep in mind as you examine various environmental issues in later chapters. In the meantime, you should think about your own worldview and discuss it with others—whose worldviews will probably be different from your own. Thinking leads to actions, and actions lead to consequences. What are the short-term and long-term consequences of your particular worldview? Are you satisfying your needs, or are your wants and desires leading you to consume in excess of what you actually need? What would be the environmental consequences if your worldview were adopted by everybody on the planet?

According to **Robert Cahn**, a 20th century environmental journalist: *The main ingredients of an environmental ethic are caring about the planet and all of its inhabitants, allowing unselfishness to control the immediate self-interest that harms others, and living each day so as to leave the lightest possible footprints on the planet.* We must develop and incorporate into our culture a long-lasting, environmentally sensitive worldview if the environment is to be sustainable for us, for other living organisms, which are linked to us through a long evolutionary history, and for future generations of both human beings and other life forms.

CONCEPT CHECK **STOP**

1. **What** is environmental ethics?
2. **What** assumptions underlie the western worldview? the deep ecology worldview?

Environmental Justice

LEARNING OBJECTIVE

1. **Define** *environmental justice.*

In the early 1970s, the Board of National Ministries of the American Baptist Churches coined the term *eco-justice* to link social and environmental ethics. At the local level, eco-justice encompasses environmental inequities faced by low-income minority communities. Many studies indicate that low-income communities and/or communities of color are more likely than others to have chemical plants, hazardous waste facilities, sanitary landfills, sewage treatment plants, and incinerators (**Figure 2.8**). A 1990 study at Clark Atlanta University, for example, found that six of Houston's eight incinerators were located in predominantly black neighborhoods. Such communities often have limited involvement in the political process and may not even be aware of their exposure to increased levels of pollutants.

Because people in low-income communities frequently lack access to sufficient health care, they may not be treated adequately for exposure to environmental contaminants. The high incidence of asthma in many minority communities, for example, may be caused or exacerbated by exposure to environmental pollutants. Many studies have examined how environmental pollutants interact with other socioeconomic factors to cause health problems. It is challenging to show to what extent a polluted environment is responsible for the disproportionate health problems of poor and minority communities. For example, a 1997 study of residents in San Francisco's polluted Bayview Hunters Point area found that hospitalization rates for chronic illnesses were the highest in the state. Researchers failed, however, to link these illnesses to an increased exposure to toxic pollutants, in part because an illness caused by exposure to pollution is clinically identical to the same illness that is caused by factors other than pollution.

In addition to their increased exposure to pollution, low-income communities may not receive equal benefits from federal cleanup programs. Several studies have reported that toxic waste sites in white communities were

A children's playground overlooks a pulp mill • Figure 2.8

Poor minority neighborhoods often have the most polluted and degraded environments. Photographed in Kingsport, Tennessee.

cleaned up faster and more thoroughly than those in Latino and African American communities.

Environmental Justice and Ethical Issues

There is an increasing awareness that environmental decisions such as where to locate a hazardous waste land-fill have important ethical dimensions. The most basic ethical dilemma centers on the rights of the poor and disenfranchised versus the rights of the rich and power-ful. Whose rights should have priority in these decisions? Is it ethically just if environmental burdens and benefits are not equally shared?

The challenge is to find and adopt solutions that respect all individuals, including those yet to be born.

| environmental justice The right of every citizen to adequate protection from environmental hazards. |

Environmental justice is a funda-mental human right in an ethical society. Although we may never completely eliminate past envi-ronmental injustices, we have a moral imperative to prevent them today so that their negative effects do not disproportion-ately affect any particular segment of society.

In response to these concerns, a growing environ-mental justice movement has emerged at the grassroots level as a strong motivator for change. Advocates are calling for special efforts to clean up hazardous sites in low-income neighborhoods, from inner-city streets to Native American reservations. On an international level, advocates of environmental justice point out that indus-trialized countries are obligated to help less-developed countries cope with climate change. These countries often suffer disproportionately from the problems caused by climate change, while it is the fossil fuel consumption in highly developed countries that is largely responsible for the changing climate.

| CONCEPT CHECK | STOP |

1. **What** is environmental justice and which communities are exposed to a disproportionate share of environmental hazards?

An Overall Plan for Sustainable Living

LEARNING OBJECTIVES

1. **Relate** poverty and population growth to carrying capacity and global sustainability.
2. **Discuss** problems related to loss of forests and declining biological diversity.
3. **Describe** the extent of food insecurity.
4. **Define** *enhanced greenhouse effect* and explain how stabilizing climate is related to energy use.
5. **Describe** at least two problems in cities in the developing world.

There is no shortage of suggestions for ways to address the world's many environmental problems. We have organized this section around the five recommendations for sustainable living presented in the 2006 book *Plan B 2.0: Rescuing a Planet Under Stress and a Civilization in Trouble*

by Lester R. Brown. If we as individuals and collectively as governments were to focus our efforts and financial support on Brown's plan, we think the quality of human life would be much improved. Brown's five recommenda-tions for sustainable living are:

1. Eliminate poverty and stabilize the human population.
2. Protect and restore Earth's resources.
3. Provide adequate food for all people.
4. Mitigate climate change.
5. Design sustainable cities.

Seriously addressing these recommendations offers hope for the kind of future we want for our children and grandchildren (**Figure 2.9**).

Environmental InSight

A plan for sustainable living • **Figure 2.9**

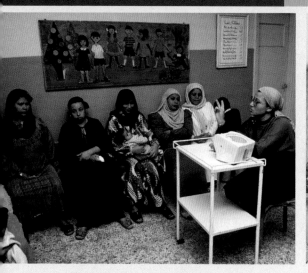

Family Planning in Egypt.
Women at a health clinic learn about family planning and birth control.

Recommendation 1: Eliminate Poverty and Stabilize the Human Population.

Restoration in Indonesia.
Mangrove trees are planted at low tide to help restore a coastal estuary.

Recommendation 2: Protect and Restore Earth's Resources.

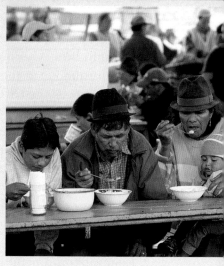

Feeding the World's People.
A family eats at a market in the Andes Mountains, Ecuador.

Recommendation 3: Provide Adequate Food for All People.

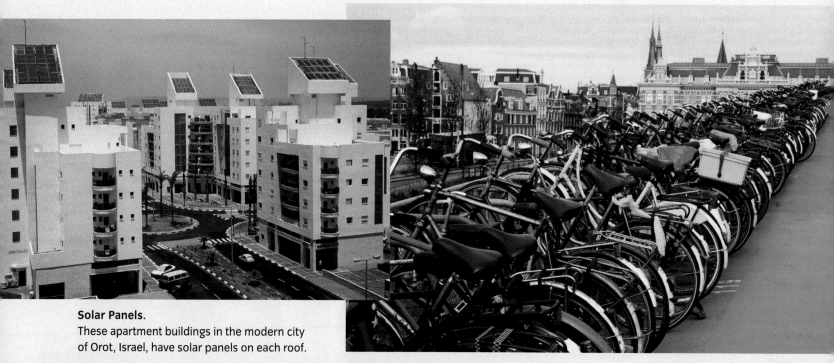

Solar Panels.
These apartment buildings in the modern city of Orot, Israel, have solar panels on each roof.

Recommendation 4: Mitigate Climate Change.

Bicycle Rack in Amsterdam.
Residents in the Netherlands ride bicycles an average of 573 mi (917 km) per year.

Recommendation 5: Design Sustainable Cities.

NATIONAL GEOGRAPHIC WILEY **RESOURCE KIT**

Recommendation 1: Eliminate Poverty and Stabilize the Human Population

The ultimate goal of economic development is to make it possible for humans throughout the world to enjoy long, healthy lives. A serious complication lies in the fact that the distribution of the world's resources is unequal. Residents of the United States are collectively the wealthiest people who have ever existed, with the highest standard of living (shared with a few other rich countries). The United States, with fewer than 5 percent of the world's people, controls about 25 percent of the world's economy but depends on other nations for this prosperity. Yet we often seem unaware of this relationship and tend to underestimate our effects on the environment that supports us.

Failing to confront the problem of poverty around the world makes it impossible to attain global sustainability. For example, most people would find it unacceptable that about 24,000 infants and children under age 5 die each day (2008 data from U.N. Children's Fund). Most of these deaths could have been prevented through access to adequate food and basic medical techniques and supplies. For us to allow so many to go hungry and to live in poverty threatens the global ecosystem that sustains us all. Everyone must have a reasonable share of Earth's

productivity. As U.S. President **Franklin Delano Roosevelt** said in his second inaugural address in 1937, "The test of our progress is not whether we add more to the abundance of those who have much; it is whether we provide enough for those who have too little."

Raising the standard of living for poor countries requires the universal education of children and the elimination of illiteracy (**Figure 2.10**). Improving the status of women is crucial because women are often disproportionately disadvantaged in poor countries. In many developing countries, women have few rights and little legal ability to protect their property, their rights to their children, and their income.

We have entered an era of global trade, within which we must establish guidelines for national, corporate, and individual behaviors. For example, the flow of money from developing countries to highly developed countries has exceeded the flow in the other direction for many years. Former West German Chancellor **Willy Brandt** termed this phenomenon "a blood transfusion from the sick to the healthy." A world that values social justice and environmental sustainability must reverse this flow. Debts from the poorest countries should be forgiven more readily than they are now, and international development assistance should be enhanced.

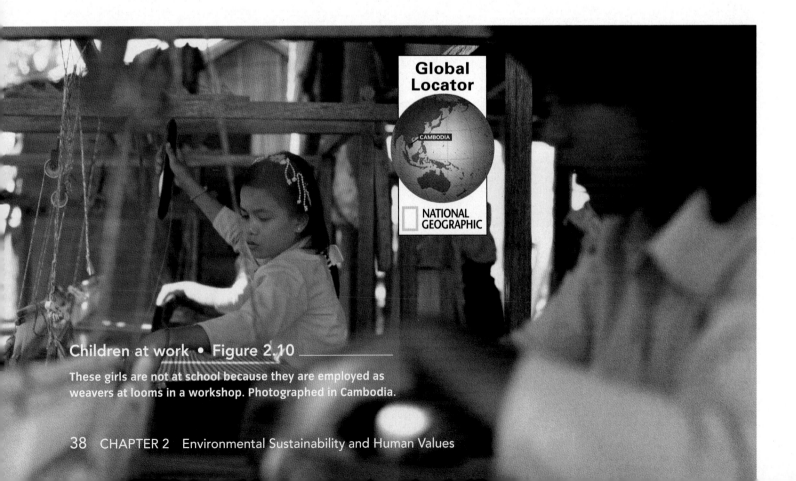

Global Locator

CAMBODIA

NATIONAL GEOGRAPHIC

Children at work • Figure 2.10

These girls are not at school because they are employed as weavers at looms in a workshop. Photographed in Cambodia.

Population growth rates are generally highest where poverty is most intense. If we pay consistent attention to overpopulation and devote the resources necessary to make family planning available for everyone, the human population will stabilize. If we do not continue to emphasize family planning measures, we simply will not achieve population stability.

To stay within Earth's carrying capacity, we must reach and sustain a stable population and reduce excessive consumption. These goals must be coupled with educational programs everywhere, so that people understand that Earth's carrying capacity is not unlimited. There is no hope for a peaceful world without overall population stability, and there is no hope for regional economic sustainability without regional population stability.

> **carrying capacity**
> The maximum population that can be sustained by a given environment or by the world as a whole.

Recommendation 2: Protect and Restore Earth's Resources

To build a sustainable society, we must preserve the natural systems that support us. The conservation of nonrenewable resources, such as oil and minerals, is obvious, although discoveries of new supplies of nonrenewable resources sometimes give the illusion that they are inexhaustible. Renewable resources such as forests, biodiversity, soils, fresh water, and fisheries must be used in ways that ensure their long-term productivity. Their capacity for renewal must be understood and respected. However, renewable resources have been badly damaged over the past 200 years. Until environmental sustainability becomes a part of economic calculations, susceptible natural resources will continue to be consumed unsustainably, driven by short-term economics.

The World's Forests Many of the world's forests are being cut, burned, or seriously altered at a frightening rate. For example, logging in Oregon, Washington, Alaska, Canada, and Siberia is destroying old-growth forests. Tropical forests are also being overexploited. Many products—hardwoods; foods such as beef, bananas, coffee, and tea; and medicines—come to the industrialized world from the tropics. As trees are destroyed, only a small fraction of them are replanted.

The pressure of rapid population growth and widespread poverty also harms the world's forests.

In many developing countries, forests have traditionally served as a "safety valve" for the poor, who, by consuming small tracts of forest on a one-time basis and moving on, find a source of food, shelter, and clothing. But now the numbers of people in developing countries are too great for their forests to support. Tropical rain forests—biologically the world's richest terrestrial areas—have been reduced to less than half their original area. Methods of forest clearing that were suitable when population levels were lower and forests had time to recover from temporary disturbances simply do not work any longer. They convert a potentially renewable resource into an unsustainable one.

Loss of Biodiversity We have a clear interest in protecting Earth's biological diversity and managing it sustainably because we obtain from living organisms all our food, most medicines, many building and clothing materials, biomass for energy, and numerous other products. In addition, organisms and the natural environment provide an array of **ecosystem services** without which we would not survive. These services include the protection of watersheds and soils, the development of fertile agricultural lands, the determination of both local climate and global climate, and the maintenance of habitats for animals and plants.

> **biological diversity** The number and variety of Earth's organisms.

Over the next few decades, we can expect human activities to cause the rate of extinction to increase to perhaps hundreds of species a day. How big a loss is this? Unfortunately, we still have limited knowledge about the world's biological diversity. An estimated five-sixths of all species have not yet been scientifically described. Some 80 percent of the species of plants, animals, fungi, and microorganisms on which we depend are found in developing countries. How will these relatively poor countries sustainably manage and conserve these precious resources? Biological diversity is an intrinsically local problem, and each nation must address it for the sake of its own people's future, as well as for the world at large. Like most other challenges of sustainable development, biological diversity can be addressed adequately only if we provide international assistance where needed, including help in training scientists and engineers from developing countries.

Biological diversity and human cultural diversity are intertwined: They are, in fact, two sides of the same coin.

Cultural diversity is Earth's variety of human communities, each with its individual languages, traditions, and identities (**Figure 2.11**). Cultural diversity enriches the collective human experience. For that reason, the U.N. Educational, Scientific, and Cultural Organization supports the protection of minorities in the context of cultural diversity.

Recommendation 3: Provide Adequate Food for All People

Globally, more than 800 million people lack access to the food needed for healthy, productive lives. This estimate, according to a 2004 report by the U.N. Food and Agriculture Organization, includes a high percentage of children. Children are particularly susceptible to food deficiencies because their brains and bodies cannot develop properly without adequate nutrition. Most malnourished people live in rural areas of the poorest developing nations. The link between poverty and **food insecurity** is inescapable.

Improving agriculture is one of the highest priorities for achieving global sustainability. In general, grain production per person has kept pace with human population growth over the past 50 years. However, expanded agricultural productivity has taken place at high environmental costs. Moreover, the global population continues to expand, putting additional pressure on food production.

Worldwide, little additional land that is not currently under cultivation is suitable for agriculture. One way to increase the productivity of agricultural land is through **multi-cropping**, or growing more than one crop per year. For example, winter wheat and summer soybean crops are grown in some areas of the United States. However,

> **food insecurity**
> The condition in which people live with chronic hunger and malnutrition.

Humans are part of the web of life • Figure 2.11

Portrait of a Yanomami father and son in Roraima State, Brazil. The Yanomami are Brazil's last large Stone Age tribe. Intrusion into isolated areas such as the Amazon Basin threatens both biological diversity and the cultures of indigenous people who have lived in harmony with nature for hundreds of generations.

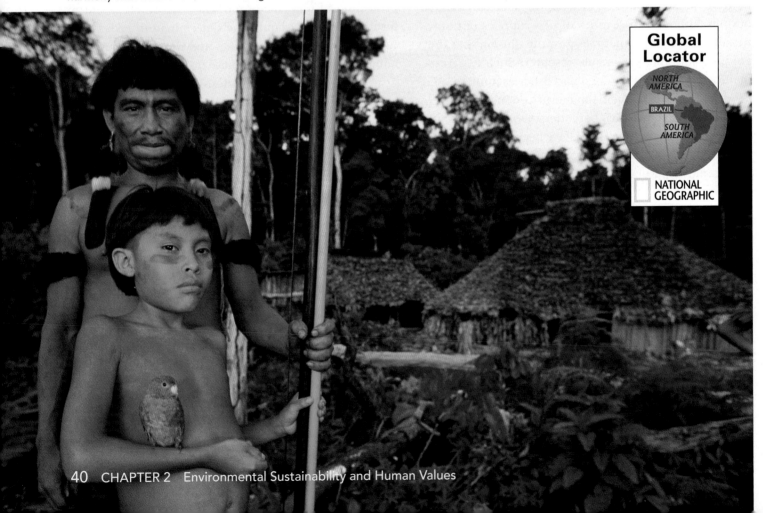

Global Locator

NORTH AMERICA

BRAZIL

SOUTH AMERICA

NATIONAL GEOGRAPHIC

Damage to soil resources • Figure 2.12

Erosion is a serious form of soil degradation. Careful stewardship of the land prevents such damage. Photographed in West Pokot, Kenya.

multi-cropping can be accomplished only in regions where water supplies are adequate for irrigation. Also, care must be taken to prevent a decline in soil fertility from such intensive use.

The negative environmental effects of agriculture, including loss of soil fertility, soil erosion, aquifer depletion, and soil, water, and air pollution, must be brought under control (**Figure 2.12**). Many strategies exist to retard the loss of topsoil, conserve water, conserve energy, and reduce the use of agricultural chemicals. For example, in **conservation tillage**, residues from previous crops are left in the soil, partially covering it and helping to hold topsoil in place.

We must develop sustainable agricultural systems that provide improved dietary standards, such as the inclusion of high-quality protein in diets in developing countries. China's expanding use of **aquaculture** is an example of efficient protein production. The carp that are raised in Chinese aquaculture are efficient at converting food into high-quality protein. In China, fish production by aquaculture now exceeds poultry production. However, aquaculture, like all other human endeavors, has negative environmental effects that must be addressed for it to be sustainable on a large scale.

Recommendation 4: Mitigate Climate Change

A widely discussed human effect on the environment is climate change caused by the **enhanced greenhouse effect**. Both highly developed and developing countries contribute to major increases in CO$_2$ in the atmosphere, as well

> **enhanced greenhouse effect** The additional warming produced by increased levels of gases that absorb infrared radiation.

as to the increasing amounts of methane, nitrous oxide, tropospheric ozone, and CFCs. The most important greenhouse gas, CO$_2$, is produced when we burn fossil fuels—coal, oil, and natural gas.

Although Earth's climate has been relatively stable during the past 10,000 years, human activities are causing it to change. The average global temperature increased by almost 1° Celsius during the 20th century; more than half of that warming occurred during the past 30 years. Precipitation patterns have shifted in many places. Climate scientists generally agree that Earth's climate will continue to change rapidly during the 21st century.

These changes will likely have serious effects because modern society has evolved and successfully adapted to conditions as they are. Keeping in mind that the change from the last ice age to the present was accompanied by an increase in global temperature of 5° Celsius puts the consequences of the present change, the most rapid of the past 10,000 years, into perspective.

We must address climate change in an aggressive and coordinated fashion, but how do we get all nations of the world to adopt the necessary approaches? Many policymakers say that we should wait until scientific knowledge of climate change is complete. This reasoning is flawed because Earth's climate system is extremely complex, and we may never completely understand it.

For example, we often say that an increase in atmospheric CO$_2$ leads to climate warming. However, the

Cascading responses of increased carbon dioxide through the environment • Figure 2.13

THE PLANNER ✓

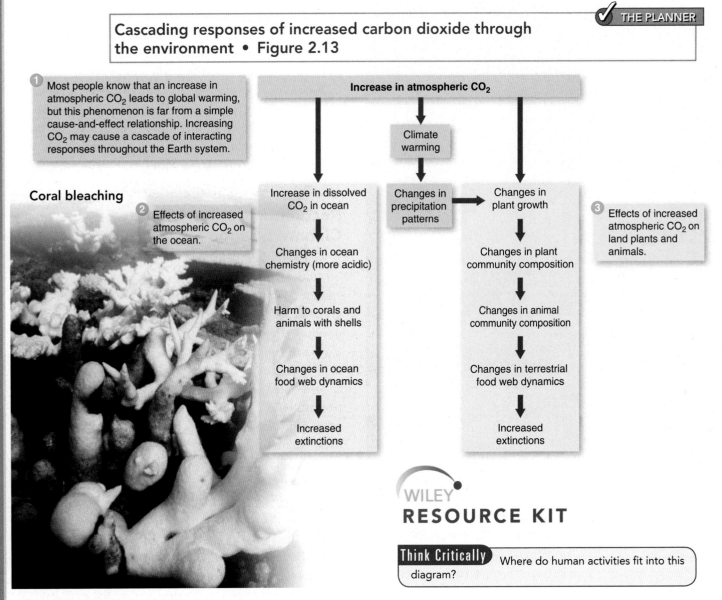

1 Most people know that an increase in atmospheric CO_2 leads to global warming, but this phenomenon is far from a simple cause-and-effect relationship. Increasing CO_2 may cause a cascade of interacting responses throughout the Earth system.

Coral bleaching

2 Effects of increased atmospheric CO_2 on the ocean.

Increase in atmospheric CO_2

Climate warming

Increase in dissolved CO_2 in ocean
↓
Changes in ocean chemistry (more acidic)
↓
Harm to corals and animals with shells
↓
Changes in ocean food web dynamics
↓
Increased extinctions

Changes in precipitation patterns → Changes in plant growth
↓
Changes in plant community composition
↓
Changes in animal community composition
↓
Changes in terrestrial food web dynamics
↓
Increased extinctions

3 Effects of increased atmospheric CO_2 on land plants and animals.

WILEY
RESOURCE KIT

Think Critically Where do human activities fit into this diagram?

---

increase in CO_2, like other human impacts, is not a simple cause-and-effect relationship but instead a cascade of interacting responses that ripple through the environment (**Figure 2.13**). We cannot begin to predict how these changes will affect humans or other organisms.

Stabilizing the climate requires a comprehensive energy plan to include phasing out fossil fuels in favor of renewable energy (such as solar and wind power), increasing energy conservation, and improving energy efficiency. Many national and local governments as well as corporations, colleges and universities, and environmentally aware individuals are setting goals to cut

carbon emissions. Other nations, however, have not recognized the urgency of the global climate problem. We need a global consensus to address climate change.

Recommendation 5: Design Sustainable Cities

At the beginning of the Industrial Revolution, in approximately 1800, only 3 percent of the world's people lived in cities, and 97 percent were rural, living on farms or in small towns. In the two centuries since then, population distribution has changed radically—toward the cities. More people live in Mexico City today than were living in all the cities of the world 200 years ago. This is a staggering difference in the way people live. Almost 50 percent of the world's population now lives in cities, and the percentage continues to grow. In industrialized countries such as the United States and Canada, almost 80 percent of the people live in cities.

City planners around the world are trying a variety of approaches to make cities more livable. Many cities are developing urban transportation systems to reduce the use of cars and the problems associated with them, such as congested roads, large areas devoted to parking, and air pollution. Urban transportation ranges from mass transit subways and light rails to pedestrian and bicycle pathways.

Investing in urban transportation in ways other than building more highways encourages commuters to use forms of transportation other than automobiles. To encourage mass transit, some cities also tax people using highways into and out of cities during business hours. When a city is built around people instead of cars—such as establishing parks and open spaces instead of highways and parking lots—urban residents gain an improved quality of life. Air pollution, including the emission of climate-warming CO_2, is substantially reduced.

Water scarcity is a major issue for many cities of the world. Some city planners think that innovative approaches must be adopted where water resources are scarce. These approaches would replace the traditional one-time water use that involves water purification before use, treatment of sewerage and industrial wastes after use, and then discharge of the treated water. For example, certain places, such as cities like Singapore, recycle some of their wastewater after it has been treated.

Effectively dealing with the problems in squatter settlements is an urgent need. Evicting squatters does not address the underlying problem of poverty. Instead, cities should incorporate some sort of plan for the eventual improvement of squatter settlements (**Figure 2.14**). Providing basic services—such as clean water to drink, transportation (so people can find gainful employment), and garbage pickup—would help improve the quality of life for the poorest of the poor.

Squatter settlement • Figure 2.14

Manila, in the Philippines, is a city of contrasts, with gleaming modern skyscrapers and abjectly poor squatter settlements.

CONCEPT CHECK	STOP

1. **What** is the global extent of poverty?

2. **What** are two ecosystem services provided by natural resources such as forests and biological diversity?

3. **What** is food insecurity?

4. **How** is stabilizing climate related to energy use?

5. **What** are two serious problems in urban environments?

CASE **STUDY**

The Loess Plateau in China

The Loess Plateau covers about 640,000 km² (247,000 mi²) in east-central China. This area is named for **loess**, the yellow-tan, fine-grained, silty soil deposited there by windstorms following the retreat of ice age glaciers. (*Loess*, pronounced "luss" in the United States, is derived from a German word meaning "loose.") Loess is a fine-grained sedimentary deposit found in many areas of the world. However, it is thickest and most extensive in China. The Loess Plateau covers much of the North China Plain and, to the west, the hilly basin of the Yellow River. It averages 75 m (250 ft) thick.

The loess, which is thick and fertile, was at one time an important resource for China. It provided a fertile agricultural soil that fed millions of people. Chinese people also dug homes in the loess; these homes were cool in summer and warm in winter, although they were prone to collapse from earthquakes.

Loess is easily eroded by wind and water, particularly when vegetation is removed from the surface. The Loess Plateau is semiarid, so water is often in short supply. Lack of water, in combination with centuries of deforestation and over-grazing, turned much of the Loess Plateau into a nonproductive desert.

In 1994 the Loess Plateau Watershed Rehabilitation Project was established to reclaim the land from encroaching desert. Portions of the Loess Plateau were reforested (see photograph). Chinese people living in the area were educated about the causes of land degradation and encouraged to keep their livestock in pens instead of allowing the animals to roam freely and overgraze the land. As portions of the Loess Plateau have slowly recovered, it is turning green again, and less silt is washing into the Yellow River.

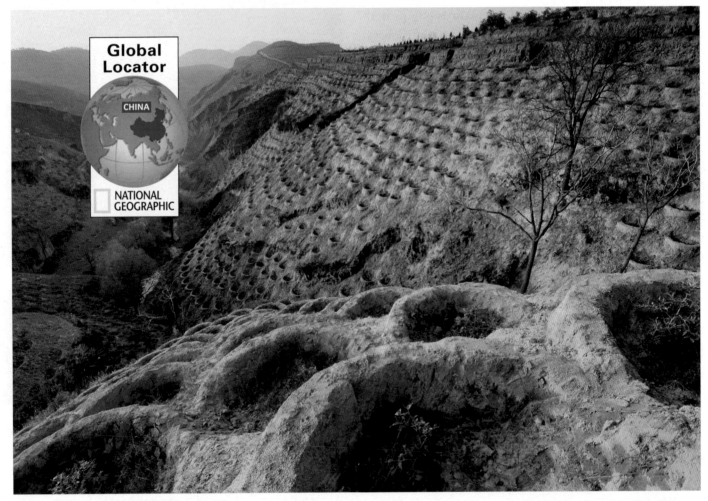

Tree seedlings have been planted in small earthworks as part of a plan to reduce erosion and restore the hills in the Loess Plateau.

Summary

1 Human Use of the Earth 28

1. **Sustainable development** is economic growth that meets the needs of the present without compromising the ability of future generations to meet their own needs. Environmentally sound decisions, economically viable decisions, and socially equitable decisions interact to promote sustainable development.

2. **Sustainable consumption** is the use of goods and services that satisfy basic human needs and improve the quality of life but also minimize the use of resources so they are available for future use.

3. **Voluntary simplicity** recognizes that individual happiness and quality of life are not necessarily linked to the accumulation of material goods.

2 Human Values and Environmental Problems 31

1. **Environmental ethics** is a field of applied ethics that considers the moral basis of environmental responsibility and how far this responsibility extends. Environmental ethicists consider how humans should relate to the natural environment.

2. **An environmental worldview** is a worldview that helps us make sense of how the environment works, our place in the environment, and right and wrong environmental behaviors. The **Western worldview** is an understanding of our place in the world based on human superiority and dominance over nature, the unrestricted use of natural

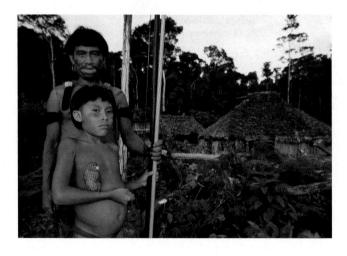

resources, and increased economic growth to manage an expanding industrial base. The **deep ecology worldview** is an understanding of our place in the world based on harmony with nature, a spiritual respect for life, and the belief that humans and all other species have equal worth.

3 Environmental Justice 35

1. **Environmental justice** is the right of every citizen, regardless of age, race, gender, social class, or other factor, to adequate protection from environmental hazards. Environmental justice is a fundamental human right in an ethical society. A growing environmental justice movement has emerged at the grassroots level.

4 An Overall Plan for Sustainable Living 36

1. Failing to confront the problem of poverty makes it impossible to attain global sustainability. To stay within Earth's **carrying capacity**, the maximum population that can be sustained indefinitely, it will be necessary to reach a stable population and reduce excessive consumption.

2. The world's forests are being cut, burned, and seriously altered for timber and other products that the global economy requires. Also, rapid population growth and poverty are putting pressure on forests. **Biological diversity**, the number and variety of Earth's organisms, is declining at an alarming rate. Humans are part of Earth's web of life and are entirely dependent on that web for survival.

3. **Food insecurity** is the condition in which people live with chronic hunger and malnutrition. Globally, more than 800 million people lack access to the food needed for healthy, productive lives.

4. The **enhanced greenhouse effect** is the additional warming produced by increased levels of gases that absorb infrared radiation. An increase in atmospheric CO_2, mostly produced when fossil fuels are burned, leads to climate warming. To stabilize climate, we must phase out fossil fuels in favor of renewable energy, increased energy conservation, and improved energy efficiency.

5. The air in cities in the developing world is badly polluted with exhaust from motor vehicles. Illegal squatter settlements proliferate in cities; the poorest inhabitants build dwellings using whatever materials they can scavenge. Squatter settlements have the worst water, sewage, and solid waste problems.

Key Terms

- biological diversity 39
- carrying capacity 39
- deep ecology worldview 32
- enhanced greenhouse effect 41
- environmental ethics 31
- environmental justice 36
- environmental worldview 32
- food insecurity 40
- sustainable consumption 29
- sustainable development 28
- Western worldview 32

What is happening in this picture?

- These fishermen are pulling up a net of jellyfish, which have proliferated, harming local fish populations. Suggest a possible reason that jellyfish swarms have become so common.

- Given that pollution and climate change are being blamed for the increase in jellyfish, suggest a plan to correct the problem. Will your plan be a quick fix, or will it take many years to address? Why?

- Where would you put "Proliferating jellyfish swarms" in Figure 2.13?

Critical and Creative Thinking Questions

1. Development is sometimes equated with economic growth. Explain the difference between sustainable development and development as an indicator of economic growth, using the figure shown to the right.

2. How are sustainable consumption and voluntary simplicity related?

3. How do the three factors shown in the figure interact to promote sustainable development?

Environmentally sound decisions do not harm environment or deplete natural resources.

Sustainable development

Economically viable decisions consider *all* costs, including long-term environmental and societal costs.

Socially equitable decisions reflect needs of society and ensure costs and benefits are shared equally by all groups.

4. Do you have an ethical obligation to protect the environment? Explain your answer.

5. State whether each of the following statements reflects the Western worldview, the deep ecology worldview, or both. Explain your answers.
   a. Species exist to be used by humans.
   b. All organisms, humans included, are interconnected and interdependent.
   c. There is a unity between humans and nature.
   d. Humans are a superior species capable of dominating other organisms.
   e. Humans should protect the environment.
   f. Nature should be used, not preserved.
   g. Economic growth will help Earth manage an expanding human population.
   h. Humans have the right to modify the environment to benefit society.
   i. All forms of life are intrinsically valuable and therefore have the right to exist.

6. What social groups generally suffer the most from environmental pollution and degradation? What social groups generally benefit from this situation?

7. Why is human population control an important part of global sustainability?

8. How is forest destruction related to declining biological diversity?

9. What is food insecurity? How does food insecurity affect the environment?

10. Discuss two ways to make cities more sustainable.

11–12. The graphs below show a computer simulation by the U.S. National Climate Assessment. In **(a)**, the level of atmospheric CO_2 is projected for the 21st century. As a result of increasing levels of CO_2 in the atmosphere, more CO_2 dissolves in ocean water, where it forms carbonic acid. In **(b)** we can see that the increasing acidity dissolves and weakens coral skeletons, which are composed of calcium carbonate. (Values in **a** and **b** are midrange projections.)

11. Why could rising CO_2 levels in the atmospheric be catastrophic to corals and other shell-forming organisms?
12. How do these graphs relate to Figure 2.13?

Environmental History, Politics, and Economics

RENEWABLE ENERGY POLICY CHALLENGES

For the past 20 years, governments have struggled to develop climate change policies. Both the costs of reducing climate change and the effects of not doing so are huge, highly uncertain, and spread out over time, space, and people. And while the scientific community agrees that human-caused climate change is happening and will worsen, many people, among them influential policymakers, are deeply skeptical.

Among the biggest issues in the climate debate is how to shift to alternative energy sources. Fossil fuels—coal, oil, and natural gas—are by far the largest source of the greenhouse gases that are changing our climate.

However, finding suitable locations for alternatives can be a daunting policy challenge.

A solar installation can require large amounts of space (see inset). In Germany, a town trying to maintain its medieval character is concerned that a proposed solar project will damage its tourism industry. Some California lawmakers object to solar panels in a wilderness area, while in Nevada, concerns about endangered desert tortoise habitat may limit installation site options. Elsewhere, aesthetic, noise, and environmental concerns threaten the launch of potential wind farm projects (see photograph). This chapter explores how environmental policymaking requires attention to ethics, economics, culture, and politics as well as to science.

NATIONAL GEOGRAPHIC

CHAPTER OUTLINE

CHAPTER PLANNER ✓

☐ Study the picture and read the opening story

☐ Scan the Learning Objectives in each section:
   p. 50 ☐   p. 51 ☐   p. 59 ☐   p. 62 ☐

☐ Read the text and study all figures and visuals. Answer any questions.

Analyze key features

☐ Process Diagram, p. 59 ☐

☐ Environmental InSight, p. 63 ☐

☐ Case Study, p. 68 ☐

☐ EnviroDiscovery, p. 58 ☐

☐ Stop: Answer the Concept Checks before you go on:
   p. 50 ☐   p. 56 ☐   p. 61 ☐   p. 67 ☐

End of Chapter

☐ Review the Summary and Key Terms.

☐ Answer What is happening in this picture?

☐ Answer the Critical and Creative Thinking Questions.

Conservation and Preservation of Resources

LEARNING OBJECTIVE

1. **Define** *conservation* and *preservation,* and distinguish between them.

Resources are any part of the natural environment used to promote the welfare of people or other species. Examples of resources include air, water, soil, forests, minerals, and wildlife. **Conservation** is the sensible and careful management of natural resources. Humans have practiced conservation of natural resources for thousands of years. More than 3000 years ago, the Phoenicians terraced hilly farmland to prevent soil erosion. More than 2000 years ago, the Greeks practiced crop rotation to maintain yields on farmlands, and the Romans practiced irrigation.

Other cultures around the world developed similar methods. Modern agriculture continues to develop conservation techniques (**Figure 3.1a**). In contrast, **preservation** involves setting aside undisturbed areas, maintaining them in a pristine state, and protecting them from human activities that might alter their "natural" state (**Figure 3.1b**).

Both conservation and preservation became pressing concerns in the early 20th century. At that time, expanding industrialization, coupled with enormous growth in the human population, began to increase pressure on the world's supply of natural resources. As the global population continues to grow, both conservation and preservation will contribute to sustainability. They will help ensure that future generations will have access to essential resources.

CONCEPT CHECK

1. **What** is conservation? preservation? how do they differ?

Conservation and preservation • Figure 3.1

a. Plowing and planting fields in curves that conform to the natural contours of the land conserves soil by reducing erosion.

b. The Arctic National Wildlife Reserve preserves caribou and other wildlife populations and their habitats.

Environmental History

LEARNING OBJECTIVES

1. **Briefly outline** the environmental history of the United States.

2. **Describe** the contributions of the following people to our understanding of the environment: John James Audubon, Henry David Thoreau, George Perkins Marsh, Theodore Roosevelt, Gifford Pinchot, John Muir, Franklin Roosevelt, Aldo Leopold, Wallace Stegner, Rachel Carson, Paul Ehrlich, and Julian Simon.

3. **Distinguish** between utilitarian conservationists and biocentric preservationists.

WILEY
RESOURCE KIT

From the establishment of the first permanent English colony at Jamestown, Virginia, in 1607, the first two centuries of U.S. history were a time of widespread environmental destruction. European settlers began to extensively exploit land, timber, wildlife, rich soil, clean water, and other resources that had been used sustainably by native peoples for thousands of years. The settlers did not recognize that the bountiful natural resources of North America would one day become scarce. During the 1700s and most of the 1800s, many Americans had a **frontier attitude**, a desire to conquer nature and put its resources to use in the most lucrative manner possible.

Protecting Forests

The great forests of the Northeast were cut down within a few generations, and, shortly after the Civil War in the 1860s, loggers began deforesting the Midwest at an alarming rate. Within 40 years, they had deforested an area the size of Europe, stripping Minnesota, Michigan, and Wisconsin of virgin forest. By 1897 the sawmills of Michigan had processed 160 billion board feet of white pine, leaving less than 6 billion board feet standing in the whole state.

During the 19th century, many U.S. naturalists began to voice concerns about conserving natural resources. **John James Audubon** (1785–1851) painted lifelike portraits of birds and other animals in their natural surroundings that aroused widespread public interest in the wildlife of North America (**Figure 3.2**). **Henry David Thoreau** (1817–1862), a prominent U.S. writer,

lived for 2 years on the shore of Walden Pond near Concord, Massachusetts. There he observed nature and contemplated how people could simplify their lives to live in harmony with the natural world. **George Perkins Marsh** (1801–1882) was a farmer, linguist, and diplomat at various times during his life. Today he is most remembered for his book *Man and Nature*, published in 1864, which

Tanagers • Figure 3.2

This portrayal is one of 500 engravings in Audubon's classic, *The Birds of America*, completed in 1844. Shown are two male Louisiana tanagers (also called western tanagers, top) and male and female scarlet tanagers (bottom).

provided one of the first discussions of humans as agents of global environmental change.

In 1875 a group of public-minded citizens formed the American Forestry Association with the intent of influencing public opinion against the wholesale destruction of America's forests. Sixteen years later, in 1891, the **Forest Reserve Act** (which was part of the General Land Law Revision Act) gave the U.S. president the authority to establish forest reserves on public (federally owned) land. Benjamin Harrison (1833–1901), Grover Cleveland (1837–1908), William McKinley (1843–1901), and Theodore Roosevelt (1858–1919) used this law to put a total of 17.4 million hectares (43 million acres) of forest, primarily in the West, out of the reach of loggers.

In 1907 angry Northwest congressmen pushed through a bill stating that national forests could no longer be created by the president but would require an act of Congress. Roosevelt signed the bill into law but only after designating 21 new national forests that totaled 6.5 million hectares (16 million acres).

Roosevelt appointed **Gifford Pinchot** (1865–1946) the first head of the U.S. Forest Service. Both Roosevelt and

> **utilitarian conservationist**
> A person who values natural resources because of their usefulness to humans but uses them sensibly and carefully.

Pinchot were **utilitarian conservationists** who viewed forests in terms of their usefulness to people—such as in providing jobs and renewable resources. Pinchot supported expanding the nation's forest reserves and managing them scientifically (for instance, harvesting trees only at the rate at which they regrow). Today, national forests are managed for multiple uses, from biological habitats to recreation to timber harvest to cattle grazing.

Establishing National Parks and Monuments

Congress established the world's first national park in 1872, after a party of Montana explorers reported on the natural beauty of the canyon and falls of the Yellowstone River. Yellowstone National Park now includes parts of Idaho, Montana, and Wyoming. In 1890 the **Yosemite National Park Bill** established the Yosemite and Sequoia national parks in California, largely in response to the efforts of a single man, naturalist and writer **John Muir** (1838–1914) (**Figure 3.3**). Muir, who as a child

President Theodore Roosevelt (left) and John Muir • Figure 3.3

Photo was taken on Glacier Point above Yosemite Valley, California.

Hetch Hetchy Valley in Yosemite • Figure 3.4

Some environmental battles involving the protection of national parks were lost. John Muir's Sierra Club fought with the city of San Francisco over its efforts to dam a river and form a reservoir in the beautiful Hetch Hetchy Valley, which lay within Yosemite National Park. In 1913 Congress approved the dam. The State of California is considering restoring Hetch Hetchy, at an estimated cost as high as $10 billion. Hetch Hetchy Valley before (a) and after (b) the dam was built.

emigrated from Scotland with his family, was a **biocentric preservationist**. Muir also founded the *Sierra Club*, a national conservation organization that is still active on a range of environmental issues.

In 1906 Congress passed the **Antiquities Act**, which authorized the president to set aside sites that had scientific, historic, or prehistoric importance. By 1916 there were 16 national parks and 21 national monuments, under the loose management of the U.S. Army. Today there are 58 national parks and

> **biocentric preservationist**
> A person who believes in protecting nature from human interference because all forms of life deserve respect and consideration.

74 national monuments under the management of the National Park Service.

Controversy over preservation battles, such as the Hetch Hetchy Valley conflict, generated a strong sentiment that the nation should better protect its national parks (**Figure 3.4**). In 1916 Congress created the National Park Service to manage the national parks and monuments for the enjoyment of the public, "without impairment." It was this clause that gave a different outcome to another battle,

Aldo Leopold • Figure 3.5 ⎯⎯⎯⎯⎯⎯⎯⎯⎯⎯

Leopold's *A Sand County Almanac* is widely considered an environmental classic.

fought in the 1950s between conservationists and dam builders over the construction of a dam within Dinosaur National Monument. No one could deny that to fill the canyon with 400 feet of water would "impair" it. This victory for conservation established the "use without impairment" clause as the firm backbone of legal protection afforded our national parks and monuments.

Conservation in the Mid-20th Century

During the Great Depression, the federal government financed many conservation projects to provide jobs for the unemployed. During his administration, **Franklin Roosevelt** (1882–1945) established the Civilian Conservation Corps, which employed 500,000 young men to plant trees, make paths and roads in national parks and forests,

build dams to control flooding, and perform other activities that protected natural resources.

During the droughts of the 1930s, windstorms carried away much of the topsoil in parts of the Great Plains, forcing many farmers to abandon their farms and search for work elsewhere. The *American Dust Bowl* alerted the United States to the need for soil conservation, and President Roosevelt formed the Soil Conservation Service in 1935.

Aldo Leopold (1886–1948), a wildlife biologist and environmental visionary, greatly influenced the conservation movement of the mid- to late 20th century (**Figure 3.5**). His textbook *Game Management*, published in 1933, supported the passage of a 1937 act in which new taxes on sporting weapons and ammunition funded wildlife management and research. Leopold also wrote about humanity's relationship with nature and the need to conserve wilderness areas in *A Sand County Almanac*, published in 1949. Leopold argued for a land ethic and the sacrifices that such an ethic requires.

Leopold profoundly influenced many American thinkers and writers, including **Wallace Stegner** (1909–1993), who penned his famous "Wilderness Essay" in 1962. Stegner's essay helped create support for the passage of the *Wilderness Act* of 1964. Stegner wrote:

> *Something will have gone out of us as a people if we ever let the remaining wilderness be destroyed; if we permit the last virgin forests to be turned into comic books and plastic cigarette cases; if we drive the few remaining members of the wild species into zoos or to extinction; if we pollute the last clean air and dirty the last clean streams and push our paved roads through the last of the silence, so that never again will Americans be free in their own country from the noise, the exhausts, the stinks of human and automotive waste . . .*
>
> *We simply need that wild country available to us, even if we never do more than drive to its edge and look in. For it can be a means of reassuring ourselves of our sanity as creatures, a part of the geography of hope.*

During the 1960s, public concern about pollution and resource quality increased, in large part due to the work of marine biologist **Rachel Carson** (1907–1964). Carson wrote about interrelationships among living organisms, including humans, and the natural environment (**Figure 3.6**). In her most famous work, *Silent Spring*, published in 1962, Carson wrote against the indiscriminate use of pesticides:

> *Pesticide sprays, dusts, and aerosols are now applied almost universally to farms, gardens, forests, and homes—nonselective*

chemicals that have the power to kill every insect, the "good" and the "bad," to still the song of birds and the leaping of fish in the streams, to coat the leaves with a deadly film, and to linger on in soil—all this though the intended target may be only a few weeds or insects. Can anyone believe it is possible to lay down such a barrage of poisons on the surface of the earth without making it unfit for all life? They should not be called "insecticides," but "biocides."

Silent Spring heightened public awareness and concern about the dangers of using DDT and other pesticides, including poisoning birds and other wildlife and contaminating human food supplies. Ultimately, the book led to restrictions on the use of certain pesticides. Around this time, the media increasingly covered environmental incidents, such as hundreds of deaths in New York City from air pollution (1963), closed beaches and fish kills in Lake Erie from water pollution (1965), and detergent foam in a Pennsylvania creek (1966).

In 1968, when the population of Earth was "only" 3.5 billion people, ecologist **Paul Ehrlich** published *The Population Bomb*. In it he described the stress that such a huge number of people impose on Earth's life support system, including global depletion of fertile soil, groundwater, and other living organisms. Ehrlich's book raised the public's awareness of the dangers of overpopulation and triggered debates about how to deal effectively with population issues.

Ehrlich's critics, in particular **Julian Simon** (1932–1998), countered that technological advances outpace the negative impacts of population growth. A decade into the 21st century, both sides of this issue have strong advocates. Ehrlich continues to point out water, climate, agriculture, and other global stresses, while many economists counter that the collapse Ehrlich predicted has not occurred.

The Environmental Movement

Until 1970 the voice of **environmentalists**, people concerned about the environment, was heard in the United States primarily through societies such as the Sierra Club and the National Wildlife Federation. There was no generally perceived **environmental movement** until the spring of 1970, when **Gaylord Nelson**, former senator of Wisconsin, urged Harvard graduate student **Denis Hayes** to organize the first nationally celebrated Earth Day.

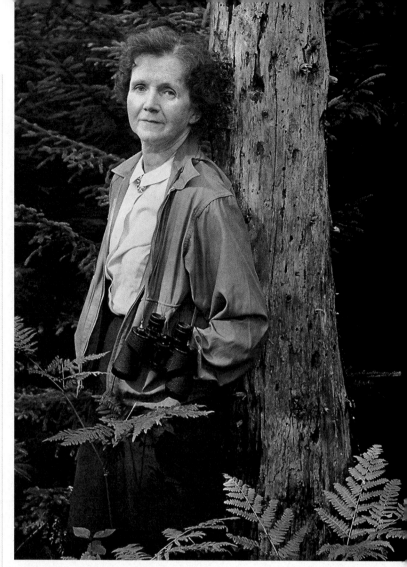

Rachel Carson • Figure 3.6

Carson's book *Silent Spring* heralded the beginning of the environmental movement.

This event awakened U.S. environmental consciousness to population growth, overuse of resources, and pollution and degradation of the environment. On Earth Day 1970, an estimated 20 million people in the United States planted trees, cleaned roadsides and riverbanks, and marched in parades to support improvements in resource conservation and environmental quality.

In the years that followed the first Earth Day, environmental awareness and the belief that individual actions could repair the damage humans were doing to Earth became a pervasive popular movement. Musicians and other celebrities popularized environmental concerns. Many of the world's religions—such as Christianity, Judaism, Islam, Hinduism, Buddhism, Taoism, Shintoism, Confucianism, and Jainism—embraced environmental themes such as protecting endangered species and controlling global climate change.

By Earth Day 1990, the movement had spread around the world, signaling the rapid growth in environmental consciousness. An estimated 200 million people in 141 nations demonstrated to increase public awareness of the importance of individual efforts ("Think globally, act locally") (**Figure 3.7**). The theme of Earth Day 2000, "Clean Energy Now," reflected the dangers of global climate change and what individuals and communities could do: Replace fossil fuel energy sources with solar electricity, wind power, and the like. However, by 2000 many environmental activists had begun to think that the individual actions Earth Day espouses, while collectively important, are not as important as pressuring governments and large corporations to make environmentally friendly decisions.

In 2009 global environmental concern was expressed as thousands of people from around the world converged on Copenhagen, Denmark, to pressure world leaders to combat climate change. **Figure 3.8** shows a timeline of selected environmental events since Earth Day 1970.

CONCEPT CHECK **STOP**

1. **How** did public perception of the environment evolve during the 20th century?

2. **What** was the environmental contribution of Rachel Carson?

3. **What** distinguishes utilitarian conservationists from biocentric preservationists?

Timeline of selected environmental events, from 1970 to the present • Figure 3.8

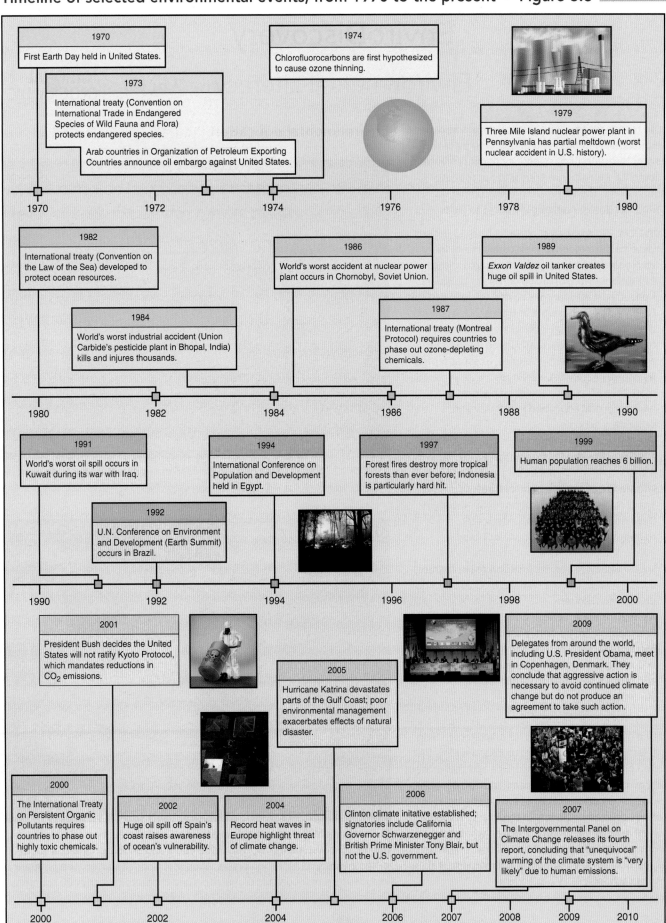

1970
First Earth Day held in United States.

1973
International treaty (Convention on International Trade in Endangered Species of Wild Fauna and Flora) protects endangered species.

Arab countries in Organization of Petroleum Exporting Countries announce oil embargo against United States.

1974
Chlorofluorocarbons are first hypothesized to cause ozone thinning.

1979
Three Mile Island nuclear power plant in Pennsylvania has partial meltdown (worst nuclear accident in U.S. history).

1970   1972   1974   1976   1978   1980

1982
International treaty (Convention on the Law of the Sea) developed to protect ocean resources.

1984
World's worst industrial accident (Union Carbide's pesticide plant in Bhopal, India) kills and injures thousands.

1986
World's worst accident at nuclear power plant occurs in Chornobyl, Soviet Union.

1987
International treaty (Montreal Protocol) requires countries to phase out ozone-depleting chemicals.

1989
Exxon Valdez oil tanker creates huge oil spill in United States.

1980   1982   1984   1986   1988   1990

1991
World's worst oil spill occurs in Kuwait during its war with Iraq.

1992
U.N. Conference on Environment and Development (Earth Summit) occurs in Brazil.

1994
International Conference on Population and Development held in Egypt.

1997
Forest fires destroy more tropical forests than ever before; Indonesia is particularly hard hit.

1999
Human population reaches 6 billion.

1990   1992   1994   1996   1998   2000

2001
President Bush decides the United States will not ratify Kyoto Protocol, which mandates reductions in CO_2 emissions.

2000
The International Treaty on Persistent Organic Pollutants requires countries to phase out highly toxic chemicals.

2002
Huge oil spill off Spain's coast raises awareness of ocean's vulnerability.

2004
Record heat waves in Europe highlight threat of climate change.

2005
Hurricane Katrina devastates parts of the Gulf Coast; poor environmental management exacerbates effects of natural disaster.

2006
Clinton climate initative established; signatories include California Governor Schwarzenegger and British Prime Minister Tony Blair, but not the U.S. government.

2009
Delegates from around the world, including U.S. President Obama, meet in Copenhagen, Denmark. They conclude that aggressive action is necessary to avoid continued climate change but do not produce an agreement to take such action.

2007
The Intergovernmental Panel on Climate Change releases its fourth report, concluding that "unequivocal" warming of the climate system is "very likely" due to human emissions.

2000   2002   2004   2006   2007   2008   2009   2010

EnviroDiscovery

Environmental Literacy

Because responses to environmental problems depend on the public's awareness and understanding of the issues and the underlying scientific concepts involved, environmental education is critical to appropriate decision making. The emphasis on environmental education has grown dramatically over the years:

- In 1990, 22 university presidents from 13 nations issued a declaration of their commitment to environmental education and research at their institutions. More than 300 university presidents from at least 40 countries have since followed suit.
- Prepared by coursework at their schools, more than 73,000 U.S. high school students in 2009 took the College Board Advanced Placement exam in Environmental Science, a test accepted by approximately 1700 colleges.
- More than 30 states require some form of environmental education in primary and secondary schools.
- The National Environmental Education Act of 1990 requires the Environmental Protection Agency to increase public awareness and knowledge of environmental issues.
- The U.N. Decade of Education for Sustainable Development (2005–2014) is dedicated to improving basic education, including public understanding about environmental sustainability. Programs focus on major themes, such as water, climate change, biodiversity, and disaster prevention.
- As of 2010 the American College and University President's Climate Commitment had more than 650 signatories. These schools agree to take actions to reduce their greenhouse gas emissions and require sustainability education for all students.

The North American Association for Environmental Education has issued guidelines for educators to help them select materials such as textbooks and films that are based on sound scientific evidence and that present a balanced perspective on environmental problems.

However, a backlash against environmental education occurred beginning in the late 1990s. Some conservative research groups criticized what they perceived as a biased presentation of environmental issues, particularly the promotion of environmental activism, in schools.

Environmental education

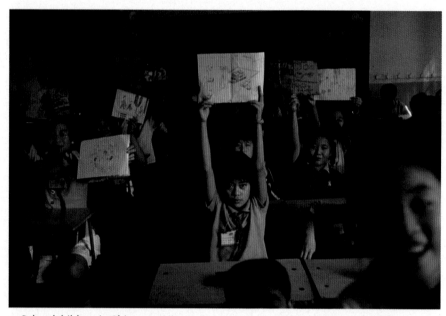

a. Schoolchildren in China participate in an environmental education exercise.

b. Environmentalist and primatologist Jane Goodall meets with Connecticut middle school students involved in the Roots and Shoots program, a youth-based environmental action organization that Goodall started. The program includes tens of thousands of members and chapters in nearly 100 countries.

Environmental Legislation

LEARNING OBJECTIVES

1. **Explain** why the National Environmental Policy Act is the cornerstone of U.S. environmental law.
2. **Describe** how environmental impact statements provide powerful protection of the environment.
3. **Define** *full cost accounting.*

Well-publicized ecological disasters, such as the 1969 oil spill off the coast of Santa Barbara, California, and overwhelming public support for the Earth Day movement, led to the **National Environmental Policy Act (NEPA)** of 1970. The Environmental Protection Agency (EPA) was created in July of the same year. A key provision of NEPA requires the federal government to consider the environmental impact of a proposed federal action, such as financing highway or dam construction, when making decisions about that action. NEPA provides the basis for developing detailed **environmental impact statements (EISs)** to accompany every federal recommendation or proposal for legislation. An EIS is a document that describes the nature and purpose of the proposal, its short- and long-term environmental impacts, and possible alternatives that would create fewer adverse effects. NEPA also requires solicitation of public comments when preparing an EIS, which generally provides a broader perspective on the proposal and its likely effects.

NEPA established the **Council on Environmental Quality** to monitor the required EISs and report directly to the president. Because this council had no enforcement powers, NEPA was originally considered innocuous, more a statement of good intentions than a regulatory policy. During the next few years, however, environmental activists took people, corporations, and the federal government to court to challenge their EISs or use them to block proposed development. The courts decreed that EISs had to thoroughly analyze the environmental consequences of anticipated projects on soil, water, and endangered species and that EISs be made available to the public (**Figure 3.9**). These rulings put sharp teeth

Environmental impact statements • Figure 3.9

① Major construction project proposed.

② Environmental Impact Statement (EIS) prepared.

③ EIS released for public review and comment.

④ Based on public debate, the project may proceed as planned, proceed with modifications, be sent back for further development, or be withdrawn.

How will the project affect wildlife habitat?

How will the landscape be altered?

How will increased soil erosion affect water quality?

Environmental Impact Statement (EIS)

Will the project harm any endangered species?

How will air quality be affected?

How will stream flow rates change?

Step ②
An EIS must answer a number of critical questions

Step ④
A major project like this solar installation near Las Vegas, Nevada, has gone through an extensive review process

NATIONAL GEOGRAPHIC

PROCESS DIAGRAM

into NEPA—particularly the provision for public scrutiny, which places intense pressure on federal agencies to respect EIS findings.

NEPA revolutionized environmental protection in the United States. Federal agencies manage federal highway construction, flood and erosion control, military projects, and many other public works. They oversee nearly one-third of the land in the United States. Federal holdings include fossil fuel and mineral reserves, millions of hectares of public grazing land, and public forests. Since 1970 very little has been done to any of them without some sort of environmental review. NEPA has also influenced environmental legislation in at least 36 states and in many other countries.

Although almost everyone agrees that NEPA has successfully reduced adverse environmental impacts of federal activities and projects, it has its critics. Some environmentalists complain that EISs are sometimes incomplete or that reports are ignored when decisions are made. Other critics think the EISs delay important projects ("paralysis by analysis") because the documents are too involved, take too long to prepare, and are often the targets of lawsuits.

Environmental Regulations

When an environmental problem becomes widely recognized, the process of environmental regulation begins with a U.S. congressperson drafting legislation. Ideally, before the legislation is drafted, the trade-offs for several proposed alternative actions are evaluated. This process, known as **full cost accounting**, is a valuable economic tool in environmental decision making.

Environmental legislation usually goes to the EPA, which translates the law's language into specific regulations. Before the EPA can enforce new regulations, several rounds of public comments allow affected parties to present their views. The EPA is required to respond to all of these comments. Then the Office of Management and Budget reviews the new regulations. Implementation and enforcement often fall to state governments, which must send the EPA details for achieving the goals of the new regulations.

> **full cost accounting** The process of evaluating and presenting to decision makers the relative benefits and costs of various alternatives.

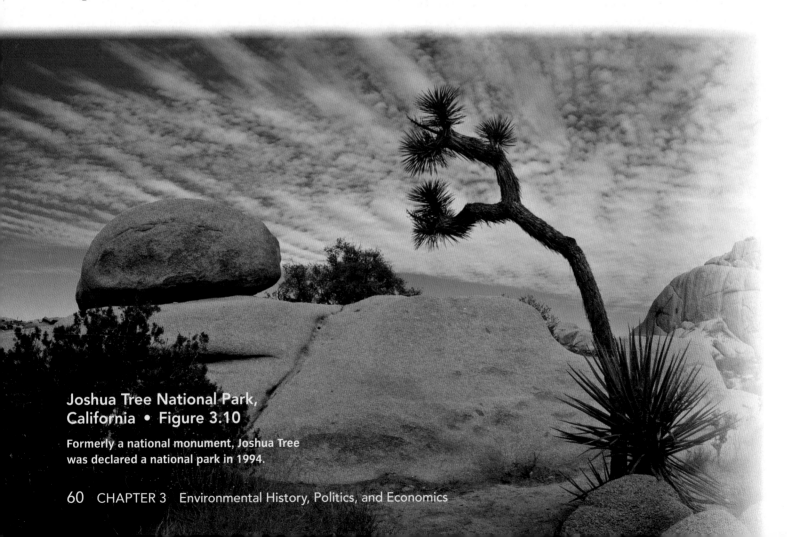

Joshua Tree National Park, California • Figure 3.10

Formerly a national monument, Joshua Tree was declared a national park in 1994.

Accomplishments of Environmental Legislation

During the period since Earth Day 1970, Congress has passed almost 40 major environmental laws that address a wide range of issues, such as endangered species, clean water, clean air, energy conservation, hazardous wastes, and pesticides. This tough interlocking mesh of laws has greatly improved environmental quality.

Despite imperfections, environmental legislation has had overall positive effects. Since 1970,

- Fifteen national parks have been established (**Figure 3.10**), and the National Wilderness Preservation System now totals more than 44 million hectares (109 million acres).

- Millions of hectares of farmland particularly vulnerable to erosion have been withdrawn from production, reducing soil erosion by more than 60 percent.

- Many endangered species are recovering, and the American alligator, California gray whale, and bald eagle have recovered enough to be removed from the endangered species list. (However, dozens of other species, such as the manatee and Kemp's ridley sea turtle, have suffered further declines or extinction since 1970.)

Although we still have a long way to go, pollution control efforts through legislation have been particularly successful. According to the EPA's *2008 Report on the Environment*:

- Emissions of six important air pollutants have dropped by more than 25 percent since 1990. (Carbon dioxide emissions, however, have continued to rise.)

- Since 1990, levels of wet sulfate, a major component of acid rain, have dropped by 20 to 35 percent.

- In 2007 almost 90 percent of the U.S. population got its drinking water from community water systems with no violations of EPA standards, up from around 75 percent in 1993 (**Figure 3.11**).

- In 2006 45 percent of municipal solid waste generated in the United States was combusted for energy recovery or recovered for composting or recycling, up from 6 percent in the 1960s.

- In 2007 the EPA considered human exposures to contamination to be under control at 93 percent of the 1968 listed hazardous waste sites.

Water treatment plant • Figure 3.11

The water supply for a town or city is treated before use so it is safe to drink. Photographed in Miami, Florida.

In the 1960s and 1970s, pollution was often obvious—witness the Cuyahoga River in Cleveland, Ohio, which burst into flames from the oily pollutants on its surface several times. Legislators, the media, and the public typically perceive things like burning rivers as serious threats that require immediate attention, without regard to the cost. As the effects of global climate change become more obvious, public pressure to develop policies to reduce greenhouse gas emissions have grown. Recognizing the high costs of historical legislation and the power of markets to drive innovative solutions, policymakers increasingly look to economics as part of the solution to environmental problems.

CONCEPT CHECK

1. **Why** is the National Environmental Policy Act the cornerstone of U.S. environmental law?

2. **What** are environmental impact statements?

3. **How** does full cost accounting influence environmental decision making?

Environmental Economics

LEARNING OBJECTIVES

1. **Explain** how economics is related to natural capital. Make sure you include sources and sinks.

2. **Give** two reasons why the national income accounts are incomplete estimates of national economic performance.

3. **Distinguish** among the following economic terms: marginal cost of pollution, marginal cost of pollution abatement, and optimum amount of pollution.

4. **Describe** various incentive-based regulatory approaches, including environmental taxes and tradable permits.

Economics is the study of how people use their limited resources to try to satisfy their unlimited wants. Economists try to understand the consequences of the ways in which people, businesses, and governments allocate their resources. Seen through an economist's eyes, the world is one large marketplace, where resources are allocated to a variety of uses, and where goods—a car, a pair of shoes, a barrel of oil—and services—a haircut, a museum tour, an education—are consumed and paid for. In a free market, supply and demand determine the price of a good (**Figure 3.12**). If something in great demand is in short supply, its price will be high. High prices encourage suppliers to produce more of a good or service, as long as the selling price is equal to the cost of producing the good or service. This interaction of demand, supply, price, and cost underlies much of what happens in the U.S. economy, from the price of a hamburger to the cycles of economic expansion (increase in economic activity) and recession (slowdown in economic activity).

Economies depend on the natural environment as *sources* for raw materials and *sinks* for waste products (**Figure 3.13**). Both sources and sinks contribute to **natural capital**. According to economists, the environment provides natural capital for human production and consumption. Resource degradation and pollution represent the

> **natural capital**
> Earth's resources and processes that sustain living organisms, including humans; includes minerals, forests, soils, water, clean air, wildlife, and fisheries.

The Hibernia oil platform on the Grand Banks in the Atlantic Ocean • Figure 3.12

When demand for crude oil goes up, economists expect that more will be pumped, and the price of a barrel of crude oil will increase.

Global Locator

GRAND BANKS

NATIONAL GEOGRAPHIC

What fossil fuel resources are extracted near where you live? How do they affect the local economy?

overuse of natural capital. *Resource degradation* is the overuse of sources, and *pollution* is the overuse of sinks; both threaten our long-term economic future.

National Income Accounts and the Environment

Much of our economic well-being flows from natural capital—such as land, rivers, the ocean, oil, timber, and the air we breathe—rather than human-made assets.

Ideally, for the purposes of economic and environmental planning, the **national income accounts** should

> **national income accounts** Measures of the total income of a nation's goods and services for a given year.

include natural resource depletion and environmental degradation. Two measures used in national income accounting are *gross domestic product (GDP)* and *net domestic product (NDP)*. Both GDP and NDP provide estimates of national economic performance that are used to make important policy decisions.

Unfortunately, current national income accounting practices provide an incomplete or inaccurate measure of income because they do not incorporate environmental factors. Two important conceptual problems exist with the way the national income accounts currently handle the economic use of natural resources and the environment: natural resource depletion and the costs and benefits of pollution control. Better accounting for environmental quality would help address whether for any given activity the benefits (both economic and environmental) exceed the costs.

Natural Resource Depletion If a manufacturing firm produces some product (output) but in the process wears out a portion of its plant and equipment, the firm's output is counted as part of GDP, but the depreciation of capital is subtracted in the calculation of NDP. Thus NDP is a measure of the net production of the economy, after a deduction for used-up capital. In contrast, when an oil company drains oil from an underground field, the value of the oil produced is counted as part of the nation's GDP,

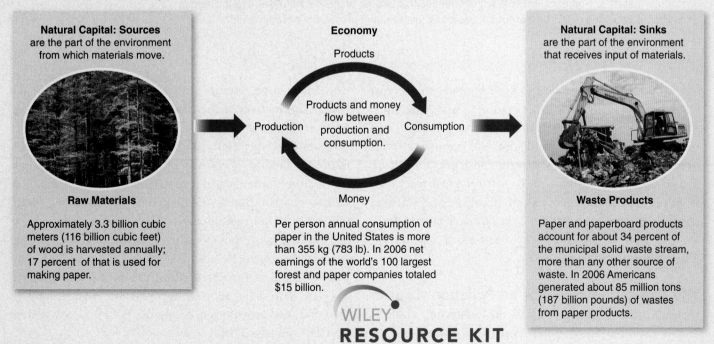

Environmental InSight
Economics and the environment • Figure 3.13

THE PLANNER

Economies Depend on Natural Capital Sources for Raw Materials and Sinks for Waste Products

Natural Capital: Sources are the part of the environment from which materials move.

Economy

Products

Products and money flow between production and consumption.

Production   Consumption

Money

Natural Capital: Sinks are the part of the environment that receives input of materials.

Raw Materials

Approximately 3.3 billion cubic meters (116 billion cubic feet) of wood is harvested annually; 17 percent of that is used for making paper.

Per person annual consumption of paper in the United States is more than 355 kg (783 lb). In 2006 net earnings of the world's 100 largest forest and paper companies totaled $15 billion.

Waste Products

Paper and paperboard products account for about 34 percent of the municipal solid waste stream, more than any other source of waste. In 2006 Americans generated about 85 million tons (187 billion pounds) of wastes from paper products.

WILEY
RESOURCE KIT

Pollution cleanup and GDP • Figure 3.14

Hazardous materials workers in protective gear remove toxic waste from a Texas beach. These cleanup costs should be added to the GDP accounts because the cleanup improves the environment.

but no offsetting deduction to NDP is made to account for the fact that nonrenewable resources were used up.

In principle, the draining of the oil field is a type of depreciation, and the oil company's net product should be accordingly reduced. The same point applies to any other natural resource that is depleted in the process of production. Natural capital is a very large part of a country's economic wealth, and we should treat it the same as human-made capital.

The Costs and Benefits of Pollution Control

Imagine that a company has the following choices: It can produce $100 million worth of output and, at the same time, dump its wastes, polluting the local river. Alternatively, if the company uses 10 percent of its workers to properly dispose of its wastes, it avoids polluting but gets only $90 million of output. Under current national income accounting rules, if the firm chooses to pollute rather than not to pollute, it will make a larger contribution to GDP ($100 million rather than $90 million) because the national income accounts attach no explicit value to a clean river. In an ideal accounting system, the economic cost of environmental degradation is subtracted in the calculation of a firm's contribution to GDP, and activities that improve the environment—because they provide real economic benefits—are added to GDP (**Figure 3.14**).

Incorporating resource depletion and pollution into national income accounting is important because GDP and related statistics are used continually in policy analyses. An increasing number of economists, government planners, and scientists support replacing GDP and NDP with a more comprehensive measure of national income accounting that includes estimates of both depletion of natural capital and the environmental cost of economic activities (**Figure 3.15**).

An Economist's View of Pollution

An important aspect of the operation of a free-market system is that the person consuming a product should pay for all the cost of producing it. However, production or consumption of a product often has an **external cost**.

> **external cost**
> A harmful environmental or social cost that is borne by people not directly involved in selling or buying a product.

A product's market price does not usually reflect an external cost—that is, the buyer or seller doesn't pay for the external cost. As a result, a market system with externalities generally does not operate in the most efficient way.

Consider the following example of an external cost. If an industry makes a product and, in so doing, also releases a pollutant into the environment, the product is bought at a price that reflects the cost of making it but not the cost of the pollutant's damage to the environment. This damage is the external cost of the product. (One common external cost of many products is air pollution released when fossil fuels are burned to transport manufacturing components or finished goods.)

Because this damage is not included in the product's price and because the consumer may not know that the pollution exists or that it harms the environment, the cost of the pollution has no impact on the consumer's decision to buy the product. As a result, consumers of the product may buy more of it than they would if its true cost, including the cost of pollution, were known or reflected in the selling price.

The failure to add the price of environmental damage to the cost of products generates a market force that encourages pollution. From the perspective of economics, then, one of the causes of the world's pollution problem is the failure to include external costs in the prices of goods.

Waterfall in Great Smoky Mountain National Park • Figure 3.15

Resources removed from pristine areas, such as this one would not be counted as a loss in standard national income accounting. Photographed in North Carolina.

We now examine industrial pollution from an economist's viewpoint, as a policymaking failure. Keep in mind, however, that lessons about the economics of industrial pollution also apply to other environmental issues (such as resource degradation) where harm to the environment is a consequence of economic activity.

How Much Pollution is Acceptable? To assign a proper price to pollution, economists first try to answer the basic question "How much pollution should we allow in our environment?" Imagine two environmental extremes: a wilderness in which no pollution is produced but neither are goods, and a "sewer" that is completely polluted from excess production of goods. A move toward a better environment often entails a cost in terms of goods.

Marginal cost of pollution • Figure 3.16

At low pollution levels, the environment may absorb the damage, so that the marginal cost of one added unit of pollution is near zero. As the level of pollution rises, the cost in terms of human health and a damaged environment increases sharply. At very high levels of pollution, the cost soars.

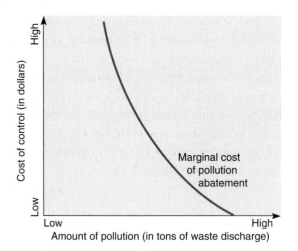

Marginal cost of pollution abatement • Figure 3.17

At high pollution levels, the marginal cost of eliminating one unit of pollution is low. As more and more pollution is eliminated from the environment, the cost of removing each additional (marginal) unit of pollution increases.

How do we, as individuals, as a country, and as part of the larger international community, decide where we want to be between the two extremes of a wilderness and a sewer? Economists analyze the marginal costs of environmental quality and of other goods to answer such questions. A **marginal cost** is the additional cost associated with one more unit of something.

The trade-off between protecting environmental quality and producing more goods involves balancing marginal costs of two kinds: (1) the external cost, in terms of environmental damage, of more pollution (the marginal cost of pollution) and (2) the cost, in terms of giving up goods, of eliminating pollution (the marginal cost of pollution abatement).

Determining the **marginal cost of pollution** involves assessing the risks associated with the pollution—for example, damage to health, property, or agriculture. (See Chapter 4 for a discussion of risk assessment.) Once the risk is known, it must be monetized. This means that injuries, deaths, loss of species, and other damages must be assigned dollar values.

Let's consider a simple example involving the marginal cost of sulfur dioxide, a type of air pollution produced during the combustion of fuels containing sulfur. Sulfur dioxide is removed from the atmosphere as acid rain, which causes damage to the environment, particularly aquatic ecosystems. Economists add up the harm of each additional unit of pollution—in this example, each ton of sulfur dioxide added to the atmosphere. As the total amount of pollution increases, the harm of each additional unit usually also increases, and as a result, the curve showing the marginal cost of pollution slopes upward, as in **Figure 3.16**.

The **marginal cost of pollution abatement** tends to rise as the level of pollution declines, as shown in **Figure 3.17**. It is relatively inexpensive to reduce automobile exhaust emissions by half, but costly devices are required to reduce the remaining emissions by half again. For this reason, the curve showing the marginal cost of pollution abatement slopes downward.

In **Figure 3.18**, the two marginal-cost curves from Figures 3.16 and 3.17 are plotted together on one graph, called a **cost-benefit diagram**. Economists use this diagram to identify the point at which the marginal cost of pollution equals the marginal cost of abatement—that is, the point where the two curves intersect. As far as economics

> **marginal cost of pollution** The added cost of an additional unit of pollution.

> **marginal cost of pollution abatement** The added cost of reducing one unit of a given type of pollution.
>
> **cost-benefit diagram** A diagram that helps policymakers make decisions about costs of a particular action and benefits that would occur if that action were implemented.

Cost–benefit diagram • Figure 3.18 _____

Economists identify the optimum amount of pollution as the amount at which the marginal cost of pollution equals the marginal cost of pollution abatement (the point at which the two curves intersect). If more pollution than the optimum is allowed, the social cost is unacceptably high. If less than the optimum amount of pollution is allowed, the pollution abatement cost is unacceptably high.

is concerned, this point represents an **optimum amount of pollution**. At this optimum, the cost to society of hav-

> **optimum amount of pollution** The amount of pollution that is economically most desirable.

ing less pollution is offset by the benefits to society of the activity creating the pollution.

There are two major objections to the economist's concept of optimum pollution. First, it is difficult to determine the true cost of environmental damage caused by pollution. The web of relationships within the environment is extremely intricate and may be more vulnerable to pollution damage than is initially obvious, sometimes with disastrous results. When cost estimates are highly uncertain, economics may lead to poor decisions. Second, many people find the notion of putting prices on lives, species, and wilderness to be unethical.

Economic Strategies for Pollution Control

Command and control regulations and incentive-based regulations are two ways that governments control pollution. To date, most pollution control efforts in the United States have involved **command and control**

> **command and control regulation** Pollution control laws that work by setting limits on levels of pollution.

regulation. Sometimes command and control laws require use of a specific pollution control method, such as the use of catalytic converters in cars to decrease polluting exhaust emissions. In other cases, a quantitative goal is set. The Clean Air Act Amendments of 1990 established a goal of a 60 percent reduction in nitrogen oxide emissions in passenger cars by the year 2003. Usually, all polluters must comply with the same rules and regulations, regardless of their particular circumstances.

Economists are concerned that command and control regulations can have excessively high costs. They argue that using economic tools can achieve the same environmental benefits at lower cost. Consequently, most economists, whether progressive or conservative, prefer **incentive-based regulation** over command and control regulation. Ideally, incentive-based regulation forces producers to internalize external cost, thereby achieving the optimum amount of pollution.

> **incentive-based regulation** Pollution control laws that work by establishing emission targets and providing industries with incentives to reduce emissions.

The two most common incentive-based regulatory approaches are environmental taxes and tradable permits. **Environmental taxes** are designed to be equal to the externality caused by a polluter. Unfortunately, this amount can be highly uncertain and so is often very difficult to set. **Tradable permit** approaches, also known as *cap and trade*, set an allowable amount of pollution and then let different companies buy and sell the right to release that pollution. Companies or individuals who can easily reduce their emissions sell some of their pollution rights to those who cannot.

CONCEPT CHECK	

1. **What** is natural capital? How is economics related to natural capital?

2. **Why** are national income accounts incomplete estimates of total national economic performance?

3. **How** are marginal cost of pollution, marginal cost of pollution abatement, and optimum amount of pollution related?

4. **How** do command and control regulation and incentive-based regulation differ regarding pollution control?

Tradable Permits and Acid Rain

Many international policy experts believe that a cap and trade system is the most promising approach to managing the problem of climate change. They argue that setting a global (or nation-by-nation) cap on greenhouse gases would encourage people to find innovative and inexpensive ways to reduce emissions. However, incentive-based environmental regulations remain less familiar than command and control regulations. The example of tradable sulfur emissions permits to reduce the effects of acid rain demonstrates how effective the approach can be.

When coal containing sulfur is burned, sulfur dioxide is created and released, causing acid rain (see Chapter 8). Through the 1970s and 1980s, Environmental Protection Agency (EPA) regulations reduced sulfur emissions primarily by mandating command and control solutions. This meant that many large coal-burning power plants had to install specific, and often very expensive, equipment. By the late 1980s, these facilities knew of less expensive options for reducing sulfur emissions but had no incentive to adopt them.

Consequently, the Clean Air Act Amendments of 1990 allowed the EPA to limit the amount of sulfur that could be emitted, with a smaller amount allowed each year, and then sell the rights to these emissions. Each year, the EPA allows the Chicago Board of Trade to auction permits to emit sulfur; companies may then buy and sell these permits as needed during the year. Industries quickly adopted a variety of technologies, such as removing sulfur before burning coal, and met the EPAs sulfur reduction goals ahead of schedule and at a lower-than-expected cost.

Tradable permits have not worked as well in all cases. Attempts to reduce water pollution have had mixed results, especially when more than one pollutant is involved. Grandfathering, or exempting older facilities, has undermined other efforts. And in the sulfur case, the EPA was able to establish clear goals and accurately measure emissions, both of which may prove a challenge for a greenhouse gas cap and trade system. Nonetheless, the success of tradable sulfur emissions suggests that incentive-based regulation has a promising future.

Coal-burning power plants in the United States, such as this one in West Virginia, emit sulfur that causes acid rain. Until the 1990s, the EPA mandated emission control technology. Since the 1990s, companies have had more flexibility in how to reduce their emissions.

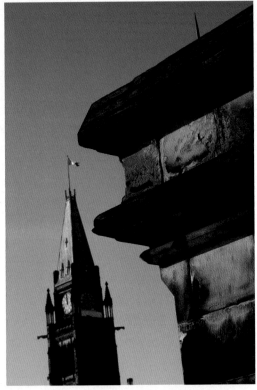

These buildings in Ottawa, Ontario, Canada, have been damaged by acid rain. This is an example of an externality caused in part by sulfur emissions from coal-burning power plants in the United States.

Summary

1 Conservation and Preservation of Resources 50

1. **Conservation** is the sensible and careful management of natural **resources**, such as air, water, soil, forests, minerals, and wildlife. **Preservation** involves setting aside undisturbed areas, maintaining them in a pristine state, and protecting them from human activities.

Sequoia national parks were established largely in response to the efforts of naturalist **John Muir**. **Franklin Roosevelt** established the Civilian Conservation Corps and the Soil Conservation Service. In *A Sand County Almanac*, **Aldo Leopold** wrote about humanity's relationship with nature. **Wallace Stegner** helped create support for the passage of the Wilderness Act of 1964. **Rachel Carson** published *Silent Spring*, alerting the public about the dangers of uncontrolled pesticide use. **Paul Ehrlich** published *The Population Bomb*, which raised the public's awareness of the dangers of overpopulation. **Julian Simon**, taking an economists perspective, challenged Ehrlich's concerns about growth.

3. A **utilitarian conservationist** is a person who values natural resources because of their usefulness to humans but uses them sensibly and carefully. A **biocentric preservationist** is a person who believes in protecting nature because all forms of life deserve respect and consideration.

2 Environmental History 51

1. The first two centuries of U.S. history were a time of widespread environmental destruction. During the 1700s and early 1800s, most Americans had a desire to conquer and exploit nature as quickly as possible. During the 19th century, many U.S. naturalists became concerned about conserving natural resources. The earliest conservation legislation revolved around protecting land—forests, parks, and monuments. By the late 20th century, environmental awareness had become a pervasive popular movement.

2. **John James Audubon**'s art aroused widespread interest in the wildlife of North America. **Henry David Thoreau** wrote about living in harmony with the natural world. **George Perkins Marsh** wrote about humans as agents of global environmental change. **Theodore Roosevelt** appointed **Gifford Pinchot** as the first head of the U.S. Forest Service. Pinchot supported expanding the nation's forest reserves and managing forests scientifically. The Yosemite and

3 Environmental Legislation 59

1. Since 1970 the federal government has addressed many environmental problems. The **National Environmental Policy Act (NEPA)** of 1970 established the **Council on Environmental Quality** to monitor required **environmental impact statements (EISs)** and report directly to the president.

2. By requiring EISs that are open to public scrutiny, NEPA initiated serious environmental protection in the United States. NEPA allows citizen suits, in which private citizens take violators, whether they are private industries or government-owned facilities, to court for noncompliance.

3. **Full cost accounting** is the process of evaluating and presenting to decision makers the relative benefits and costs of various alternatives.

4 Environmental Economics 62

1. **Economics** is the study of how people use their limited resources to try to satisfy their unlimited wants. Economies depend on the natural environment as sources for raw materials and sinks for waste products. Both sources and sinks contribute to **natural capital**, which is Earth's resources and processes that sustain living organisms, including humans. Natural capital includes minerals, forests, soils, water, clean air, wildlife, and fisheries.

2. **National income accounts** are measures of the total income of a nation's goods and services for a given year. An **external cost** is a harmful environmental or social cost that is borne by people not directly involved in buying or selling a product. National income accounts are incomplete estimates of national economic performance because they do not include both natural resource depletion and the environmental costs of economic activities. Many economists, government planners, and scientists support more comprehensive income accounting that includes these estimates.

3. From an economic point of view, the appropriate amount of pollution is a trade-off between harm to the environment and inhibition of development. The **marginal cost of pollution** is the added cost of an additional unit of pollution. The **marginal cost of pollution abatement** is the added cost of reducing one unit of a given type of pollution. Economists think the use of resources for pollution abatement should increase only until the cost of abatement equals the cost of the pollution damage. This results in the **optimum amount of pollution**— the amount of pollution that is economically most desirable.

4. **Incentive-based regulations** take advantage of economic markets to reduce environmental damage. **Environmental taxes** require polluters to pay an amount equal to the harm they cause. **Tradable permit** systems limit the total amount of a pollutant that can be released, allowing people to buy and sell rights to emit and reduce emissions as inexpensively as possible.

Key Terms

- biocentric preservationist 53
- command and control regulation 67
- cost–benefit diagram 66
- external cost 65
- full cost accounting 60
- incentive-based regulation 67
- marginal cost of pollution 66
- marginal cost of pollution abatement 66
- national income accounts 63
- natural capital 62
- optimum amount of pollution 67
- utilitarian conservationist 52

What is happening in this picture?

- This photo was taken in Hong Kong in 1990. What event is taking place?

- Note the ages of the individuals in this photo. How do their ages contribute to growing environmental awareness?

Research how Earth Day was celebrated in 2010. What issues did people focus on?

Critical and Creative Thinking Questions

1. Is a ban on logging in a national park an example of conservation or preservation? Explain.

2. Explain why policymaking for renewable energy projects requires attention to ethics, economics, culture, and politics as well as to science.

3. Describe how writers influenced environmental history in the 19th and 20th centuries.

4. List at least three issues that would be included in a national income account that incorporates full cost accounting.

5. The National Environmental Policy Act (NEPA) is sometimes called the "Magna Carta of environmental law." What is meant by such a comparison?

6. How would a utilitarian conservationist approach each of the problems listed in question 6? a biocentric preservationist?

7. Do you think a global cap on carbon dioxide would benefit developed countries more than developing countries? Explain.

8. In the graph shown below, is the amount of pollution indicated by the vertical dashed line more or less than the economically optimum amount of pollution? Explain your answer.

9. If you were an economist examining the previous graph, would you recommend increasing or decreasing pollution abatement measures? Why?

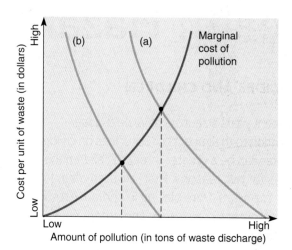

10. The graph above shows two curves, labeled a and b, that represent marginal cost of pollution abatement. In this hypothetical situation, technological innovations were developed between 2003 and 2006 that lowered the abatement cost. Which curve corresponds to 2003 and which to 2006? Explain your answer.

11. If you were a member of Congress, what legislation would you introduce to deal with each of the following problems?
    • Toxins from a major sanitary landfill are polluting your state's groundwater.
    • Acid rain from a coal-burning power plant in a nearby state is harming the trees in your state. Loggers and foresters are upset.
    • There is a high incidence of cancer in the area of your state where heavy industry is concentrated.

THE PLANNER

Risk Analysis and Environmental Health Hazards

4

PESTICIDES AND CHILDREN

In recent years scientists have paid closer attention to the impacts of pesticides on children. Pesticides can be a greater threat to children than to adults for two reasons. First, children often face greater *exposure*, from playing in contaminated areas and putting their hands and other objects into their mouths. Second, children's developing bodies can exhibit greater *response* from a given amount of pesticide than do less sensitive adults. Pesticide exposures have a range of effects, including cancers and mental or physical disabilities.

Research indicates that pesticide exposure can affect the development of intelligence and motor skills in young children. A study published in *Environmental Health Perspectives* compared two groups of rural Yaqui Indian preschoolers in Mexico. These two nearly identical groups differed mainly in their exposure to pesticides: One group lived in a farming community where pesticides were used frequently (see photograph), and the other lived in an area where pesticides were rarely used. When asked to draw a person, most of the 17 children from the low-pesticide area drew recognizable stick figures (see part **a** of inset), whereas most of the 34 children from the high-pesticide area drew meaningless lines and circles (see part **b**). Additional tests of simple mental and physical skills revealed similar striking differences between the two groups of children.

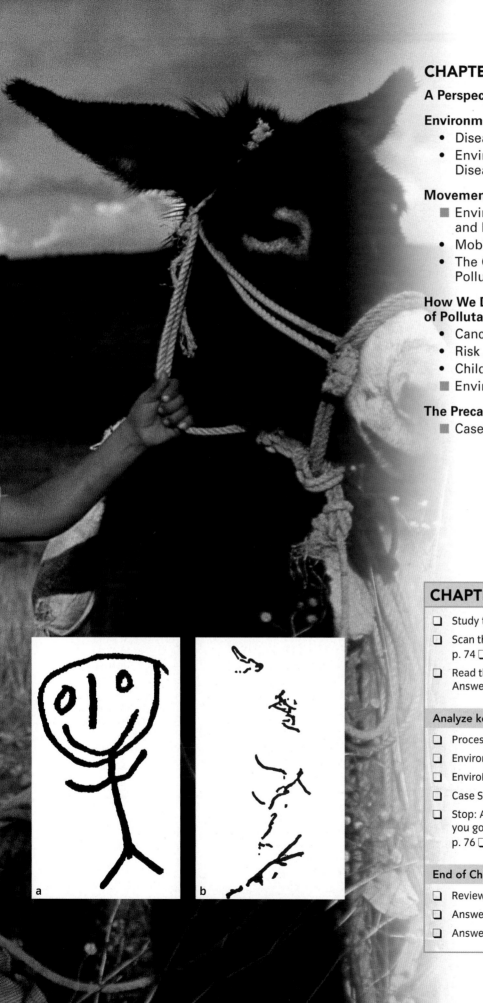

CHAPTER OUTLINE

CHAPTER PLANNER ✔

- ❏ Study the picture and read the opening story.
- ❏ Scan the Learning Objectives in each section:
  p. 74 ❏   p. 77 ❏   p. 81 ❏   p. 85 ❏   p. 90 ❏
- ❏ Read the text and study all figures and visuals. Answer any questions.

Analyze key features

- ❏ Process Diagram, p. 75 ❏
- ❏ Environmental InSight, p. 82 ❏
- ❏ EnviroDiscovery, p. 88 ❏
- ❏ Case Study, p. 92 ❏
- ❏ Stop: Answer the Concept Checks before you go on:
  p. 76 ❏   p. 81 ❏   p. 84 ❏   p. 89 ❏   p. 91 ❏

End of Chapter

- ❏ Review the Summary and Key Terms.
- ❏ Answer What is happening in this picture?
- ❏ Answer the Critical and Creative Thinking Questions.

A Perspective on Risks

LEARNING OBJECTIVES

1. **Define** *risk* and *risk assessment*.

2. **Explain** how risk assessment helps us manage potential health threats.

Threats to our health, particularly from toxic chemicals in the environment, make big news. Many of these stories are more sensational than factual. Human health in highly developed counties is generally better today than at any previous time in our history, although life expectancy in some of the poorest U.S. countries has begun to decline. This does not mean that you should ignore chemicals that humans introduce into the environment. Nor should you discount all the stories that the news media sometimes sensationalize, since they can identify serious health threats that we can manage only if we are aware of them.

Risk is inherent in all our actions and in everything in our environment. All of us take risks every day of our lives. Walking on stairs involves a small risk, but a risk nonetheless because sometimes people die from falls on stairs. Using household appliances leads to some risk of electrocution

> **risk** The probability of harm (such as injury, disease, death, or environmental damage) occurring under certain circumstances.

when wires are faulty or people operate appliances unsafely. Driving or riding in a car or flying in a jet has risks that are easier for most of us to recognize. Yet few of us hesitate to get in a car or board a plane because of the associated risk. In order to successfully manage risks, we must have a sense of their causes, likelihoods, and effects (**Figure 4.1**).

Each of us uses intuition, habit, and experience to make many decisions regarding risk every day. However, environmental and health risks often affect many individuals, and the best choices cannot always be made on an intuitive or routine level. **Risk management** is the

Lifetime probability of death by selected causes • Figure 4.1 ——————

These 2008 data from the National Safety Council are for U.S. residents. Note that few of these risks apply to everyone. For example, only motorcyclists can die in motorcycle accidents.

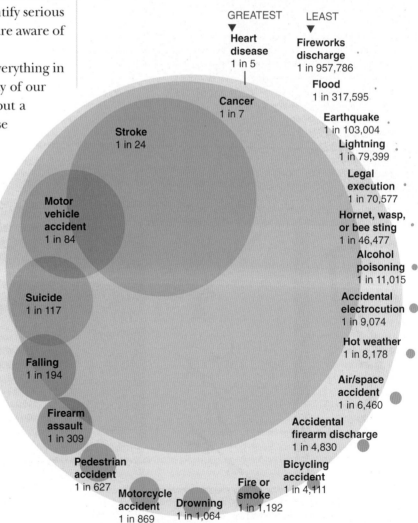

GREATEST ▼   LEAST ▼

Heart disease 1 in 5

Cancer 1 in 7

Stroke 1 in 24

Motor vehicle accident 1 in 84

Suicide 1 in 117

Falling 1 in 194

Firearm assault 1 in 309

Pedestrian accident 1 in 627

Motorcycle accident 1 in 869

Drowning 1 in 1,064

Fire or smoke 1 in 1,192

Fireworks discharge 1 in 957,786

Flood 1 in 317,595

Earthquake 1 in 103,004

Lightning 1 in 79,399

Legal execution 1 in 70,577

Hornet, wasp, or bee sting 1 in 46,477

Alcohol poisoning 1 in 11,015

Accidental electrocution 1 in 9,074

Hot weather 1 in 8,178

Air/space accident 1 in 6,460

Accidental firearm discharge 1 in 4,830

Bicycling accident 1 in 4,111

Four steps for risk assessment • Figure 4.2

① Hazard identification
Can exposure to a substance cause adverse health effects such as cancer or brain defects?

② Dose-response assessment
What is the relationship between the amount of exposure (dose) and the severity of the adverse health effects (response)? Very low doses may have no effect or delayed effects, while higher doses may have delayed effects, immediate effects, or both.

③ Exposure assessment
How much, how long, and how often are individuals exposed to the substances in question? Exposure assessment requires understanding where people live, what they eat or drink, as well as how different substances travel in air, water, and food.

④ Risk characterization
Information about dose-response and exposures are combined to provide a detailed description of the likelihood and extent of adverse health effects. Risk characterization indicates that Mexican-Americans, who make up a disproportionate fraction of agricultural workers in the United States, are at higher risk from pesticides than are other groups (see graph).

Risk management
Once risks have been characterized, this information can help inform decisions about risk. Importantly, the process of assessing and managing risks does not just go in one direction, from hazard identification to assessment to management. Rather, each step feeds back information to the others. For example, managers may find that a risk characterization does not describe the risks most important to them, or may identify a new risk needing further analysis.

Interpreting Data
What is the ratio of the DDT blood level for Mexican-Americans to that of the rest of the population of the United States?

Blood level of DDT (ng/g) — General population, Mexican-Americans

Agricultural workers have a greater than average exposure to chemicals such as pesticides.

NATIONAL GEOGRAPHIC

process of identifying, assessing, and reducing risks. The four steps involved in **risk assessment** for adverse health effects are summarized in **Figure 4.2**. Once a risk assessment is performed, its results are combined with relevant political, social, and economic considerations to determine how we can best avoid, reduce, or eliminate a particular risk and, if so, what we should do. This evaluation includes the development and implementation of laws to regulate hazardous substances.

risk assessment
The quantitative and qualitative characterization of risks so that they can be compared, contrasted and managed.

Risk assessment estimates the probability that an event will occur and lets us set priorities and manage risks in an appropriate way. As an example, consider a person who smokes a pack of cigarettes a day and drinks well water containing traces of the cancer-causing chemical trichloroethylene (in amounts permitted under Environmental Protection Agency [EPA] limits). Without knowledge of risk assessment, this person might buy bottled water in an attempt to reduce

Lung cancer • Figure 4.3

Cancer was diagnosed in the right lung of this 73-year-old woman (shown by the red areas) after years of heavy smoking, a high-risk behavior. Data suggest that smoking is on the decline among younger people, although the poor and some minority groups have disproportionately high smoking rates.

Percentage of U.S. students (grades 9–12) who smoked cigarettes on 20 or more of the past 30 days • Table 4.1

Year	Total	Females	Males
2005	9.3	9.3	9.3
2003	9.7	9.7	9.6
2001	13.8	12.9	14.9
1999	16.8	15.6	17.9
1997	16.7	15.7	17.6
1995	16.1	15.9	16.3
1993	13.8	13.5	14.0
1991	12.7	12.4	13.0

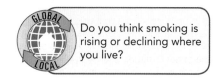

Do you think smoking is rising or declining where you live?

his or her chances of getting cancer. Based on risk assessment calculations, the annual risk of death from smoking is 0.00059, or 5.9×10^{-4}, whereas the annual risk from drinking water with EPA-accepted levels of trichloroethylene is 0.000000002, or 2.0×10^{-9}. This means that this person is almost *300,000 times* more likely to get cancer from smoking than to get it from ingesting such low levels of trichloroethylene (**Figure 4.3**). Knowing this, the person in our example would, we hope, stop smoking.

One of the most perplexing dilemmas of risk assessment is that people often accept substantial risks but get extremely upset about minor risks. The average life expectancy of smokers is more than eight years shorter than that of nonsmokers, and almost one-third of all smokers die from diseases that the habit causes or exacerbates. Yet many people can get more upset over a one-in-a-million chance of getting cancer from pesticide residues on food than they do over the relationship between smoking and cancer. Perhaps part of the reason for this attitude is that behaviors such as diet, smoking, and exercise are parts of our lives that many people see as choices (**Table 4.1**). Risks over which most of us have no control, such as pesticide residues or nuclear wastes, tend to evoke more fearful responses.

CONCEPT CHECK

1. **What** are risk and risk assessment?

2. **What** are the four steps of risk assessment?

Environmental Health Hazards

LEARNING OBJECTIVES

1. **Define** *toxicology* and *epidemiology*.
2. **Explain** why public water supplies are monitored for fecal coliform bacteria despite the fact that most strains of *E. coli* do not cause disease.
3. **Describe** the link between environmental changes and emerging diseases, such as swine flu.

The human body is exposed to many kinds of chemicals in the environment. Both natural and synthetic chemicals are in the air we breathe, the water we drink, and the food we eat. All chemicals, even "safe" chemicals such as sodium chloride (table salt), are toxic if exposure is high enough. For example, a 1-year-old child will die from ingesting about 2 tablespoons of table salt; table salt is also harmful to people with heart or kidney disease. Chemicals with adverse effects are known as **toxicants**.

Toxicology (a) studies the effects of toxicants on living organisms, (b) studies the mechanisms that cause toxicity, and (c) develops ways to prevent or minimize adverse effects. (Developing appropriate handling or exposure guidelines for specific toxicants is one of these ways.)

Epidemiology involves studying how chemicals (toxicants), biological agents (disease), and physical hazards (accidents, radiation) affect the health of human populations. Epidemiologists study large groups of people and investigate a range of possible causes and types of diseases and injuries. The effects of toxicants following exposure can be immediate (*acute toxicity*) or prolonged (*chronic toxicity*). Symptoms of **acute toxicity** range from dizziness and nausea to death. Acute toxicity occurs immediately to several days following a single exposure. In comparison, **chronic toxicity** generally produces damage following long-term, low-level exposure to a toxicant. Toxicologists know far less about chronic toxicity than they do about acute toxicity, partly because the symptoms of chronic toxicity often mimic

toxicology The study of toxicants, chemicals with adverse effects on health.

epidemiology The study of the effects of chemical, biological, and physical agents on the health of human populations.

acute toxicity Adverse effects that occur within a short period after high-level exposure to a toxicant.

chronic toxicity Adverse effects that occur after a long period of low-level exposure to a toxicant.

those of other chronic diseases associated with risky lifestyle patterns, poor nutrition, and aging. Also, it is difficult to isolate a causative agent from among the multiple toxicants we are routinely exposed to.

Disease-Causing Agents in the Environment

Disease-causing agents are *infectious* organisms, such as bacteria, viruses, protozoa, and parasitic worms that cause diseases. Typhoid, cholera, bacterial dysentery, polio, and infectious hepatitis are some of the most common bacterial or viral diseases that are transmissible through contaminated food and water. Diseases such as these are considered environmental health hazards. Other human diseases, such as acquired immunodeficiency syndrome (AIDS), are not transmissible through the environment and aren't discussed here.

The vulnerability of water supplies to waterborne disease-causing agents was dramatically demonstrated in 2000, when the first waterborne outbreak in North America of a deadly strain of *Escherichia coli* occurred in Ontario, Canada. Several people were killed, and several thousand became sick. Prior to this outbreak, this deadly *E. coli* strain had been transmitted almost exclusively through contaminated food.

The largest outbreak of a waterborne disease ever recorded in the United States occurred in 1993, when a microorganism (*Cryptosporidium*) contaminated the water supply in the greater Milwaukee area. About 370,000 people developed diarrhea. These and similar outbreaks raise concerns about the safety of our drinking water. Because sewage-contaminated water is an environmental threat to public health, periodic tests are made for the presence of sewage in our drinking water supplies. The best indicator of sewage-contaminated water is the presence of the common intestinal bacterium *E. coli* because it doesn't appear in the environment except from human and animal

Some human diseases transmitted by polluted water • Table 4.2

Disease	Type of organism	Symptoms
Cholera	Bacterium	Severe diarrhea, vomiting; fluid loss of as much as 20 quarts per day causes cramps and collapse
Dysentery	Bacterium	Infection of the colon causes painful diarrhea with mucus and blood in the stools; abdominal pain
Enteritis	Bacterium	Inflammation of the small intestine causes general discomfort, loss of appetite, abdominal cramps, and diarrhea
Typhoid	Bacterium	Early symptoms include headache, loss of energy, fever; later, a pink rash appears, along with (sometimes) hemorrhaging in the intestines
Infectious hepatitis	Virus	Inflammation of liver causes jaundice, fever, headache, nausea, vomiting, severe loss of appetite, muscle aches, and general discomfort
Poliomyelitis	Virus	Early symptoms include sore throat, fever, diarrhea, and aching in limbs and back; when infection spreads to spinal cord, paralysis and atrophy of muscles occur
Cryptosporidiosis	Protozoon	Diarrhea and cramps that last up to 22 days
Amoebic dysentery	Protozoon	Infection of the colon causes painful diarrhea with mucus and blood in the stools; abdominal pain
Schistosomiasis	Fluke	Tropical disorder of the liver and bladder causes blood in urine, diarrhea, weakness, lack of energy, repeated attacks of abdominal pain
Ancylostomiasis	Hookworm	Severe anemia, sometimes symptoms of bronchitis

feces. Tests such as those for *E. coli* are used to indicate the possible presence of various disease-causing agents (**Table 4.2**). Although most strains of coliform bacteria found in sewage do not cause disease, testing for these bacteria is a reliable way to indicate the likely presence of **pathogens** in water.

pathogen An agent (usually a microorganism) that causes disease.

The **fecal coliform test** assesses whether *E. coli* is present in water (**Figure 4.4**). A small sample of water is passed through a filter to trap the bacteria, which are then transferred to a petri dish that contains nutrients. After an incubation period, the number of greenish colonies present indicates the number of *E. coli*. Safe drinking water should contain no more than one coliform bacterium per 100 mL of water (about ½ cup); safe swimming water should have no more than

200 per 100 mL of water; and general recreational water (for boating) should have no more than 2000 per 100 mL. In contrast, raw sewage may contain several million coliform bacteria per 100 mL of water. Water pollution and purification are discussed further in Chapter 10.

Environmental Changes and Emerging Diseases

Human health has improved significantly in developed countries over the past several decades, but environmental factors remain a significant cause of human disease in many areas of the world. Epidemiologists have established links between human health and human activities that alter the environment.

Fecal coliform test • Figure 4.4

This test indicates the likely presence of disease-causing agents in water. A water sample is first passed through a filtering apparatus. **a.** The filter disk is then placed on a medium that supports coliform bacteria for 24 hours. **b.** After incubation, the number of bacterial colonies is counted. each colony of *Escherichia coli* arose from a single coliform bacterium in the original water sample.

b

A 1997 World Health Organization report concluded that about 25 percent of disease and injury worldwide is related to human-caused environmental changes. The environmental component of human health is sometimes direct and obvious, as when people drink unsanitary water and contract dysentery, a waterborne disease that causes diarrhea. Diarrhea causes 4 million deaths worldwide each year, mostly in children.

The health effects of many human activities are complex and often indirect. For example, a recent study suggests that cholera, which in 2009 caused more than 4000 deaths in Zimbabwe, is a greater problem as temperatures increase due to climate change. The disruption of natural environments may give disease-causing agents an opportunity to thrive. Cutting down forests, building dams, and expanding agriculture may bring humans into contact with new or previously rare disease-causing agents by increasing the population and distribution of disease-carrying organisms such as mosquitoes (**Figure 4.5**). Social factors may also contribute

Road clearing in the Amazon rain forest • Figure 4.5

The drainage ditches that will be added to each side of the road will hold standing water where mosquito larvae thrive.

Global ocator

to disease epidemics. Highly concentrated urban populations promote the rapid spread of infectious organisms among large numbers of people (**Figure 4.6**). Global travel also has the potential to contribute to the rapid spread of disease as infected individuals move easily from one place to another.

Consider malaria, a disease that mosquitoes transmit to humans. Each year, between 200 million and 500 million people worldwide contract malaria, resulting in more than 1 million deaths. About 60 different species of *Anopheles* mosquito transmit the parasites that cause malaria. Each mosquito species thrives in its own unique combination of environmental conditions (such as elevation, amount of precipitation, temperature, relative humidity, and availability of surface water).

Another recent concern is pandemic influenza (flu). A **pandemic** disease reaches nearly every part of the world and has the potential to infect almost every person. Avian influenza is a strain of influenza virus that is common in birds. It tends to be difficult for humans to contract because it is usually transferred from bird to human but not from human to human. It is extremely potent once contracted and has a high fatality rate.

In late spring 2009 a strain of the swine flu appeared in Mexico, and by early summer was pandemic, killing thousands of people worldwide. Flu and other diseases generally spread more easily between related species: Humans are more closely related to pigs than to birds. (AIDS is thought to have originated from human contact with diseased monkeys.)

The international response to the 2009 swine flu outbreak continues to be controversial. Some argue that, since far fewer deaths occurred than predicted under

Crowds at an open-air market in Dhaka, Bangladesh • Figure 4.6

The development of cities, and the concentration of people living in them, permit the rapid spread of infectious disease-causing agents.

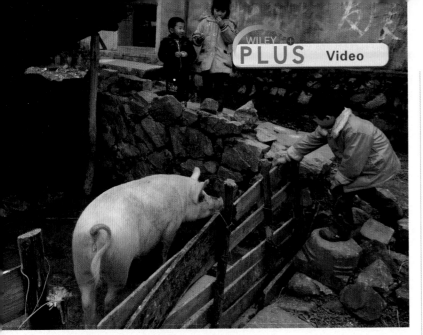

People and livestock in close proximity • Figure 4.7

These children in rural Jiangxi, China, live and play near where pigs are raised. Humans and livestock, including chickens and pigs, share enough genetic similarities that some diseases can be transferred from species to species. This includes the swine flu epidemic of 2009, which started in rural Mexico, and spread around the world in a matter of months.

worst-case assumptions, we over reacted. Others think that we should have done more, since we did not know that the strain would be relatively mild.

Understanding and controlling an influenza pandemic requires study of the environment that allows the virus to survive and travel, as well as cooperation among many governments and individuals. The virus often originates in areas that have dense populations of domestic animals, especially chickens and pigs, raised in small cages or close to human households (**Figure 4.7**). In the past decade, large numbers of domestic pigs, cows, and poultry have been killed and burned to prevent or stop disease outbreaks.

CONCEPT CHECK STOP

1. **What** is the difference between toxicology and epidemiology?
2. **Why** is the fecal coliform test performed on public drinking water supplies?
3. **How** is the incidence of swine flu related to human activities that alter the environment?

Movement and Fate of Toxicants

LEARNING OBJECTIVES

1. **Distinguish** among persistence, bioaccumulation, and biological magnification of toxicants.
2. **Discuss** the mobility of persistent toxicants in the environment.
3. **Describe** the purpose of the Stockholm Convention on Persistent Organic Pollutants.

Some chemically stable toxicants are particularly dangerous because they resist degradation and readily move around in the environment. These include certain pesticides, radioactive isotopes, heavy metals such as mercury, flame retardants such as PBDEs (polybrominated diphenyl ethers), and industrial chemicals such as PCBs (polychlorinated biphenyls).

Mercury is the only metal that is liquid at room temperature. As such, it moves readily through air, water, and land. Mercury can cause brain and nerve damage, loss of hair, and even death.

The effects of the pesticide **DDT (dichlorodiphenyltrichloroethane)** on many bird species demonstrate the problem. Falcons, pelicans, bald eagles, and many other birds are sensitive to traces of DDT in their tissues. Substantial evidence indicates that DDT causes these birds to lay eggs with thin, fragile shells that usually break during incubation, causing the chicks' deaths. After 1972, the year DDT was banned in the United States, the reproductive success of many birds began to slowly improve.

The impact of DDT on birds is the result of (1) its persistence, (2) bioaccumulation, and (3) biological magnification. **Persistence** means that the substance is extremely stable and may take many years to break down

Biological magnification of DDT on a Long Island salt marsh

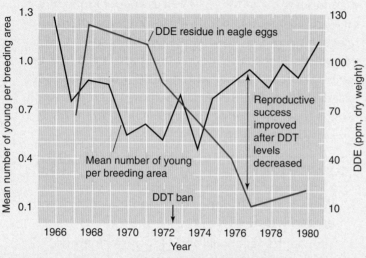

*DDT is converted to DDE in the birds' bodies

a. A comparison of the number of successful bald eagle offspring with the level of DDT residues in their eggs.

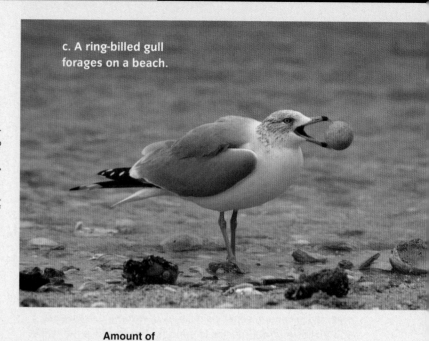

c. A ring-billed gull forages on a beach.

Effects of DDT on birds

b. A bald eagle feeds its chicks.

Amount of DDT in tissue		Trophic level
75.5 ppm		Tertiary consumer (ring-billed gull)
2.07 ppm		Secondary consumer (Atlantic needlefish)
0.28 ppm		Secondary consumer (American eel)
0.16 ppm	1,510,000 times increase	Primary consumer (shrimp)
0.04 ppm		Producer and primary consumer (algae and other plankton)
0.00005 ppm		Water

d. Note how the level of DDT, expressed as parts per million, increased in the tissues of various organisms as DDT moved through the food chain from producers to consumers (bottom to top of figure). The ring-billed gull at the top of the food chain had approximately 1.5 million times more DDT in its tissues than the water contained.

Interpreting Data
In which year was the DDE level in eagle eggs the highest? What was the level in that year?

NATIONAL GEOGRAPHIC

into a less toxic form. When an organism can't metabolize (break down) or excrete a toxicant, it is simply stored, usually in fatty tissues. Over time, the organism may accumulate high concentrations of the toxicant.

The buildup of a persistent toxicant in an organism is **bioaccumulation** (**Figure 4.8a** and **b**). Organisms at the top of the food chain tend to store greater concentrations of bioaccumulated toxicants in their bodies than those lower on the food chain. As an example of **biological magnification**, consider a food chain studied in a Long Island salt marsh that was sprayed with DDT over several years for mosquito control: algae and plankton shrimp → American eel → Atlantic needlefish → ring-billed gull (**Figure 4.8c** and **d**). All top carnivores, from fishes to humans, are at risk of health problems from biological magnification. Scientists therefore test pesticides to

> **biological magnification** The increase in toxicant concentrations as a toxicant passes through successive levels of the food chain.

ensure that they do not persist and accumulate in the environment.

Mobility in the Environment

Persistent toxicants tend to move through the soil, water, and air, sometimes long distances. For example, pesticides applied to agricultural lands may be washed into rivers and streams by rain, harming aquatic life (**Figure 4.9**). If the pesticide level in their aquatic ecosystem is high enough, plants and animals may die. At lower pesticide levels, aquatic life may still suffer from symptoms of chronic toxicity such as bone degeneration in fishes. These symptoms may, for example, decrease fishes' competitiveness or increase their chances of being eaten by predators.

Mobility of persistent toxicants is also a risk for humans. In 1994 the Environmental Working Group, a private organization, analyzed five common herbicides (weed-killing chemicals) found in drinking water.

Mobility of pesticides in the environment • Figure 4.9

The intended pathway of pesticides in the environment is shown in the tan column to the right of the figure, and the actual pathways are shown in the blue column to the left.

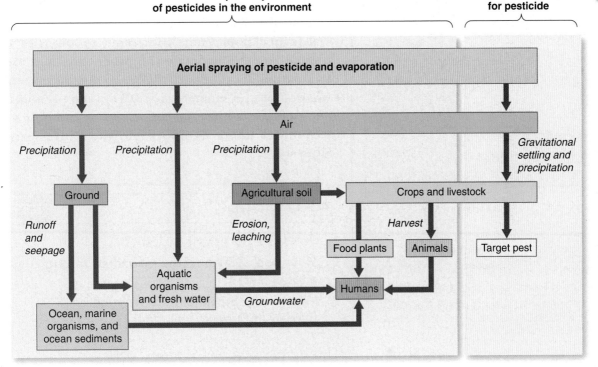

It concluded that 3.5 million people in the Midwest face a slightly elevated cancer risk because of exposure to the herbicides. The EPA has since mandated a reduction in use of the five herbicides.

The Global Ban of Persistent Organic Pollutants

persistent organic pollutants (POPs) Persistent toxicants that bioaccumulate in organisms and travel through air and water to contaminate sites far from their source.

The **Stockholm Convention on Persistent Organic Pollutants**, which was adopted in 2001, is an important U.N. treaty that seeks to protect human health and the environment from the 12 most toxic persistent organic pollutants (POPs) on Earth (**Table 4.3**). Some POPs disrupt the endocrine system

Persistent organic pollutants: The "dirty dozen" • Table 4.3

Persistent organic pollutant	Use
Aldrin	Insecticide
Chlordane	Insecticide
DDT (dichlorodiphenyl-trichloroethane)	Insecticide
Dieldrin	Insecticide
Endrin	Rodenticide and insecticide
Heptachlor	Fungicide
Hexachlorobenzene	Insecticide; fire retardant
Mirex™	Insecticide
Toxaphene™	Insecticide
PCBs (polychlorinated biphenyls)	Industrial chemicals
Dioxins	By-products of certain manufacturing processes
Furans (dibenzofurans)	By-products of certain manufacturing processes

Mosquito net • Figure 4.10

This net was treated with DDT to kill malaria-transmitting mosquitoes. Photographed in Nigeria.

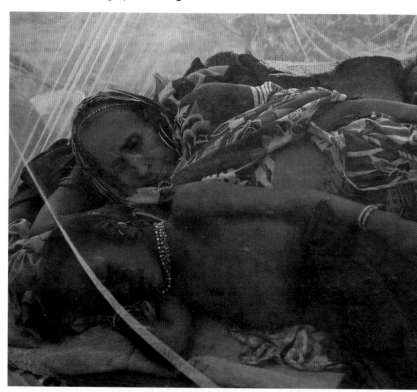

(discussed later in this chapter), cause cancer, or adversely affect the developmental processes of organisms.

The Stockholm Convention requires countries to develop plans to eliminate the production and use of intentionally produced POPs. A notable exception to this requirement is that DDT is still produced and used to control malaria-carrying mosquitoes in countries where no affordable alternatives exist (**Figure 4.10**).

CONCEPT CHECK **STOP**

1. **What** is a persistent toxicant?
2. **How** does DDT become magnified through a food chain?
3. **What** is the Stockholm Convention on Persistent Organic Pollutants?

How We Determine the Health Effects of Pollutants

LEARNING OBJECTIVES

1. **Describe** how a dose–response curve is used to determine the health effects of environmental pollutants.

2. **Describe** the most common method of determining whether a chemical causes cancer.

3. **Distinguish** among additive, synergistic, and antagonistic interactions in chemical mixtures.

4. **Explain** why children are particularly susceptible to toxicants.

We measure toxicity by the dose at which adverse effects are produced. A **dose** of a toxicant is the amount that enters the body of an exposed organism. The **response** is the type and amount of damage that exposure to a particular dose causes. A dose may cause death (*lethal dose*) or cause harm but not death (*sub-lethal dose*). Lethal doses, which are usually expressed in milligrams of toxicant per kilogram of body weight, vary depending on the organism's age, sex, health, and metabolism, as well as on how the dose was administered (all at once or over a period of time). The lethal doses, for humans, of many toxicants are known through records of homicides and accidental poisonings.

One way to determine acute toxicity is to administer different-sized doses to populations of laboratory animals, measure the responses, and use these data to predict the chemical effects on humans (**Figure 4.11**). The dose that is lethal to 50 percent of a population of test animals is called the **lethal dose–50 percent**, or **LD_{50}**. It is usually reported in milligrams of chemical toxicant per kilogram of body weight. An inverse relationship exists between the size of the LD_{50} and the acute toxicity of a chemical: The smaller the LD_{50}, the more toxic the chemical, and, conversely, the greater the LD_{50}, the less toxic the chemical (**Table 4.4**). The LD_{50} is determined for new synthetic chemicals—thousands are produced each year—as a way of estimating their toxic potential. It is generally assumed that a chemical with a low LD_{50} for several species of test animals is also very toxic in humans.

The **effective dose–50 percent**, or **ED_{50}**, measures a wide range of biological responses, such as stunted

Laboratory rat • Figure 4.11

The results of chemicals eaten or inhaled by laboratory animals are extrapolated to humans.

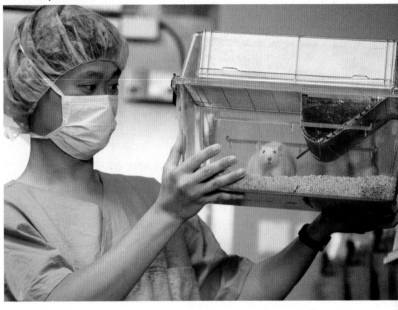

LD_{50} values for selected chemicals • Table 4.4

Chemical	LD_{50} (mg/kg)*
Aspirin	1,750
Ethanol	1,000
Morphine	500
Caffeine	200
Heroin	150
Lead	20
Cocaine	17.5
Sodium cyanide	10
Nicotine	2
Strychnine	0.8

*Administered orally to rats.

development in the offspring of a pregnant animal, reduced enzyme activity, or onset of hair loss. The ED_{50} is the dose that causes 50 percent of a population to exhibit whatever response is under study.

To develop a **dose–response curve**, scientists first test the effects of high doses and then work their way down to a **threshold level**, the maximum dose that has no measurable effect (or, alternatively, the minimum dose that produces a measurable effect) (**Figure 4.12**). Scientists assume that doses lower than the threshold level are safe.

A growing body of evidence, however, suggests that for certain toxicants there is no safe dose. A threshold does not exist for these chemicals, and even the smallest amount causes a measurable response.

> **dose–response curve** In toxicology, a graph that shows the effects of different doses on a population of test organisms.

Cancer-Causing Substances

Because cancer is so feared, for many years it was the main effect evaluated in chemical risk assessment. Environmental contaminants are linked to many serious health concerns, including other diseases, birth defects, damage to the immune system, reproductive problems, and damage to the nervous system or other body systems. We focus here on

risk assessment as it relates to cancer, but non-cancer hazards are assessed in similar ways.

The most common method of determining whether a chemical causes cancer is to expose groups of laboratory animals, such as rats, to various large doses and count how many animals develop cancer at the different levels. This method is indirect and uncertain, however. For one thing, although humans and rats are both mammals, they are different organisms and may respond differently to exposure to the same chemical. (Even rats and mice often respond differently to the same toxicant.)

Another problem is that lab rats are exposed to massive doses of the suspected **carcinogen** relative to their body size, whereas humans are usually exposed to much lower amounts. Researchers must use large doses to cause cancer in a small group of laboratory animals within a reasonable amount of time. Otherwise, such tests would take years, require thousands of test animals, and be prohibitively expensive to produce enough data to have statistically significant results.

> **carcinogen** Any substance (for example, chemical, radiation, virus) that causes cancer.

Risk assessment assumes that we can extrapolate (work backward) from the huge doses of chemicals and

Dose–response curves • Figure 4.12 _____

a. This hypothetical dose–response curve demonstrates two assumptions: first, that the biological response increases as the dose is increased; second, that harmful responses occur only above a certain threshold level.

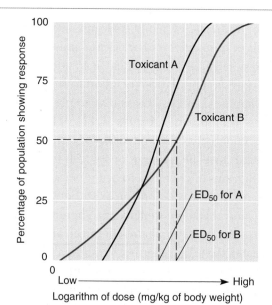

b. Dose–response curves for two hypothetical toxicants, A and B. In this example, A has a lower effective dose–50 percent (ED_{50}) than B. However, at lower doses, B is more toxic than A.

the high rates of cancer they cause in rats to determine the expected rates of cancer in humans exposed to lower amounts of the same chemicals. However, even if we are reasonably sure that exposure to high doses of a chemical causes the same effects for the same reasons in both rats and humans, we cannot assume that these same mechanisms work at low doses in humans. The body metabolizes small and large doses of a chemical in different ways. For example, enzymes in the liver may break down carcinogens in small quantities, but an excessive amount of carcinogen might overwhelm the liver enzymes.

In short, extrapolating from one species to another and from high doses to low doses is uncertain and may overestimate or underestimate a toxicant's danger. However, animal carcinogen studies provide valuable information: A toxicant that does not cause cancer in laboratory animals at high doses is not likely to cause cancer in humans at levels found in the environment or in occupational settings.

Scientists do not currently have a reliable way to determine whether exposure to small amounts of a substance causes cancer in humans. However, the EPA is planning to change how toxic chemicals are evaluated and regulated. Toxicologists are developing methods to provide direct evidence of the risk involved in exposure to low doses of cancer-causing chemicals (**Figure 4.13**).

Epidemiological evidence, including studies of human groups accidentally exposed to high levels of suspected carcinogens, is also used to determine whether chemicals are carcinogenic. For example, in 1989 epidemiologists in Germany established a direct link between cancer and a group of persistent organic pollutants called *dioxins* (see Table 4.3 on page 84). They observed the incidence of cancer in workers exposed to high concentrations of dioxins during an accident at a chemical plant in 1953 and found unexpectedly high levels of cancers in their digestive and respiratory tracts.

Risk Assessment of Chemical Mixtures

Humans are frequently exposed to various combinations of chemical compounds, in the air we breathe, the food we eat, and the water we drink. For example, cigarette smoke contains a mixture of chemicals, as does automobile exhaust. Cigarette smoke is a mixture of air pollutants that includes hydrocarbons, carbon dioxide, carbon monoxide, particulate matter, nicotine, cyanide, and a small amount of radioactive materials that come from the fertilizer used to grow the tobacco plants.

Measuring low doses of a toxicant (dioxin) in human blood serum • Figure 4.13 _____

Scientists are developing increasingly sophisticated methods of biomonitoring to analyze human tissues and fluids. Photographed at the Centers for Disease Control and Prevention's Environmental Health Laboratory.

The vast majority of toxicology studies have been performed on single chemicals rather than chemical mixtures, and for good reason. Mixtures of chemicals interact in a variety of ways, increasing the level of complexity in risk assessment, a field already complicated by many uncertainties. Moreover, there are simply too many chemical mixtures to evaluate.

Chemical mixtures interact by **additivity**, **synergy**, or **antagonism**. When a chemical mixture is *additive*, the effect is exactly what you would expect, given the individual effects of each component of the mixture. If a chemical with a toxicity level of 1 is mixed with a different chemical, also with a toxicity level of 1, the combined effect of exposure to the mixture is 2. A chemical mixture that is *synergistic* has a greater combined effect than expected; two chemicals, each with a toxicity level of 1, might have a combined toxicity of 3. An *antagonistic* interaction in a chemical mixture results in a smaller combined effect than expected; for example, two chemicals, each with a toxicity level of 1, might have a combined effect of 1.3.

If toxicological studies of chemical mixtures are lacking, how is risk assessment for chemical mixtures assigned? Toxicologists use *additivity* to assign risk to mixtures—that is, they add the known effects of each compound in the mixture. Such an approach sometimes overestimates or underestimates the actual risk involved, but it is the best method currently available. The alternative—waiting for years or decades until numerous studies have been designed, funded, and completed—is unreasonable.

Children and Chemical Exposure

Children weigh less than adults, tend to interact more with their environments, and are undergoing rapid internal changes as they grow. They are also less aware

EnviroDiscovery

Smoking: A Significant Risk

Tobacco use is the single largest cause of preventable death. Smoking causes serious diseases such as lung cancer, emphysema, and heart disease and is responsible for the premature deaths of nearly half a million people in the United States each year. Cigarette smoking annually causes about 120,000 of the 140,000 deaths from lung cancer in the United States. Smoking also contributes to heart attacks and strokes and to cancers of the bladder, mouth, throat, pancreas, kidney, stomach, voice box, and esophagus.

Passive smoking, which is nonsmokers' chronic breathing of smoke from cigarette smokers, also increases the risk of cancer. Passive smokers suffer more respiratory infections, allergies, and other chronic respiratory diseases than other nonsmokers. Passive smoking is particularly harmful to infants and young children, pregnant women, the elderly, and people with chronic lung disease.

There is good news and bad news about smoking. The good news is that fewer people in highly developed nations such as the United States are smoking. Recent research indicates that bans on smoking in public buildings and restaurants have reduced heart disease–related emergency room visits by as much as 17 percent.

The bad news is that more and more people are taking up the habit in China, Brazil, Pakistan, and other developing nations. Tobacco companies in the United States promote smoking abroad, and a substantial portion of the U.S. tobacco crop is exported. The World Health Organization (WHO) estimates that worldwide, 5 million people die each year of smoking-related causes. In an attempt to establish a global ban on tobacco advertising, WHO developed the Framework Convention on Tobacco Control. The treaty went into effect in 2005 but as of early 2010, the United States was not among the 167 parties that had joined.

NATIONAL GEOGRAPHIC

◀ **A sugarcane worker smokes during a break in his long, hard day**

Cigarette use continues to increase in many developing countries, even as consumption of tobacco decreases in the United States. Photographed in the Philippines.

Air pollution and respiratory disease in children • Figure 4.14

A Honduran mother gives oxygen to her baby, who suffers from environmentally linked respiratory disease. Farmers nearby burn land to prepare for the planting season; the resulting smoke triggers breathing problems, mostly in children and the elderly.

of potential risks from exposures. Consequently, they are often more susceptible than adults to the effects of chemicals. Consider a toxicant with an LD_{50} of 100 mg/kg. A potentially lethal dose for a child who weighs 11.3 kg (25 lb) is $100 \times 11.3 = 1130$ mg, which is equal to a scant ¼ teaspoon if the chemical is a liquid. In comparison, the potentially lethal dose for an adult who weighs 68 kg (150 lb) is 6,800 mg, or about 2 teaspoons. We must therefore protect children from exposure to environmental chemicals.

Children and Pollution Consider the toxicants in air pollution. Air pollution is a greater health threat to children than it is to adults (**Figure 4.14**). Lungs continue to develop throughout childhood, and air pollution restricts lung development. In addition, a child has a higher metabolic rate than an adult and therefore needs more oxygen. To obtain this oxygen, a child breathes more air—about two times as much air per pound of body weight as

an adult. This means that a child also breathes more air pollutants into the lungs. A 1990 study in which autopsies were performed on 100 Los Angeles children who died for reasons unrelated to respiratory problems found that more than 80 percent had subclinical lung damage, which is lung disease in its early stages, before clinical symptoms appear. (Los Angeles has some of the worst air quality in the world.)

CONCEPT CHECK **STOP**

1. **What** is a dose–response curve?

2. **What** is one way that scientists determine whether a chemical causes cancer? What are two problems with this method?

3. **What** are the three ways that chemical mixtures interact?

4. **Why** are children particularly susceptible to toxicants?

The Precautionary Principle

LEARNING OBJECTIVE

1. **Discuss** the precautionary principle as it relates to the introduction of new technologies or products.

You've probably heard the expression "An ounce of prevention is worth a pound of cure." This statement is the heart of the **precautionary principle** that many politicians and environmental activists advocate. According to the precautionary principle, we should not introduce new technology, practice, or material until it is demonstrated that (a) the risks are small and (b) the benefits outweigh the risks. The precautionary principle puts the burden of proof on the developers of the new technology or

> **precautionary principle** The idea that new technologies, practices, or materials should not be adopted until there is strong evidence that they will not adversely affect human or environmental health.

substance. The precautionary principle might require that all new industrial chemicals be tested for acute and chronic toxicity before they can be sold.

The precautionary principle is also applied to existing technologies when new evidence suggests that they are more dangerous than originally thought. For example, when observations and experiments suggested that lead added to gasoline as an anti-knock ingredient was contaminating soil, particularly in inner cities near major highways, the precautionary principle led to the phase-out of leaded gasoline (**Figure 4.15**).

To many people the precautionary principle is just common sense, given that science and risk assessment often cannot provide definitive answers to policymakers

Children in the South Bronx of New York City • Figure 4.15

In addition to air pollution, these children are probably exposed to lead in the soil (from leaded gasoline) and in paint from old buildings. Children with even low levels of lead in their blood may suffer from partial hearing loss, hyperactivity, attention deficit, lowered IQ, and learning disabilities.

Precautionary principle or economic protectionism? • Figure 4.16

Is Europe's ban of U.S. beef the result of its concern over the safety of hormone-treated beef or an excuse to support the European beef industry? Photographed on a range in Wyoming.

dealing with environmental and public health problems. The developers of a new technology or substance must prove that it is safe instead of society proving that it is harmful after it has already been introduced. However, the precautionary principle does not require that developers provide absolute proof that their product is safe; such proof would be impossible to provide.

Certain laws and decisions in many European Union nations have incorporated the precautionary principle, and some laws in the United States have a precautionary tone. In 2000, Christine Todd Whitman, then governor of New Jersey, said in a speech to the National Academy of Sciences,

> *Policy makers need to take a precautionary approach to environmental protection. . . . We must acknowledge that uncertainty is inherent in managing natural resources, recognize it is usually easier to prevent environmental damage than to repair it later, and shift the burden of proof away from those advocating protection toward those proposing an action that may be harmful.*

The precautionary principle has generated much controversy. Some scientists fear that the precautionary principle challenges the role of science and endorses making decisions without the input of science. Some critics contend that its imprecise definition reduces trade and limits technological innovations. For example, several European countries made precautionary decisions to ban beef from the United States and Canada because these countries use growth hormones to make the cattle grow faster. Europeans contend that the growth hormone might harm humans eating the beef, but the ban, in effect since 1989, is widely viewed as protecting the European beef industry (**Figure 4.16**). Another international controversy in which the precautionary principle is involved is the cultivation of genetically modified foods (discussed further in Chapter 14).

CONCEPT CHECK STOP

1. **What** is the precautionary principle? What are two criticisms of the precautionary principle?

Endocrine Disrupters

Mounting evidence suggests that dozens of widely used industrial and agricultural chemicals are **endocrine disrupters**, which interfere with the normal actions of the endocrine system (the body's hormones) in humans and animals. These chemicals include chlorine-containing industrial compounds known as PCBs and dioxins, the heavy metals lead and mercury, pesticides such as DDT, and certain plastics and plastic additives.

Hormones are chemical messengers that organisms produce to regulate their growth, reproduction, and other important biological functions. Some endocrine disrupters mimic the *estrogens*, a class of female sex hormones. Other endocrine disrupters mimic *androgens* (male hormones such as testosterone) or *thyroid hormones*. Like hormones, endocrine disrupters are active at very low concentrations and therefore may cause significant health effects at relatively low doses.

Many endocrine disrupters appear to alter the reproductive development of various animal species. A chemical spill in 1980 contaminated Lake Apopka, Florida's third largest lake, with DDT and other agricultural chemicals that have known estrogenic properties. In the years following the spill, male alligators had low levels of testosterone and elevated levels of estrogen. The mortality rate for eggs in this lake was extremely high, which reduced the alligator population for many years (see photo). Fortunately, a 2006 study indicates that Lake Apopka's alligator population is recovering.

Humans may also be at risk from endocrine disrupters. The number of reproductive disorders, infertility cases, and hormonally related cancers (such as testicular cancer and breast cancer) appears to be increasing. However, we cannot make definite connections between environmental endocrine disrupters and human health problems at this time because of the limited number of human studies. Complicating such assessments is the fact that humans are also exposed to *natural*, hormone-mimicking substances in the plants we eat.

For example, soy-based foods such as bean curd and soymilk contain natural estrogens.

Congress amended the *Food Quality Protection Act* and the *Safe Drinking Water Act* in 1996 to require the Environmental Protection Agency to develop a plan and establish priorities to test thousands of chemicals for their potential to disrupt endocrine systems. Chemicals testing positive are tested further to determine what specific damage, if any, they cause to reproduction and other biological functions. These tests, which may take decades to complete, should reveal the level of human and animal exposure to endocrine disrupters and the effects of this exposure.

Lake Apopka alligators

A young American alligator hatches from eggs that University of Florida researchers took from Lake Apopka, Florida. Many of the young alligators that hatch have abnormalities in their reproductive systems. This young alligator may not leave any offspring.

Summary

1 A Perspective on Risks 74

1. A **risk** is the probability of harm (such as injury, disease, death, or environmental damage) occurring under certain circumstances. **Risk assessment** is the quantitative and qualitative characterization of risks that allows us to compare, contrast, and manage them.

2. Risk assessment characterizes the dose–response relationship between exposures to hazards and the effects of those exposures. These characterizations can be used to help inform decisions about how best to avoid, reduce, or eliminate risks.

2 Environmental Health Hazards 77

1. **Toxicology** is the study of **toxicants**, chemicals that have adverse effects on health. **Epidemiology** is the study of the effects of chemical, biological, and physical agents on the health of human populations.

2. While most strains of coliform bacteria do not cause disease, the **fecal coliform test** is a reliable way to indicate the likely presence of **pathogens**, or disease-causing agents, in water.

3. About 25 percent of disease and injury worldwide is related to human-caused environmental changes. The environmental component of human health is sometimes direct, as when people drink unsanitary water and contract a waterborne disease. The health effects of other human activities are complex and indirect, as when climate change allows disease-causing agents to prosper.

3 Movement and Fate of Toxicants 81

1. Some toxicants exhibit **persistence**—they are extremely stable in the environment and may take many years to break down into less toxic forms. **Bioaccumulation** is the buildup of a persistent toxicant in an organism's body. **Biological magnification** is the increase in toxicant concentration as a toxicant passes through successive levels of the food chain.

2. Persistent toxicants do not stay where they are applied but tend to move through the soil, water, and air, sometimes long distances. For example, pesticides applied to agricultural lands may wash into rivers and streams, harming fishes.

3. The Stockholm Convention on Persistent Organic Pollutants requires countries to eliminate the production and use of the 12 worst **persistent organic pollutants (POPs)**. POPs are a group of persistent toxicants that bioaccumulate in organisms and travel thousands of kilometers through air and water, contaminating sites far removed from their source.

4 How We Determine the Health Effects of Pollutants 85

1. A **dose–response curve** is a graph that shows the effect of different doses on a population of test organisms. Scientists test the effects of high doses and work their way down to a **threshold level**, the maximum dose that has no measurable effect. It is assumed that doses lower than the threshold level will not have an effect on the organism and are safe.

2. A **carcinogen** is any substance (for example, chemical, radiation, virus) that causes cancer. The most common method of determining whether a chemical is carcinogenic is to expose laboratory animals such as rats to large doses of that chemical and see if they develop cancer. It is assumed that we can extrapolate from high doses of chemicals and the high rates of cancer they cause in rats to determine the rates of cancer expected in humans exposed to lower amounts of the same chemicals.

3. When a chemical mixture is **additive**, the effect is exactly what you would expect, given the combined individual effects of each component of the mixture. A chemical mixture that is **synergistic** has a greater combined effect than expected. An **antagonistic** interaction in a chemical mixture results in a smaller combined effect than expected.

4. Because they weigh less than adults, are growing, are more often exposed to toxicants in the environment, and don't know to avoid exposures, children are generally more susceptible to chemicals. The potentially lethal dose for a child is considerably less than the potentially lethal dose for an adult. Air pollution restricts a child's lung development. Also, a child has a higher metabolic rate than an adult and therefore breathes more air and more air pollutants into the lungs.

5 The Precautionary Principle 90

1. The **precautionary principle** is the ideal that new technologies, practices, or materials should not be adopted until there is strong evidence that they will not adversely affect human or environmental health.

Key Terms

- acute toxicity 77
- biological magnification 83
- carcinogen 86
- chronic toxicity 77

- dose–response curve 86
- epidemiology 77
- pathogen 78
- persistant organic pollutants (POPs) 84

- precautionary principle 90
- risk 74
- risk assessment 75
- toxicology 77

What is happening in this picture?

- Why do we use animal testing to determine whether a new pesticide causes cancer?

- This mouse developed cancer after exposure to high levels of a toxicant. What uncertainties are associated with extrapolating this result to low levels of exposure of that toxicant in humans?

- Chemical manufacturers have sometimes paid human subjects to be exposed to low doses of new chemicals, but the EPA currently has a moratorium on such testing. Suggest a possible reason for the moratorium. Do you consider such testing ethical? Why or why not?

Critical and Creative Thinking Questions

1. Which risk—an extremely small amount of a cancer-causing chemical in drinking water or smoking cigarettes—tends to generate the greatest public concern? Explain why this view can be counterproductive.

2. Should public policymakers be more concerned with public risk perception or with risks as calculated by experts? Explain your answer.

3. Describe what you would expect to find in a risk characterization.

4. What is the Stockholm Convention on Persistent Organic Pollutants?

5. Is the absence of scientific certainty about the health effects of an environmental pollutant synonymous with the absence of risk? Explain your answer.

6. Would you support the United States adopting the precautionary policy in all of its legislation? Why or why not?

7. Explain how toxicology and epidemiology contribute to risk assessment.

8. Describe how a persistent pesticide might move around in the environment.

9. Why is air pollution a greater threat to children than it is to adults?

10. Distinguish among persistence, bioaccumulation, and biological magnification.

11. How do acute and chronic toxicity differ?

12–14. The figure to the right shows the organisms sampled in the Long Island salt marsh study of DDT (also see Figure 4.8 on page 82).

12. If DDT is sprayed on land to control insects, how does it get into the bodies of aquatic species?

13. Why does the Atlantic needlefish (5) contain more DDT in its body than an American eel (4)?

14. How does high concentration of DDT cause reproductive failure in birds at the top of the food chain?

THE PLANNER

How Ecosystems Work

LAKE VICTORIA'S ECOLOGICAL IMBALANCE

Africa's Lake Victoria, the world's second largest freshwater lake, was once home to about 400 species of small, colorful fishes known as cichlids (pronounced "SICK'lids"; see inset). These fishes thrived throughout the lake ecosystem and provided protein for the millions of humans living nearby.

The Nile perch was introduced into the lake in the early 1960s to stimulate the local economy. By 1985, most of the catch was Nile perch, which was increasing in number because it had an abundant food supply—the cichlids.

Today, more than half of the cichlids and other native fish species in Lake Victoria are extinct. Local fishermen, who once caught and ate hundreds of different types of fishes, now catch only a few types. Because most of the algae-eating cichlids disappeared, the algal population has increased explosively. Decomposing dead algae use up dissolved oxygen in the water. The bottom zone of the lake, once filled with cichlids, is empty because it contains too little dissolved oxygen.

Many scientists doubt that the depleted lake can sustain the Nile perch fishery for long. In the meantime, conservation efforts such as the American Zoo and Aquarium Association's Lake Victoria Species Survival Program work to preserve in captivity cichlids that are now absent from Lake Victoria.

WILEY
RESOURCE KIT

CHAPTER OUTLINE

CHAPTER PLANNER ✔

- ❑ Study the picture and read the opening story.
- ❑ Scan the Learning Objectives in each section:
  p. 98 ❑   p. 101 ❑   p. 106 ❑   p. 113 ❑   p. 116 ❑
- ❑ Read the text and study all figures and visuals. Answer any questions.

Analyze key features

- ❑ Process Diagram, p. 104 ❑   p. 105 ❑   p. 107 ❑
  p. 108 ❑   p. 109 ❑   p. 111 ❑   p. 112 ❑
- ❑ What a Scientist Sees, p. 115 ❑
- ❑ Environmental InSight, p. 117 ❑
- ❑ EnviroDiscovery, p. 118 ❑
- ❑ Case Study, p. 122 ❑
- ❑ Stop: Answer the Concept Checks before you go on:
  p. 100 ❑   p. 106 ❑   p. 113 ❑   p. 115 ❑   p. 121 ❑

End of Chapter

- ❑ Review the Summary and Key Terms.
- ❑ Answer What is happening in this picture?
- ❑ Answer the Critical and Creative Thinking Questions.

What Is Ecology?

LEARNING OBJECTIVES

1. **Define** *ecology*.
2. **Distinguish** among the following ecological levels: population, community, ecosystem, landscape, and biosphere.

In the 19th century the German biologist Ernst Haeckel first developed the concept of ecology and devised its name—*eco* from the Greek word for "house" and *logy* from the Greek word for "study." Thus, **ecology** literally means "the study of one's house." The environment—one's house—consists of two parts: the *biotic* (living) environment, which includes all organisms, and the *abiotic* (nonliving, or physical) environment, which includes such physical factors as living space, temperature, sunlight, soil, wind, and precipitation.

> **ecology** The study of the interactions among organisms and between organisms and their abiotic environment.

The focus of ecology is local or global, specific or generalized, depending on what questions the scientist is trying to answer. One ecologist might determine the temperature or light requirements of a single oak, another might study all the organisms that live in a forest where the oak is found, and another might examine how nutrients flow between the forest and surrounding communities.

Ecology is the broadest field within the biological sciences. It is linked to every other biological discipline and to other fields as well. Geology and earth science are extremely important to ecology, especially when ecologists examine the physical environment of Earth. Chemistry and physics are also important. Humans are biological organisms, and all our activities have a bearing on ecology. Even economics and politics have profound ecological implications, as discussed in Chapter 3.

Ecologists are most interested in the levels of biological organization including and above the level of the individual organism. Individuals of the same species occur in **populations**. (A *species* is a group of similar organisms whose members freely interbreed with one another in the wild to produce fertile offspring; members of one species generally don't interbreed with other species of organisms.) A population ecologist might study a population of walruses or a population of marsh grass.

> **population** A group of organisms of the same species that live together in the same area at the same time.

Populations are organized into **communities**. The number and kinds of species that live within a community, along with their relationships with one another, characterize the community. A community ecologist might study how organisms interact with one another—including feeding relationships (who eats whom)—in an alpine meadow community or in a tidal pool (**Figure 5.1**).

Ecosystem is a more inclusive term than community. An ecosystem includes all the biotic interactions of a community as well as the interactions between organisms and their abiotic environment. In an ecosystem, all the biological, physical, and chemical components of an area form an extremely complicated interacting network of energy flow and materials cycling. An ecosystem ecologist might examine how energy, nutrient composition, or water affects the organisms living in a desert community or a coastal bay ecosystem.

> **community** A natural association that consists of all the populations of different species that live and interact together within an area at the same time.
>
> **ecosystem** A community and its physical environment.

The ultimate goal of ecosystem ecologists is to understand how ecosystems function. This is not a simple task, but it is important because ecosystem processes collectively regulate the global cycles of water, carbon, nitrogen, sulfur, and phosphorus that are essential to the survival of humans and all other organisms. As humans increasingly alter ecosystems for their own uses, the natural functioning of ecosystems is changed, and we must learn whether these changes will affect the *sustainability* of our life-support system.

Landscape ecology is a subdiscipline in ecology that studies ecological processes that operate over larger areas. Landscape ecologists

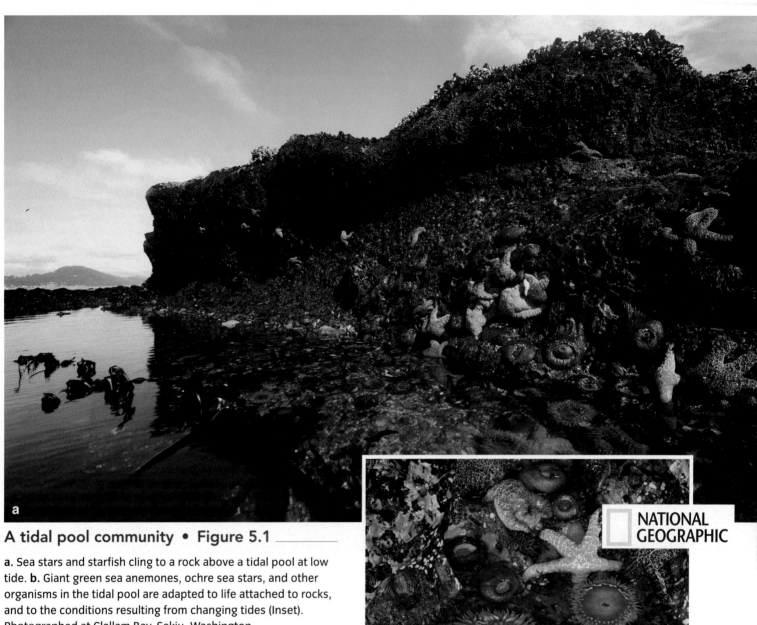

A tidal pool community • Figure 5.1

a. Sea stars and starfish cling to a rock above a tidal pool at low tide. **b.** Giant green sea anemones, ochre sea stars, and other organisms in the tidal pool are adapted to life attached to rocks, and to the conditions resulting from changing tides (Inset). Photographed at Clallam Bay, Sekiu, Washington.

NATIONAL GEOGRAPHIC

examine the connections among ecosystems found in a particular region. Consider, for example, a **landscape** consisting of a forest ecosystem and a pond ecosystem located adjacent to the forest. One possible connection between these two ecosystems is the great blue heron, which eats fish, frogs, insects, crustaceans, and snakes along the shallow water of the pond but often builds nests and raises its young in the secluded treetops of the nearby

landscape A region that includes several interacting ecosystems.

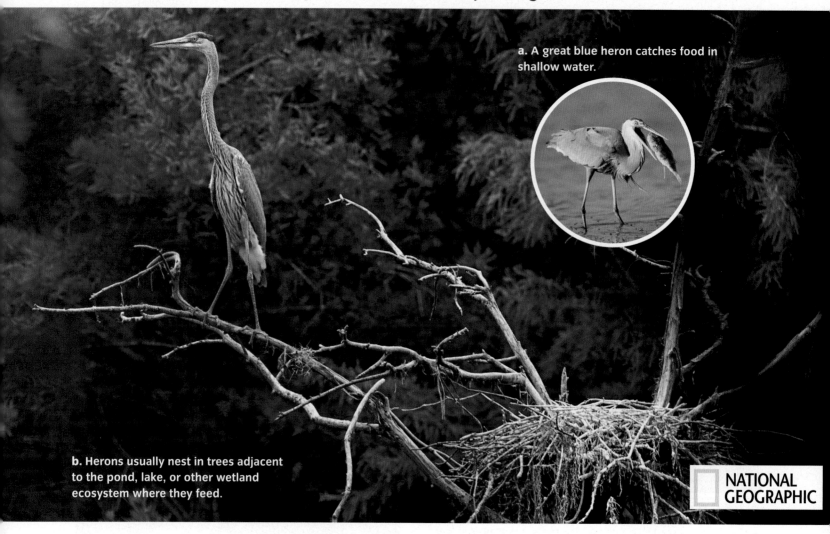

a. A great blue heron catches food in shallow water.

b. Herons usually nest in trees adjacent to the pond, lake, or other wetland ecosystem where they feed.

NATIONAL GEOGRAPHIC

forest (**Figure 5.2**). Landscapes, then, are based on larger land areas that include several ecosystems.

The organisms of the **biosphere**—Earth's communities, ecosystems, and landscapes—depend on one another and on the other realms of Earth's physical environment: the atmosphere, hydrosphere, and lithosphere. The *atmosphere* is the gaseous envelope surrounding Earth; the *hydrosphere* is Earth's supply of water—liquid and frozen, fresh and salty; and the *lithosphere* is the soil and rock of Earth's crust. Ecologists who study the biosphere examine the global interrelationships among Earth's atmosphere, land, water, and organisms.

biosphere The layer of Earth that contains all living organisms.

The biosphere teems with life. Where do these organisms get the energy to live? And how do they harness this energy? We now examine the importance of energy to organisms, which survive only as long as the environment continuously supplies them with energy. We will revisit energy as it relates to human endeavors in many chapters throughout this text.

CONCEPT CHECK STOP

1. **What** is the definition of *ecology*?

2. **What** is the difference between an ecosystem and a landscape? between a community and an ecosystem?

The Flow of Energy Through Ecosystems

LEARNING OBJECTIVES

1. **Define** *energy*, and state the first and second laws of thermodynamics.
2. **Distinguish** among producers, consumers, and decomposers.
3. **Summarize** how energy flows through a food web.

Energy is the capacity or ability to do work. Organisms require energy to grow, move, reproduce, and maintain and repair damaged tissues. Energy exists as stored energy—called **potential energy**—and as **kinetic energy**, the energy of motion. We can think of potential energy as an arrow on a drawn bow (**Figure 5.3**). When the string is released, this potential energy is converted to kinetic energy as the motion of the bow propels the arrow. Similarly, the grass a bison eats has chemical potential energy, some of which

> **first law of thermodynamics**
> A physical law which states that energy cannot be created or destroyed, although it can change from one form to another.

is converted to kinetic energy and heat as the bison runs across the prairie. Thus, energy changes from one form to another.

The First and Second Laws of Thermodynamics

Thermodynamics is the study of energy and its transformations. Two laws about energy apply to all things in the universe: the first and second laws of thermodynamics. According to the **first law of thermodynamics**, an organism may utilize energy by converting it from one form to another, but the total energy content of the organism and its surroundings is always the same. An organism can't create the energy it requires to live. Instead, it must capture energy from the environment to use for biological work, a process that involves the transformation of energy from one form

---

Potential and kinetic energy • Figure 5.3 _____

Potential energy is stored in the drawn bow (**a**) and is converted to kinetic energy (**b**) as the arrow speeds toward its target. Photographed in Athens, Greece, during the 2004 Summer Olympics.

photosynthesis
The biological process that captures light energy and transforms it into the chemical energy of organic molecules, which are manufactured from carbon dioxide and water.

to another. In **photosynthesis**, for example, plants absorb the radiant energy of the sun and convert it into the chemical energy contained in the bonds of sugar molecules (**Figure 5.4**). Later, an animal that eats the plant may transform some of the chemical energy into the mechanical energy of muscle contraction, enabling the animal to walk, run, slither, fly, or swim.

As each energy transformation occurs, some of the energy is changed to heat that is released into the cooler surroundings. No organism can ever use this energy again for biological work; it is "lost" from the biological point of view. However, it isn't gone from a thermodynamic point of view because it still exists in the surrounding physical environment. The use of food to enable you to walk or run doesn't destroy the chemical energy that was once present in the food molecules. After you have performed the task of walking

Capturing energy from the environment • Figure 5.4 _____

The sun powers photosynthesis, producing chemical energy stored in the leaves and seeds of this umbrella tree. Photographed in Hanging Rock State Park, North Carolina.

or running, the energy still exists in your surroundings as heat.

According to the **second law of thermodynamics**, the amount of usable energy available to do work in the universe decreases over time. The second law of thermodynamics is consistent with the first law—that is, the total amount of energy in the universe isn't decreasing with time. However, the total amount of energy in the universe available to do biological work is decreasing over time.

second law of thermodynamics
A physical law which states that when energy is converted from one form to another, some of it is degraded into heat, a less usable form that disperses into the environment.

Less usable energy is more diffuse, or disorganized, than more usable energy. *Entropy* is a measure of this disorder or randomness. Organized, usable energy has low entropy, whereas disorganized energy such as heat has high entropy. Another way to explain the second law of thermodynamics is that entropy, or disorder, in a system tends to increase over time. As a result of the second law of thermodynamics, no process that requires an energy conversion is ever 100 percent efficient because much of the energy is dispersed as heat, resulting in an increase in entropy. For example, an automobile engine, which converts the chemical energy of gasoline to mechanical energy, is between 20 and 30 percent efficient: Only 20 to 30 percent of the original energy stored in the chemical bonds of the gasoline molecules is actually transformed into mechanical energy, or work.

Producers, Consumers, and Decomposers

The organisms of an ecosystem are divided into three categories, based on how they obtain nourishment: producers, consumers, and decomposers. Virtually all ecosystems contain representatives of all three groups, which interact extensively with one another, both directly and indirectly.

Plants and other photosynthetic organisms are **producers** and manufacture large organic molecules from simple inorganic substances, generally carbon dioxide and water, usually using the energy of sunlight. Producers are potential food resources for other organisms because they incorporate the chemicals they manufacture into their own bodies. Plants are the most significant producers on land, and algae and certain types of bacteria are important producers in aquatic environments.

Animals are **consumers**—they consume other organisms as a source of food energy and bodybuilding materials. Consumers that eat producers are *primary consumers*, or *herbivores*. Grasshoppers, deer, moose, and rabbits are examples of primary consumers (**Figure 5.5a**). *Secondary consumers* eat primary consumers, whereas *tertiary consumers* eat secondary consumers. Both secondary and tertiary consumers are *carnivores* that eat other animals. Lions, spiders, and lizards are examples of carnivores (**Figure 5.5b**). Other consumers, called *omnivores*, eat a variety of organisms. Bears, pigs, and humans are examples of omnivores.

Some consumers, called detritus feeders, consume **detritus**, organic matter that includes animal carcasses, leaf litter, and feces (**Figure 5.5c**). Detritus feeders, such as snails, crabs, clams, and worms, are abundant in aquatic environments. Earthworms, termites, beetles, snails, and millipedes are terrestrial (land-dwelling) detritus feeders. Detritus feeders work together with microbial decomposers to destroy dead organisms and waste products.

Bacteria and fungi are important examples of **decomposers**, organisms that break down dead organisms and waste products (**Figure 5.5d**). Decomposers release simple inorganic molecules, such as carbon dioxide and mineral salts, which producers can then reuse.

Consumers and decomposers • Figure 5.5

a. The moose is an herbivore, or a primary consumer. The chemical energy stored in grasses transfers to the moose cow as it eats.

c. A Sally Lightfoot crab forages for detritus on volcanic rock in the Galápagos Islands.

d. The mushrooms growing on a dead beech tree are reproductive structures; the invisible branching, threadlike body of the mushroom grows in the tree trunk, decomposing dead organic material.

b. A Madagascar day gecko (a tertiary consumer) feeds on a spider (a secondary consumer). Both the gecko and the spider are carnivores.

Energy flow through a food chain • Figure 5.6

Much of the energy acquired by a given level of a food chain is used and escapes into the surrounding environment as heat. This energy, as the second law of thermodynamics stipulates, is unavailable to the next level of the food chain.

2 Energy flows linearly—in a one-way direction—through ecosystems.

1 Energy enters ecosystem from the sun.

3 Energy exits as heat loss.

| **First trophic level:** Producers | **Second trophic level:** Primary consumers | **Third trophic level:** Secondary consumers | **Fourth trophic level:** Tertiary consumers | **Decomposers** |

Energy from sun

Heat   Heat   Heat   Heat   Heat

Think Critically Why do you suppose that, within a particular ecosystem, the population size of a top consumer like a hawk is much smaller than the population size of a primary consumer like a seed-eating rat? Express your answer in terms of energy flow.

WILEY
RESOURCE KIT

The Path of Energy Flow in Ecosystems

In an ecosystem, energy flow occurs in **food chains**, in which energy from food passes from one organism to the next in a sequence. When a food chain is diagrammed, it consists of a series of arrows, each of which points from the species that is consumed to the species that uses it as food (**Figure 5.6**). Each level, or "link," in a food chain is a **trophic level**. (The Greek *tropho* means "nourishment.") An organism is assigned a trophic level based on the number of energy transfer steps to that level.

Producers form the first trophic level, primary consumers form the second trophic level, secondary consumers form the third trophic level, and so on. At

energy flow
The passage of energy in a one-way direction through an ecosystem.

every step in a food chain are decomposers, which respire organic molecules in the carcasses and body wastes of all members of the food chain.

Simple food chains rarely occur in nature because few organisms eat just one kind of organism. Plants, for example, are eaten by a variety of insects, birds, and mammals; and most of these herbivores are consumed by several different predators. Thus, the flow of energy through an ecosystem typically takes place in accordance with a range of food choices for each organism involved. In an ecosystem of average complexity, numerous alternative pathways are possible. An owl eating a rabbit is a different energy pathway than an owl eating a snake. A **food web**, a complex of interconnected food chains in an ecosystem, is a more realistic model of the flow of energy and materials through ecosystems (**Figure 5.7**). A food web helps us visualize feeding relationships that indicate how a community is organized.

Food web at the edge of an eastern U.S. deciduous forest • Figure 5.7

This food web is greatly simplified compared to what actually happens in nature. Many species aren't included, and numerous links in the web aren't shown.

Key
1. Pitch pine
2. White oak
3. Barred owl
4. Gray squirrel
5. Eastern chipmunk
6. Eastern cottontail
7. Red fox
8. White-tailed deer
9. Red-tailed hawk
10. Eastern bluebird
11. Red-winged blackbird
12. Blackberry
13. American robin
14. Red-headed woodpecker
15. Red clover
16. Bacteria
17. Worms and ants
18. Moths
19. Deer mouse
20. Spiders
21. Insect larvae
22. Insects
23. Fungi

Think Critically Compare and explain the very different locations of decomposers and producers.

The most important thing to remember about energy flow in ecosystems is that it is *linear*, or *one way*. Energy moves along a food chain or food web from one organism to the next, as long as it isn't used for biological work. Once an organism uses energy, it is lost as heat (recall the second law of thermodynamics) and is unavailable for any other organism in the ecosystem.

Organisms at each step of a food chain use a large amount of the potential energy available to them. Because this energy is ultimately lost into the environment as heat, the number of steps in any food chain is limited.

The longer the food chain, the less energy is available for organisms at the higher trophic levels.

CONCEPT CHECK

1. **What** is the first law of thermodynamics? the second?

2. **Why** is a balanced ecosystem unlikely to contain only producers and consumers? only consumers and decomposers? Explain your answers.

3. **How** does energy move through a food web?

The Cycling of Matter in Ecosystems

LEARNING OBJECTIVE

1. **Diagram and explain** the carbon, hydrologic, nitrogen, sulfur, and phosphorus cycles.

In contrast to energy flow, matter, the material of which organisms are composed, moves in numerous cycles from one part of an ecosystem to another—from one organism to another and from living organisms to the abiotic environment and back again. We call these cycles of matter **biogeochemical cycles** because they involve biological, geological, and chemical interactions. Five different biogeochemical cycles of matter—carbon, hydrologic, nitrogen, sulfur, and phosphorus—are representative of all biogeochemical cycles. These five cycles are particularly important to organisms, for these materials make up the chemical compounds of cells. Humans affect all of these cycles on both local and global scales; we conclude the chapter with an example of this human influence.

The Carbon Cycle

Proteins, carbohydrates, and other molecules that are essential to living organisms contain carbon, so organisms must have carbon available to them. Carbon makes up approximately 0.04 percent of the atmosphere as a gas, carbon dioxide (CO_2). It is present in the ocean in several chemical forms, such as carbonate (CO_3^{2-}) and bicarbonate (HCO_3^-), and in sedimentary rocks such as limestone, which consists primarily of calcium carbonate ($CaCO_3$). The global movement of carbon between organisms and the abiotic environment—including the atmosphere, ocean, and sedimentary rock—is known as the *carbon cycle* (**Figure 5.8**).

During photosynthesis, plants, algae, and certain bacteria remove carbon (as CO_2) from the air and fix (incorporate) it into chemical compounds such as sugar. Plants use sugar to make other compounds. Thus, photosynthesis incorporates carbon from the abiotic environment into the biological compounds of producers. Those compounds are usually used as fuel for cellular respiration by the producer that made them, by a consumer that eats the producer, or by a decomposer that breaks down the remains of the producer or consumer. During respiration, sugar is broken down to carbon dioxide that is returned to the atmosphere. A similar carbon cycle occurs in aquatic ecosystems, involving carbon dioxide dissolved in the water.

Sometimes the carbon in biological molecules isn't recycled back to the abiotic environment for quite a while. For example, a large amount of carbon is stored in the wood of trees, where it may stay for several

The carbon cycle • Figure 5.8

The movement of carbon between the abiotic environment (the atmosphere and ocean) and living organisms is known as the carbon cycle. Because proteins, carbohydrates, and other living molecules contain carbon, the process is essential to life.

Think Critically How would removing a large forest affect the carbon cycle? What would be the additional effects if the wood from those trees were used as fuel?

WILEY
RESOURCE KIT

hundred years or even longer. Coal, oil, and natural gas, called *fossil fuels* because they formed from the remains of ancient organisms, are vast deposits of carbon compounds—the end products of photosynthesis that occurred millions of years ago. In *combustion*, organic molecules in wood, coal, oil, and natural gas are burned, with accompanying releases of heat, light, and carbon dioxide.

The thick deposits of shells of marine organisms contain carbon. These shells settle to the ocean floor and are eventually cemented together to form the sedimentary rock limestone. The crust is dynamically active, and over millions of years, sedimentary rock on the bottom of the seafloor may lift to form land surfaces. The summit of Mt. Everest, for example, is composed of sedimentary rock.

The Hydrologic Cycle

In the *hydrologic cycle* water continuously circulates from the ocean to the atmosphere to the land and back to the ocean. It provides a renewable supply of purified water for terrestrial organisms. This cycle results in water distributed among the ocean, the land, and the atmosphere (**Figure 5.9**).

Water may evaporate from land and reenter the atmosphere directly. Alternatively, it may flow in rivers and streams to lakes or the ocean. The movement of water from land to rivers, lakes, wetlands, and the ocean is *runoff*, and the area of land where runoff drains is a *watershed*. Regardless of its physical form—solid, liquid, or vapor—or location, every molecule of water moves through the hydrologic cycle repeatedly.

PROCESS DIAGRAM

The hydrologic cycle • Figure 5.9

In the hydrologic cycle, water moves among the ocean, the atmosphere, the land, and back to the ocean in a continuous process that supports life.

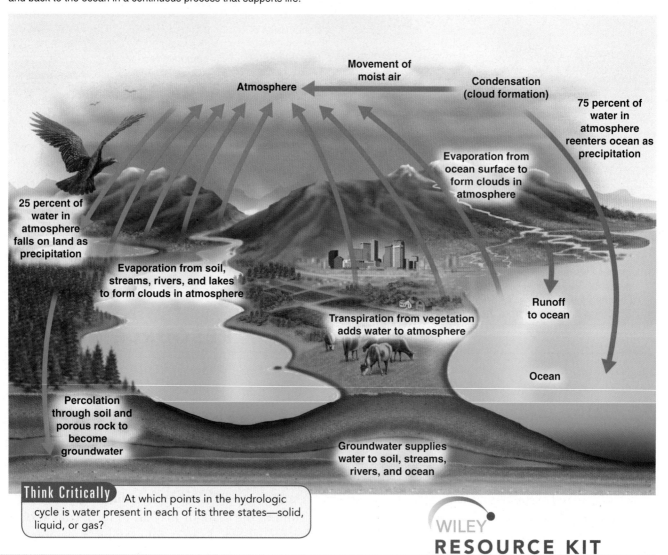

Movement of moist air

Atmosphere

Condensation (cloud formation)

75 percent of water in atmosphere reenters ocean as precipitation

Evaporation from ocean surface to form clouds in atmosphere

25 percent of water in atmosphere falls on land as precipitation

Evaporation from soil, streams, rivers, and lakes to form clouds in atmosphere

Transpiration from vegetation adds water to atmosphere

Runoff to ocean

Ocean

Percolation through soil and porous rock to become groundwater

Groundwater supplies water to soil, streams, rivers, and ocean

Think Critically At which points in the hydrologic cycle is water present in each of its three states—solid, liquid, or gas?

WILEY
RESOURCE KIT

The Nitrogen Cycle

Nitrogen is critical for all organisms because it is an essential part of biological molecules such as proteins and nucleic acids (for example, DNA). At first glance, a shortage of nitrogen for organisms appears impossible. The atmosphere is 78 percent nitrogen gas (N_2). But atmospheric nitrogen is so stable that it does not readily combine with other elements. Atmospheric nitrogen must first be broken apart before the nitrogen atoms combine with other elements to form proteins and nucleic acids.

There are five steps in the *nitrogen cycle*, in which nitrogen cycles between the abiotic environment and organisms: nitrogen fixation, nitrification, assimilation, ammonification, and denitrification (**Figure 5.10**).

The nitrogen cycle • Figure 5.10

 THE PLANNER

The movement of nitrogen between the abiotic environment (primarily the atmosphere) and living organisms is known as the nitrogen cycle. The five steps of the nitrogen cycle are nitrogen fixation, nitrification, assimilation, ammonification, and denitrification.

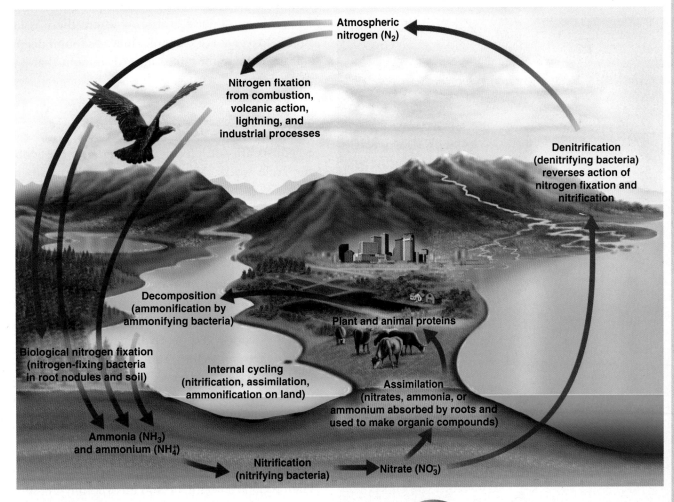

Atmospheric nitrogen (N_2)

Nitrogen fixation from combustion, volcanic action, lightning, and industrial processes

Denitrification (denitrifying bacteria) reverses action of nitrogen fixation and nitrification

Decomposition (ammonification by ammonifying bacteria)

Plant and animal proteins

Biological nitrogen fixation (nitrogen-fixing bacteria in root nodules and soil)

Internal cycling (nitrification, assimilation, ammonification on land)

Assimilation (nitrates, ammonia, or ammonium absorbed by roots and used to make organic compounds)

Ammonia (NH_3) and ammonium (NH_4^+)

Nitrification (nitrifying bacteria)

Nitrate (NO_3^-)

Think Critically How might a worldwide decline in nitrifying bacteria alter the nitrogen cycle?

WILEY
RESOURCE KIT

Bacteria are the only organisms involved in each of these steps except assimilation.

Nitrogen-fixing bacteria carry out *nitrogen fixation* in soil and aquatic environments. The process gets its name from the fact that nitrogen is fixed into a form that organisms can use, ammonia (NH_3). Volcanic activity, lightning, and human activities—combustion and industrial processes—also fix considerable nitrogen because all supply enough energy to break apart atmospheric nitrogen.

Nitrogen-fixing bacteria split atmospheric nitrogen and combine the resulting nitrogen atoms with hydrogen. Some nitrogen-fixing bacteria, *Rhizobium*, live inside swellings, or nodules, on the roots of legumes such as beans or peas and some woody plants (**Figure 5.11a**). In moist environments, photosynthetic bacteria called *cyanobacteria* perform most of the nitrogen fixation (**Figure 5.11b**).

During *nitrification*, soil bacteria convert ammonia to nitrate (NO_3^-). The process of nitrification furnishes these bacteria, called nitrifying bacteria, with energy. In *assimilation*, plants absorb ammonia or nitrate through their roots and convert the nitrogen into plant compounds such as proteins. Animals assimilate nitrogen when they consume plants or other animals and convert the proteins into animal proteins.

Ammonification occurs when organisms produce nitrogen-containing waste products such as urine. These substances, plus the nitrogen compounds that occur in dead organisms, are decomposed, releasing the nitrogen into the abiotic environment as ammonia. The bacteria that perform this process are called ammonifying bacteria. Other bacteria perform *denitrification*, in which nitrate is converted back to nitrogen gas. Denitrifying bacteria typically live and grow where there is little or no free oxygen. For example, they are found deep in the soil near the water table, an environment that is nearly oxygen free.

Nitrogen fixation • Figure 5.11

a. Bacteria carry out nitrogen fixation in the nodules of a pea plant's roots.

b. *Nostoc*, a cyanobacterium that fixes nitrogen, grows here on a mossy bank. This particular species forms colonies that range in size from a pinhead to a potato.

The Sulfur Cycle

Scientists are still piecing together how the global sulfur cycle works. Most sulfur is underground in sedimentary rocks and minerals, which over time erode to release sulfur-containing compounds into the ocean (**Figure 5.12**). Sulfur gases enter the atmosphere from natural sources in both the ocean and land. Sea spray delivers sulfates (SO_4^{2-}) into the air, as do forest fires and dust storms. Volcanoes release both hydrogen sulfide (H_2S), a poisonous gas that smells like rotten eggs, and sulfur oxides (SO_x). Hydrogen sulfide reacts with oxygen to form sulfur oxides, and sulfur oxides react with water to form sulfuric acid (H_2SO_4). Although sulfur gases make up a minor part of the atmosphere, the total movement of sulfur to and from the atmosphere is substantial.

A tiny fraction of global sulfur is present in living organisms, where it is an essential component of proteins. Plant roots absorb sulfate and incorporate the sulfur into plant proteins. Animals assimilate sulfur when they consume plant proteins and convert them to animal proteins.

The sulfur cycle • Figure 5.12

The largest sources of sulfur on Earth are sedimentary rock and the ocean. In the sulfur cycle, sulfur compounds are incorporated into organisms and move among them, the atmosphere, the ocean, and land.

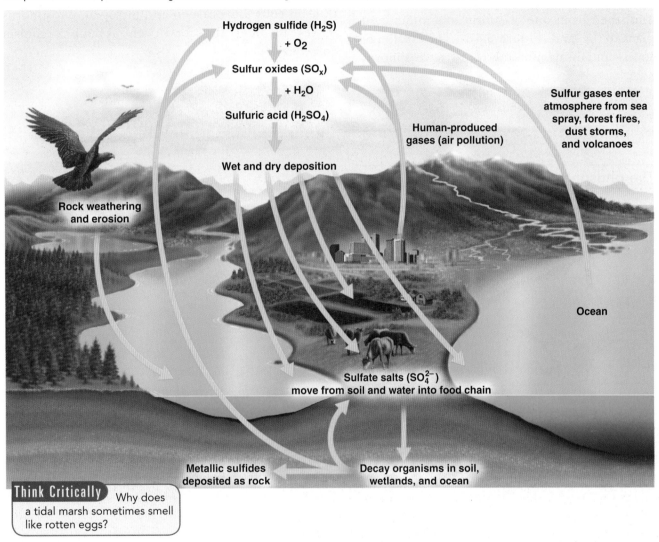

Hydrogen sulfide (H₂S)

+ O₂

Sulfur oxides (SOₓ)

+ H₂O

Sulfuric acid (H₂SO₄)

Sulfur gases enter atmosphere from sea spray, forest fires, dust storms, and volcanoes

Human-produced gases (air pollution)

Wet and dry deposition

Rock weathering and erosion

Ocean

Sulfate salts (SO₄²⁻) move from soil and water into food chain

Metallic sulfides deposited as rock

Decay organisms in soil, wetlands, and ocean

Think Critically Why does a tidal marsh sometimes smell like rotten eggs?

In the ocean, certain marine algae release a compound that bacteria convert to dimethyl sulfide (DMS). DMS is released into the atmosphere, where it helps condense water into droplets in clouds and may affect weather and climate. Atmospheric DMS is converted to sulfate, most of which is deposited into the ocean.

As in the nitrogen cycle, bacteria drive the sulfur cycle. In freshwater wetlands, tidal flats, and flooded soils, which are oxygen deficient, certain bacteria convert sulfates to hydrogen sulfide gas, which is released into the atmosphere, or to metallic sulfides, which are deposited as rock. In the absence of oxygen, other bacteria perform a type of photosynthesis that uses hydrogen sulfide instead of water. Where oxygen is present, different bacteria oxidize sulfur compounds to sulfates.

The Cycling of Matter in Ecosystems **111**

The Phosphorus Cycle

Unlike the biogeochemical cycles just discussed, the phosphorus cycle doesn't have an atmospheric component. Phosphorus cycles from the land into living organisms, then from one organism to another, and finally back to the land (**Figure 5.13**). The erosion of phosphorus-containing minerals releases phosphorus into the soil, where plant roots absorb it in the form of inorganic phosphates. Phosphates are used in biological molecules such as nucleic acids and ATP, a compound that is important in energy transfer reactions in cells. Like carbon and nitrogen, phosphorus moves through the food web as one organism consumes another.

The phosphorus cycle • Figure 5.13

THE PLANNER

Phosphorus moves from the land through aquatic and terrestrial communities, between organisms in these communities, and back to the land in a process known as the phosphorus cycle. Unlike other biogeochemical cycles, the phosphorus cycle does not involve the atmosphere.

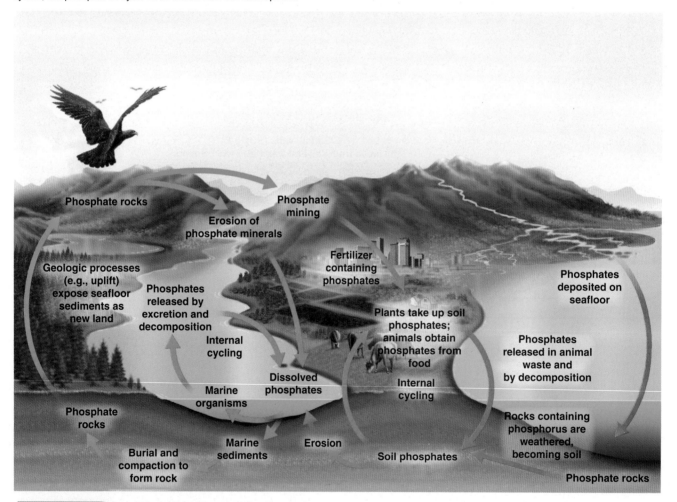

Phosphate rocks

Erosion of phosphate minerals

Phosphate mining

Geologic processes (e.g., uplift) expose seafloor sediments as new land

Phosphates released by excretion and decomposition

Internal cycling

Fertilizer containing phosphates

Plants take up soil phosphates; animals obtain phosphates from food

Internal cycling

Phosphates deposited on seafloor

Phosphates released in animal waste and by decomposition

Phosphate rocks

Marine organisms

Dissolved phosphates

Rocks containing phosphorus are weathered, becoming soil

Burial and compaction to form rock

Marine sediments

Erosion

Soil phosphates

Phosphate rocks

Think Critically If the phosphates present in fertilizer on a field are too abundant to be taken up by plants and animals, how are those excess phosphates likely to move in the phosphorus cycle?

Phosphorus cycles through aquatic communities in much the same way that it does through terrestrial communities. Dissolved phosphorus enters aquatic communities as algae and plants absorb and assimilate it; plankton and larger organisms obtain phosphorus when they consume the algae and plants. A variety of fishes and molluscs eat plankton in turn. Ultimately, decomposers release inorganic phosphorus into the water, where it is available for aquatic producers to use again.

Phosphate can be lost from biological cycles. Some phosphate is carried from the land by streams and rivers to the ocean, where it can be deposited on the seafloor and remain for millions of years.

A small portion of the phosphate in the aquatic food web finds its way back to land. A few fishes and aquatic invertebrates are eaten by seabirds, which may defecate on land where they roost. The manure of seabirds, called guano, contains large amounts of phosphate and nitrate. Once on land, these minerals may be absorbed by the roots of plants. The phosphate contained in guano may enter terrestrial food webs in this way, although the amounts involved are small.

CONCEPT CHECK STOP

1. **What** are the differences and similarities among the five biogeochemical cycles, particularly in the roles organisms play in them?

Ecological Niches

LEARNING OBJECTIVES

1. **Describe** the factors that contribute to an organism's ecological niche.
2. **Explain** the concept of resource partitioning.

You have seen that a diverse assortment of organisms inhabits each community and that these organisms obtain nourishment in a variety of ways. You have also considered energy flow and biogeochemical cycles. Now let's examine the way of life of a given species in its ecosystem. An ecological description of a species typically includes whether it is a producer, consumer, or decomposer. However, we need other details to provide a complete picture.

Every organism is thought to have its own role, or **ecological niche**, within the structure and function of an ecosystem. The ecological niche describes the place and function of a species within a complex system of biotic and abiotic factors.

An ecological niche is difficult to define precisely because it takes into account all aspects of the organism's existence—all physical, chemical, and biological factors the organism needs to survive, remain healthy, and reproduce. Among other things,

> **ecological niche**
> The totality of an organism's adaptations, its use of resources, and the lifestyle to which it is fitted.

the niche includes the local environment in which an organism lives—its **habitat**. An organism's niche also encompasses what it eats, what organisms eat it, what organisms it competes with, and how the abiotic components of its environment, such as light, temperature, and moisture, interact with and influence it. A complete description of an organism's ecological niche involves numerous dimensions to explain when, where, and how an organism makes its living.

An organism's potential ecological niche may be much broader than it actually is in nature. Put differently, an organism may be capable of using much more of its environment's resources or of living in a wider assortment of habitats than it actually does. The potential, idealized ecological niche of an organism is its **fundamental niche**, but various factors, such as competition with other species, usually exclude it from part of its fundamental niche. The lifestyle an organism actually pursues and the resources it actually uses make up its **realized niche**.

An example helps illustrate the difference between fundamental and realized niches. The green anole, a lizard native to Florida and other southeastern states,

The green anole's realized niche • Figure 5.14

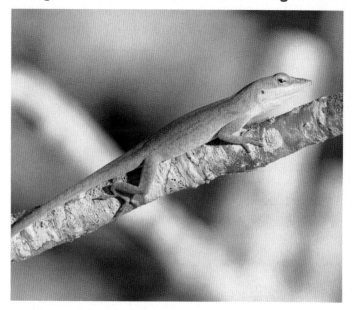

a. The green anole is native to Florida.

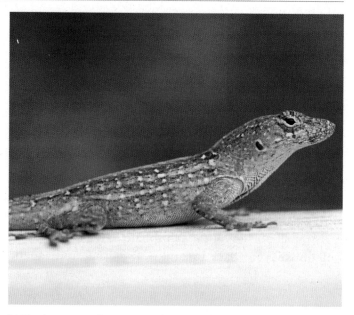

b. The brown anole was introduced in Florida.

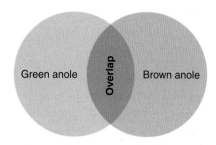

Green anole | Overlap | Brown anole

c. The fundamental niches of the two lizards initially overlapped.

Green anole | Brown anole

d. The brown anole out-competed the green anole, restricting its realized niche.

perches on trees, shrubs, walls, or fences during the day and waits for insect and spider prey (**Figure 5.14a**). In the past, these little lizards were widespread in Florida.

Several years ago, a related species, the brown anole, was introduced from Cuba into southern Florida and quickly became common (**Figure 5.14b**). Suddenly the green anoles became rare, apparently driven out of their habitat by competition from the slightly larger brown anoles. Careful investigation revealed that green anoles were still present but were now confined largely to the wetland vegetation and to the leafy crowns of trees, where they were less obvious.

The habitat portion of the green anole's fundamental niche includes all of the places where it originally lived in Florida: trunks and crowns of trees, exterior house walls,

and many other locations. Where they became established, brown anoles drove green anoles out from all but wetlands and tree crowns, so the green anoles' realized niche—the areas where it could survive—became smaller (**Figure 5.14c** and **d**). Natural communities consist of numerous species, and the interactions among species produce the realized niche of each.

When two species are similar—as are green and brown anoles—their ecological niches may appear to overlap. However, many ecologists think no two species indefinitely occupy the same niche in the same community. Resource partitioning is one way some species avoid or at least reduce niche overlap. **Resource partitioning** is the reduction in competition for environmental resources such as food among coexisting species as a result of the niche of each species differing from

WHAT A SCIENTIST SEES

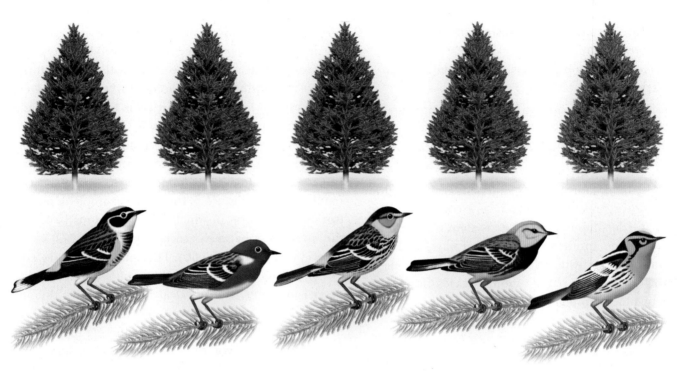

Yellow-rumped warbler   Bay-breasted warbler   Cape May warbler   Black-throated green warbler   Blackburnian warbler

Resource Partitioning

Robert MacArthur's study of five American warbler species is a classic example of resource partitioning. Although it initially appeared that the niches of the species were nearly identical, MacArthur determined that individuals of each species spend most of their feeding time in different portions of spruces and other conifer trees. They also move in different directions through the canopy, consume different combinations of insects, and nest at slightly different times. The photo shows a male black-throated green warbler in a spruce tree.

 Interpreting Data
Which two warblers are least likely to overlap when feeding?

the niches of others in one or more ways. Evidence of resource partitioning in animals is well documented and includes studies in tropical forests of Central and South America that demonstrate little overlap in the diets of fruit-eating birds, primates, and bats that coexist in the same habitat. Although fruits are the primary food for several hundred bird, primate, and bat species, the wide variety of fruits available has allowed fruit eaters to specialize, thereby reducing competition. Resource partitioning may also include timing of feeding, location of feeding, nest sites, and other aspects of an organism's ecological niche (see *What a Scientist Sees*).

CONCEPT CHECK

1. **What** are three aspects of an organism's ecological niche?

2. **What** is resource partitioning?

Interactions Among Organisms

LEARNING OBJECTIVES

1. **Distinguish** among mutualism, commensalism, and parasitism.
2. **Define** *predation* and describe predator–prey relationships.
3. **Define** *competition* and distinguish between intraspecific and interspecific competition.
4. **Discuss** an example of a keystone species.

No organism exists independently of other organisms. The producers, consumers, and decomposers of an ecosystem interact with one another in a variety of ways, and each forms associations with other organisms. Three main types of interactions occur among species in an ecosystem: symbiosis, predation, and competition.

Symbiosis

In **symbiosis**, one species usually lives in or on another species. The partners in a symbiotic relationship may benefit, be unaffected, or be harmed by the relationship.

> **symbiosis** An intimate relationship or association between members of two or more species; includes mutualism, commensalism, and parasitism.

Symbiosis is the result of **coevolution**, the interdependent evolution of two interacting species. Flowering plants and their animal pollinators are an excellent example of coevolution. Bees, beetles, hummingbirds, bats, and other animals transport pollen from one plant to another. During the millions of years over which these associations developed, flowering plants evolved several ways to attract animal pollinators. One of the rewards for the pollinator is food—nectar (a sugary solution) and pollen. Plants have a variety of ways to get the pollinator's attention, most involving showy petals and scents.

As plants acquire specialized features to attract pollinators, animals coevolve specialized body parts and behaviors to aid pollination and obtain nectar and pollen as a reward. Coevolution is responsible for the hairy bodies of bumblebees, which catch and hold sticky pollen for transport from one flower to another. Coevolution is also responsible for the long, curved beaks of certain Hawaiian birds that insert their beaks into tubular flowers to obtain nectar (**Figure 5.15**).

The thousands, or even millions, of symbiotic associations that result from coevolution fall into three categories: mutualism, commensalism, and parasitism (summarized in **Figure 5.16a**). One example of **mutualism**, an association in which both organisms benefit, is the interaction between acacia ants and the bull's horn acacia plant (**Figure 5.16b**). The ants make hollow nests out of thorns at the base of the plant's leaves and gain special nutrients from the leaf tips. In return, the ants effectively protect the plant from invertebrate and vertebrate herbivores and clear away competing plants. Both ant and acacia depend on this association for survival.

Commensalism is a symbiotic relationship in which one species benefits and the other is neither harmed nor helped. One example of commensalism is the relationship between a tropical tree and its *epiphytes*, smaller plants such as mosses, orchids, and ferns that live attached to the bark of the tree's branches (**Figure 5.16**).

Figure 5.15 • Coevolution

This Hawaiian honeycreeper uses its gracefully curved bill to sip nectar from the long, tubular flowers of the lobelia.

Environmental InSight

Symbiotic relationships
- Figure 5.16

	Organism 1	Organism 2	Characteristic of relationship
Mutualism	Benefits	Benefits	Each organism depends on the other
Commensalism	Benefits	Not affected	Only one organism depends on the other
Parasitism	Benefits	Harmed	Host harmed, rarely killed; host usually much larger than parasite

a. Categories of Symbiosis.

Interpreting Data

A bee pollinates a plant species while gathering material to make nectar. What type of symbiosis exists between plant and bee?

b. Mutualism. Most common in Central America, the acacia ant gains shelter and nutrients from the acacia plant, in turn protecting the plant from predators, Photographed in Costa Rica.

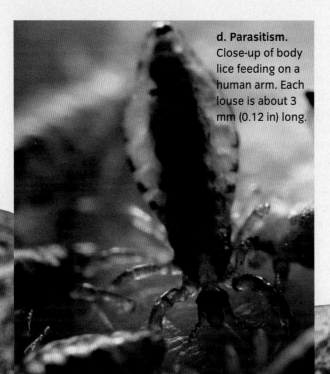

d. Parasitism. Close-up of body lice feeding on a human arm. Each louse is about 3 mm (0.12 in) long.

c. Commensalism. Epiphytes are small plants that attach to the branches and trunks of larger trees. Photographed in the Fiji Islands.

NATIONAL GEOGRAPHIC

An epiphyte anchors itself to a tree but typically doesn't obtain nutrients or water directly from the tree. Its location on the tree enables it to obtain adequate light, water (as rainfall dripping down the branches), and required nutrient minerals (which rainfall washes out of the tree's leaves). The epiphyte benefits from the association, whereas the tree is apparently unaffected.

Parasitism is a symbiotic relationship in which one species (the *parasite*) benefits at the expense of the other (the *host*). Parasitism is a successful lifestyle; more than 100 parasites live in or on the human species (**Figure 5.16d**).

EnviroDiscovery

Bee Colonies Under Threat

Since late 2006, many U.S. beekeepers have experienced major losses in their bee colonies, 30 to 90 percent of total individuals. Similar bee disappearances have occurred in other countries, resulting in losses of millions of bees worldwide. These sudden declines, now known in the United States as colony collapse disorder (CCD), are thought to be triggered by a complex mix of factors. Researchers are investigating three major potential causes of CCD:

- The negative effects of pesticides

- Damage caused by pathogens or parasites, such as *Varroa* and tracheal mites

- Deaths resulting from viruses, including the Israeli acute-paralysis virus (IAPV), which can be spread by mites

Bees are necessary for the pollination of a variety of important crops, many of which—nearly 100—are potentially threatened by these bee declines. The demand on bees to pollinate crop species has increased even as the number of bee colonies maintained by keepers has dropped. Some agricultural researchers believe that the added stress placed on bee colonies as they are transported to carry out pollination has increased the susceptibility of these colonies to health threats by compromising bees' immune systems. Recent research indicates that improving hive hygiene can help prevent CCD.

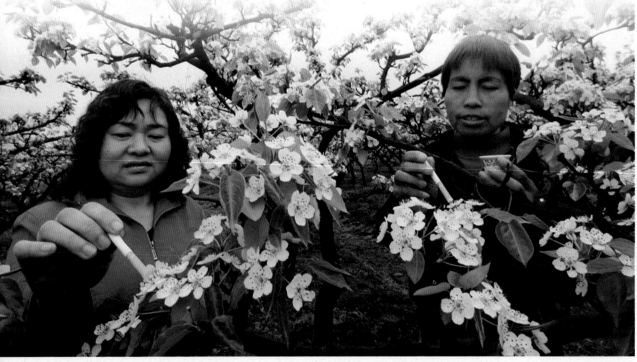

Two fruit growers hand pollinate pear trees in Yongchuan, Chongqing, China, a region where local bee populations have vanished.

A parasite, usually much smaller than its host, obtains nourishment from its host, but although a parasite may weaken its host, it rarely kills it quickly. (A parasite would have a difficult life if it kept killing off its hosts!) Some parasites, such as ticks, live outside the host's body; other parasites, such as tapeworms, live within the host.

Predation

Predation includes both animals eating other animals (for example, herbivore–carnivore interactions) and animals eating plants (producer–herbivore interactions). Predation has resulted in an "arms race," with the coevolution of predator strategies—more efficient ways to catch prey—and prey strategies—better ways to escape the predator. An efficient predator exerts a strong selective force on its prey, and over time the prey species may evolve some sort of countermeasure that reduces the probability of its being captured. The

> **predation** The consumption of one species (the prey) by another (the predator).

countermeasure that the prey acquires in turn may act as a strong selective force on the predator.

The cheetah is the world's fastest animal and can sprint at 110 km (68 mi) per hour for short distances (**Figure 5.17a**). Orcas (commonly known as killer whales) hunt in packs and often herd salmon or tuna into a cove so that they are easier to catch. Any trait that increases hunting efficiency, such as the speed of a cheetah or the intelligence of orcas, favors predators that pursue their prey. Ambush is another effective way to catch prey. The goldenrod spider is the same color as the white or yellow flowers in which it hides (**Figure 5.17b**). This camouflage prevents unwary insects that visit the flower for nectar from noticing the spider until it is too late.

Many potential animal prey, such as woodchucks, run to their underground burrows to escape predators. Others have mechanical defenses, such as the barbed quills of a porcupine and the shell of a pond turtle. Some animals live in groups—a herd

WILEY
RESOURCE KIT

Predation • Figure 5.17

a. The cheetah sprints at high speed to catch prey. Photographed in the Okavango Delta, Botswana, Africa.

b. The goldenrod spider employs camouflage to ambush its prey.

of antelope, colony of honeybees, school of anchovies, or flock of pigeons. This social behavior decreases the likelihood of a predator catching an individual unaware because the group has so many eyes, ears, and noses watching, listening, and smelling for predators (**Figure 5.18a**).

Avoiding predators • Figure 5.18

a. Adult meerkats stand guard at their burrow. If one of the sentries spies a predator such as an eagle or a hawk, it will alert the other meerkats, and all will scramble into their burrows. Photographed in the Kalahari Desert, South Africa.

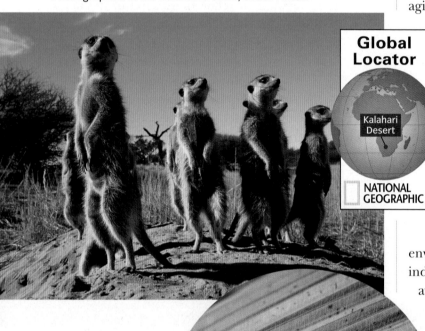

Global Locator

Kalahari Desert

NATIONAL GEOGRAPHIC

b. When at rest among forest leaves, the aptly named dead leaf moth is difficult for a predator to spot. Photographed in Panama.

How might behaviors that animals use to avoid predators be the same around the world? What predator avoidance behaviors might animals in your environment show that are different than animals in the African desert?

Chemical defenses are common among animal prey. The South American poison arrow frog has poison glands in its skin and bright warning colors that experienced predators avoid. Some animals blend into their surroundings and so hide from predators. Certain insects resemble twigs, tree trunks, or leaves so closely you would not guess that they are animals until they move (**Figure 5.18b**).

Plants possess adaptations that protect them from being eaten. The presence of spines, thorns, tough leathery leaves, or even thick wax on leaves discourages foraging herbivores from grazing. Other plants produce an array of protective chemicals that are unpalatable or even toxic to herbivores. The nicotine found in tobacco is so effective at killing insects that it is an ingredient in many commercial insecticides.

Competition

Competition occurs when two or more individuals attempt to use an essential common resource such as food, water, shelter, living space, or sunlight. Resources are often in limited supply in the environment, and their use by one individual decreases the amount available to others. If a tree in a dense forest grows taller than surrounding trees, it absorbs more of the incoming sunlight. Less sunlight is available for nearby trees that the taller tree shades. Competition occurs among individuals within a population (*intraspecific competition*) and between different species (*interspecific competition*).

competition The interaction among organisms that vie for the same resources in an ecosystem (such as food or living space).

Competition isn't always a straightforward, direct interaction. Consider a variety of flowering plants that live in a young pine forest and compete with conifers for such resources as soil moisture and soil nutrient minerals. Their relationship is more involved than simple competition. The flowers produce nectar that some insect species consume; these insects also prey on needle-eating insects, reducing the number of insects feeding on pines. It is therefore difficult to assess the overall effect of flowering plants on pines, and scientists have many questions. If the flowering plants were removed from the community, would the pines grow faster because they were no longer competing for necessary resources? Or would the increased presence of needle-eating insects (caused by fewer omnivorous insects) inhibit pine growth?

Short-term experiments in which one competing plant species is removed from a forest community have in several instances demonstrated improved growth for the remaining species. However, few studies have tested the long-term effects on forest species of removing one competing species. These long-term effects may be subtle, indirect, and difficult to assess. They may reduce or negate the negative effects of competition for resources.

Keystone species • Figure 5.19

a. The gray wolf is considered a keystone species in its ecosystem.

b. Elk populations may grow exponentially when wolves, one of their key predators, disappear from an ecosystem.

Keystone Species

Certain species are more crucial to the maintenance of their ecosystem than others. Such **keystone species** are vital in determining an ecosystem's species composition and how the ecosystem functions. The fact that other species depend on or are greatly affected by the keystone species is revealed when the keystone species is removed. Keystone species are usually not the most abundant species in the ecosystem. Although present in relatively small numbers, keystone species exert a profound influence on the entire ecosystem because they often affect the available amount of food, water, or some other resource.

Identifying and protecting keystone species are crucial goals of conservation biologists because if a keystone species disappears from an ecosystem, other organisms may become more common or more rare, or they may even disappear. One example of a keystone species is a top predator such as the gray wolf (**Figure 5.19**). Where wolves were hunted to extinction, the populations of deer, elk, and other herbivores increased explosively. As these herbivores overgrazed the vegetation, plant species that couldn't tolerate such grazing pressure disappeared. Smaller animals such as insects were lost from

the ecosystem because the plants they depended on for food had become less abundant. Thus, the disappearance of the wolf resulted in the ecosystem having considerably less biological diversity.

Some scientists think we should abandon the concept of keystone species because it is problematic. For one thing, most of the information about keystone species is anecdotal. Scientists have performed few long-term studies to identify keystone species and to determine the nature and magnitude of their effects on the ecosystems they inhabit.

CONCEPT CHECK

1. **What** is one example of mutualism? of parasitism?

2. **What** is one example of a predator–prey interaction?

3. **What** is the difference between interspecific and intraspecific competition?

4. **How** does a keystone species affect its ecosystem?

Global Climate Change: How Does It Affect the Carbon Cycle?

Within the scientific community, the question is no longer whether human-caused climate change will occur but at what rate and with what effects, and what, if anything, can we do about it. The biggest culprit in climate change is an increase in atmospheric carbon dioxide (CO_2), which is generated primarily through burning fossil fuels such as coal, oil, and natural gas, as well as through the clearing and burning of forests. During the past two centuries, the level of carbon dioxide in the atmosphere has climbed dramatically, increasing more than 20 percent just over the past 50 years. Atmospheric CO_2 allows solar radiation to pass through but does not allow heat to radiate into space. Instead, the heat is radiated back to Earth's surface. As the CO_2 accumulates, it may trap enough heat to warm the planet.

Robert Socolow and Stephen Pacala of the Princeton University Carbon Mitigation Initiative (CMI) suggest that somewhere just below a doubling of atmospheric carbon from preindustrial levels—600 billion tons in the early 1800s—lies the amount of atmospheric carbon that, if accumulated, will lead to the most dangerous of consequences. They propose that to keep future atmospheric carbon below this doubling would require generating 7 billion fewer tons of carbon each year by 2056 than are currently expected.

Many people despair of ever finding a solution to this enormous challenge. Socolow and Pacala, however, along with other CMI researchers, propose a "stabilization wedges" approach to solving the carbon emissions dilemma. In this framework, carbon reductions can be thought of in terms of "wedges," each of which would result in a 1-billion-ton-per-year reduction by 2056. A combination of any seven wedges would put us on a path to avoid the critical doubling of CO_2. Socolow and Pacala identify fifteen technologies in five categories, any one of which could serve as one of the seven wedges. Four of these wedges are:

- *Increase the fuel economy of 2 billion cars from 30 to 60 mpg*. Two billion cars are expected to be on the world's roads by 2056, each traveling an average of 10,000 miles per year. If the typical car operates at 60 mpg, there will be 1 billion fewer tons of carbon generated each year than if they operate at 30 mpg *(see photo)*.

- *Install carbon capture and storage devices at 800 large coal-fired power plants*. Currently, the CO_2 produced from burning coal is released to the atmosphere. If 90 percent of the carbon released each year is instead captured and stored, 1 billion fewer tons will be released to the atmosphere.

- *Stop all tropical deforestation and double the current rate of planting new forests*. Deforestation worldwide currently releases 2 billion tons of carbon to the atmosphere each year. However, it is expected to slow to 1 billion tons per year without intervention. To achieve a 1-billion-ton wedge below the expected amount would require significant efforts in both reducing deforestation and increasing reforestation.

- *Add twice today's nuclear energy output to displace coal*. Nuclear energy currently accounts for about 17 percent of the world's electricity. Increasing the amount of nuclear power alone will not affect the amount of CO_2 generated. However, replacing existing coal-fired power plants with nuclear power facilities would reduce greenhouse gases. Similarly, we could replace coal with solar or wind power.

We say more about global climate change in Chapter 9.

Hybrid vehicles such as this model of the Ford Escape allow consumers to lower their gasoline consumption and thus reduce their CO_2 emissions.

Summary

1 What Is Ecology? 98

1. **Ecology** is the study of the interaction among organisms and between organisms and their abiotic environment.

2. A **population** is a group of organisms of the same species that live together in the same area at the same time. A **community** is a natural association that consists of all the populations of different species that live and interact together within an area at the same time. An **ecosystem** is a community and its physical environment. A **landscape** is a region that includes several interacting ecosystems. The **biosphere** is the layer of Earth that contains all living organisms.

2 The Flow of Energy Through Ecosystems 101

1. **Energy** is the capacity or ability to do work. According to the **first law of thermodynamics**, energy can be neither created nor destroyed, although it can change from one form to another. As a result of the **second law of thermodynamics**, when energy is converted from one form to another, some of it is degraded into heat, a less usable form that disperses into the environment.

2. A **producer** manufactures large organic molecules from simple inorganic substances. A **consumer** cannot make its own food and uses the bodies of other organisms as a source of energy and bodybuilding materials. **Decomposers** are microorganisms that break down dead organic material and use the decomposition products to supply themselves with energy.

3. **Energy flow** is the passage of energy in a one-way direction through an ecosystem, from producers to consumers to decomposers.

3 The Cycling of Matter in Ecosystems 106

1. **Biogeochemical cycles** are the processes by which matter cycles from the living world to the nonliving, physical environment and back again. Carbon dioxide is the important gas of the carbon cycle; carbon enters the living world through photosynthesis and returns to the abiotic environment when organisms respire. The hydrologic cycle continuously renews the supply of water and involves an exchange of water among the land, the atmosphere, and organisms. There are five steps in the nitrogen cycle: nitrogen fixation, nitrification, ammonification, assimilation, and denitrification. In the sulfur cycle, sulfur compounds whose natural sources are the ocean and rock are incorporated by organisms into proteins and move between organisms, the atmosphere, the ocean, and land. The phosphorus cycle has no biologically important gaseous compounds; phosphorus erodes from rock and is absorbed by plant roots.

4 Ecological Niches 113

1. An **ecological niche** is the totality of an organism's adaptations, its use of resources, and the lifestyle to which it fits. An organism's ecological niche includes its **habitat**, its distinctive lifestyle, and its role in the community.

2. **Resource partitioning** is the reduction in competition for environmental resources, such as food, that occurs among coexisting species as a result of the niche of each species differing from the niches of other species in one or more ways.

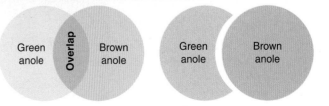

5 Interactions Among Organisms 116

1. **Symbiosis**, an intimate relationship or association between members of two or more species, is the result of **coevolution**, the interdependent evolution of two interacting species. **Mutualism** is a symbiotic relationship in which both species benefit. **Commensalism** is a symbiotic relationship in which one species benefits and the other species is neither harmed nor helped. **Parasitism** is a symbiotic relationship in which one species (the parasite) benefits at the expense of the other (the host).

2. **Predation** is the consumption of one species (the prey) by another (the predator). With coevolution between predator and prey, the predator evolves more efficient ways to catch prey (such as pursuit and ambush), and the prey evolves better ways to escape the predator (such as flight, association in groups, and camouflage).

3. **Competition** is the interaction among organisms that vie for the same resources in an ecosystem (such as food or living space). Competition occurs among individuals within a population (intraspecific competition) and between species (interspecific competition).

4. A **keystone species** is crucial in determining the nature and structure of the entire ecosystem in which it lives. Though present in relatively small numbers, keystone species have disproportionate effects on ecosystems.

Key Terms

- biosphere 100
- community 98
- competition 120
- ecological niche 113
- ecology 98

- ecosystem 98
- energy flow 104
- first law of thermodynamics 101
- landscape 99
- photosynthesis 102

- population 98
- predation 119
- second law of thermodynamics 102
- symbiosis 116

What is happening in this picture?

- This caterpillar has inflated its thorax to make its body look like the head of a snake. Suggest a possible reason the caterpillar does this.

- Note the two large spots. What do they resemble? Why would this animal have such conspicuous spots?

- If a hungry bird saw this caterpillar, do you think it would have second thoughts about eating it? Why or why not?

Critical and Creative Thinking Questions

1. To function, ecosystems require inputs of energy. Where does this energy come from?

2. After an organism uses energy, what happens to the energy?

3. What is a biogeochemical cycle? Why is the cycling of matter essential to the continuance of life?

4. What types of resources might two organisms compete over if those resources are scarce? How might interspecific competition affect two species' ecological niches?

5. How are food chains important in biogeochemical cycles?

6. The many cichlid species in Lake Victoria obtained food in a variety of ways—some grazed on algae, others preyed on insects, others consumed dead organic matter, and so on. Explain how the cichlids are an example of resource partitioning.

7. In both parasitism and predation, one organism benefits at the expense of another. What is the difference between the two relationships?

8. Some biologists think protecting keystone species would help preserve biological diversity in an ecosystem. Do you agree? Explain your answer.

9. How does the role of humans in the carbon cycle influence global climate change? How might your role in the carbon cycle compare to that of a young person on a remote South American farm that uses animal labor rather than machines?

10. Describe how the close-up image below of an alpine meadow represents a community.

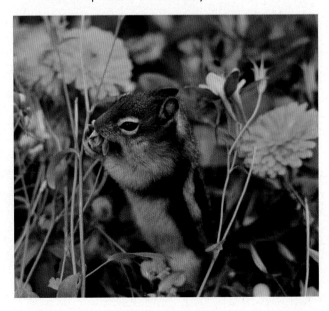

11. Ecologists investigating interactions of two species at a study site first counted individuals of Species A and then removed all Species B individuals. Six months later, the ecologists again counted individuals of Species A. Viewing their results as graphed below, what is the likely ecological interaction between Species A and Species B? Explain your answer.

The figure below shows the components of a simple food chain. Use it to answer questions 12–14.

12. Identify the producers, consumers, and decomposers in the food chain. How many trophic levels are represented?

13. Describe or indicate the flow of food and energy within this system.

14. Which forms of energy are present within this chain?

THE PLANNER

Ecosystems and Evolution

THE FLORIDA EVERGLADES

The Everglades, a "river of grass" in the southernmost part of Florida, is a vast expanse of predominantly sawgrass wetlands dotted with small islands of trees.

It is a haven for wildlife, including alligators (see photograph), snakes, panthers, otters, raccoons, and thousands of birds.

The Everglades today is about half its original size of 1.6 million hectares (4 million acres) and suffers from many serious environmental problems. Wading bird populations have dropped 93 percent since 1930, and the area is now home to 50 endangered or threatened species.

More than 70 years of engineering projects aimed at protecting the human population from storm-related flooding have reduced the quantity of water flowing into the Everglades, restricting the natural recharging process there. Flood-control measures created dry spaces that were then converted to agricultural or residential use, fragmenting wildlife habitat and polluting the water that does enter.

The Everglades will never return completely to its original condition because there are now too many cities and sugar plantations in the region. However, state and federal governments are working together on the massive Comprehensive Everglades Restoration Plan to eventually restore a more natural water flow to the area, repel invasions of foreign species, and reestablish native species.

126

CHAPTER OUTLINE

CHAPTER PLANNER ✓

- ❑ Study the picture and read the opening story.
- ❑ Scan the Learning Objectives in each section:
  p. 128 ❑   p. 142 ❑   p. 147 ❑   p. 151 ❑
- ❑ Read the text and study all figures and visuals. Answer any questions.

Analyze key features

- ❑ Environmental InSight, p. 129 ❑   p. 150 ❑
- ❑ National Geographic Map, pp. 130–131 ❑
- ❑ EnviroDiscovery, p. 138 ❑
- ❑ What a Scientist Sees, p. 143 ❑
- ❑ Process Diagram, p. 149 ❑   p. 152 ❑   p. 153 ❑
- ❑ Case Study, p. 154
- ❑ Stop: Answer the Concept Checks before you go on:
  p. 141 ❑   p. 147 ❑   p. 151 ❑   p. 153 ❑

End of Chapter

- ❑ Review the Summary and Key Terms.
- ❑ Answer What is happening in this picture?
- ❑ Answer the Critical and Creative Thinking Questions.

Earth's Major Biomes

LEARNING OBJECTIVES

1. **Define** *biome* and discuss how biomes are related to climate.

2. **Briefly describe** the nine major terrestrial biomes, giving attention to the climate, soil, and characteristic organisms of each.

Earth has many different climates, based primarily on temperature and precipitation differences. Characteristic organisms have adapted to each climate within large regions called biomes. Each **biome** encompasses many interacting ecosystems (**Figure 6.1**). In terrestrial ecology, a biome is considered the next level of ecological organization above community, ecosystem, and landscape.

> **biome** A large, relatively distinct terrestrial region with similar climate, soil, plants, and animals, regardless of where it occurs in the world.

Near the poles, temperature is generally the overriding climate factor defining a biome, whereas in temperate and tropical regions, precipitation is more significant than temperature, as shown in **Figure 6.2**. Other *abiotic* factors to which certain biomes are sensitive include extreme temperatures as well as rapid temperature changes, fires, floods, droughts, and strong winds. Elevation also affects biomes: Changes in vegetation with increasing elevation resemble the changes in vegetation observed in going from warmer to colder climates. These differences across biomes can be further defined by types of vegetation present and land use patterns (**Figure 6.3** on pages 130 and 131).

The world's terrestrial biomes • Figure 6.1

Although sharp boundaries are shown in this highly simplified map, biomes actually grade together at their boundaries. Use the legend below to identify the locations of the different biomes.

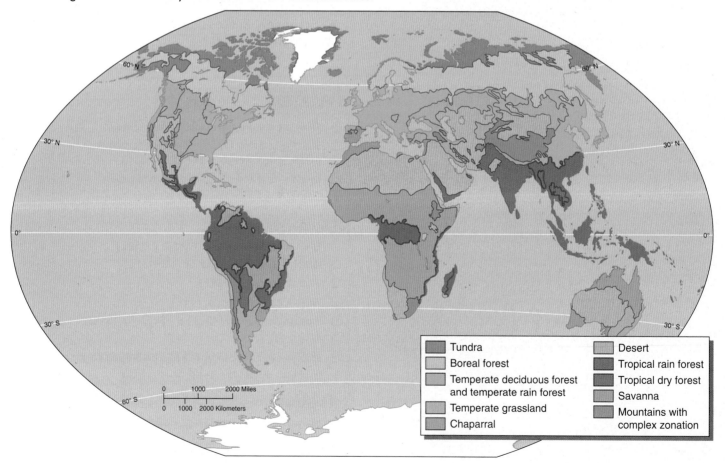

Legend:
- Tundra
- Boreal forest
- Temperate deciduous forest and temperate rain forest
- Temperate grassland
- Chaparral
- Desert
- Tropical rain forest
- Tropical dry forest
- Savanna
- Mountains with complex zonation

Two climate factors, temperature and precipitation, have a predominant effect on biome distribution.

Alaska Tundra. At higher latitudes, temperature is more important than precipitation in shaping biomes, as mean annual temperatures decline poleward.

Costa Rica Tropical Rain Forest. In temperate and tropical zones, precipitation is more important than temperature in shaping biomes.

Cold • Arctic

DECREASING TEMPERATURE

INCREASING LATITUDE

Tundra

Boreal forest

Temperate deciduous forest | Temperate grassland | Chaparral | Temperate desert

Hot

Tropical rain forest | Dry tropical forest | Savanna | Moist tropical desert | Dry tropical desert

Tropics

Wet — DECREASING PRECIPITATION → Dry

Arizona Desert. Biomes differ in the relative amounts of precipitation they receive and in the seasonal distribution of precipitation.

Earth's Major Biomes **129**

A diversity of land cover types • Figure 6.3

Similar vegetation types can occur at many different locations.

GLOBAL LAND COVER COMPOSITION
Three characteristics underlie these categories: life-form (woody, herbaceous, or bare); leaf type (needle or broad); and leaf duration (evergreen or deciduous).

EVERGREEN NEEDLE-LEAVED FOREST
Tree height exceeds 5 m (16 ft); more than 60 percent of the land is canopied by forest. Typical example: boreal (northern) region. On tree plantations, trees are logged for paper and building products.

EVERGREEN BROAD-LEAVED FOREST
More than 60 percent of the land is covered by a forest canopy, with trees over 5 m (16 ft). Such forests dominate in the tropics and are home to great biodiversity, when not cleared for mechanized farms, ranches, and tree plantations.

Land Cover
Forest
- Evergreen needle-leaf
- Evergreen broad-leaf
- Deciduous needle-leaf
- Deciduous broad-leaf
- Mixed forest
- Woodland

Grassland or shrubland
- Wooded grassland
- Closed shrubland
- Opened shrubland
- Grassland

- Cropland
- Barren (desert and polar ice)
- Built up

DECIDUOUS NEEDLE-LEAVED FOREST
A forest canopy covers more than 60 percent of the land; tree height exceeds 5 m (16 ft). This class is dominant only in Siberia, taking the form of larch forests.

DECIDUOUS BROAD-LEAVED FOREST
More than 60 percent of the land is covered by a forest canopy; tree height exceeds 5 m (16 ft). In temperate regions, much of this forest has been converted to cropland.

MIXED FOREST
Both needle and deciduous types of trees appear. Mixed forest is largely found between temperate deciduous and boreal evergreen forests.

WOODLAND
Land has herbaceous or woody understory; trees exceed 5 m (16 ft) and may be deciduous or evergreen. Highly degraded in long-settled human environments, such as West Africa.

WOODED GRASSLAND
Woody or herbaceous understories are punctuated by trees. Examples are African savanna as well as open boreal border land between trees and tundra.

CLOSED SHRUBLAND
Found where prolonged or dry seasons limit plant growth. This cover is dominated by bushes or shrubs not exceeding 5 m (16 ft). Tree canopy is less than 10 percent.

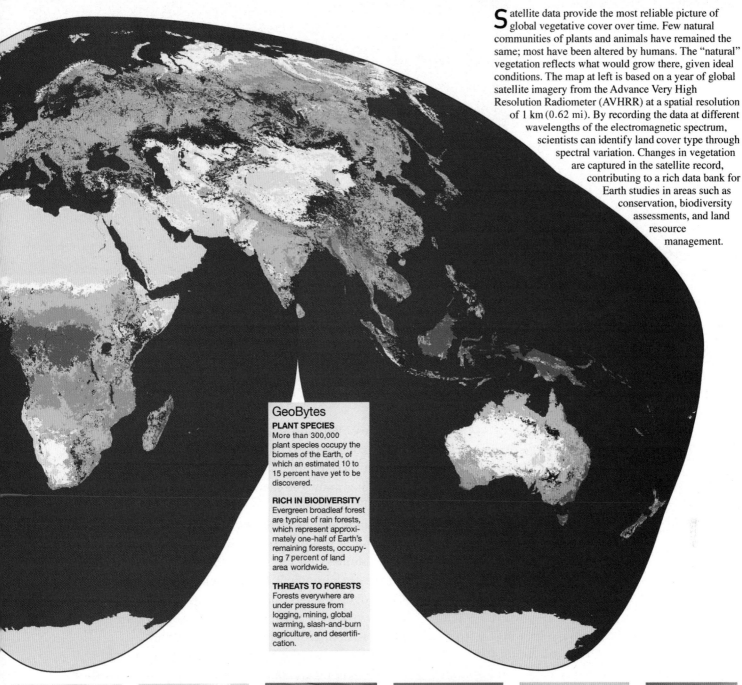

Satellite data provide the most reliable picture of global vegetative cover over time. Few natural communities of plants and animals have remained the same; most have been altered by humans. The "natural" vegetation reflects what would grow there, given ideal conditions. The map at left is based on a year of global satellite imagery from the Advance Very High Resolution Radiometer (AVHRR) at a spatial resolution of 1 km (0.62 mi). By recording the data at different wavelengths of the electromagnetic spectrum, scientists can identify land cover type through spectral variation. Changes in vegetation are captured in the satellite record, contributing to a rich data bank for Earth studies in areas such as conservation, biodiversity assessments, and land resource management.

GeoBytes

PLANT SPECIES
More than 300,000 plant species occupy the biomes of the Earth, of which an estimated 10 to 15 percent have yet to be discovered.

RICH IN BIODIVERSITY
Evergreen broadleaf forest are typical of rain forests, which represent approximately one-half of Earth's remaining forests, occupying 7 percent of land area worldwide.

THREATS TO FORESTS
Forests everywhere are under pressure from logging, mining, global warming, slash-and-burn agriculture, and desertification.

OPEN SHRUBLAND
Shrubs are dominant not exceeding 2 m (6.5 ft) in height. They can be evergreen or deciduous. This type occures in semiarid or severely cold areas.

GRASSLAND
Occurring in a wide range of habitats, this landscape has continuous herbaceous cover. The American Plains and central Russia are the premier examples.

CROPLAND
Crop-producing fields constitute over 80 percent of the land. Temperate regions are home to large areas of mechanized farming; in the developing world, plots are small.

BARREN (DESERT)
The land never has more than 10 percent vegetated cover. True deserts, such as the Sahara, as well as areas succumbing to desertification are examples.

BUILT UP
This land cover type was mapped using digital population data from around the world. It represents the most densely inhabited areas.

BARREN (POLAR ICE)
Permanent snow cover characterizes this class, the greatest examples of which are in the polar regions, as well as on high elevations in Alaska and the Himalayas.

Arctic tundra • Figure 6.4

Because of the tundra's short growing season and permafrost, only small, hardy plants grow in the northernmost biome that encircles the Arctic Ocean. Photographed in the Yukon Territory, Canada. The Ogilvie Mountains are in the background. Climate graph shows monthly temperatures and precipitation for Fort Yukon, Alaska.

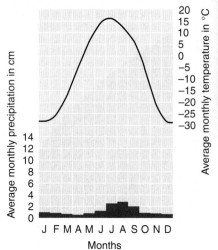

Interpreting Data
During how many months each year is the average monthly temperature in the Alaskan tundra at or above freezing?

Tundra

tundra The treeless biome in the far north that consists of boggy plains covered by lichens and mosses; it has harsh, cold winters and extremely short summers.

Tundra (or *arctic tundra*) occurs in the extreme northern latitudes where the snow melts seasonally (**Figure 6.4**). The Southern Hemisphere has no equivalent of the arctic tundra because it has no land in the corresponding latitudes. A similar ecosystem located in the higher elevations of mountains, above the tree line, is called *alpine tundra*.

Although the arctic tundra's growing season is short, the days are long. Above the Arctic Circle, the sun does not set at all for many days in midsummer, although the amount of light at midnight is one-tenth that at noon. There is little precipitation, and most of the yearly 10 to 25 cm (4 to 10 in) of rain or snow falls during summer months.

Most tundra soils formed when glaciers began retreating after the last ice age, about 17,000 years ago. These soils are usually nutrient poor and have little *detritus*, such as dead leaves and stems, animal droppings, or remains of organisms. Although the tundra's surface soil thaws during summer, beneath it lies a layer of *permafrost*, permanently frozen ground that varies in depth and thickness. Permafrost impedes drainage, so the thawed upper zone of soil is usually waterlogged during summer. Limited precipitation, combined with low temperatures, flat topography (or surface features), and the layer of permafrost, produces a landscape of broad, shallow lakes and ponds, sluggish streams, and bogs.

Tundra supports relatively few species compared to other biomes, but the species that do occur there often exist in great numbers. Mosses, lichens, grasses, and grasslike sedges are the dominant plants. Stunted trees and shrubs grow only in sheltered locations. As a rule, tundra plants seldom grow taller than 30 cm (12 in).

Animals adapted to live year-round in the tundra include lemmings, voles, weasels, arctic foxes, snowshoe hares, ptarmigan, snowy owls, and musk oxen. In summer, caribou migrate north to the tundra to graze on sedges, grasses, and dwarf willow. Dozens of bird species also migrate north in summer to nest and feed on abundant insects. Mosquitoes, blackflies, and deerflies survive winter as eggs or pupae and appear in great numbers during summer weeks.

Tundra recovers slowly from even small disturbances. Oil and natural gas exploration and military use have caused damage to tundra likely to persist for hundreds of years (see the Chapter 17 case study).

Boreal Forest

Just south of the tundra is the **boreal forest**, or northern coniferous forest, which stretches across North America and Eurasia (**Figure 6.5**). There is no biome in the Southern Hemisphere comparable to the boreal forest. Winters in the boreal forest are extremely cold and severe, although not as harsh as those in the tundra. Boreal forest receives little precipitation, perhaps 50 cm (20 in) per year, and its soil is typically acidic and mineral poor, with a thick surface layer of partly decomposed pine and spruce needles. Where permafrost occurs, it is found deep under the surface. Boreal forest has numerous ponds and lakes dug by ice sheets during the last ice age.

Black and white spruces, balsam fir, eastern larch, and other conifers (cone-bearing evergreens) dominate the boreal forest. Conifers have many drought-resistant adaptations, such as needle-like leaves whose minimal surface area prevents water loss by evaporation. Such

boreal forest A region of coniferous forest (such as pine, spruce, and fir) in the Northern Hemisphere; located just south of the tundra. Also called *taiga*.

an adaptation helps conifers withstand the drought of the northern winter, when roots cannot absorb water through the frozen ground. Being evergreen, conifers resume photosynthesis as soon as warmer temperatures return.

The animal life of the boreal forest consists of some larger species such as caribou, which migrate from the tundra for winter; wolves; brown and black bears; and moose. However, most boreal mammals are medium sized to small, including rodents, rabbits, and smaller predators such as lynx, sable, and mink. Birds are abundant in the summer but migrate to warmer climates for winter. Insects are plentiful, but few amphibians and reptiles occur except in the southern boreal forest.

Currently, boreal forest is the world's top source of industrial wood and wood fiber. Extensive logging, gas and oil exploration, mining, and farming have contributed to loss of boreal forest.

Boreal forest • Figure 6.5

These coniferous forests occur in cold regions of the Northern Hemisphere adjacent to the tundra. Photographed in Saskatchewan, Canada. Climate graph shows monthly temperatures and precipitation for Fort Smith, Northwest Territories, Canada.

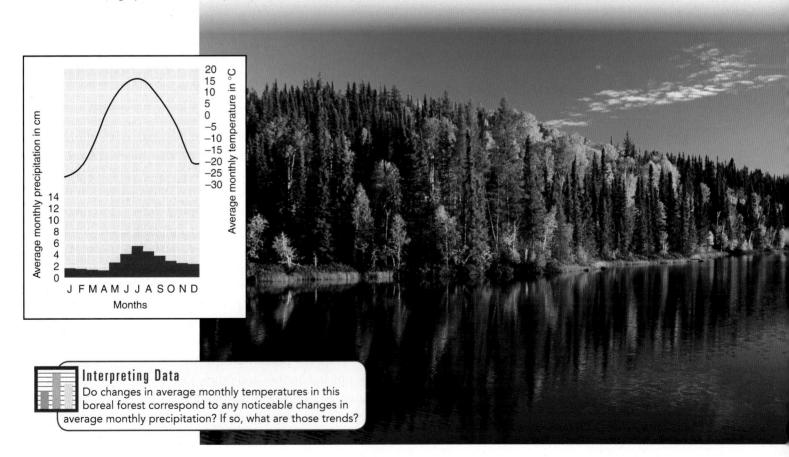

Interpreting Data
Do changes in average monthly temperatures in this boreal forest correspond to any noticeable changes in average monthly precipitation? If so, what are those trends?

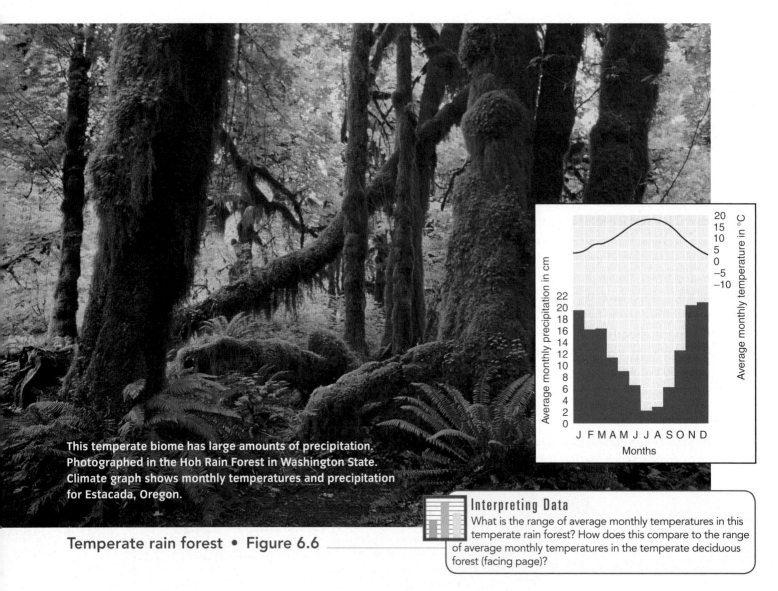

This temperate biome has large amounts of precipitation. Photographed in the Hoh Rain Forest in Washington State. Climate graph shows monthly temperatures and precipitation for Estacada, Oregon.

Temperate rain forest • Figure 6.6

Interpreting Data

What is the range of average monthly temperatures in this temperate rain forest? How does this compare to the range of average monthly temperatures in the temperate deciduous forest (facing page)?

Temperate Rain Forest

A coniferous **temperate rain forest** occurs on the northwest coast of North America. Similar vegetation exists in southeastern Australia and in southern South America. Annual precipitation in this biome is high—more than 127 cm (50 in)—and is augmented by condensation of water from dense coastal fogs. The proximity of temperate rain forest to the coastline moderates its temperature so that the seasonal fluctuation is narrow; winters are mild, and summers are cool. Temperate rain forest has relatively nutrient-poor soil, though its organic content may be high. Cool temperatures slow the activity of bacterial and fungal decomposers. Thus, needles and large fallen branches and trunks accumulate on the ground as litter that takes many years to decay and release nutrient minerals to the soil.

temperate rain forest A coniferous biome with cool weather, dense fog, and high precipitation.

The dominant vegetation in the North American temperate rain forest is large evergreen trees such as western hemlock, Douglas fir, western red cedar, Sitka spruce, and western arborvitae (**Figure 6.6**). Temperate rain forests are rich in epiphytes, smaller plants that grow on the trunks and branches of large trees. Epiphytes in this biome are mainly mosses, club mosses, lichens, and ferns, all of which also carpet the ground. Squirrels, wood rats, mule deer, elk, numerous bird species, and several species of amphibians and reptiles are common temperate rainforest animals.

The temperate rain forest is a rich wood producer, supplying lumber and pulpwood. Issues surrounding logging of old-growth temperate rain forests of the Pacific Northwest are discussed in the Chapter 13 case study, "The Tongass Debate over Clear-Cutting."

Temperate Deciduous Forest

Hot summers and cold winters characterize the **temperate deciduous forest**. Typically, the soil of a temperate deciduous forest consists of a topsoil rich in organic material and a deep, clay-rich lower layer. Broad-leaved hardwood trees that lose their leaves seasonally, such as oak, hickory, and beech, dominate the temperate deciduous forests of the northeastern and mideastern United States (**Figure 6.7**). In the southern areas of the temperate deciduous forest, the number of broad-leaved evergreen trees, such as magnolia, increases. The trees of the temperate deciduous forest form a dense canopy that overlies saplings and shrubs.

Temperate deciduous forests originally contained a variety of large mammals, such as puma, wolves, and bison, which are now absent. Other more common animals include deer, bears, and many small mammals and birds.

temperate deciduous forest
A forest biome that occurs in temperate areas where annual precipitation ranges from about 75 cm to 126 cm (30 to 50 in).

In Europe and North America, logging and land clearing for farms, tree plantations, and cities destroyed much of the original temperate deciduous forest. Where it has regenerated, temperate deciduous forest is often in a seminatural state that humans have modified for recreation, livestock foraging, timber harvest, and other uses. Many forest organisms have successfully reestablished themselves in these returning forests.

Worldwide, deciduous forests were among the first biomes converted to agricultural use. In Europe and Asia, many soils that originally supported deciduous forests have been cultivated by traditional agricultural methods for thousands of years without a substantial loss in fertility. During the 20th century, widely adopted intensive agricultural practices, along with overgrazing and deforestation, contributed to the degradation of some agricultural lands.

Temperate deciduous forest • Figure 6.7

The broad-leaved trees that dominate this biome are deciduous and shed their leaves before winter. Photographed in Germany Valley, West Virginia. Climate graph shows monthly temperatures and precipitation for Nashville, Tennessee.

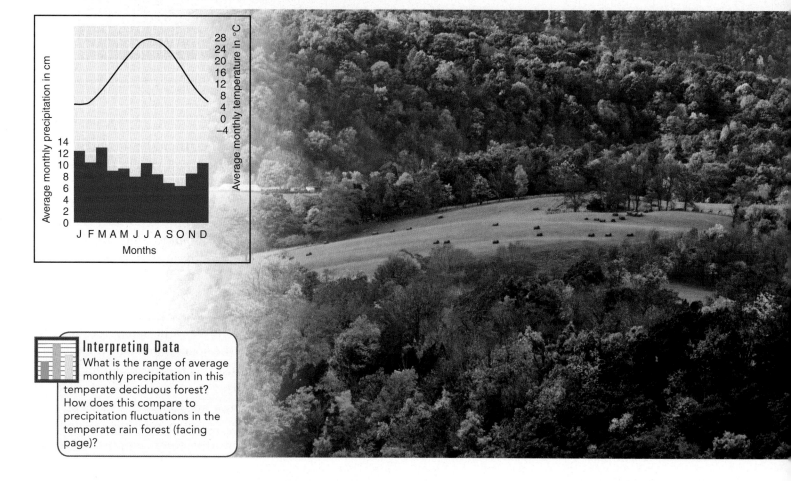

Interpreting Data
What is the range of average monthly precipitation in this temperate deciduous forest? How does this compare to precipitation fluctuations in the temperate rain forest (facing page)?

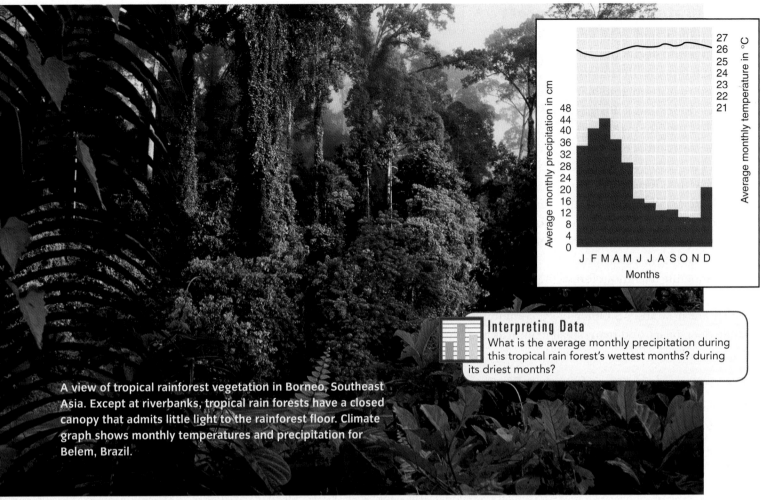

Average monthly precipitation in cm: 0 4 8 12 16 20 24 28 32 36 40 44 48

Average monthly temperature in °C: 21 22 23 24 25 26 27

Months: J F M A M J J A S O N D

Interpreting Data
What is the average monthly precipitation during this tropical rain forest's wettest months? during its driest months?

A view of tropical rainforest vegetation in Borneo, Southeast Asia. Except at riverbanks, tropical rain forests have a closed canopy that admits little light to the rainforest floor. Climate graph shows monthly temperatures and precipitation for Belem, Brazil.

Tropical rain forest • Figure 6.8

Tropical Rain Forest

tropical rain forest
A lush, species-rich forest biome that occurs where the climate is warm and moist throughout the year.

Tropical rain forest occurs where temperatures are warm throughout the year and precipitation occurs almost daily. The annual precipitation in a tropical rain forest is typically between 200 and 450 cm (80 to 180 in). Tropical rain forest commonly occurs in areas with ancient, highly weathered, mineral-poor soil. Little organic matter accumulates in such soils; because temperatures are high year-round, bacteria, fungi, and detritus-feeding ants and termites decompose organic litter quite rapidly. Roots quickly absorb nutrient minerals from the decomposing material. Tropical rain forests are found in Central and South America, Africa, and Southeast Asia.

Of all the biomes, the tropical rain forest is unexcelled in species richness and variety (**Figure 6.8**). No single species dominates this biome. The trees are typically evergreen flowering plants. A fully developed tropical rain forest has at least three distinct stories, or layers, of vegetation. The topmost story, or emergent layer, consists of the crowns of very tall trees, some 50 m (164 ft) or more in height, which are exposed to direct sunlight. The middle story, or canopy, which reaches a height of 30 to 40 m (100 to 130 ft), forms a continuous layer of leaves that lets in very little sunlight to support the smaller plants in the sparse understory. Only 2 to 3 percent of the light bathing the forest canopy reaches the forest understory. The vegetation of tropical rain forests is not dense at ground level except near stream banks or where a fallen tree has opened the canopy. Tropical rainforest trees support extensive communities of epiphytic plants such as ferns, mosses, orchids, and bromeliads.

Not counting bacteria and other soil-dwelling organisms, about 90 percent of tropical rainforest organisms are adapted to live in the canopy. Rain forests shelter the most abundant and varied insects, reptiles, and amphibians on Earth. The birds, often brilliantly colored, are also varied. Some are specialized to consume fruit (parrots, for example) and others to consume nectar (hummingbirds and sunbirds, for example). Most rainforest mammals, such as sloths and monkeys, are adapted to live only in the trees and rarely climb down to the ground, although some large, ground-dwelling mammals, including elephants, are also found in rain forests.

Human population growth and industrial expansion in tropical countries threaten the survival of tropical rain forests. (See Chapter 13 for more discussion of the ecological impacts of rainforest destruction.)

Chaparral

Some hilly temperate environments have mild winters with abundant rainfall combined with very dry summers (**Figure 6.9**). Such *Mediterranean climates*, as they are called, occur not only in the area around the Mediterranean Sea but also in the North American Southwest, southwestern and southern Australia, central Chile, and southwestern South Africa. On the mountain slopes of southern California, this Mediterranean-type biome is known as **chaparral**.

Chaparral soil is thin and often not very fertile. Wildfires occur naturally in this environment and are particularly frequent in late summer and autumn.

Chaparral vegetation looks strikingly similar in different parts

> **chaparral** A biome with mild, moist winters and hot, dry summers; vegetation is typically small-leaved evergreen shrubs and small trees.

Chaparral • Figure 6.9

Chaparral vegetation consists mainly of drought-resistant evergreen shrubs and small trees. Hot, dry summers and mild, rainy winters characterize the chaparral. Photographed on Mount Tamalpais in the Marin Hills, California. Climate graph shows monthly temperatures and precipitation for Culver City, California.

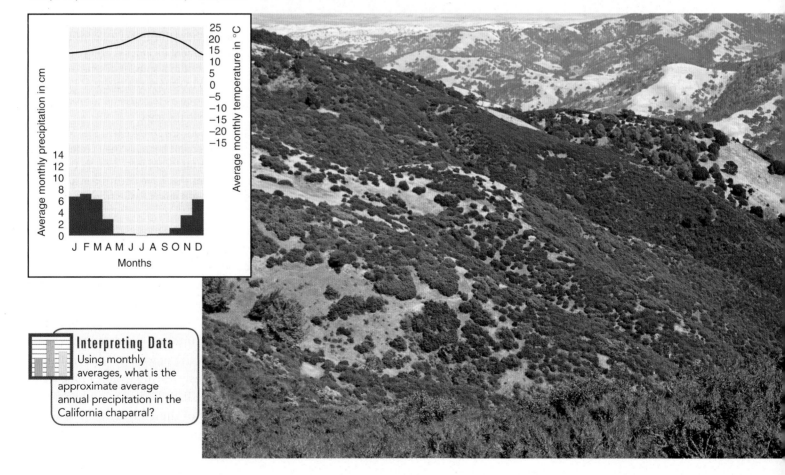

Interpreting Data
Using monthly averages, what is the approximate average annual precipitation in the California chaparral?

EnviroDiscovery

NATIONAL GEOGRAPHIC

Using Goats to Fight Fires

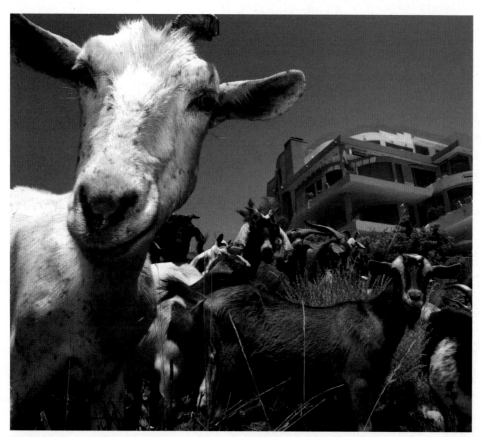

Goats prefer woody and weedy species, such as those common to chaparral.

California has about 6,000 wildfires each year, and they are becoming increasingly expensive and dangerous to manage because many people are building homes and living in the fire-vulnerable chaparral. Yet the topography of chaparral is so steep that firefighters often cannot use mechanized equipment but must transport equipment to fires with helicopters. Afraid that prescribed burns will get out of control, local governments are increasingly trying an effective, low-tech method to reduce the fuel load: During the 6-month fire season, goats are clearing hills around Oakland, Berkeley, Monterey, and Malibu.

A herd of 350 goats can clear an entire acre of heavy brush in about a day, but their use entails advance organization and support. Before goats can remove hazardous dry fuels from surrounding hillsides, botanists must fence off small trees and rare or endangered plants, to keep the goats from eating those plants. Goats are confined with installed electric fencing, and the goatherds typically use dogs to help herd the goats. Responsible management includes grazing the goats for the optimal amount of time—long enough to reduce the threat of fire but not so long as to cause unnecessary erosion in the area.

Goats are an excellent tool for fire management because they preferentially browse woody shrubs and thick undergrowth—exactly the fuel that causes disastrous fires. Fires that have occurred in areas after goats have browsed there are much easier to contain.

of the world, even though the individual species differ by location. A dense thicket of evergreen shrubs—often short, drought-resistant pine or scrub oak trees that grow 1 to 3 m (3 to 10 ft) tall—usually dominates chaparral. These plant species have evolved adaptations that equip them to live where precipitation is seasonal. During the rainy winter season, the environment may be lush and green, and during the hot, dry summer, the plants lie dormant. The hard, small, leathery leaves of trees and shrubs resist water loss.

Many plants are also specifically fire adapted and grow best in the months following a fire. Such growth is possible because fire releases into the soil the nutrient minerals present in the aboveground parts of the plants that burned. The seeds and underground parts of plants that survive fire make use of the newly availabile nutrient minerals and sprout vigorously during winter rains. Mule deer, wood rats, chipmunks, lizards, and many species of birds are common animals of the chaparral. (For more on the role fire plays in nature and on how humans have disrupted this role, see the case study on page 154.)

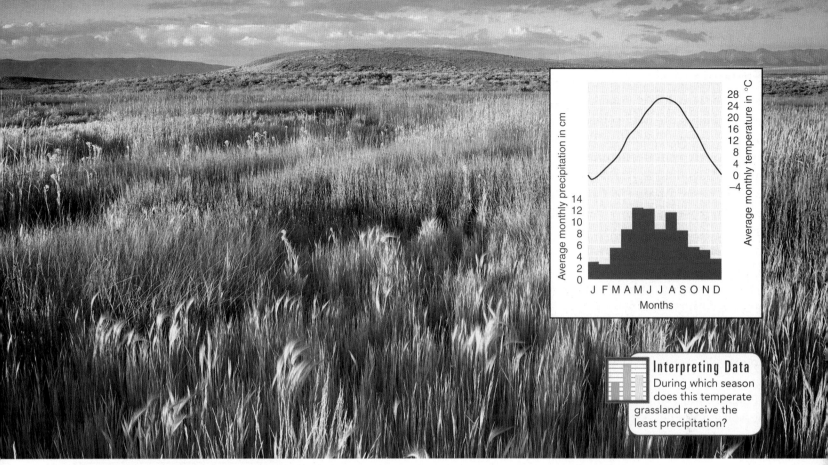

Average monthly precipitation in cm

Average monthly temperature in °C

J F M A M J J A S O N D
Months

Interpreting Data
During which season does this temperate grassland receive the least precipitation?

Temperate grassland • Figure 6.10

This short-grass prairie contains a mixture of grasses and other herbaceous flowering plants. Photographed in Arapaho National Wildlife Refuge, Colorado. Climate graph shows monthly temperatures and precipitation for Lawrence, Kansas.

Temperate Grassland

Summers are hot, winters are cold, and rainfall is often uncertain in temperate grassland. Average annual precipitation ranges from 25 to 75 cm (10 to 30 in). Grassland soil has considerable organic material because the aboveground portions of many grasses die off each winter and contribute to the organic content of the soil, while the roots and rhizomes (underground stems) survive underground. Many grasses are sod formers—that is, their roots and rhizomes form a thick, continuous underground mat.

Moist temperate grasslands, also known as *tall-grass prairies*, occur in the United States in parts of Illinois, Iowa, Minnesota, Nebraska, Kansas, and other midwestern states. Trees grow sparsely except near rivers and streams, but grasses taller than a person grow in great profusion in the deep, rich soil. Periodic wildfires help to maintain grasses as the dominant vegetation in grasslands.

More than 90 percent of the North American grassland encountered by European settlers has been

temperate grassland A grassland with hot summers, cold winters, and less rainfall than is found in the temperate deciduous forest biome.

converted to farmland, and the remaining prairie is highly fragmented. Today, the tall-grass prairie is considered North America's rarest biome. Tall-grass prairie formerly supported large herds of grazing animals, such as bison and pronghorn elk. The principal predators were wolves, although in sparser, drier areas coyotes took their place. Smaller animals included prairie dogs and their predators (foxes, black-footed ferrets, and various birds of prey), grouse, reptiles such as snakes and lizards, and great numbers of insects.

Short-grass prairies are temperate grasslands that receive less precipitation than moist temperate grasslands but more precipitation than deserts. In the United States, short-grass prairies occur in parts of Montana, Wyoming, South Dakota, and other midwestern states (**Figure 6.10**). Grasses that grow knee high or lower dominate short-grass prairies. Plants grow less abundantly than in the moister grasslands, and bare soil is occasionally exposed.

Savanna

Savanna, a tropical grassland, is found in areas of low rainfall or, more commonly, in areas of intense seasonal rainfall with prolonged dry periods. Temperatures in tropical savannas vary little throughout the year. Precipitation is the overriding climate factor: Annual precipitation is 85 to 150 cm (34 to 60 in). Savanna soil is somewhat low in essential nutrient minerals, in part because it is heavily leached during rainy periods—that is, nutrient minerals filter out of the topsoil. Although the African savanna is best known, savanna also occurs in South America, western India, and northern Australia.

Savanna has wide expanses of grasses interrupted by occasional trees like the acacia, which bristles with thorns to provide protection against herbivores. Both trees and grasses have fire-adapted features, such as extensive underground root systems, that enable them to survive seasonal droughts as well as periodic fires.

> **savanna** A tropical grassland with widely scattered trees or clumps of trees.

Spectacular herds of herbivores such as antelope, giraffe, elephants, wildebeest, and zebra occur in the African savanna (**Figure 6.11**). Large predators, such as lions and hyenas, kill and scavenge the herds. In areas of seasonally varying rainfall, the herds and their predators may migrate annually.

Savanna in many places is being converted into rangeland for cattle and other domesticated animals. The problem is particularly serious in Africa, where human populations are growing rapidly.

Desert

Desert consists of dry areas found in both temperate (cold deserts) and subtropical or tropical regions (warm deserts). The low water vapor content of the desert

> **desert** A biome in which the lack of precipitation limits plant growth; deserts are found in both temperate and tropical regions.

Savanna • Figure 6.11

Tropical grasslands such as this one, with widely scattered acacia trees, support large herds of grazing animals and their predators. Photographed in Masai Mara National Reserve, Kenya. Climate graph shows monthly temperatures and precipitation for Lusaka, Zambia.

Interpreting Data

In terms of ranges shown across the year, how does the pattern of average monthly precipitation differ from the pattern of average monthly temperature in this savanna?

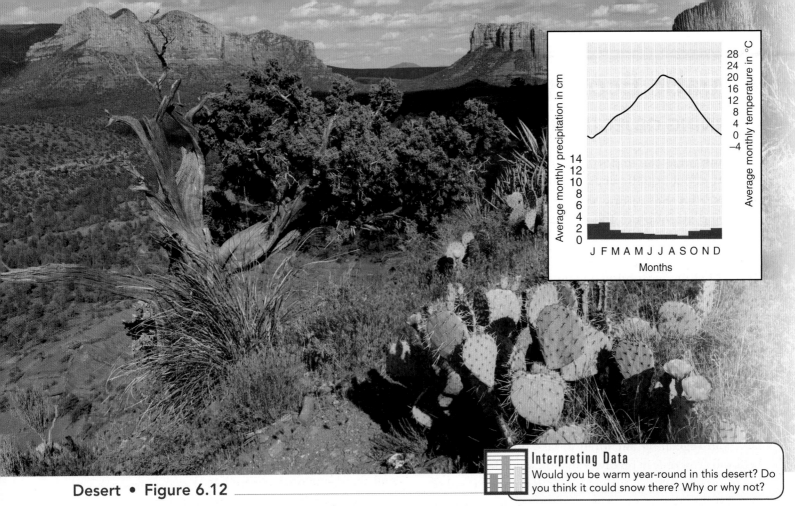

Desert • Figure 6.12

This desert landscape includes tough-leaved yuccas and spine-covered prickly pear cacti. Desert inhabitants are strikingly adapted to the demands of their environment. Photographed near Sedona, Arizona. Climate graph shows monthly temperatures and precipitation for Reno, Nevada.

Interpreting Data
Would you be warm year-round in this desert? Do you think it could snow there? Why or why not?

atmosphere results in daily temperature extremes of heat and cold, so that a major change in temperature occurs in each 24-hour period. Desert environments vary greatly depending on the amount of precipitation they receive, which is generally less than 25 cm (10 in) per year. As a result of sparse vegetation, desert soil is low in organic material but is often high in mineral content, particularly salts.

Plant cover is sparse in deserts, and much of the soil is exposed. Plants in North American deserts include cacti, yuccas, Joshua trees, and sagebrush (**Figure 6.12**). Desert plants are adapted to conserve water and as a result tend to have few, small, or no leaves. Cactus leaves are modified into spines, which discourage herbivores. Other desert plants shed their leaves for most of the year, growing only during the brief moist season.

Desert animals are typically small. During the heat of the day, they remain under cover or return to shelter periodically, emerging at night to forage or hunt. In addition to desert-adapted insects and arachnids (such as tarantulas and scorpions), there are a few

desert-adapted amphibians (frogs and toads) and many reptiles, such as the desert tortoise, Gila monster, and Mojave rattlesnake. Desert mammals in North America include rodents such as kangaroo rats, as well as mule deer and jackrabbits. Birds of prey, especially owls, live on the rodents and jackrabbits, and even the scorpions. During the driest months of the year, many desert animals tunnel underground, where they remain inactive.

In many areas, human development encroaches on deserts. Use of off-road vehicles damages vegetation, and expansion of farms, cities, and residential areas places severe demands on limited groundwater.

CONCEPT CHECK   STOP

1. **What** is a biome?

2. **How** do you distinguish between temperate rain forest and tropical rain forest? between savanna and desert?

Aquatic Ecosystems

LEARNING OBJECTIVES

1. **Summarize** the important environmental factors that affect aquatic ecosystems.
2. **Describe** the various aquatic ecosystems, giving attention to the environmental characteristics of each.

The most fundamental division in aquatic ecology is probably between freshwater and saltwater environments. Salinity, which is the concentration of dissolved salts (such as sodium chloride) in a body of water, affects the kinds of organisms present in aquatic ecosystems, as does the amount of dissolved oxygen. Water greatly interferes with the penetration of light, so floating aquatic organisms that photosynthesize must remain near the water's surface, and vegetation anchored to lake floors or streambeds will grow only in relatively shallow water. In addition, low levels of essential nutrient minerals limit the number and distribution of organisms in certain aquatic environments. In this section, we discuss freshwater ecosystems only; because the immense marine environment is so critical to the environmental well-being of Earth, we devote an entire chapter to it (see Chapter 11).

Aquatic ecosystems contain three main ecological categories of organisms: free-floating plankton, strongly swimming nekton, and bottom-dwelling benthos. **Plankton** are usually small or microscopic organisms. They tend to drift or swim feebly, so, for the most part, they are carried about at the mercy of currents and waves. **Nekton** are larger, more strongly swimming organisms such as fishes, turtles, and whales. **Benthos** are bottom-dwelling organisms that fix themselves to one spot (sponges and oysters), burrow into the sand (worms and clams), or simply walk about on the bottom (crawfish and aquatic insect larvae).

Freshwater Ecosystems

Freshwater ecosystems include lakes and ponds (standing-water ecosystems), rivers and streams (flowing-water ecosystems), and marshes and swamps (freshwater wetlands).

Specific abiotic conditions and characteristic organisms distinguish each freshwater ecosystem. Although freshwater ecosystems occupy only about 2 percent of Earth's surface, they play an important role in the hydrologic cycle: They help recycle precipitation that flows into the ocean as surface runoff. (See Chapter 5 for a detailed explanation of the hydrologic cycle.) Large bodies of fresh water help moderate daily and seasonal temperature fluctuations on nearby land regions, and freshwater habitats provide homes for many species.

Zonation is characteristic of **standing-water ecosystems**. A large lake has three zones: the littoral, limnetic, and profundal zones (see *What a Scientist Sees*). The *littoral zone* is a shallow-water area along the shore of a lake or pond. Emergent vegetation, such as cattails and bur reeds, as well as several deeper-dwelling aquatic plants and algae, live in the littoral zone. The *limnetic zone* is the open water beyond the littoral zone—that is, away from the shore. The limnetic zone extends down as far as sunlight penetrates to permit photosynthesis. The main organisms of the limnetic zone are microscopic plankton. Larger fishes also spend most of their time in the limnetic zone, although they may visit the littoral zone to feed and reproduce. The deepest zone, the *profundal zone*, is beneath the limnetic zone of a large lake; smaller lakes and ponds typically lack a profundal zone. Because light does not penetrate effectively to this depth, plants and algae do not live there. Detritus drifts into the profundal zone from the littoral and limnetic zones; bacteria decompose this detritus. This marked zonation is accentuated by **thermal stratification**, in which the temperature changes sharply with depth.

Flowing-water ecosystems are highly variable. The surrounding environment changes greatly between a river's source and

> **standing-water ecosystem** A body of fresh water surrounded by land and whose water does not flow; a lake or a pond.

> **flowing-water ecosystem** A freshwater ecosystem such as a river or stream in which water flows in a current.

Zonation in a Large Lake

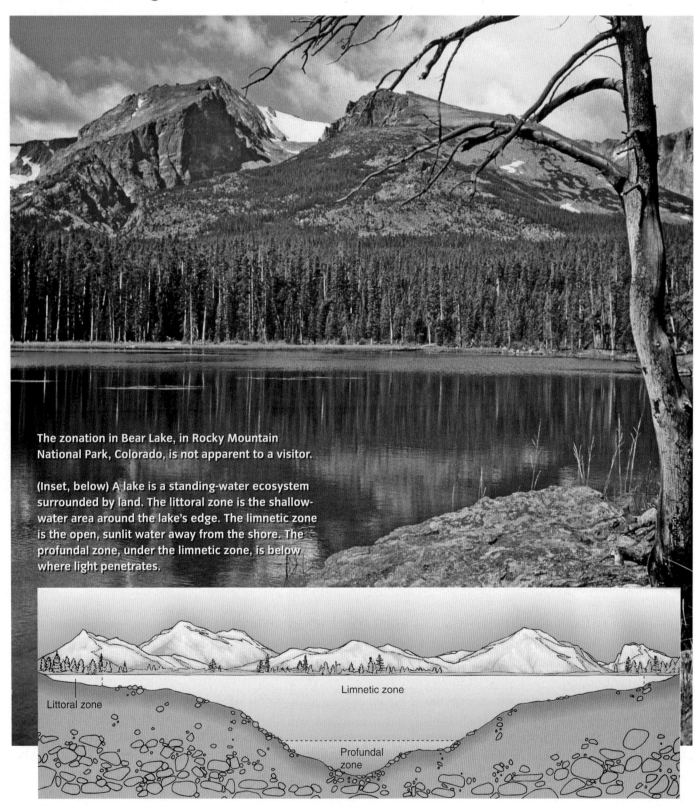

The zonation in Bear Lake, in Rocky Mountain National Park, Colorado, is not apparent to a visitor.

(Inset, below) A lake is a standing-water ecosystem surrounded by land. The littoral zone is the shallow-water area around the lake's edge. The limnetic zone is the open, sunlit water away from the shore. The profundal zone, under the limnetic zone, is below where light penetrates.

Littoral zone

Limnetic zone

Profundal zone

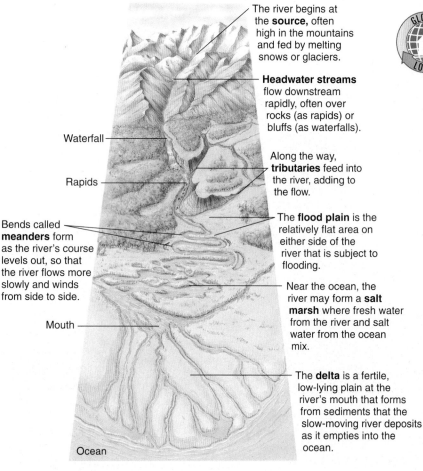

The river begins at the **source,** often high in the mountains and fed by melting snows or glaciers.

Headwater streams flow downstream rapidly, often over rocks (as rapids) or bluffs (as waterfalls).

Along the way, **tributaries** feed into the river, adding to the flow.

Waterfall

Rapids

Bends called **meanders** form as the river's course levels out, so that the river flows more slowly and winds from side to side.

The **flood plain** is the relatively flat area on either side of the river that is subject to flooding.

Near the ocean, the river may form a **salt marsh** where fresh water from the river and salt water from the ocean mix.

Mouth

The **delta** is a fertile, low-lying plain at the river's mouth that forms from sediments that the slow-moving river deposits as it empties into the ocean.

Ocean

a. A river flows from its source to the ocean.

b. Aerial view of meanders in the Tambopata River, Peru.

Many of the world's busiest cities are located at river mouths, or along rivers, on the flood plain. Can you name examples in your region, or anywhere else in the world?

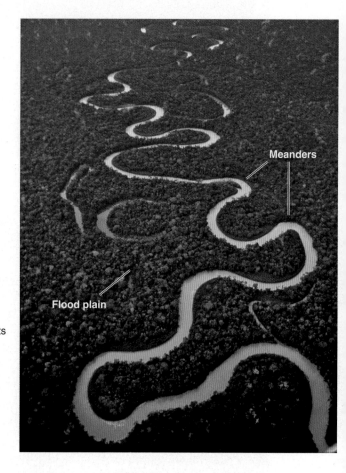

Meanders

Flood plain

---

its mouth (**Figure 6.13**). Certain parts of the stream's course are shaded by forest, while other parts are exposed to direct sunlight. Groundwater may well up through sediments on the bottom in one particular area, making the water temperature cooler in summer or warmer in winter than in adjacent parts of the stream or river. The kinds of organisms found in flowing water vary greatly from one stream to another, depending primarily on the strength of the current. In streams with fast currents, some inhabitants have adaptations such as suckers, with which they attach themselves to rocks to prevent being swept away. Some stream inhabitants have flattened bodies to slip under or between

freshwater wetlands Lands that shallow fresh water covers for at least part of the year; wetlands have a characteristic soil and water-tolerant vegetation.

rocks. Other inhabitants such as fish are streamlined and muscular enough to swim in the current.

Human activities such as pollution and dam construction have adverse impacts on rivers and streams. These activities damage wildlife habitat and threaten water supplies and fisheries.

Freshwater wetlands include **marshes,** dominated by grasslike plants, and **swamps,** dominated by woody trees or shrubs (**Figure 6.14**). Wetland soils are waterlogged for variable periods and are therefore anaerobic (without oxygen). They are rich in accumulated organic materials, partly because anaerobic conditions discourage decomposition.

ecosystem services Important environmental benefits, such as clean air to breathe, clean water to drink, and fertile soil in which to grow crops, that the natural environment provides.

Wetlands provide excellent wildlife habitat for migratory waterfowl and other bird species, as well as for beaver, otters, muskrats, and game fish. In addition to providing unique wildlife habitat, wetlands serve other important environmental functions, known as **ecosystem services**. When rivers flood their banks, wetlands are capable of holding or even absorbing the excess water, thereby helping to control flooding. The floodwater then drains slowly back into the rivers, providing a steady flow of water throughout the year. Wetlands also serve as groundwater recharging areas. One of their most important roles is to trap and hold pollutants in the flooded soil, thereby cleansing and purifying the water.

Although wetlands are afforded some legal protection, they are still threatened by pollution, development, agriculture, and dam construction. (See Chapter 10 for more on threats to freshwater ecosystems.)

Freshwater swamp • Figure 6.14

Freshwater swamps are inland areas covered by water and dominated by trees, such as baldcypress. Photographed in the Atchafalaya National Wildlife Refuge, Louisiana.

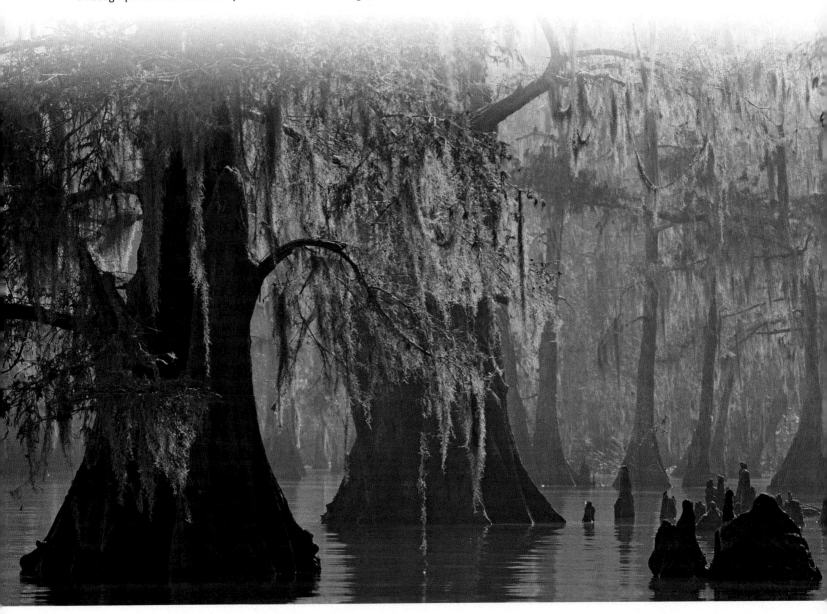

Brackish Ecosystems: Estuaries

estuary A coastal body of water, partly surrounded by land, with access to the open ocean and a large supply of fresh water from a river.

Where the ocean meets the land, there may be one of several kinds of ecosystems: a rocky shore, a sandy beach, an intertidal mud flat, or a tidal *estuary*. Water levels in an estuary rise and fall with the tides; salinity fluctuates with tidal cycles, the time of year, and precipitation. Salinity also changes gradually within the estuary, from fresh water at the river entrance, to *brackish* (somewhat salty) water, to salty ocean water at the mouth of the estuary. Because estuaries undergo significant daily, seasonal, and annual variations in physical factors such as temperature, salinity, and depth of light penetration, estuarine organisms must have a high tolerance for changing conditions.

Temperate estuaries usually feature **salt marshes**, shallow wetlands in which salt-tolerant grasses grow (**Figure 6.15a**). Salt marshes perform many ecosystem services, including providing biological habitats, trapping sediment and pollution, supplying groundwater, and buffering storms by absorbing their energy, which prevents flood damage elsewhere.

Mangrove forests, the tropical equivalent of salt marshes, cover perhaps 70 percent of tropical coastlines (**Figure 6.15b**). Like salt marshes, mangrove forests provide valuable ecosystem services. Their interlacing roots are breeding grounds and nurseries for several

Estuaries • Figure 6.15

b. Brain coral growing around red mangrove roots in Tunicate Cove, Belize. Mangrove roots grow into deeper water as well as into mudflats that are exposed at low tide. Many animals live among the complex root systems of mangrove forests.

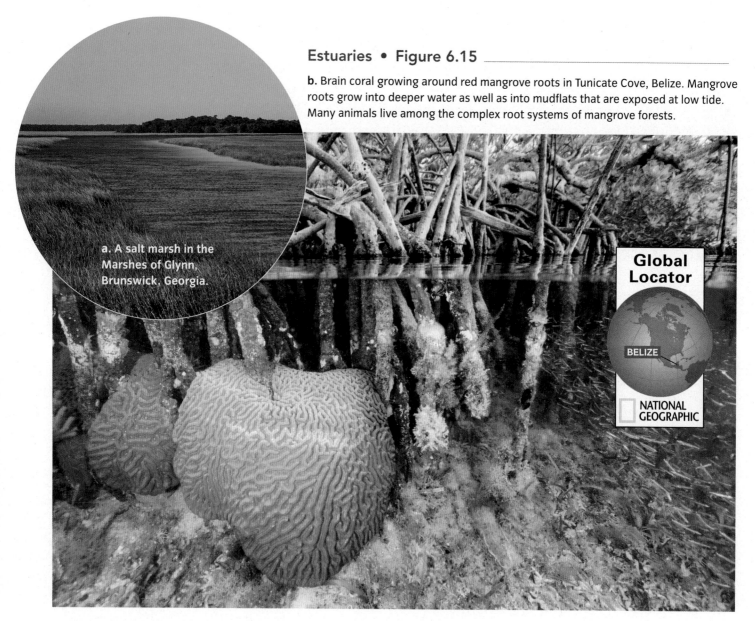

a. A salt marsh in the Marshes of Glynn, Brunswick, Georgia.

Global Locator

BELIZE

NATIONAL GEOGRAPHIC

commercially important fishes and shellfish, such as mullet, spotted sea trout, crabs, and shrimp. Mangrove branches are nesting sites for many species of birds, such as pelicans, herons, egrets, and roseate spoonbills. Mangrove roots stabilize the submerged soil, thereby preventing coastal erosion and providing a barrier against the ocean during storms.

Both salt marsh and mangrove forest ecosystems have experienced significant losses due to coastal development. Salt marshes have been polluted and turned into dumping grounds; mangrove forests have been logged and used as aquaculture sites.

CONCEPT CHECK	STOP

1. **Which** environmental factors shape flowing-water ecosystems? standing-water ecosystems?

2. **How** do the characteristics of a freshwater wetland differ from those of an estuary? How does a mangrove swamp differ from a salt marsh?

Population Responses to Changing Conditions over Time: Evolution

LEARNING OBJECTIVES

1. **Define** *evolution.*
2. **Explain** how evolution by natural selection consists of four observations of the natural world.
3. **Describe** various types of evidence that supports evolution.

S cientists think all of Earth's remarkable variety of organisms descended from earlier species by a process known as **evolution**. The concept of evolution dates back to the time of Aristotle, but **Charles Darwin** (1809–1882), a 19th-century naturalist, proposed the mechanism of evolution that today's scientific community still accepts (**Figure 6.16**). As you will see, the environment plays a crucial role in Darwin's theory of evolution.

> **evolution**
> The cumulative genetic changes in populations that occur during successive generations.

It occurred to Darwin that in a population, inherited traits favorable to survival in a given environment tended to be preserved over successive generations, whereas unfavorable traits were eliminated. The result is *adaptation,* an evolutionary modification that improves the chance of survival and reproductive success of a species in a given environment. Eventually the accumulation of many adaptive modifications might result in a new species.

Darwin proposed the theory of evolution by natural selection in his monumental book *The Origin of Species by Means of Natural Selection,* which was published in 1859. Since that time, scientists have accumulated an enormous body of observations and experiments that support Darwin's theory. Although biologists still do not agree completely on some aspects of the evolutionary process, the concept that evolution by natural selection has taken place and is still occurring is now well documented.

Portrait of a young Charles Darwin • Figure 6.16

Natural Selection

Evolution occurs through the process of **natural selection**. As favorable traits increase in frequency in successive generations, and as unfavorable traits decrease or disappear, the collection of characteristics of a given population changes. Natural selection is the process by which successful traits are passed on to the next generation and unsuccessful ones are weeded out. It consists of four observations about the natural world:

> **natural selection**
> The tendency of better-adapted individuals—those with a combination of genetic traits best suited to environmental conditions—to survive and reproduce, increasing their proportion in the population.

1. *Surplus reproductive capacity.* Each species produces more offspring than will survive to maturity. Natural populations have the reproductive potential to increase their numbers continuously over time (**Figure 6.17**).

2. *Heritable variation.* The individuals in a population exhibit variation. Each individual has a unique combination of traits, such as size, color, and ability to tolerate harsh environments. Some traits improve the chances of an individual's survival and reproductive success, whereas others do not. It is important to remember that the variation necessary for evolution by natural selection must be inherited so that it can be passed to offspring.

3. *Limits on population growth, or a struggle for existence.* There is only so much food, water, light, growing space, and so on available to a population, and organisms compete with one another for the limited resources available to them. Because there are more individuals than the environment can support, not all of an organism's offspring will survive to reproductive age. Other limits on population growth include predators and diseases.

4. *Differential reproductive success.* Individuals that possess the most favorable combination of characteristics (those that make individuals better adapted to their environment) are more likely than others to survive, reproduce, and pass their traits to the next generation. Sexual reproduction is the key to natural selection: The best-adapted individuals are those that

Surplus reproductive capacity • Figure 6.17

If each breeding pair of elephants were to produce six offspring that lived and reproduced, in 750 years a single pair of elephants would have given rise to more than 15 million elephants! Yet elephants have not overrun the planet. Photographed in Kenya.

reproduce most successfully, whereas less-fit individuals die prematurely or produce fewer or inferior offspring. Over time, enough changes may accumulate in geographically separated populations (often with slightly different environments) to produce new species (**Figure 6.18**).

One premise on which Darwin based his theory of evolution by natural selection is that individuals transmit traits to the next generation. However, Darwin could not explain *how* this occurs or *why* individuals within a population vary. Beginning in the 1930s and 1940s, biologists combined the principles of genetics with Darwin's theory of natural selection. The resulting unified explanation of evolution is known as the **modern synthesis** (where *synthesis* refers to a combination of parts of previous theories). The modern synthesis explains Darwin's observation of variation among offspring in terms of **mutation**, or changes in DNA. Mutations provide the genetic variability on which natural selection acts during evolution.

A vast body of evidence supports evolution, most of which is beyond the scope of this text. This evidence includes observations from the fossil record, comparative anatomy, biogeography (the study of the geographic locations of organisms), and molecular biology

Darwin's finches • Figure 6.18

Charles Darwin was a ship's naturalist on a 5-year voyage around the world. During an extended stay in the Galápagos Islands off the coast of Ecuador, he studied the plants and animals of each island, including 14 species of finches.

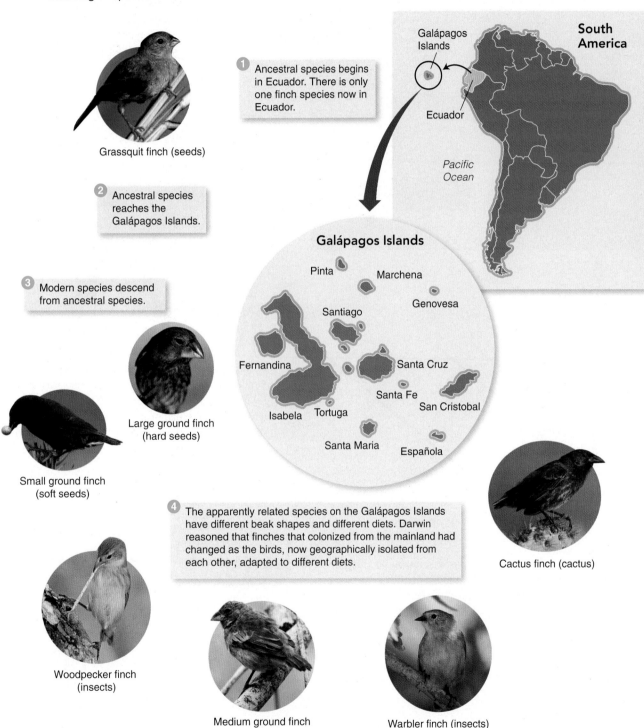

1 Ancestral species begins in Ecuador. There is only one finch species now in Ecuador.

Grassquit finch (seeds)

2 Ancestral species reaches the Galápagos Islands.

3 Modern species descend from ancestral species.

Large ground finch (hard seeds)

Small ground finch (soft seeds)

South America

Galápagos Islands

Ecuador

Pacific Ocean

Galápagos Islands

Pinta   Marchena

Genovesa

Santiago

Fernandina   Santa Cruz

Santa Fe

Isabela   Tortuga   San Cristobal

Santa Maria   Española

4 The apparently related species on the Galápagos Islands have different beak shapes and different diets. Darwin reasoned that finches that colonized from the mainland had changed as the birds, now geographically isolated from each other, adapted to different diets.

Cactus finch (cactus)

Woodpecker finch (insects)

Medium ground finch (moderate seeds)

Warbler finch (insects)

Think Critically Which of the species shown have similar beaks? Is this reflected in their diet similarities?

NATIONAL GEOGRAPHIC

WILEY **RESOURCE KIT**

Evidence for evolution • **Figure 6.19**

a. The Fossil Record.
Fossils deposited in rock layers, which can be dated, show how organisms evolved over time. These fish fossils from Liaoning Province, China, date from 120 million years ago.

b. Comparative Anatomy.
Similarities among organisms demonstrate how they are related. These similarities among four vertebrate limbs illustrate that, while proportions of bones have changed in relation to each organism's way of life, the forelimbs have the same basic bone structure.

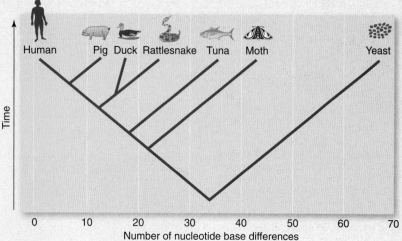

c. Molecular Biology.
The organisms pictured here all share a particular enzyme, but in the course of evolution, mutations have resulted in changes in the gene that codes for that enzyme. This diagram shows the nucleotide base differences in this gene among humans and other organisms. Note that organisms thought to be more closely related to humans have fewer differences than organisms that are more distantly related to humans.

Interpreting Data
Based on the diagram, which organism shown is most closely related to the duck?

RESOURCE KIT

(**Figure 6.19**, on facing page). In addition, evolutionary hypotheses are tested experimentally.

On the basis of these kinds of evidence, virtually all biologists accept the principles of evolution by natural selection, although they don't agree on all the details. They try to better understand certain aspects of evolution, such as the role of chance and how quickly new species evolve. As discussed in Chapter 1, science is an ongoing process, and information obtained in the future may require modifications to certain parts of the theory of evolution by natural selection.

CONCEPT CHECK STOP

1. **What** is evolution?
2. **What** four observations are the basis of natural selection?
3. **Which** types of evidence support evolution?

Community Responses to Changing Conditions over Time: Succession

LEARNING OBJECTIVES

1. **Define** *ecological succession.*
2. **Distinguish** between primary and secondary succession.

A community of organisms does not spring into existence full blown. By means of **ecological succession**, a given community develops gradually through a sequence of species. Certain organisms colonize an area; over time, others replace them, and eventually the replacements are themselves replaced by still other species. Ecologists first studied succession in three diverse ecosystems: an abandoned field, a northern freshwater bog, and sand dunes (**Figure 6.20**).

ecological succession The process of community development over time, which involves species in one stage being replaced by different species.

The actual mechanisms that underlie succession are not clear. In some cases, it may be that a resident species modified the environment in some way, thereby making it more suitable for a later species to colonize. It is also possible that prior residents lived there in the first place because there was little competition from other species. Later, as more invasive species arrived, the original species were displaced.

Ecologists initially thought that succession inevitably led to a stable and persistent community, known as a *climax community,* such as a forest. But more recently, this traditional view has fallen out of favor. The apparent stability of a "climax" forest is probably the result of how long trees live relative to the human life span. It is now recognized that mature climax communities are not in a state of stable equilibrium but rather in a state of continual disturbance. Over time, a mature community changes in species composition

Ecological succession on sand dunes • Figure 6.20

Sand dunes along Lake Michigan were one of the first sites where ecological succession was studied.

and in the relative abundance of each species, despite the fact that it retains an overall uniform appearance.

Succession is usually described in terms of the changes in the plant species growing in a given area, although each stage of the succession may also have its own kinds of animals and other organisms. Ecological succession is measured on the scale of tens, hundreds, or thousands of years, not the millions of years involved in the evolutionary time scale.

Primary Succession

Primary succession is the change in species composition over time in a previously uninhabited environment (**Figure 6.21**). No soil exists when primary succession begins. Bare rock surfaces, such as recently formed volcanic lava and rock scraped clean by glaciers, are examples of sites where primary succession may take place. Details vary from one site to another, but on bare rock, lichens are often the most important element in the *pioneer community*, which is the initial community that develops during primary succession. Lichens secrete acids that help break apart the rock, beginning the process of soil formation. Over time, mosses and drought-resistant ferns may replace the lichen community, followed in turn by tough grasses and herbs. Once enough soil accumulates, low shrubs may replace the grasses and herbs; over time, forest trees in several distinct stages would replace the shrubs. Primary succession on bare rock from a pioneer community to a forest community often occurs in this sequence: lichens → mosses → grasses → shrubs → trees.

The concept of succession was developed in the 1880s by Henry Cowles, who studied the process as it occurred on sand dunes along the shores of Lake Michigan (see Figure 6.20). As in many other lake and ocean shore areas, the Lake Michigan sand dune environment is severe, with temperatures ranging from high during the day to low at night. Few plants could tolerate these stresses and the low nutrient content of the sand making up the dunes. As Cowles observed, grasses are common pioneer

PROCESS DIAGRAM

Primary succession on glacial moraine • Figure 6.21

 THE PLANNER

During the past 200 years, glaciers have retreated in Glacier Bay, Alaska. Although these photos were not taken in the same area, they show some of the stages of primary succession on glacial moraine (rocks, gravel, and sand that a glacier deposits).

① After a glacier's retreat, lichens initially colonize the barren landscape, followed by mosses and small shrubs.

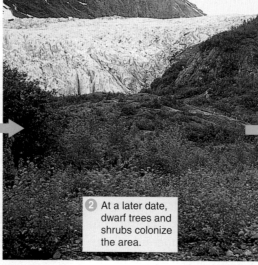

② At a later date, dwarf trees and shrubs colonize the area.

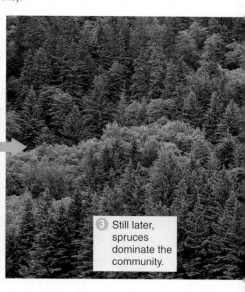

③ Still later, spruces dominate the community.

Think Critically Why is it that spruce trees can't grow directly on the rocks deposited by glaciers?

Secondary succession on an abandoned field in North Carolina • Figure 6.22

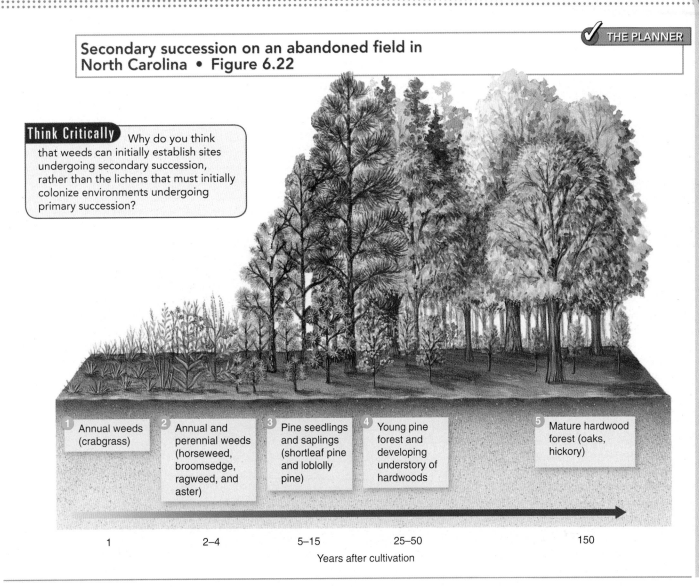

Think Critically Why do you think that weeds can initially establish sites undergoing secondary succession, rather than the lichens that must initially colonize environments undergoing primary succession?

1 Annual weeds (crabgrass)

2 Annual and perennial weeds (horseweed, broomsedge, ragweed, and aster)

3 Pine seedlings and saplings (shortleaf pine and loblolly pine)

4 Young pine forest and developing understory of hardwoods

5 Mature hardwood forest (oaks, hickory)

1   2–4   5–15   25–50   150

Years after cultivation

plants on Great Lakes dunes, helping to stabilize the dune surface. Mat-forming shrubs then invade to further stabilize the dune, followed by a succession of tree species over the course of many years. Primary succession on sand dunes around the Great Lakes might proceed in this sequence: grasses → shrubs → poplars (cottonwoods) → pine trees → oak trees.

Secondary Succession

Secondary succession is the change in species composition that takes place after some disturbance destroys the existing vegetation; soil is already present (**Figure 6.22**). Abandoned farmland and an open area caused by a forest fire are common examples of sites where secondary succession occurs. Biologists have studied secondary succession

on abandoned farmland extensively. Although it takes more than 100 years for secondary succession to occur at a single site, a single researcher can study old-field succession in its entirety by observing different sites undergoing succession in the same general area. The biologist may examine county tax records to determine when each field was abandoned. Secondary succession on abandoned farmland in the southeastern United States proceeds in this sequence: crabgrass → horseweed, broomsedge, and other weeds → pine trees → hardwood trees.

CONCEPT CHECK **STOP**

1. **What** is ecological succession?

2. **How** does primary succession differ from secondary succession?

Wildfires

A wildfire is any unexpected—and unwanted—fire that burns in grass, shrub, and forest areas. Whether started by lightning or by humans, wildfires are an important environmental force in many geographic areas, especially places with wet seasons followed by dry seasons, such as chaparral. Vegetation that grows during the wet season dries to tinder during the dry season. After fire ignites the dry organic material, wind spreads the fire through the area.

At the peak of the wildfire season in the American West, an area prone to wildfires, hundreds of new wildfires can break out each day. In southern California alone, devastating wildfires in 2007, 2008, and 2009 consumed well over 380,000 hectares (950,000 acres).

Fires have several effects on the environment. First, combustion frees minerals locked in dry organic matter. The ashes left by fire are rich in potassium, phosphorus, calcium, and other nutrient minerals essential for plant growth. Thus, vegetation flourishes after a fire. Second, fire removes plant cover and exposes the soil, which stimulates the germination of seeds that require bare soil and the growth of shade-intolerant plants. Third, fire increases soil erosion because it removes plant cover, leaving soil more vulnerable to wind and water.

Fires were a part of the natural environment long before humans appeared, and many terrestrial ecosystems have adapted to fire. Grasses adapted to wildfire have underground stems and buds. After fire kills the aboveground parts, the untouched underground parts send up new sprouts. Fire-adapted trees such as bur oak and ponderosa pine have thick, fire-resistant bark; others, such as jack pine, depend on fire for successful reproduction because the fire's heat opens the cones and releases the seeds.

Human interference also affects the frequency and intensity of wildfires, even when the goal is fire prevention. When fire is excluded from a fire-adapted ecosystem, organic litter accumulates. As a result, when a fire does occur, it burns hotter and is much more destructive than ecologically helpful. Decades of fire suppression in the West are partly responsible for the massively destructive fires that have occurred there in recent years. *Prescribed burning* is an ecological management tool that allows for controlled burning to reduce organic litter and suppress fire-sensitive trees in fire-adapted areas.

A deer flees a wildfire near Malibu, California, November 2007.

A NASA satellite image of the southern California wildfires, October 24, 2007. Red pixels were added to indicate fires.

Summary

1 Earth's Major Biomes 128

1. A **biome** is a large, relatively distinct terrestrial region with characteristic climate, soil, plants, and animals, regardless of where it occurs; a biome encompasses many interacting ecosystems. Near the poles, temperature is generally the overriding climate factor in determining biome distribution, whereas in temperate and tropical regions, precipitation is more significant.

2. **Tundra** is the treeless biome in the far north that consists of boggy plains covered by lichens and small plants such as mosses; it has harsh, very cold winters and extremely short summers. **Boreal forest** is a region of coniferous forest in the Northern Hemisphere, located just south of the tundra. **Temperate rain forest** is a coniferous biome with cool weather, dense fog, and high precipitation. **Temperate deciduous forest** is a forest biome that occurs in temperate areas where annual precipitation ranges from about 75 cm to 126 cm (30–50 in). **Tropical rain forest** is a lush, species-rich forest biome that occurs where the climate is warm and moist throughout the year. **Chaparral** is a biome with mild, moist winters and hot, dry summers; vegetation is typically small-leafed evergreen shrubs and small trees. **Temperate grassland** is grassland with hot summers, cold winters, and less rainfall than is found in the temperate deciduous forest biome. **Savanna** is tropical grassland with widely scattered trees or clumps of trees. **Desert** is a biome in which the lack of precipitation limits plant growth; deserts are found in both temperate and tropical regions.

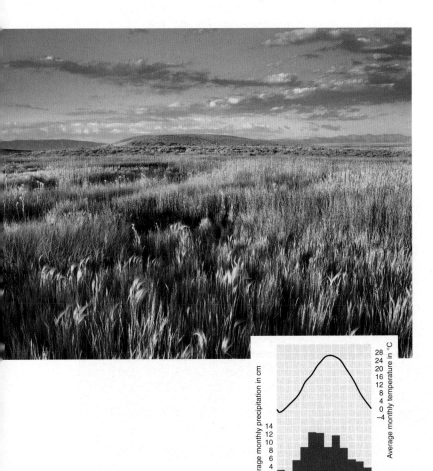

2 Aquatic Ecosystems 142

1. In aquatic ecosystems, important environmental factors include salinity, amount of dissolved oxygen, and availability of light for photosynthesis.

2. Freshwater ecosystems include standing-water, flowing-water, and freshwater wetlands. A **standing-water ecosystem** is a body of fresh water surrounded by land and whose water does not flow, such as a lake or pond. A **flowing-water ecosystem** is a freshwater ecosystem such as a river or stream in which the water flows in a current. **Freshwater wetlands** are marshes and swamps—lands that are covered by shallow fresh water at least part of the year; wetlands have a characteristic soil and water-tolerant vegetation. An **estuary** is a coastal body of water, partly surrounded by land, with access to the open ocean and a large supply of fresh water from a river. Water in an estuary is brackish rather than truly fresh. Temperate estuaries usually contain **salt marshes**, whereas tropical estuaries are lined with **mangrove forests**.

3 Population Responses to Changing Conditions over Time: Evolution 147

1. **Evolution** is the cumulative genetic changes in populations that occur during successive generations.

2. **Natural selection** is the tendency of better-adapted individuals—those with a combination of genetic traits best suited to environmental conditions—to survive and reproduce, increasing their proportion in the population. Natural selection is based on four observations established by Charles Darwin: (1) Each species produces more offspring than will survive to maturity. (2) The individuals in a population exhibit inheritable variation in their traits. (3) Organisms compete with one another for the resources needed to survive. (4) Individuals with the most favorable combination of traits are most likely to survive and reproduce, passing their genetic traits to the next generation.

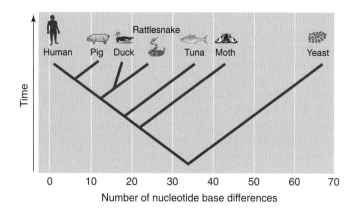

Number of nucleotide base differences

3. Scientific evidence supporting evolution comes from the fossil record, comparative anatomy, biogeography, and molecular biology.

4 Community Responses to Changing Conditions over Time: Succession 151

1. **Ecological succession** is the process of community development over time, which involves species in one stage being replaced by different species.

2. **Primary succession** is the change in species composition over time in an environment that was not previously inhabited by organisms; examples include bare rock surfaces, such as recently formed volcanic lava and rock scraped clean by glaciers. **Secondary succession** is the change in species composition that takes place after some disturbance destroys the existing vegetation; soil is already present. Examples include abandoned farmland and open areas caused by forest fires.

Key Terms

- biome 128
- boreal forest 133
- chaparral 137
- desert 140
- ecological succession 151
- ecosystem services 145

- estuary 146
- evolution 147
- flowing-water ecosystem 142
- freshwater wetlands 144
- natural selection 148
- savanna 140

- standing-water ecosystem 142
- temperate deciduous forest 135
- temperate grassland 139
- temperate rain forest 134
- tropical rain forest 136
- tundra 132

What is happening in this picture?

- This picture shows expensive homes built in the chaparral of the Santa Monica Mountains. Based on what you have learned in this chapter, what environmental problem might threaten these homes?

- Sometimes people have removed the chaparral vegetation to prevent fires from damaging their homes. Where that has occurred, the roots no longer hold the soil in place. What could happen when the winter rains come?

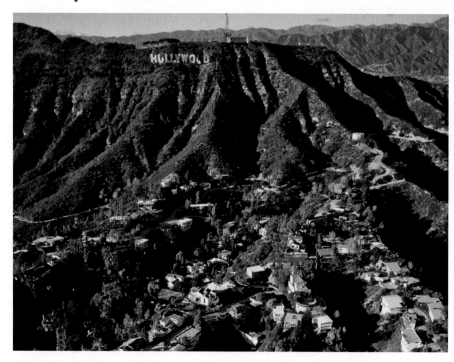

Critical and Creative Thinking Questions

1. What two climate factors are most important in determining an area's characteristic biome?

2. In which biome do you live? Where would you place your biome in the figure below? How would that compare with your placement of the biome in northern Siberia or the biome dominating northern Africa and Saudi Arabia?

3. What environmental factors are most important in determining the kinds of organisms found in aquatic environments?

4. Distinguish between freshwater wetlands and estuaries and between flowing-water and standing-water ecosystems.

5. Name and compare temperate and tropical estuaries. What types of plants are characteristic of each?

6. During the mating season, male giraffes slam their necks together in fighting bouts to determine which male is stronger and can therefore mate with females. Explain how the long necks of giraffes may have evolved, using Darwin's theory of evolution by natural selection.

7. Explain why evolution, by definition, cannot take place within one individual and during that individual's life span.

8. Describe the process and stages of ecological succession.

9. Which type of ecological succession might occur in a forest where a stand of trees has been cleared? Explain your choice by comparing primary and secondary succession.

10. Although most salamanders have four legs, the aquatic salamander shown below resembles an eel. It lacks hind limbs and has very tiny forelimbs. Propose a hypothesis to explain how these salamanders evolved according to Darwin's theory of natural selection.

11. How could you test the hypothesis you proposed in question 10? What type of evidence might you produce?

12. Which biome discussed in this chapter is depicted by the information in the graph below? Explain your answer.

Human Population Change and the Environment

INDIA'S POPULATION PRESSURES

India is the world's second most populous nation, with a mid-2009 population of 1.17 billion. In the 1950s, it became the first country to establish government-sponsored family planning. India did not experience immediate results from its efforts to control population growth, in part because of the diverse cultures, religions, and customs in different regions of the country. Indians speak 15 main languages and more than 700 dialects, which makes communicating a program of family planning education difficult.

In recent years, India has attempted to integrate economic development and family planning projects. Adult literacy and population education programs have been combined. Multimedia advertisements and education promote voluntary birth control, and contraceptives are more available. India has emphasized that improving health services lowers infant and child mortality rates. These efforts have had an effect: The average number of children born per Indian woman declined from 5.3 in 1980 to 2.7 in 2009.

Despite these gains, population pressure has caused the deterioration of India's environment in the past few decades, and 76 percent of Indians live below the official poverty level (less than US$2 a day). India's large population exacerbates its poverty, environmental degradation, and economic underdevelopment.

WILEY
RESOURCE KIT

CHAPTER OUTLINE

CHAPTER PLANNER ✓

Population Ecology

LEARNING OBJECTIVES

1. **Define** *population ecology*.
2. **Explain** the four factors that produce changes in population size.
3. **Define** *biotic potential* and *carrying capacity*.

ndividuals of a given species are part of a larger organization called a *population*. Populations exhibit characteristics that are distinct from those of the individuals in them. Some of the features characteristic of populations but not of individuals are birth and death rates, growth rates, and age structure. Studying populations of other species provides insight into some of the processes that affect the growth of human populations. Understanding human population change is important because the size of the human population is central to most of Earth's environmental problems and their solutions.

> **population ecology** The branch of biology that deals with the number of individuals of a particular species found in an area and why those numbers increase or decrease over time.

Scientists who study **population ecology** try to determine the processes common to all populations (**Figure 7.1**). Population ecologists study how a population responds to its environment—such as how individuals in a given population compete for food or other resources, and how predation, disease, and other environmental pressures affect that population. Environmental pressures such as these prevent populations—whether of bacteria or maple trees or giraffes—from increasing indefinitely.

What we learn about one population helps us make predictions about other populations • Figure 7.1

At first glance, the two populations shown here appear to have little in common, but they share many characteristics.

a. A population of blue columbines in Yankee Boy Basin, Colorado. Populations of other flowers are in the background.

b. A herd of impala survey their surroundings. Photographed in Tanzania.

How Do Populations Change in Size?

Populations of organisms, whether sunflowers, eagles, or humans, change over time. On a global scale, this change is due to two factors: the rate at which individual organisms produce offspring (the birth rate) and the rate at which individual organisms die (the death rate) (**Figure 7.2a**). In humans, the birth rate (b) is usually

> **growth rate (r)**
> The rate of change (increase or decrease) of a population's size, expressed in percentage per year.

expressed as the number of births per 1000 people per year and the death rate (d) as the number of deaths per 1000 people per year. The **growth rate (r)** of a population is the birth rate (b) minus the death rate (d):

$$r = b - d$$

Growth rate is also referred to as *natural increase* in human populations.

If more individuals in a population are born than die, the growth rate is more than zero, and population size increases. If more individuals in a population die faster than are born, the growth rate is less than zero, and population size decreases. If the growth rate is equal to zero, births and deaths match, and population size is stationary, despite continued reproduction and death.

In addition to birth and death rates, **dispersal**—movement from one region or country to another—affects local populations. There are two types of dispersal: **immigration (i)**, in which individuals enter a population and increase its size, and **emigration (e)**, in which

individuals leave a population and decrease its size. The growth rate (r) of a local population must take into account birth rate (b), death rate (d), immigration (i), and emigration (e) (**Figure 7.2b**):

$$r = (b - d) + (i - e)$$

Maximum Population Growth

Different species have different **biotic potentials** (also called *intrinsic rates of increase*). Several factors influence the biotic potential of a species: the age at which reproduction begins, the fraction of the life span during which an individual can reproduce, the number of reproductive periods per lifetime, and the number of offspring produced during each period of reproduction. These factors, called *life history*

> **biotic potential**
> The maximum rate at which a population could increase under ideal conditions.

characteristics, determine whether a particular species has a large or a small biotic potential.

Generally, larger organisms, such as blue whales and elephants, have the smallest biotic potentials, whereas microorganisms have the greatest biotic potentials. Under ideal conditions (that is, in an environment with unlimited resources), certain bacteria reproduce by dividing in half every 30 minutes. At this rate of growth, a single bacterium increases to a population of more than 1 million in just 10 hours and exceeds 1 billion in 15 hours. If you plot population number versus time, the graph takes on the characteristic J shape of

Factors that interact to change population size • Figure 7.2

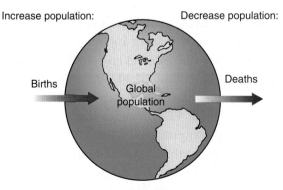

a. On a global scale, the change in a population is due to the number of births and deaths.

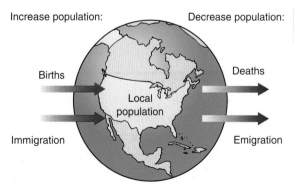

b. In local populations, such as the population of the United States, the number of births, deaths, immigrants, and emigrants affect population size.

Exponential population growth • Figure 7.3

a. *Streptococcus* bacterium in the process of dividing.

Time (hours)	Number of bacteria
0	1
1	4
2	16
3	64
4	256
5	1,024
6	4,096
7	16,384
8	65,536
9	262,144
10	1,048,576

b. When bacteria divide at a constant rate, their number increases exponentially.

c. When bacterial numbers are graphed, the curve of exponential population growth has a characteristic J shape.

exponential population growth (**Figure 7.3**). When a population grows exponentially, the larger the population gets, the faster it grows. Regardless of species, whenever a population grows at its biotic potential, population size plotted versus time gives the same J-shaped curve. The only variable is time.

> **exponential population growth**
> The accelerating population growth that occurs when optimal conditions allow a constant reproductive rate.

Environmental Resistance and Carrying Capacity

Certain populations may exhibit exponential population growth for a short period. However, organisms don't reproduce indefinitely at their biotic potentials because the environment sets limits, which are collectively called **environmental resistance**. Examples of environmental resistance include limited food, water, shelter, and other essential resources, as well as increased disease and predation.

Using the earlier example, we find that bacteria never reproduce unchecked for an indefinite period because they run out of food and living space, and poisonous body wastes accumulate in their vicinity. With crowding, bacteria become more susceptible to parasites (high population densities facilitate the spread of infectious organisms such as viruses among individuals) and predators (high population densities increase the likelihood of a predator catching an individual). As the environment deteriorates, bacterial birth rate declines and death rate increases. The environmental conditions might worsen to a point where the death rate exceeds the birth rate, and as a result, the population decreases. Thus, the environment controls population size: As the population increases, so does environmental resistance, which limits population growth.

Over longer periods, the rate of population growth may decrease to nearly zero. This leveling out occurs at or near the environment's **carrying capacity (K)**. In nature, carrying capacity is dynamic and changes in response to environmental changes. An extended drought, for example, might decrease the amount of vegetation growing in an area, and this change, in turn, would lower the carrying capacity for deer and other herbivores in that environment.

G.F. Gause, a Russian ecologist who conducted experiments in the 1930s, grew a population of *Paramecium* in a test tube (**Figure 7.4a**). He supplied a limited amount of food daily and replenished the media to eliminate the buildup of wastes. Under these conditions, the population increased exponentially at first, but then its growth rate declined to zero, and the population size leveled off.

When a population influenced by environmental resistance is graphed over a long period, the curve has

> **carrying capacity (K)** The largest population a particular environment can support sustainably (long term), if there are no changes in that environment.

an S shape (**Figure 7.4b**). The curve shows the population's initial exponential increase (note the curve's J shape at the start, when environmental resistance is low). Then the population size levels out as it approaches the carrying capacity of the environment. Although the S curve is an oversimplification of how most populations change over time, it fits some populations studied in the laboratory, as well as a few studied in nature.

A population rarely stabilizes at *K* (carrying capacity), as shown in Figure 7.4, but its size may temporarily rise higher than *K*. It will then drop back to, or below, the carrying capacity. Sometimes a population that overshoots *K* will experience a *population crash*, an abrupt decline from high to low population density when resources are exhausted. Such an abrupt change is commonly observed in bacterial cultures, zooplankton, and other populations whose resources are exhausted.

Population growth as carrying capacity is approached • Figure 7.4

a. *Paramecium* is a unicellular microorganism.

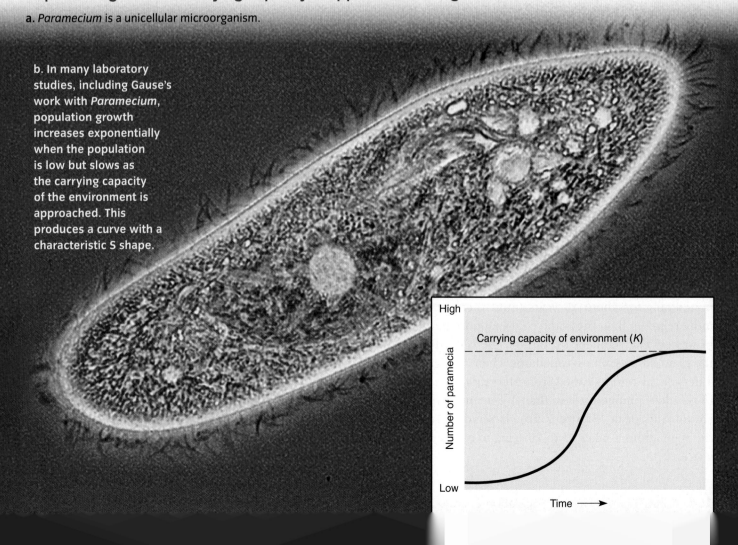

b. In many laboratory studies, including Gause's work with *Paramecium*, population growth increases exponentially when the population is low but slows as the carrying capacity of the environment is approached. This produces a curve with a characteristic S shape.

High

Carrying capacity of environment (*K*)

Number of paramecia

Low

Time ⟶

A population crash • Figure 7.5

a. A herd of reindeer on one of the Pribilof Islands off the coast of Alaska.

b. Graph of the reindeer population originally introduced to one of the Pribilof Islands in 1911. Note the population crash, which followed the peak population attained in 1935.

The availability of winter forage largely determines the carrying capacity for reindeer, which live in cold northern habitats. In 1911, a small herd of reindeer was introduced on one of the Pribilof Islands in the Bering Sea (**Figure 7.5a**). The herd's population increased exponentially for about 25 years, until there were many more reindeer than the island could support, particularly in winter. The reindeer overgrazed the vegetation until the plant life was almost wiped out. Then, in slightly over a decade, as reindeer died from starvation, the number of reindeer plunged to less than 1 percent of the population at its peak (**Figure 7.5b**). If reindeer overgraze the vegetation, it takes 15 to 20 years to recover. During that period, the carrying capacity for reindeer is greatly reduced.

CONCEPT CHECK **STOP**

1. **What** is population ecology?

2. **How** do each of the following affect population size: birth rate, death rate, immigration, and emigration?

3. **How** do biotic potential and/or carrying capacity produce the J-shaped and S-shaped population growth curves?

Human Population Patterns

LEARNING OBJECTIVES

1. **Summarize** the history of human population growth.

2. **Identify** Thomas Malthus, relate his ideas on human population growth, and explain why he may or may not have been wrong.

3. **Explain** why it is impossible to precisely determine how many people Earth can support—that is, Earth's carrying capacity for humans.

N ow that you have examined some of the basic concepts of population ecology, let's apply those concepts to the human population. **Figure 7.6** shows the increase in human population. Reexamine Figure 7.3 and compare the two curves. The characteristic J curve of exponential population growth shown in Figure 7.6 reflects the decreasing amount of time it has taken to add each

additional billion people to our numbers. It took tens of thousands of years for the human population to reach 1 billion, a milestone that took place around 1800. It took 130 years to reach 2 billion (in 1930), 30 years to reach 3 billion (in 1960), 15 years to reach 4 billion (in 1975), 12 years to reach 5 billion (in 1987), and 12 years to reach 6 billion (in 1999). The United Nations projects that the human population will reach 7 billion by 2012. Population experts predict that the population will level out during the 21st century, possibly forming an S curve as observed in some other species.

One of the first people to recognize that the human population can't increase indefinitely was **Thomas Malthus** (1766–1834), a British economist. He pointed out that human population growth is not always desirable—a view contrary to the beliefs of his day and to those of many people even today. Noting that human population can increase faster than its food supply, he warned that the inevitable consequences of population growth would be famine, disease, and war. Since Malthus's time, the human population has increased from about 1 billion to more than 6 billion. On the surface, it seems that Malthus was wrong. Our population has grown dramatically because geographic expansion and scientific advances have allowed food production to keep pace with population growth. Malthus's ideas may ultimately be proved correct, however, because we don't know whether this increased food production is sustainable. Have we achieved this increase in food production at the environmental cost of reducing the planet's ability to meet the needs of future populations?

Our world population was 6.8 billion in 2009, an increase of about 105 million from 2008. This increase was not due to a rise in the birth rate (b). In fact, the world birth rate has declined slightly during the past 200 years. The population growth is due instead to a dramatic *decrease* in the death rate (d), which has occurred primarily because greater food production, better medical care, and improvements in water quality and sanitation practices have increased life

Human population growth • Figure 7.6

Compare this figure to Figure 7.5b. Do you think what happened to the reindeer on the Pribilof Islands could happen to Earth's human population? Why or why not? (Black Death refers to a devastating disease, probably bubonic plague, that decimated Europe and Asia in the 14th century.)

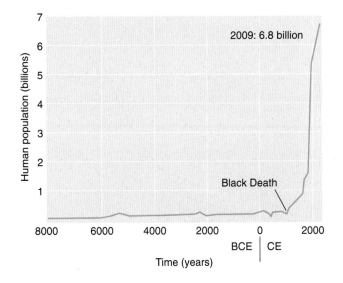

Advances in global health • Figure 7.7

A child in Bangladesh receives a dose of oral polio vaccine. At one time, polio killed or crippled millions of children each year. Polio is still endemic (constantly present) in Nigeria, India, Afghanistan, and Pakistan, and it sometimes spreads from those countries to other countries.

expectancy for a great majority of the global population (**Figure 7.7**).

Projecting Future Population Numbers

The human population has reached a turning point. Although our numbers continue to increase, the world growth rate (*r*) has declined slightly over the past several years, from a peak of 2.2 percent per year in the mid-1960s to the current growth rate of 1.2 percent per year. Population experts at the United Nations and the World Bank project that the growth rate will continue to decrease slowly until zero population growth is attained toward the end of the 21st century. Exponential growth of the human population will end, and the S curve may replace the J curve.

> **zero population growth** The state in which the population remains the same size because the birth rate equals the death rate.

The United Nations periodically publishes population projections for the 21st century. The latest (2004) U.N. figures forecast that the human population will total between 7.7 billion (their "low" projection) and 10.6 billion (their "high" projection) in the year 2050,

with 9.1 billion thought to be most likely (**Figure 7.8**). The estimates vary depending on fertility changes, particularly in less developed countries, because that is where almost all of the growth will take place.

Population projections must be interpreted with care because they vary depending on what assumptions are made. In projecting that the world population will be 7.7 billion (their low projection) in the year 2050, U.N. population experts assume that the average number of children born to each woman in all countries will have declined to 1.5 in the 21st century. The average number of children born to each woman on Earth is currently 2.6. If the decline to 1.5 doesn't occur, our population could be significantly higher. If the average number of children born to each woman declines to only 2.5 instead of 1.5, the 2050 population will be 10.6 billion (the U.N. high

World population projections to 2050 • Figure 7.8

In 2004 the United Nations made three projections, each based on different fertility rates.

Environmental degradation on a cattle ranch in Brazil • Figure 7.9

Part of the rain forest in the background was cleared for a cattle pasture. After a few years, the pasture became unproductive, and erosion degraded the land further. Photographed in Amazonas State in the Amazon River Basin.

projection). Small differences in fertility, then, produce large differences in population forecasts.

The main unknown factor in any population growth scenario is Earth's *carrying capacity*. Most published estimates of how many people Earth can support range from 4 billion to 16 billion. For example, in 2004, environmental economists in the Netherlands performed a detailed analysis of 69 recent studies of Earth's carrying capacity for humans. Based on current technology, they estimated that 7.7 billion is the upper limit of human population that the world can support. The medium and high U.N. projections for 2050 exceed this value (see Figure 7.8).

These estimates vary widely depending on what assumptions are made about standard of living, resource consumption, technological innovations, and waste generation. If we want all people to have a high level of material well-being equivalent to the lifestyles in highly developed countries, then Earth will support far fewer humans than if everyone lives just above the subsistence level. Unlike with other organisms, environmental constraints aren't the exclusive determinant of Earth's carrying capacity for humans. Human choices and values must be factored into the assessment.

What will happen to the human population when it approaches Earth's carrying capacity? Optimists suggest that a decrease in the birth rate will stabilize the human population. Some experts take a more pessimistic view and predict that our ever-expanding numbers will cause widespread environmental degradation and make Earth uninhabitable for humans as well as other species (**Figure 7.9**). These experts contend that a massive wave of human suffering and death will occur. This view doesn't mean we will go extinct as a species, but it projects severe hardship for many people. Some experts think the human population has already exceeded the carrying capacity of the environment, a potentially dangerous situation that threatens our long-term survival as a species.

Human population trends are summarized in **Figure 7.10** on pages 168 to 169.

CONCEPT CHECK

1. **How** would you describe human population growth for the past 200 years?

2. **Who** was Thomas Malthus, and what were his views on human population growth?

3. **When** determining Earth's carrying capacity for humans, why is it not enough to just consider human numbers?

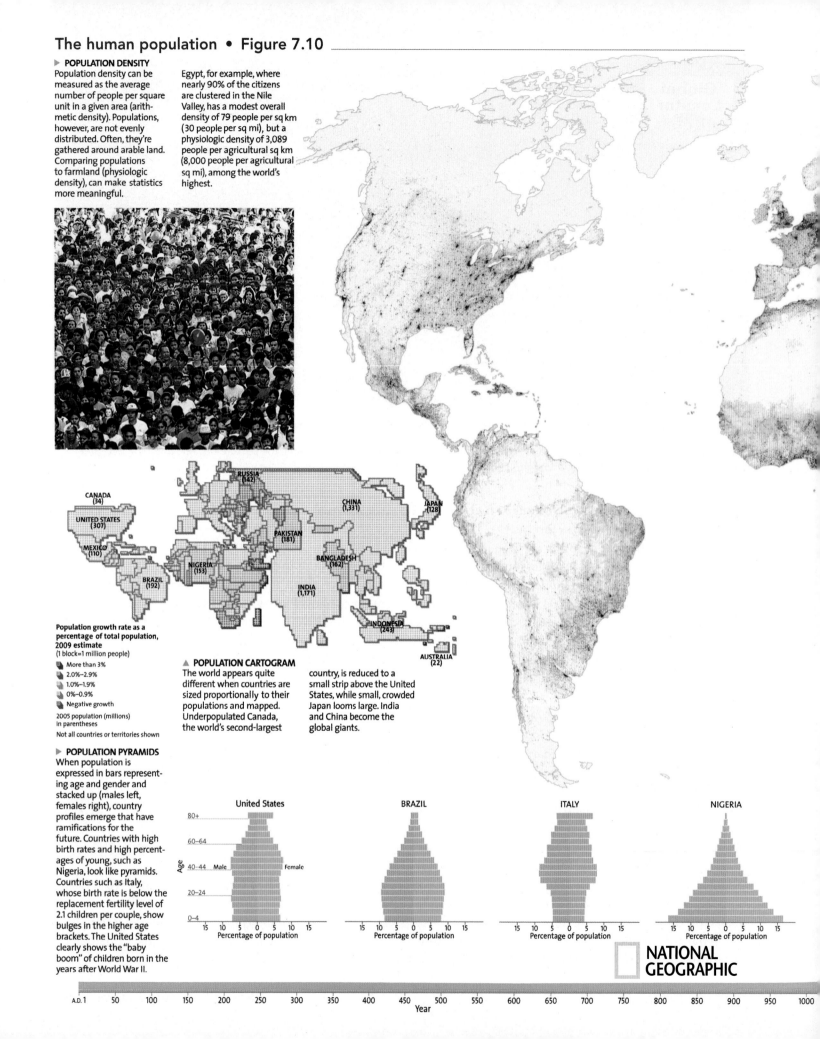

▶ **POPULATION DENSITY**

Population density can be measured as the average number of people per square unit in a given area (arithmetic density). Populations, however, are not evenly distributed. Often, they're gathered around arable land. Comparing populations to farmland (physiologic density), can make statistics more meaningful.

Egypt, for example, where nearly 90% of the citizens are clustered in the Nile Valley, has a modest overall density of 79 people per sq km (30 people per sq mi), but a physiologic density of 3,089 people per agricultural sq km (8,000 people per agricultural sq mi), among the world's highest.

Population growth rate as a percentage of total population, 2009 estimate
(1 block=1 million people)

- More than 3%
- 2.0%–2.9%
- 1.0%–1.9%
- 0%–0.9%
- Negative growth

2005 population (millions) in parentheses

Not all countries or territories shown

▲ **POPULATION CARTOGRAM**

The world appears quite different when countries are sized proportionally to their populations and mapped. Underpopulated Canada, the world's second-largest country, is reduced to a small strip above the United States, while small, crowded Japan looms large. India and China become the global giants.

Cartogram labels:
CANADA (34)
UNITED STATES (307)
MEXICO (110)
BRAZIL (192)
RUSSIA (142)
CHINA (1,331)
JAPAN (128)
PAKISTAN (181)
NIGERIA (153)
BANGLADESH (162)
INDIA (1,171)
INDONESIA (243)
AUSTRALIA (22)

▶ **POPULATION PYRAMIDS**

When population is expressed in bars representing age and gender and stacked up (males left, females right), country profiles emerge that have ramifications for the future. Countries with high birth rates and high percentages of young, such as Nigeria, look like pyramids. Countries such as Italy, whose birth rate is below the replacement fertility level of 2.1 children per couple, show bulges in the higher age brackets. The United States clearly shows the "baby boom" of children born in the years after World War II.

United States
Age: 80+, 60–64, 40–44 Male / Female, 20–24, 0–4
Percentage of population: 15 10 5 0 5 10 15

BRAZIL
Percentage of population: 15 10 5 0 5 10 15

ITALY
Percentage of population: 15 10 5 0 5 10 15

NIGERIA
Percentage of population: 15 10 5 0 5 10 15

NATIONAL GEOGRAPHIC

A.D. 1 | 50 | 100 | 150 | 200 | 250 | 300 | 350 | 400 | 450 | 500 | 550 | 600 | 650 | 700 | 750 | 800 | 850 | 900 | 950 | 1000
Year

Geographers approach the study of human populations, or demography, from a spatial perspective, asking why density, distribution, resources, births, deaths, and migrations vary from place to place. Earth's population, now at 6.8 billion, grows by 105 million per year, or 1.2% annually. The bulk of the increase occurs in developing countries in Asia, Africa, and Latin America. Physiologic density—the number of people per unit of agricultural land—shows concentrations in Asia, in particular in China and India; in Europe, from Britain into Russia; along the eastern seaboard of the United States; and in West Africa in Nigeria and along the Nile Valley.

GeoBytes

COUNTING HEADS
Most governments conduct a population census every 10 years, although the process is expensive and accuracy is difficult to attain for most.

PACKED NEIGHBORHOODS
The most densely populated place in the world is Macau, a Chinese Special Administrative Region, with 21,346 people per sq km (8,239 people per sq mi). The least dense is Greenland.

IOWA VS. BANGLADESH
Bangladesh's population density is up to 1,127 per sq km (435 per sq mi) in an area the size of Iowa. In Iowa, the figure is fewer than 145 per sq km (55 per sq mi).

SHEER NUMBERS
The most populated country is the People's Republic of China, with 1.3 billion. The least populated jurisdiction is Pitcairn Island in the Pacific, with 46 people.

Population density, 2005

People per square km	People per square mi
More than 500	More than 195
150–500	60–195
25–149	10–59
1–24	1–9
Less than 1	Less than 1

Locate your home or school on the world map showing population density. What is the approximate population density per square km in this area? Are there other areas of the world with greater population densities? Where?

GLOBAL LOCAL

INDIA

| 15 | 10 | 5 | 0 | 5 | 10 | 15 |
Percentage of population

LAOS

| 15 | 10 | 5 | 0 | 5 | 10 | 15 |
Percentage of population

AUSTRALIA

| 15 | 10 | 5 | 0 | 5 | 10 | 15 |
Percentage of population

REGIONAL POPULATION GROWTH
Earth's population has burgeoned since 1800, from approximately 1 billion to today's 6.8 billion. Africa is sustaining high fertility rates (average number of children per woman) and is projected to contain 21% of the world's population by 2050.

- Asia
- Africa
- Latin America
- Europe
- North America
- Australia & Oceania

Projected growth

Number of people (in billions)

| 1100 | 1150 | 1200 | 1250 | 1300 | 1350 | 1400 | 1450 | 1500 | 1550 | 1600 | 1650 | 1700 | 1750 | 1800 | 1850 | 1900 | 1950 | 2000 | 2050 |

Year

Demographics of Countries

LEARNING OBJECTIVES

1. **Define** *demographics* and describe the demographic transition.

2. **Explain** how highly developed and developing countries differ in population characteristics such as infant mortality rate, total fertility rate, replacement-level fertility, and age structure.

World population figures illustrate overall trends but don't describe other important aspects of the human population story, such as population differences from country to country (**Table 7.1**). **Demographics** provides information on the populations of various countries. Recall from Chapter 1 that countries are classified into two main groups—highly developed and developing—based on population growth rates, degree of industrialization, and relative prosperity.

demographics The applied branch of sociology that deals with population statistics.

Highly developed countries have the lowest birth rates in the world. Some countries, such as Germany, have birth rates just below those needed to sustain their populations and are declining slightly in numbers. Highly developed countries also have low **infant mortality rates** (**Figure 7.11a**). The infant mortality rate of the United States is 6.5, compared with a world rate of 52. Highly developed countries have longer life expectancies (78 years in the United States versus 68 years worldwide).

infant mortality rate The number of deaths of infants under age 1 per 1000 live births.

Per person GNI PPP is a country's gross national income (GNI) in purchasing power parity (PPP) divided by its population. It indicates the amount of goods and services an average citizen of that particular country could buy in the United States. There is a high average per person GNI PPP in the United States—$46,970—versus the worldwide figure of $10,090.

In *moderately developed countries*, birth rates and infant mortality rates are higher than those of highly developed countries, but they are declining. Moderately developed countries have a medium level of industrialization, and their average per person GNI PPPs are lower than those of highly developed countries. *Less developed countries* have the shortest life expectancies, the lowest average per person GNI PPPs, the highest birth rates, and the highest infant mortality rates in the world (**Figure 7.11b**).

Replacement-level fertility is usually given as 2.1 children. The number is greater than 2.0 because some infants and children die before they reach reproductive age. Worldwide, the **total fertility rate (TFR)** is currently 2.6, well above the replacement level.

replacement-level fertility The number of children a couple must produce to "replace" themselves.

total fertility rate (TFR) The average number of children born to each woman.

The Demographic Transition

Demographers recognize four demographic stages based on their observations of Europe as it became industrialized and urbanized (**Figure 7.11c**). These stages converted Europe from relatively high birth and death rates to relatively low birth and death rates. All highly developed and moderately developed countries with more

The world's 10 most populous countries • Table 7.1

Country	2009 Population (in millions)	Population density (per square kilometer)
China	1331	139
India	1171	356
United States	307	32
Indonesia	243	128
Brazil	192	22
Pakistan	181	227
Bangladesh	162	1127
Nigeria	153	165
Russia	142	8
Japan	128	338

Demographics of countries • **Figure 7.11**

a. Infant Mortality Rates in Highly Developed Countries. Nurses care for newborn infants in Israel, a highly developed country with an infant mortality rate of 3.6.

b. Infant Mortality Rates in Developing Countries. This infant, just born in the less developed country of Bangladesh, is underweight and therefore at risk. The infant mortality rate in Bangladesh is 48.

NATIONAL GEOGRAPHIC

WILEY RESOURCE KIT

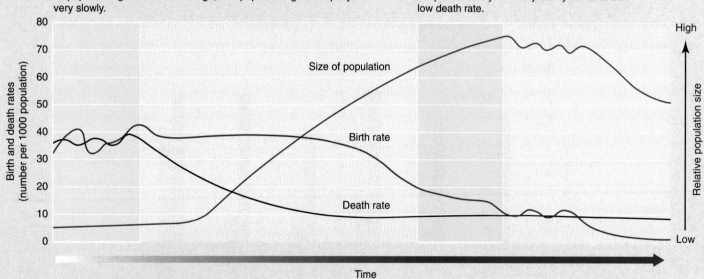

Stage 1 Preindustrial

Women have many children, but infant mortality rate is high, so population grows very slowly.

Stage 2 Transitional

Lowered death rate from improved health care and more reliable food and water supplies. Birth rate is still high, and population grows rapidly.

Stage 3 Industrial

Decline in birth rate slows population growth despite relatively low death rate.

Stage 4 Postindustrial

People tend to desire smaller families and take steps to limit family size. Population grows very slowly or not at all.

Size of population

Birth rate

Death rate

Birth and death rates (number per 1000 population)

Relative population size

Time

High

Low

c. The Demographic Transition. Demographers have identified four stages through which a population progresses as its society becomes industrialized.

Interpreting Data

Is population increasing or decreasing in Stage 3 of the demographic transition? Why?

advanced economies have gone through this **demographic transition**, and demographers assume that the same progression will occur in less developed countries as they industrialize.

Why has the population stabilized in more than 30 highly developed countries in the fourth (postindustrial) demographic stage? The reasons are complex. Declining birth rate is associated with an improvement in living standards. It is difficult to say whether improved socioeconomic conditions have resulted in a decrease in birth rate or whether a decrease in birth rate has resulted in improved socioeconomic conditions. Perhaps both are true. Another reason for the decline in birth rate in highly developed countries is the increased availability of family planning services. Other socioeconomic factors that influence birth rate are increased education, particularly of women, and *urbanization* of society (discussed later in this chapter).

Once a country reaches the fourth demographic stage, is it correct to assume that the country will continue to have a low birth rate indefinitely? We don't know. Low birth rates may be a permanent response to the socioeconomic factors of an industrialized, urbanized society. On the other hand, low birth rates may be a response to socioeconomic factors, such as the changing roles of women in highly developed countries. Unforeseen changes in the socioeconomic status

> **demographic transition** The process whereby a country moves from relatively high birth and death rates to relatively low birth and death rates.

of women and men in the future may again change birth rates. No one knows for sure.

The population in many developing countries is beginning to approach stabilization (**Figure 7.12**). For example, the TFR in Brazil in 1960 was 6.7 children per woman. Today it is 2.0. Worldwide, the TFR in developing countries has decreased from an average of 6.1 children per woman in 1970 to 2.7 today.

Although fertility rates in these countries have declined, many still exceed replacement-level fertility. Consequently, populations in these countries are still increasing. Even when fertility rates equal replacement-level fertility, population growth will still continue for some time. To understand why this is so, let's examine the age structure of various countries.

Age Structure of Countries

A population's **age structure** helps predict future population growth. The number of males and the number of females at each age, from birth to death, are represented in an *age structure diagram*. Each diagram is divided vertically in half, the left side representing the males in a population and the right side the females. The bottom third of each diagram

> **age structure** The number and proportion of people at each age in a population.

Fertility changes in selected developing countries • Figure 7.12

Since the 1960s, fertility levels have dropped dramatically in many developing countries.

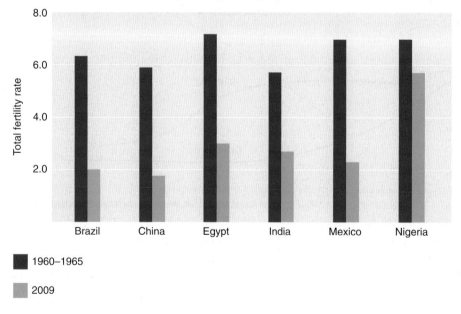

Shown are countries with **a** rapid (Ethiopia), **b** slow (United States), and **c** no growth (Italy) or declining population growth.

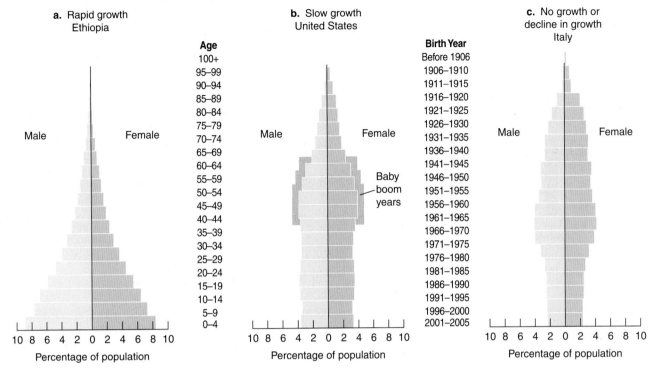

a. Rapid growth
Ethiopia

b. Slow growth
United States

c. No growth or decline in growth
Italy

Age
100+
95–99
90–94
85–89
80–84
75–79
70–74
65–69
60–64
55–59
50–54
45–49
40–44
35–39
30–34
25–29
20–24
15–19
10–14
5–9
0–4

Male   Female

Male   Female   Baby boom years

Male   Female

Birth Year
Before 1906
1906–1910
1911–1915
1916–1920
1921–1925
1926–1930
1931–1935
1936–1940
1941–1945
1946–1950
1951–1955
1956–1960
1961–1965
1966–1970
1971–1975
1976–1980
1981–1985
1986–1990
1991–1995
1996–2000
2001–2005

10 8 6 4 2 0 2 4 6 8 10
Percentage of population

10 8 6 4 2 0 2 4 6 8 10
Percentage of population

10 8 6 4 2 0 2 4 6 8 10
Percentage of population

represents prereproductive humans (between 0 and 14 years of age); the middle third, reproductive humans (15 to 44 years); and the top third, postreproductive humans (45 years and older). The widths of these segments are proportional to the population sizes: A broader width implies a larger population. The overall shape of an age structure diagram indicates whether the population is increasing, stable, or shrinking.

The age structure diagram of a country with a high growth rate, based on a high fertility rate—for example, Ethiopia or Guatemala—is shaped like a pyramid (**Figure 7.13a**). The largest percentage of the population is in the prereproductive age group (0 to 14 years of age), so the probability of future population growth is great. A positive **population growth momentum** exists because when all these children mature, they will become the parents of the next generation, and this group of parents will be larger than the previous group. Even if the fertility rate of such a country has declined to replacement level (that is, if couples are having smaller families than their parents did), the population will continue to grow for some time. Population growth momentum explains how a population's present age distribution affects its future growth.

The more tapered bases of the age structure diagrams of countries with slowly growing, stable, or declining populations indicate that a smaller proportion of the population will become the parents of the next generation (**Figure 7.13b** and **c**). The age structure diagram of a stable population (neither growing nor shrinking) demonstrates that the numbers of people at prereproductive and reproductive ages are approximately the same. A larger percentage of the population is older—that is, postreproductive—than in a rapidly increasing population. Many countries in Europe have stable populations.

In a shrinking population, the prereproductive age group is smaller than either the reproductive or postreproductive age group. Russia, Ukraine, and Bulgaria are examples of countries with slowly shrinking populations.

Worldwide, 27 percent of the human population is under age 15. When these people enter their reproductive years, they have the potential to cause a large increase in the growth rate. Even if the birth rate doesn't increase, the growth rate will increase simply because there are more people reproducing.

Most of the world population increase since 1950 has taken place in developing countries as a result of the younger age structure and the higher-than-replacement-level fertility rates of their populations. In 1950, 67 percent of the world's population was in

Percentages of prereproductive and elderly populations for various regions of the world • Figure 7.14

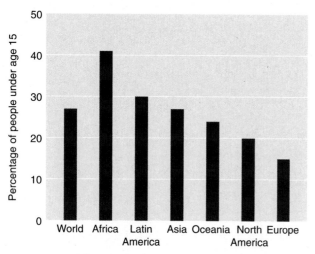

a. Percentages of the population under age 15 in 2009. The higher this percentage, the greater the potential for population growth.

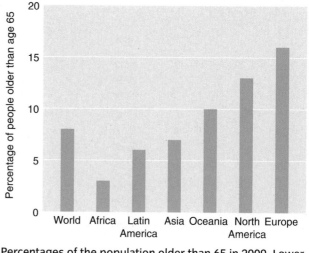

b. Percentages of the population older than 65 in 2009. Lower fertility rates lead to aging populations.

developing countries in Africa, Asia (minus Japan), and Latin America. Between 1950 and 2000, the world's population more than doubled in size, but most growth occurred in developing countries. As a reflection of this occurrence, the current number of people in developing countries has increased to 82 percent of the world population. Most of the population increase during the 21st century will take place in developing countries, largely as a result of their younger age structures (**Figure 7.14a**). These countries will have economic difficulty supporting such growth.

Declining fertility rates have profound social and economic implications because as fertility rates drop, the percentage of the population that is elderly increases (**Figure 7.14b**). An aging population has a higher percentage of people who are chronically ill or disabled, and these people require more health care and other social services. An aging population reduces a country's productive workforce, increases its tax burden, and strains its social security, health, and pension systems. However, in an aging population, there is often a reduction in the rate of violent crimes. Governments with growing elderly populations may offer incentives to the elderly to work longer before retiring.

CONCEPT CHECK

1. **What** is the demographic transition?
2. **What** is infant mortality rate? How does it vary in highly developed and developing countries?

Stabilizing World Population

LEARNING OBJECTIVES

1. **Relate** total fertility rates to each of the following: cultural values, social and economic status of women, the availability of family planning services, and government policies.
2. **Explain** the link between education and total fertility rates.

Dispersal used to be a solution for overpopulation, but not today. As a species, we humans have expanded our range throughout Earth, and few habitable areas remain with the resources to adequately support a major increase in human population. Nor is increasing the death rate an acceptable means of regulating population size. Clearly, reducing the birth rate is the way to control our expanding population. Cultural traditions, women's social and economic status, family planning, and government policies all influence total fertility rate (TFR).

Culture and Fertility

The values and norms of a society—what is considered right and important and what is expected of a person—constitute a part of that society's **culture**. Gender—that is, varying roles men and women are expected to fill—is an important part of **culture**. Different societies have different gender expectations (**Figure 7.15**). With respect to fertility and culture, a couple is expected to have the number of children traditional in their society.

High TFRs are traditional in many cultures. The motivations for having many babies vary from culture to culture, but a major reason for high TFRs is that infant and child mortality rates are high. For a society to endure, it must produce enough children who survive to reproductive age. If infant and child mortality rates are high, TFRs must be high to compensate. Although infant and child mortality rates are decreasing, it will take longer for culturally embedded fertility levels to decline. Parents must have confidence that the children they already have will survive before they stop having additional babies. Another reason for the lag in fertility decline is cultural: Changing anything traditional, including large family size, usually takes a long time.

Higher TFRs in some developing countries are also due to the important economic and societal roles of

Varying roles of men and women • Figure 7.15

a. In parts of Latin America, men do the agricultural work. This Argentinian man is harvesting grapes.

b. In sub-Saharan Africa, women do most of the agricultural work in addition to caring for their children. Photographed in South Africa.

children. In some societies, children usually work in family enterprises such as farming or commerce, contributing to the family's livelihood. The International Labour Organization estimates that, worldwide, about 218 million children under age 15 work full time (household chores are not counted as labor). More than 95 percent of these children live in developing countries (**Figure 7.16**; also see Figure 2.10). More than half of child laborers—some 126 million—do hazardous work such as mining and construction. These child laborers often suffer from chronic health problems caused by the dangerous, unhealthy conditions to which they are exposed. Children who work full time do not have childhoods, nor do they receive education.

When child laborers become adults, they provide support for their aging parents. In contrast, children in highly developed countries have less value as a source of labor because they attend school and because less human labor is required in an industrialized society. Furthermore, highly developed countries provide many social services for the elderly, so the burden of their care doesn't fall entirely on offspring.

Many cultures place a higher value on male children than on female children. In these societies, a woman who bears many sons achieves a high status; thus, the social pressure to have male children keeps the TFR high.

Working child • Figure 7.16

This Indian girl works on glass bangles at a workshop that uses child labor.

Religious values are another aspect of culture that affects TFRs. Several studies done in the United States point to differences in TFRs among Catholics, Protestants, and Jews. In general, Catholic women have a higher TFR than either Protestant or Jewish women, and women who don't follow any religion have the lowest TFRs of all. The observed differences in TFRs may not be the result of religious differences alone. Other variables, such as ethnicity (certain religions are associated with particular ethnic groups) and residence (certain religions are associated with urban or with rural living), complicate any generalizations that might be made.

The Social and Economic Status of Women

Gender inequality exists in most societies; in these societies, women don't have the same rights, opportunities, or privileges as men. Gender disparities include the lower political, social, economic, and health status of women compared to men. For example, more women than men live in poverty, particularly in developing countries. In most countries, women are not guaranteed equality in legal rights, education, employment and earnings, or political participation.

Because sons are more highly valued than daughters, girls are often kept at home to work rather than being sent to school (**Figure 7.17a**). In most developing countries, a higher percentage of women are illiterate than men (**Figure 7.17b**). Fewer women than men attend secondary school (high school). In some African countries only 2 to 5 percent of girls are enrolled in secondary school. Worldwide, some 90 million girls aren't given the opportunity to receive a primary (elementary school) education. Laws, customs, and lack of education often limit women to low-skilled, low-paying jobs. In such societies, marriage is usually the only way for a woman to achieve social influence and economic security.

Evidence suggests that the single most important factor affecting high TFRs may be the low status of women in many societies. An effective population growth reduction

Gender discrimination • Figure 7.17

a. Nigerian students. Note the number of boys versus girls. Why do you think the school has many more boys than girls? Where are all the girls?

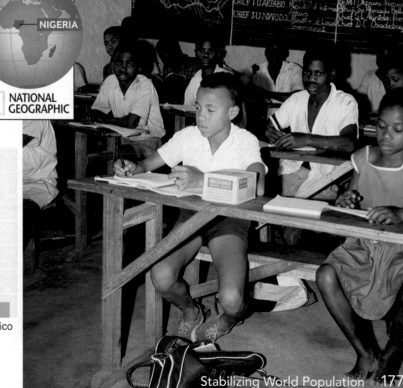

Global Locator

NIGERIA

NATIONAL GEOGRAPHIC

b. Illiteracy percentage of men and women in selected developing countries. A higher percentage of women than men are illiterate.

strategy, then, is to improve the social and economic status of women.

Let's examine how marriage age and educational opportunities, especially for women, affect fertility. The average age at which women marry affects the TFR; in turn, the laws and customs of a given society affect marriage age. Women who marry are more apt to bear children than women who don't marry, and the earlier a woman marries, the more children she is likely to have.

In nearly all societies women with more education tend to marry later and have fewer children. Providing women with educational opportunities delays their first childbirth, thereby reducing the number of childbearing years and increasing the amount of time between generations. Education provides greater career opportunities and may change women's lifetime aspirations. The *What a Scientist Sees* feature on the following page discusses this. In the United States, it isn't uncommon for a woman to give birth to her first child in her thirties or forties, after establishing a career.

Education increases the probability that women will know how to control their fertility. It also provides knowledge to improve the health of the women's families, which results in a decrease in infant and child mortality. A study in Kenya showed that 10.9 percent of children born to women with no education died by age 5, as compared with 7.2 percent of children born to women with a primary education, and 6.4 percent of children born to women with a secondary education. Education also increases women's career options and provides ways of achieving status besides having babies.

Education may also have an indirect effect on TFR. Children who are educated have a greater chance of improving their living standards, partly because they have more employment opportunities. Parents who recognize this may be more willing to invest in the education of a few children than in the birth of many children whom they can't afford to educate. The ability of better-educated people to earn more money may be one reason smaller family size is associated with increased family income.

Family Planning Services

Socioeconomic factors may encourage people to want smaller families, but fertility reduction won't become a reality without the availability of health and family planning services. The governments of most countries recognize the importance of educating people about basic maternal and child health care. Developing countries that have significantly lowered their TFRs credit many of these results to effective family planning programs. Prenatal care and proper birth spacing make women healthier. In turn, healthier women give birth to healthier babies, leading to fewer infant deaths.

Family planning services provide information on reproductive physiology and contraceptives, as well as the actual contraceptive devices, to people who wish to control the number of children they have or to space out their children's births. Family planning programs are most effective when they are designed with sensitivity to local social and cultural beliefs. Family planning services don't try to force people to limit their family sizes; rather, they attempt to convince people that small families (and the contraceptives that promote small families) are acceptable and desirable.

Contraceptive use is strongly linked to lower TFRs. Research has shown that 90 percent of the decrease in fertility in 31 developing countries was a direct result of increased knowledge and availability of contraceptives (**Figure 7.18**). In highly developed countries, where TFRs are at replacement levels or lower, an average of 68 percent of married women of reproductive age use contraceptives. Fertility declines are occurring in

Access to contraceptives • Figure 7.18

A birth control vendor explains condoms to women at the Adjame market in Côte d'Ivoire, a West African country with a TFR of 5.0. Currently, only 7 percent of Ivory Coast women aged 15 to 49 use modern methods of contraception.

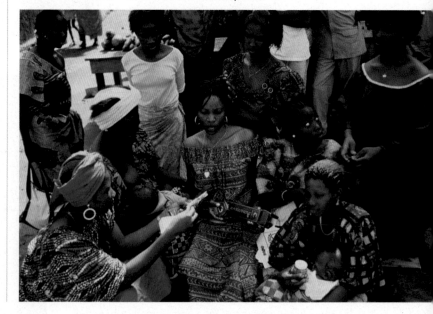

WHAT A SCIENTIST SEES

Education and Fertility

a. Teen mothers gather during lunch at Lincoln High School in Nebraska. A demographer looking at this scene would see the possibility of lower TFRs for the women in this photograph because they are continuing their education.

NATIONAL GEOGRAPHIC

b. Demographers know that the total number of children a woman has during her life (TFR) is affected by the amount of education she has received. The bar graph shows TFRs for 35- to 44-year-old women in the United States by level of education. A similar trend—in which more education leads to lower TFRs—also occurs among women in developing countries.

EnviroDiscovery

Microcredit Programs

Microcredit programs extend small loans ($50 to $500) to very poor people to help them establish businesses that generate income. The poor use these loans for a variety of projects. Some have purchased used sewing machines to make clothing faster than sewing by hand. Others have opened small grocery stores after purchasing used refrigerators to store food so that it does not spoil.

The **Foundation for International Community Assistance (FINCA)** is a not-for-profit agency that administers a global network of microcredit banks. FINCA uses *village banking*, in which a group of very poor neighbors guarantees one another's loans, administers group lending and saving activities, and provides mutual support. These village banks give autonomy to local people.

FINCA primarily targets women because an estimated 70 percent of the world's poorest people are women. FINCA believes that the best way to alleviate the effects of poverty and hunger on children is to provide their mothers with a means of self-employment. A woman's status in the community is raised as she begins earning income from her business (see photo).

Microcredit.
This Bangladeshi woman feeds chickens at her poultry farm. She received her first microcredit loan to buy a few chickens and has built the farm into a thriving business.

---

developing countries where contraceptives are readily available. Since the 1970s, use of contraceptives in East Asia and many areas of Latin America increased significantly, and these regions experienced corresponding declines in birth rates.

Family planning centers provide information and services primarily to women. As a result, in the male-dominated societies of many developing countries, such services may not be as effective as they could otherwise be. Polls of women in developing countries reveal that many who say they don't want additional children still don't practice any form of birth control. When asked why they don't use birth control, these women frequently respond that their husbands or in-laws want additional children.

Government Policies and Fertility

The involvement of governments in childbearing and child rearing is well established. Laws determine the minimum age at which people may marry and the amount of compulsory education they receive. Governments may allot portions of their budgets to family planning services, education, health care, old-age security, or incentives for smaller or larger family size. The tax structure, including additional charges or allowances based on family size, also influences fertility.

In recent years, the governments of at least 78 developing countries—41 in Africa, 19 in Asia, and 18 in Latin America and the Caribbean—have taken measures to limit population growth. Most countries sponsor family planning projects, which are integrated with health care, education, economic development, and efforts to improve women's status.

CONCEPT CHECK

1. **What** is family planning? Is family planning effective in reducing fertility rates?

2. **What** is the relationship between fertility rate and educational opportunities for women?

Population and Urbanization

LEARNING OBJECTIVES

1. **Define** *urbanization* and describe trends in the distribution of people in rural and urban areas.

2. **Describe** some of the problems associated with rapid growth rates in large urban areas.

3. **Explain** how compact development makes a city more livable.

The geographic distribution of people in rural areas, towns, and cities significantly influences the social, environmental, and economic aspects of population growth. During recent history, the human population has become increasingly urbanized. **Urbanization** involves the movement of people from rural to urban areas as well as the transformation of rural areas into urban areas. When Europeans first settled in North America, the majority of the population consisted of farmers in rural areas. Today, approximately 79 percent of the U.S. population lives in cities.

> **urbanization** A process whereby people move from rural areas to densely populated cities.

How many people does it take to make an urban area or city? The answer varies from country to country. According to the U.S. Bureau of the Census, a location with 2500 or more people qualifies as an urban area. One important distinction between rural and urban areas isn't how many people live there but how people make a living. Most people residing in rural areas have occupations that involve harvesting natural resources—such as fishing, logging, and farming. In urban areas, most people have jobs that are not connected directly with natural resources. Cities have traditionally provided more jobs than rural areas because cities are sites of industry, economic development, and educational and cultural opportunities.

Cities are urban ecosystems, and scientists study the effects of humans on the urban environment. This research focuses on the ecological effects of human settlement rather than the interactions among humans themselves. Study of urban ecosystems is complicated because the flow of energy, water, and other resources into and out of the city is linked to the flow of money and the human population (**Figure 7.19**). Often political power is connected to better environmental quality of specific (wealthy) neighborhoods.

Every city is unique in terms of size, climate, culture, and economic development. Although there is no such thing as a typical city, certain traits are common to city populations in general. One basic characteristic of city populations is their far greater heterogenity with respect to race, ethnicity, religion, and socioeconomic status than populations in rural areas. People living in urban areas tend to be younger than those living in the surrounding countryside. The young age structure of cities is due to the influx of many young adults from rural areas.

The city as a dynamic ecosystem • Figure 7.19

The human population in an urban environment requires inputs from the surrounding countryside and produces outputs that flow into surrounding areas. Not shown in this figure is the internal cycling of materials and energy within the urban system.

Natural capital (inputs)

Energy (fuel)
Clean water
Clean air
Food
Raw and refined materials for construction and industry
Business and consumer products

Products and wastes (outputs)

Waste heat, greenhouse gases
Wastewater, water pollution
Air pollution
Solid waste
Goods, services

Brownfield of vacant warehouses and stores • Figure 7.20

Urban and rural areas often have different proportions of males and females. Cities in developing nations tend to have more males. In cities in Africa, for example, males migrate to the city in search of employment, whereas females tend to remain in the country and tend their farms and children. Cities in highly developed countries often have a higher ratio of females to males. Women in rural areas often have little chance of employment after they graduate from high school, so they move to urban areas.

Environmental Problems of Urban Areas

Growing urban areas affect land-use patterns. Suburban sprawl that encroaches into former forest, wetland, desert, or agricultural land destroys or fragments wildlife habitat. Portions of Chicago, Boston, New York City, and New Orleans, for example, are former wetlands. Most cities have blocks and blocks of **brownfields**—areas of abandoned, vacant factories, warehouses, and residential sites that may be contaminated from past uses (**Figure 7.20**). Meanwhile, the suburbs continue to expand outward, swallowing natural areas and farmland.

Reuse of brownfields is complicated because many have environmental contaminants that must be cleaned up before redevelopment can proceed. Nonetheless, brownfields represent an important potential land resource. Pittsburgh is known for its redevelopment of brownfields that were once steel mills and meatpacking centers. Residential and commercial sites now occupy several of these former brownfields (**Figure 7.21**).

Most workers in U.S. cities have to commute dozens of miles through traffic-congested streets from the suburbs where they live to downtown areas where they work. Because development is so spread out in the suburbs, having automobiles is a necessity to accomplish everyday chores. This dependence on motor vehicles as our primary means of transportation increases air pollution and causes other environmental problems.

The high density of automobiles, factories, and commercial enterprises in urban areas causes a buildup of airborne emissions, including particulate matter (dust), sulfur oxides, carbon oxides, nitrogen oxides, and volatile organic compounds. Urban areas in developing nations have the worst air pollution in the world. Although we have made progress in reducing air pollution in highly

Brownfield redevelopment in Pittsburgh • Figure 7.21

Pittsburgh's first brownfield redevelopment project, Washington's Landing at Herr's Island, today features upscale housing, recreation, and commerce.

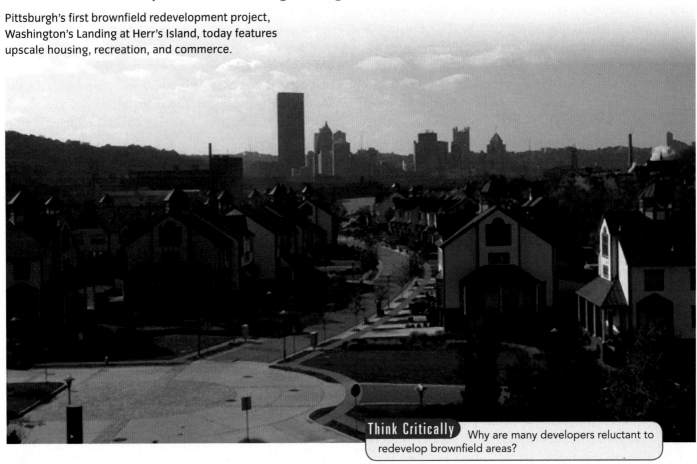

Think Critically Why are many developers reluctant to redevelop brownfield areas?

developed nations, the atmosphere in many cities often contains higher levels of pollutants than are acceptable based on health standards.

Cities affect water flow because they cover the rainfall-absorbing soil with buildings and paved roads. Storm sewage systems are built to handle the runoff from rainfall, which is polluted with organic wastes (garbage, animal droppings, and such), motor oil, lawn fertilizers, and heavy metals. In most cities across the United States, urban runoff is cleaned up in sewage treatment plants before being discharged into nearby waterways. In many cities, however, high levels of precipitation can overwhelm the sewage treatment plants, resulting in the release of untreated urban runoff. When this occurs, the polluted runoff contaminates water far beyond the boundaries of the city.

Environmental Benefits of Urbanization

Although our previous discussion may suggest that the concentration of people in cities has a harmful effect overall on the environment, urbanization does have the potential to provide tangible environmental benefits that in many cases outweigh its environmental problems. A well-planned city actually benefits the environment by reducing pollution and preserving rural areas.

One solution to urban growth is **compact development**, which uses land efficently. Dependence on motor vehicles and their associated pollution are reduced as people walk, cycle, or take public transit such as buses or light rails to work and shopping districts. With compact development, fewer parking lots and highways are needed, so there is more room for parks, open space, housing, and businesses. Compact development makes a city more livable and attractive to people.

Portland, Oregon, provides a good example of compact development. Although Portland is still grappling with many issues, the city government has developed effective land-use policies that

> **compact development**
> Design of cities in which tall, multiple-unit residential buildings are close to shopping and jobs, and all are connected by public transportation.

dictate where and how growth will occur. The city looks inward to brownfields rather than outward to the suburbs for new development sites. From 1975 to 2005, Portland's population grew 50 percent, from 0.9 million to 1.8 million, yet the urbanized area increased about 2 percent. In contrast, from 1975 to 2005, the population of Chicago grew 22 percent, and its urbanized area increased more than 50 percent due to sprawl.

Although the automobile is still the primary means of transportation in Portland, the city's public transportation system is an important part of its regional master plan. Public transportation incorporates light-rail lines, bus routes (many of which feature buses arriving every 15 minutes), bicycle lanes, and walkways. Employers are encouraged to provide bus passes to their employees instead of paying for parking. The emphasis on public transportation has encouraged commercial and residential growth along light-rail lines and bus routes instead of in suburbs.

Urbanization Trends

Urbanization is a worldwide phenomenon. Currently, more than 50 percent of the world population lives in urban areas with populations of 2000 or greater (**Figure 7.22**). The percentage of people living in cities compared with rural settings is greater in highly developed countries than in developing countries. In 2009, urban inhabitants comprised 75 percent of the total population of highly developed countries but only 44 percent of the total population of developing countries.

Although proportionately more people still live in rural settings in developing countries, urbanization is increasing rapidly, by both natural increase (more births than deaths) and immigration from rural areas. Each week the world's cities increase by approximately one million people!

Currently, most urban growth in the world is occurring in developing countries, whereas highly developed countries are experiencing little urban growth. As a result of the greater urban growth of developing nations, most of the world's largest cities are in developing countries. In 1975, 4 of the world's 10 largest cities—Mexico City, São Paulo, Buenos Aires, and Calcutta—were in developing countries. In 2005, 8 of the world's 10 largest cities—Mexico City, São Paulo, Mumbai (Bombay), Delhi, Shanghai, Kolkata (Calcutta), Jakarta, and Buenos Aires—were in developing countries. By 2015, 8 of the world's 10 largest cities will still be in developing countries, although Dhaka, Bangladesh, will have made the list, replacing Buenos Aires (**Figure 7.23**).

According to the United Nations, almost 400 cities worldwide have a population of at least 1 million inhabitants, and 284 of these cities are in developing countries. The number and size of **megacities** (cities with more than 10 million inhabitants) has also increased. In some places, separate urban areas have merged into **urban agglomerations**, urbanized core regions, each of which consists of several adjacent cities or megacities and their surrounding developed suburbs. An example is the Tokyo-Yokohama-Osaka-Kobe agglomeration in Japan, which is home to about 50 million people.

However, according to the U.N. Population Division, most of the world's urban population still lives in small or medium-sized cities with populations of less than 1 million.

It is useful to compare urbanization in highly developed countries with that in developing countries. Consider the United States as representative of highly developed nations. Here, most of the migration to cities occurred during the past 150 years, when an increased need for industrial

The worldwide shift from rural to urban areas, 1950–2030 • Figure 7.22

In 2008, a significant milestone was reached as over 50 percent of the world's population now lives in urban areas.

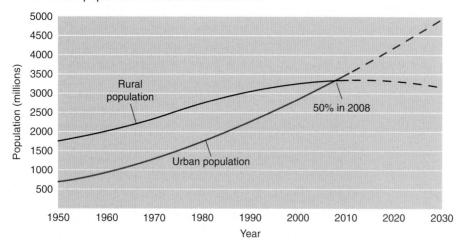

The world's 10 largest cities in 2015 • Figure 7.23

In 2015, 8 of the 10 largest cities will be in developing countries: Mumbai (Bombay), Mexico City, São Paulo, Delhi, Shanghai, Kolkata (Calcutta), Dhaka, and Jakarta.

2015 projections

labor coincided with a decreased need for agricultural labor. The growth of U.S. cities over such a long period was typically slow enough to allow important city services such as water purification, sewage treatment, education, and adequate housing to keep pace with the influx of people from rural areas.

The challenge of meeting a fast-growing city's water needs • Figure 7.24

A woman in Lagos, Nigeria, fills water from a communal tank as another woman walks by with fruit balanced on her head. Lack of access to safe water and basic sanitation services is a problem for many urban residents, particularly the poor, in less developed countries.

In contrast, the fast pace of urban growth in developing nations has outstripped the limited capacity of many cities to provide basic services. It has overwhelmed their economic growth (although cities still offer more job possibilities than rural areas). Consequently, cities in developing nations are generally faced with more serious challenges than cities in highly developed countries. These challenges include substandard housing (slums and squatter settlements); poverty; exceptionally high unemployment; heavy pollution; and inadequate or nonexistent water, sewage, and waste disposal (**Figure 7.24**). Rapid urban growth also strains school, medical, and transportation systems.

CONCEPT CHECK   STOP

1. **Which** countries are the most urbanized? the least urbanized? Which countries have the highest rates of urbanization today?

2. **What** are some of the problems caused by rapid urban growth in developing countries?

3. **How** does compact development affect city living?

Urban Planning in Curitiba, Brazil

Livable cities aren't restricted to highly developed countries. Curitiba, a Brazilian city of more than 2.9 million people, provides a good example of *compact development* in a moderately developed country. Curitiba's city officials and planners have had notable successes in public transportation, traffic management, land-use planning, waste reduction and recycling, and community livability.

The city developed an inexpensive, efficient mass transit system that uses clean, modern buses that run in high-speed bus lanes. High-density development was largely restricted to areas along the bus lines, encouraging population growth where public transportation was already available. About 72 percent of commuters use mass transportation.

Since the 1970s, Curitiba's population has more than doubled, yet traffic has declined by 30 percent. Curitiba doesn't rely on automobiles as much as comparably sized cities do, so it has less traffic congestion and significantly cleaner air, both of which are major goals of compact development. Instead of streets crowded with vehicular traffic, the center of Curitiba is a *calcadao*, or "big sidewalk," that consists of 49 downtown blocks of pedestrian walkways connected to bus stations, parks, and bicycle paths.

Over several decades, Curitiba purchased and converted flood-prone properties along rivers in the city to a series of interconnected parks crisscrossed with bicycle paths. This move reduced flood damage and increased the per person amount of "green space" from 0.5 m² (5.4 ft²) in 1950 to 50 m² (540 ft²) today, a significant accomplishment considering Curitiba's rapid population growth during the same period.

Another example of Curitiba's creativity is its labor-intensive Garbage Purchase program, in which poor people exchange filled garbage bags for bus tokens, surplus food (eggs, butter, rice, and beans), or school notebooks. This program encourages garbage pickup from the unplanned shantytowns (which garbage trucks can't access) that surround the city. Curitiba supplies more services to these unplanned settlements than most cities do. It tries to provide water, sewer, and bus service for them.

These changes didn't happen overnight. Urban planners can carefully reshape most cities over several decades to make better use of space and to reduce dependence on motor vehicles. City planners and local and regional governments are increasingly adopting measures to provide the benefits of compact development in the future.

Curitiba, Brazil

- ▬ Express routes
- ▬ Interdistrict routes
- ▬ Direct routes
- ▬ Feeder bus routes
- ▬ Workers' routes

City center

a. Curitiba's bus network, arranged like the spokes of a wheel, has concentrated development along the bus lines, saving much of the surrounding countryside from development.

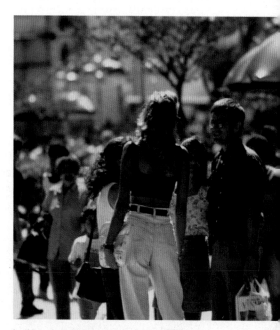

b. The downtown area of Curitiba has open terraces lined with shops and restaurants.

Summary

1 Population Ecology 160

1. **Population ecology** is the branch of biology that deals with the number of individuals of a particular species found in an area and how and why those numbers change over time.

2. The **growth rate (r)** is the rate of change (increase or decrease) of a population's size, expressed in percentage per year. On a global scale, growth rate is due to the **birth rate (b)** and the **death rate (d)**: $r = b - d$. **Emigration (e)**, the number of individuals leaving an area, and **immigration (i)**, the number of individuals entering an area, also affect a local population's growth rate.

3. **Biotic potential** is the maximum rate a population could increase under ideal conditions. **Exponential population growth** is the accelerating population growth that occurs when optimal conditions allow a constant reproductive rate for limited periods. Eventually, the growth rate decreases to around zero or becomes negative because of **environmental resistance**, unfavorable environmental conditions that prevent organisms from reproducing indefinitely at their biotic potential. The **carrying capacity (K)** is the largest population a particular environment can support sustainably (long term) if there are no changes in that environment.

2 Human Population Patterns 165

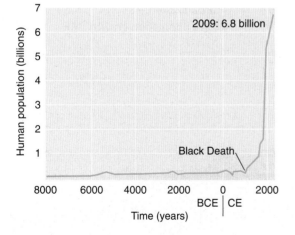

1. It took thousands of years for the human population to reach 1 billion (around 1800). Since then, the population has grown exponentially. The United Nations projects that the population will reach 7 billion by 2012. Although our numbers continue to increase, the growth rate (r) has declined slightly over the past several years. The population should reach **zero population growth**, in which it remains the same size because the birth rate equals the death rate, toward the end of the 21st century.

2. **Thomas Malthus** was a British economist who said that the human population increases faster than its food supply, resulting in famine, disease, and war. Malthus' ideas appear to be erroneous because the human population has grown from about 1 billion in his time to more than 6 billion today, and food production has generally kept pace with population. But Malthus may ultimately be proved correct because we don't know whether our increase in food production is sustainable.

3. Estimates of Earth's carrying capacity for humans vary widely depending on what assumptions are made about standard of living, resource consumption, technological innovations, and waste generation. In addition to natural environmental constraints, human choices and values determine Earth's carrying capacity for humans.

3 Demographics of Countries 170

1. **Demographics** is the applied branch of sociology that deals with population statistics. As a country becomes industrialized, it goes through a demographic transition as it moves from relatively high birth and death rates to relatively low birth and death rates.

2. The **infant mortality rate** is the number of deaths of infants under age 1 per 1000 live births. The **total fertility rate (TFR)** is the average number of children born to each woman.

Replacement-level fertility is the number of children a couple must produce to "replace" themselves. **Age structure** is the number and proportion of people at each age in a population. A country can have replacement-level fertility and still experience population growth if the largest percentage of the population is in the prereproductive years. In contrast to developing countries, highly developed countries have low infant mortality rates, low total fertility rates, and an age structure in which the largest percentage of the population isn't in the prereproductive years.

4 Stabilizing World Population 174

1. Four factors are most responsible for high total fertility rates: high infant and child mortality rates, the important economic and societal roles of children in some cultures, the low status of women in many societies, and a lack of health and family planning services. The single most important factor affecting high TFRs is the low status of women. The governments of many developing countries are trying to limit population growth.

2. Education of women decreases the total fertility rate, in part by delaying the first childbirth. Education increases the likelihood that women will know how to control their fertility. Education also increases women's career options, which provide ways of achieving status besides having babies.

5 Population and Urbanization 181

1. **Urbanization** is the process whereby people move from rural areas to densely populated cities. In developing nations, most people live in rural settings, but their rates of urbanization are rapidly increasing.

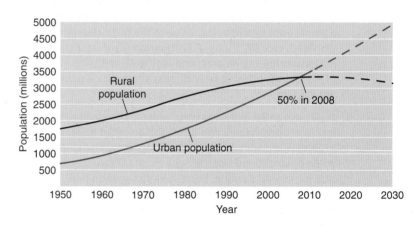

2. Rapid urbanization makes it difficult to provide city dwellers with basic services such as housing, water, sewage, and transportation systems.

3. **Compact development** is the design of cities so that tall, multiple-unit residential buildings are close to shopping and jobs, and all are connected by public transportation.

Key Terms

What is happening in this picture?

- The photo shows people—mainly displaced rural workers—picking through trash at the Smoky Mountain Dump in Manila, Philippines. They are looking mainly for scraps of plastic and metal that they can sell. What valuable environmental service does such scavenging provide?

- Do you think we in the United States will ever mine our landfills? Why or why not?

Critical and Creative Thinking Questions

1. How does the study of population ecology help us understand why some populations grow, some remain stable, and others decline?

2. The growth rates of various populations are usually expressed in percentages. Why are percentages advantageous in comparing growth rates?

3. The human population has grown as we have increased our global carrying capacity. In your opinion, can the global carrying capacity continue to increase? Explain your answer.

4. Why has human population growth, which increased exponentially for centuries, started to decline in the past few decades?

5. Malthus originally suggested that the population of England would collapse because it could not continue to increase its production of food. Why did this not happen?

6. What is carrying capacity? Do you think carrying capacity applies only to other organisms and not to people? Why?

7. What can the governments of developing countries do to help their countries experience the demographic transition?

8. If you were to draw an age structure diagram for Poland, with a total fertility rate of 1.3, which of the following overall shapes would the diagram have? Explain why a country like Poland faces a population decline even if its fertility rate were to start increasing today.

9. Explain the rationale behind this statement: It is better for highly developed countries to spend millions of dollars on family planning in developing countries now than to have to spend billions of dollars on relief efforts later.

10. Which factor do you think would have a larger effect on total fertility rate: the increased education of men or of women? Explain your answer.

11. Urbanization varies from one country to another (see figure). Which countries are mostly urban? Which countries are mostly rural?

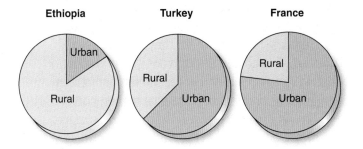

12. What are two serious problems associated with the rapid growth of large urban areas? Explain why they are serious.

13. In compact cities, motor vehicle use is reduced. What are some alternatives to motor vehicles?

14. Should the rapid increase in world population be of concern to the average citizen in the United States? Why or why not?

8 Air and Air Pollution

LONG-DISTANCE TRANSPORT OF AIR POLLUTION

P ersistent toxic compounds are found in the Yukon (in northwestern Canada) and in other pristine arctic regions, far from where they were originally produced. This occurs through the *global distillation effect*, in which chemicals enter the atmosphere in warm regions and move to areas at higher, cooler latitudes, where they are deposited on the surface. Chemicals concentrate in the body fat of animals at the top of food chains, including humans (see Chapter 4). When an Inuit woman consumes a single bite of raw whale skin, she ingests more toxic PCBs than scientists think should be consumed in a week (*see* photograph). Five times as much PCB is found in the breast milk of Inuit women than in the milk of women who live in southern Canada.

Atmospheric conditions also cause pollutants from Asia to move east across the Pacific Ocean. In 1998 a Chinese dust storm produced a visible cloud of particulate matter that transferred toxic metals from ore smelters in Manchuria to the United States.

Around the world, the air we breathe can be contaminated with a variety of pollutants. Because air pollution causes many health and environmental problems, most highly developed nations and many developing nations have policies and regulations limiting emissions from transportation, industry, and even households.

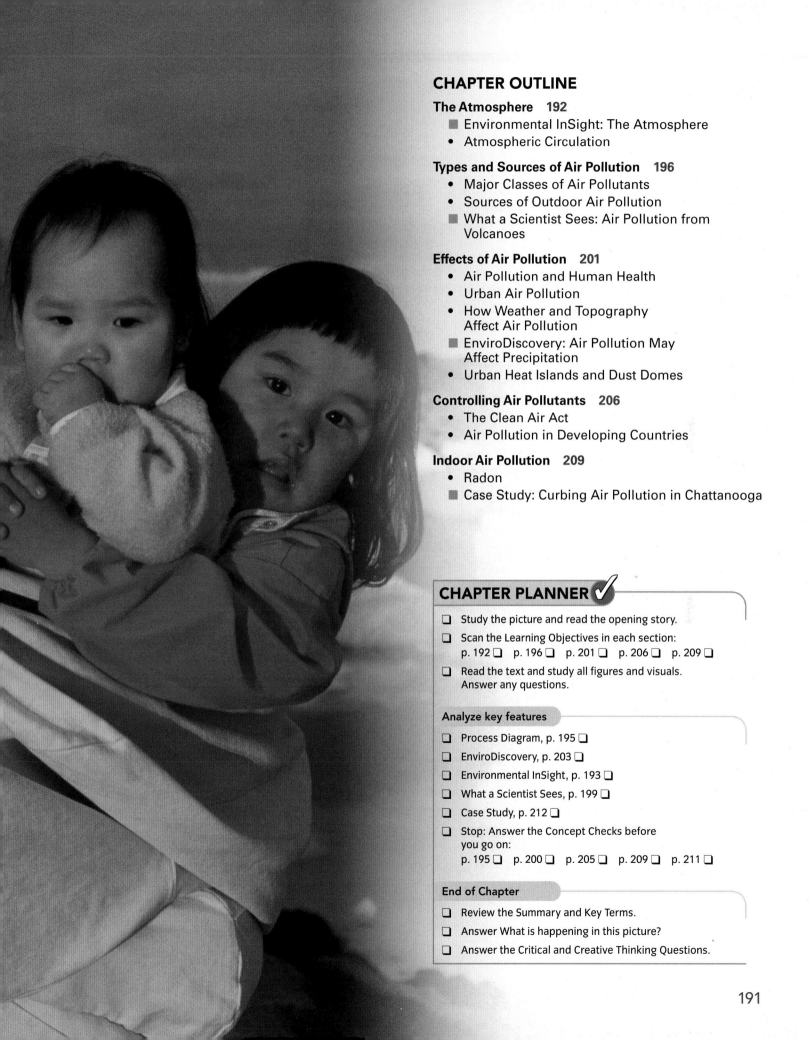

CHAPTER OUTLINE

CHAPTER PLANNER ✓

- ❑ Study the picture and read the opening story.
- ❑ Scan the Learning Objectives in each section:
  p. 192 ❑   p. 196 ❑   p. 201 ❑   p. 206 ❑   p. 209 ❑
- ❑ Read the text and study all figures and visuals. Answer any questions.

Analyze key features

- ❑ Process Diagram, p. 195 ❑
- ❑ EnviroDiscovery, p. 203 ❑
- ❑ Environmental InSight, p. 193 ❑
- ❑ What a Scientist Sees, p. 199 ❑
- ❑ Case Study, p. 212 ❑
- ❑ Stop: Answer the Concept Checks before you go on:
  p. 195 ❑   p. 200 ❑   p. 205 ❑   p. 209 ❑   p. 211 ❑

End of Chapter

- ❑ Review the Summary and Key Terms.
- ❑ Answer What is happening in this picture?
- ❑ Answer the Critical and Creative Thinking Questions.

The Atmosphere

LEARNING OBJECTIVES

1. **Define** *atmosphere* and list the major gases comprising the atmosphere.
2. **Briefly describe** the four major concentric layers of the atmosphere.
3. **Define** *Coriolis effect.*

Oxygen and nitrogen are the predominant gases in the **atmosphere**, accounting for about 99 percent of dry air (**Figure 8.1**). Other gases make up the remaining 1 percent. In addition, water vapor and trace amounts of air pollutants are present in the air. The atmosphere becomes less dense as it extends outward into space.

> **atmosphere** The gaseous envelope surrounding Earth.

Ulf Merbold, a German space shuttle astronaut, felt differently about the atmosphere after viewing it in space (**Figure 8.2**): "For the first time in my life, I saw the horizon as a curved line. It was accentuated by a thin seam of dark blue light—our atmosphere. Obviously, this wasn't the 'ocean' of air I had been told it was so many times in my life. I was

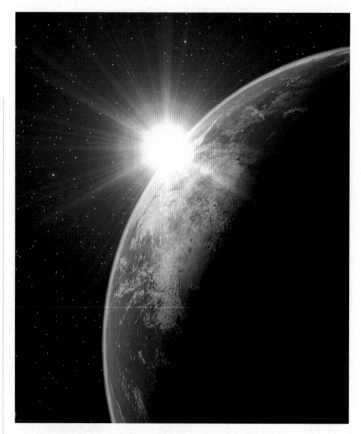

The atmosphere • Figure 8.2 _____

The "ocean of air" is a thin blue layer that separates the planet from the blackness of space.

terrified by its fragile appearance." The atmosphere is composed of four major concentric layers—the troposphere, stratosphere, mesosphere, and thermosphere (**Figure 8.3**). These layers vary in altitude and temperature, depending on the latitude and season.

The atmosphere performs several valuable **ecosystem services**. First, it protects Earth's surface from most of the sun's ultraviolet (UV) radiation and x-rays, and from lethal exposures to cosmic rays from space. Life as we know it would cease to exist without this shielding. Second, atmospheric greenhouse gases absorb some of the heat reradiated from Earth's surface, which keeps the lower atmosphere within the range of temperatures that support life.

Organisms depend on the atmosphere for existence, but they also maintain and, in certain instances, modify its composition. Atmospheric oxygen is thought to have increased to its present level as a result of billions of years of photosynthesis. A balance between oxygen-producing photosynthesis and oxygen-using cellular respiration maintains the current level of oxygen.

Composition of the atmosphere • Figure 8.1 _____

Nitrogen and oxygen form most of the atmosphere. Air also contains water vapor and various pollutants (methane, ozone, dust particles, microorganisms, and chlorofluorocarbons [CFCs]).

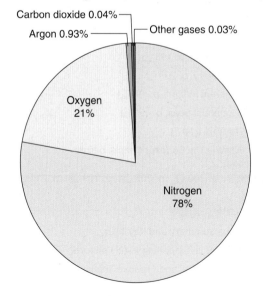

Carbon dioxide 0.04%

Argon 0.93%

Other gases 0.03%

Oxygen 21%

Nitrogen 78%

b. A Thunderstorm in New Mexico. During a lightning flash, a negative charge moves from the bottom of the cloud to the ground, followed by an upward-moving charge along the same channel. The expansion of air around the lightning strike produces sound waves, or thunder.

c. An Aurora in the Northern Hemisphere. Electrically charged particles from the sun collide with the gas molecules in the thermosphere, releasing energy visible as light of different colors.

a. Layers of Atmosphere.

Thermosphere
Extends to 480 km (300 mi)
Gases in extremely thin air absorb x-rays and short-wave radiation, raising the temperature to 1000°C (1800°F) or more. The thermosphere is important in long-distance communication because it reflects outgoing radio waves back to Earth without the use of satellites. Auroras occur here.

Mesosphere
Extends to 80 km (50 mi)
Directly above the stratosphere, temperatures drop to the lowest in the atmosphere—as low as −138°C (−216°F). Meteors often burn up from friction with air molecules in the mesosphere.

Stratosphere
Extends to 50 km (30 mi)
Steady wind occurs but no turbulence; commercial jets fly here. Contains a layer of ozone that absorbs much of the Sun's damaging ultraviolet (UV) radiation. Temperature increases with increasing altitude because absorption of UV radiation by ozone layer heats the air.

Troposphere
Average thickness: 12 km (7.5 mi)
16 km (10 mi) thick at equator
8 km (5 mi) thick at poles
Layer of atmosphere closest to Earth's surface. Temperature decreases with increasing altitude. Weather, including turbulent wind, storms, and most clouds, occurs here.

Ozone layer

NATIONAL GEOGRAPHIC   WILEY RESOURCE KIT

Atmospheric Circulation

Variations in the amount of solar energy that reaches different areas on Earth cause differences in temperature, which then drive the circulation of the atmosphere. The very warm regions near the equator heat the air, which expands and rises (**Figure 8.4**). As this warm air rises, it cools, and then it sinks again. Much of it recirculates almost immediately to the same areas it has left. The remainder of the heated air splits and flows toward the poles. The air chills enough to sink to the surface at about 30 degrees north and south latitudes. This descending air splits and flows over the surface.

Similar upward movements of warm air and its subsequent flow toward the poles also occur at higher latitudes farther from the equator. At the poles, the air cools, sinks, and flows back toward the equator, generally beneath the currents of warm air that simultaneously flow toward the poles. These constantly moving currents transfer heat from the equator toward the poles and cool the land over which they pass on their return. This continuous circulation moderates temperatures over Earth's surface.

In addition to these global circulation patterns, the atmosphere features smaller-scale horizontal movements, or **winds**. The motion of wind, with its eddies, lulls, and turbulent gusts, is difficult to predict. It results partly from fluctuations in atmospheric pressure and partly from the planet's rotation.

The gases that constitute the atmosphere have weight and exert a pressure—about 1013 millibars (14.7 lb per in^2) at sea level. Air pressure is variable, depending on altitude, temperature, and humidity. Winds tend to blow from areas of high atmospheric pressure to areas of low pressure, and the greater the difference between the high- and low-pressure areas, the stronger the wind.

Atmospheric circulation and heat exchange • Figure 8.4

a. In atmospheric convection, heating of the ground surface heats the air, producing an updraft of less dense, warm air. The convection process ultimately causes air currents that mix warmer and cooler parts of the atmosphere.

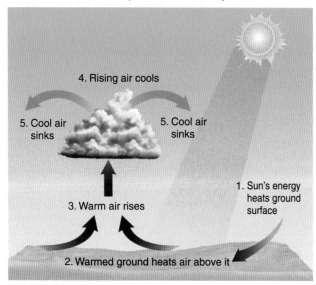

4. Rising air cools

5. Cool air sinks

5. Cool air sinks

1. Sun's energy heats ground surface

3. Warm air rises

2. Warmed ground heats air above it

Polar easterlies

60°N

Westerlies

30°N

Trade winds

0°

Trade winds

30°S

Westerlies

60°S

Polar easterlies

b. Atmospheric circulation transports heat from the equator to the poles (left side of figure). The greatest solar energy input occurs at the equator, heating air most strongly in that area. The air rises, travels toward the poles, and cools in the process so that much of it descends again at around 30 degrees latitude in both hemispheres. At higher latitudes, the patterns of air circulation are more complex.

What direction do winds usually come from where you live? Is this consistent with what you would expect based on your latitude?

The Coriolis effect • Figure 8.5

 THE PLANNER

Viewed from the North Pole, the Coriolis effect appears to deflect ocean currents and winds to the right. From the South Pole, the deflection appears to be to the left.

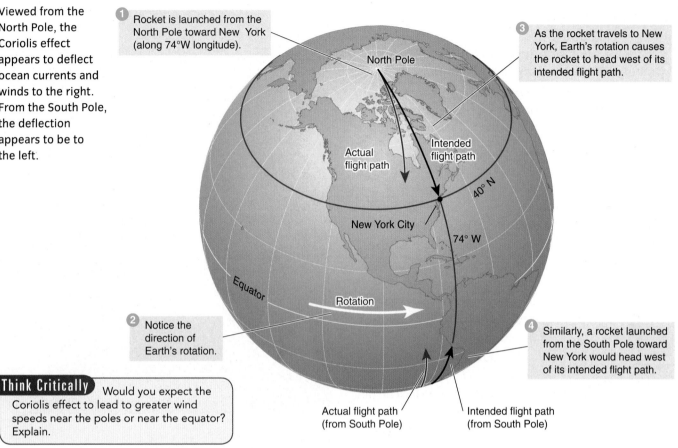

1 Rocket is launched from the North Pole toward New York (along 74°W longitude).

3 As the rocket travels to New York, Earth's rotation causes the rocket to head west of its intended flight path.

North Pole

Intended flight path

Actual flight path

40° N

New York City

74° W

Equator

Rotation

2 Notice the direction of Earth's rotation.

4 Similarly, a rocket launched from the South Pole toward New York would head west of its intended flight path.

Actual flight path (from South Pole)

Intended flight path (from South Pole)

Think Critically Would you expect the Coriolis effect to lead to greater wind speeds near the poles or near the equator? Explain.

As a result of the Coriolis effect, Earth's rotation from west to east also influences the direction of wind.

Coriolis effect
The tendency of moving air or water to be deflected from its path and swerve to the right in the Northern Hemisphere and to the left in the Southern Hemisphere.

To visualize the Coriolis effect, imagine that a rocket is launched from the North Pole toward New York (**Figure 8.5**).

The atmosphere has three **prevailing winds**—major surface winds that blow more or less continually (see Figure 8.4). Prevailing winds from the northeast near the North Pole, or from the southeast near the South Pole, are called *polar easterlies*. Winds that blow in the middle latitudes from the southwest in the Northern Hemisphere or from the northwest in the Southern Hemisphere are called *westerlies*. Tropical winds from the northeast in the Northern Hemisphere or from the southeast in the Southern Hemisphere are called *trade winds*.

CONCEPT CHECK

1. **What** gases make up the atmosphere?
2. **What** two layers of the atmosphere are closest to Earth's surface? How do they differ from one another?
3. **What** is the Coriolis effect, and how does it influence atmospheric circulation?

Types and Sources of Air Pollution

LEARNING OBJECTIVES

1. **Define** *air pollution* and distinguish between primary and secondary air pollutants.

2. **List** the seven major classes of air pollutants and describe their characteristics and sources.

Air pollution can come from natural sources, such as smoke from a forest fire ignited by lightning or gases from an erupting volcano. However, human activities release many kinds of substances into the atmosphere and contribute greatly to global air pollution. Some of these substances are harmful when they are inhaled or settle on land and surface waters, and some substances are harmful because they alter the chemistry of the atmosphere.

Although many different air pollutants exist, we focus on the seven most important classes from a regulatory perspective: particulate matter, nitrogen oxides, sulfur oxides, carbon oxides, hydrocarbons, ozone, and air toxics.

Air pollutants are often divided into two categories, primary and secondary (**Figure 8.6**). The major **primary air pollutants** are carbon oxides, nitrogen oxides, sulfur dioxide, particulate matter, and hydrocarbons.

Ozone, sulfur trioxide, and several acids are called **secondary air pollutants** because they are formed from chemical reactions that take place in the atmosphere.

> **air pollution**
> Various chemicals (gases, liquids, or solids) present in the atmosphere in high enough levels to harm humans, other organisms, or materials.

> **primary air pollutants** Harmful chemicals that enter directly into the atmosphere due to either human activities or natural processes.

> **secondary air pollutants** Harmful chemicals that form in the atmosphere when primary air pollutants react chemically with one another or with natural components of the atmosphere.

Major Classes of Air Pollutants

Particulate matter consists of dusts and mists—thousands of different solid and liquid particles suspended in the atmosphere. Particulate matter includes soil particles, soot, lead, asbestos, sea salt, and sulfuric

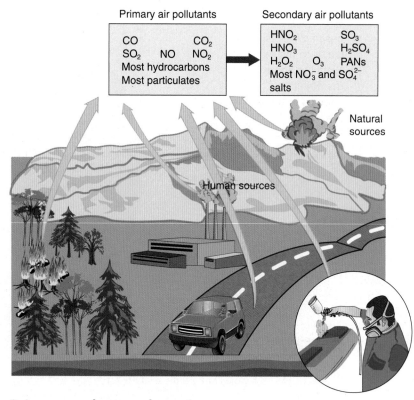

Primary and secondary air pollutants • Figure 8.6

Primary air pollutants are emitted, unchanged, from a source directly into the atmosphere, whereas secondary air pollutants are produced from chemical reactions involving primary air pollutants.

acid droplets. Some particulate matter has toxic or carcinogenic effects.

Particulate matter scatters and absorbs sunlight, reducing visibility. Urban areas receive less sunlight than rural areas, partly as a result of greater quantities of particulate matter in the air. Particulate matter can corrode metals, erode buildings and sculptures when the air is humid, and soil clothing and draperies.

All particulate matter eventually settles out of the atmosphere, but microscopic particles can remain suspended in the atmosphere for weeks or even years. Trace amounts of hundreds of different chemicals bind to these microscopic particles; inhaling the particles introduces the chemicals, some of which are toxic, into the human body. Microscopic particles are considered more dangerous than larger particles because they are inhaled more deeply into the lungs.

The Environmental Protection Agency (EPA) samples microscopic particulate matter at 1,000 locations around the United States because its composition varies with location and season. Particulate matter includes:

Pollutant	Category	Characteristics
Dust particles	Primary	Solid particles
Lead (Pb)	Primary	Solid particles
Sulfuric acid (H_2SO_4)	Secondary	Liquid droplets

Nitrogen oxides are gases produced by chemical interactions between nitrogen and oxygen when a source of energy, such as fuel combustion, produces high temperatures. Collectively known as NO_x, nitrogen oxides consist mainly of nitric oxide (NO), nitrogen dioxide (NO_2), and nitrous oxide (N_2O). Nitrogen oxides inhibit plant growth and, when inhaled, aggravate health problems such as asthma. They are involved in the production of *photochemical smog* (discussed later in the chapter) and acid deposition (see Chapter 9). Nitrous oxide is associated with global warming, and it depletes ozone in the stratosphere (again, see Chapter 9). Nitrogen oxides cause metals to corrode and textiles to fade and deteriorate. Nitrogen oxides include:

Pollutant	Category	Characteristics
Nitrogen dioxide (NO_2)	Primary	Reddish-brown gas
Nitric oxide (NO)	Primary	Colorless gas

Sulfur oxide gases result from chemical interactions between sulfur and oxygen. Sulfur dioxide (SO_2), a colorless, nonflammable gas with a strong, irritating odor, is emitted as a primary air pollutant. Sulfur trioxide (SO_3) is a secondary air pollutant that forms when sulfur dioxide reacts with oxygen in the air. Sulfur trioxide, in turn, reacts with water to form another secondary air pollutant, sulfuric acid. Sulfur oxides play a major role in acid deposition, and they corrode metals and damage stone and other materials. Sulfuric acid and other sulfur oxides damage plants and irritate the respiratory tracts of humans and other animals. Sulfur oxides include:

Pollutant	Category	Characteristics
Sulfur dioxide (SO_2)	Primary	Colorless gas with strong odor
Sulfur trioxide (SO_3)	Secondary	Reactive colorless gas

Carbon oxides are the gases carbon monoxide (CO) and carbon dioxide (CO_2). Carbon monoxide is a colorless, odorless, and tasteless gas produced in larger quantities than any other atmospheric pollutant except carbon dioxide. Carbon monoxide is poisonous and reduces the blood's ability to transport oxygen. Carbon dioxide, also colorless, odorless, and tasteless, is associated with global warming. Carbon oxides are:

Pollutant	Category	Characteristics
Carbon monoxide (CO)	Primary	Colorless, odorless gas
Carbon dioxide (CO_2)	Primary	Colorless, odorless gas

Hydrocarbons are a diverse group of organic compounds that contain only the elements hydrogen and carbon. Small hydrocarbon molecules, such as methane (CH_4), are gaseous at room temperature. Methane is colorless and odorless and is the principal component of natural gas. (The odor of natural gas comes from sulfur compounds deliberately added so that humans can detect the gas's presence.) Medium-sized hydrocarbons, such as benzene (C_6H_6), are liquids at room temperature, although many are volatile and may evaporate easily. The largest hydrocarbons, such as the waxy fuel paraffin, are solids at room temperature. The many different hydrocarbons have a variety of effects on human and animal health. Some cause no adverse effects, some injure the respiratory tract, and others cause cancer. All except methane contribute to the production of photochemical smog.

Methane is linked to global warming. Hydrocarbons include:

Pollutant	Category	Characteristics
Methane (CH_4)	Primary	Colorless, odorless gas
Benzene (C_6H_6)	Primary	Liquid with sweet smell

Ozone (O_3) is a form of oxygen considered a pollutant in one part of the atmosphere but an essential component of another. In the stratosphere, oxygen reacts with solar UV radiation to form ozone. Stratospheric ozone protects Earth's surface from receiving harmful levels of solar UV radiation. Unfortunately, certain human-made pollutants, such as chlorofluorocarbons (CFCs), react with stratospheric ozone, breaking it down into molecular oxygen (O_2). As a result of this breakdown, more solar UV reaches Earth's surface.

Unlike stratospheric ozone, ozone in the troposphere—the layer of atmosphere closest to Earth's surface—is a human-made air pollutant. (Ground-level, or tropospheric, ozone does not replenish the ozone depleted from the stratosphere because tropospheric ozone breaks down to form oxygen long before it drifts up to the stratosphere.) Ozone in the troposphere is a secondary air pollutant formed when sunlight triggers reactions between nitrogen oxides and volatile hydrocarbons. The most harmful component of photochemical smog, ozone reduces air visibility and causes health problems. Ozone also reduces plant vigor, and chronic ozone exposure (of long duration) lowers crop yields (**Figure 8.7**). Chronic exposure to ozone is one possible contributor to forest decline, and ground-level ozone is associated with global warming. As discussed:

Pollutant	Category	Characteristics
Ozone (O_3)	Secondary	Pale blue gas with irritating odor

Hazardous air pollutants (HAPs), or **air toxics**, include hundreds of other air pollutants—such as chlorine (Cl_2), lead, hydrochloric acid, formaldehyde, radioactive substances, and fluorides. HAPs are present in very low concentrations, although it is possible to have high local concentrations of specific pollutants. They are potentially harmful and may pose long-term health risks to people

Ozone damage • Figure 8.7

A scientist measures the effects of ozone on the growth and productivity of several plants. Plants exposed to ozone generally exhibit damaged leaves, reduced root growth, and reduced productivity. Photographed in Birmensdorf, Switzerland.

who live and work around chemical factories, incinerators, or other facilities that produce or use them. To limit the release of more than 180 HAPs, the Clean Air Act Amendments of 1990 (discussed later in this chapter) regulate the pollutant emissions of both large and small businesses. Hazardous air pollutants include:

Pollutant	Category	Characteristics
Chlorine (Cl_2)	Primary	Yellow-green gas
Formaldehyde	Primary	Colorless gas with pungent odor

Sources of Outdoor Air Pollution

Not all air pollution is human generated. Throughout Earth's history, volcanoes have released particulate matter and sulfur oxides (see *What a Scientist Sees*). Plants can also contribute to air pollution, producing a variety of hydrocarbons in response to heat. The hydrocarbon isoprene, for example, may protect leaves from high temperatures. However, isoprene and other hydrocarbons are volatile and evaporate into the air, where they interact with other substances to affect atmospheric chemistry.

WHAT A SCIENTIST SEES

Air Pollution from Volcanoes

Volcanoes occur where the hot magma inside Earth reaches the surface. Active volcanoes can release large quantities of pollutants.

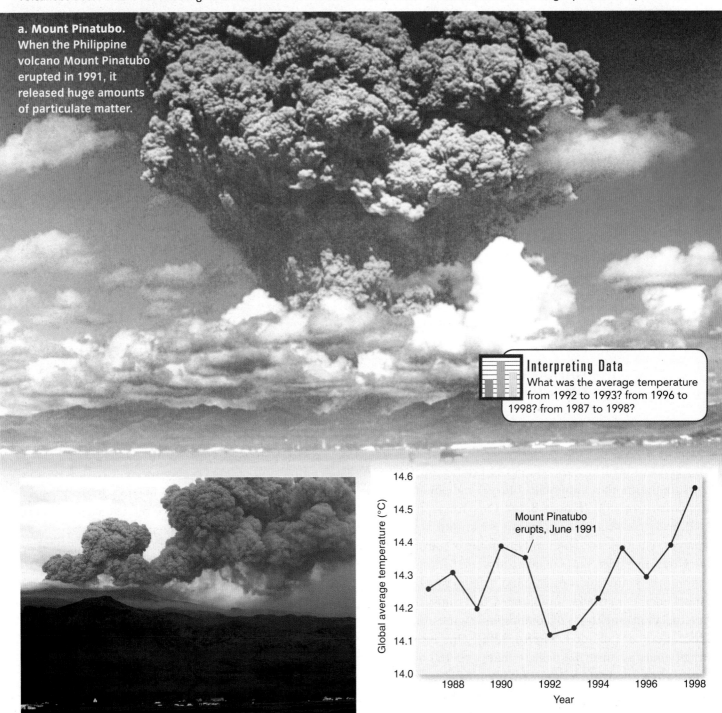

a. Mount Pinatubo. When the Philippine volcano Mount Pinatubo erupted in 1991, it released huge amounts of particulate matter.

Interpreting Data
What was the average temperature from 1992 to 1993? from 1996 to 1998? from 1987 to 1998?

Mount Pinatubo erupts, June 1991

c. The Eyjafjallajokull Volcano. In 2010, the Eyjafjallajokull volcano in Iceland erupted. The resulting ash cloud disrupted air traffic for days.

b. Global Average Temperature, 1987 to 1998. Climate scientists observed that the years following Mount Pinatubo's eruption were cooler than previous and subsequent years. This brief cooling period temporarily interrupted a longer-term warming trend (Chapter 9).

Sources of primary air pollutants • Figure 8.8

b. Stationary source of air pollution. Ash, nitrogen oxides, sulfur oxides, and hazardous air pollutants are released from smokestacks at this coal-fired electric power plant in Tennessee.

Transportation
57%

Miscellaneous
10%

Industrial
processes
12%

Fuel
combustion
(electric power
plants)
21%

a. Transportation and industrial fuel combustion (such as electric power plants) are major contributors of pollutants.

NATIONAL GEOGRAPHIC

c. Mobile source of air pollution. Diesel trucks on the New Jersey Turnpike produce particulate matter and other kinds of air pollution.

The two main human sources of primary air pollutants are transportation (mobile sources) and power plants (stationary sources) (**Figure 8.8a**). Cars, trucks, tractors, and heavy construction equipment are known as mobile sources. They release significant quantities of nitrogen oxides, carbon oxides, particulate matter, and hydrocarbons during the combustion of gasoline or diesel fuel. While diesel engines in trucks, buses, trains, and ships consume less fuel than other types of combustion engines, they produce more air pollution (**Figure 8.8b**). One heavy-duty truck emits as much particulate matter as 150 automobiles, whereas one diesel train engine produces, on average, 10 times the particulate matter of a diesel truck.

Electric power plants and other industrial facilities, known as *stationary sources*, emit most of the particulate matter and sulfur oxides released in the United States; they also emit sizable amounts of nitrogen oxides, hydrocarbons, and carbon oxides (**Figure 8.8c**).

The combustion of fossil fuels, especially coal, is responsible for most of these emissions. The top three industrial sources of toxic air pollutants are the chemical industry, the metals industry, and the paper industry.

Around the world, burning forests to allow crop planting and burning the remains of a previous year's crops creates large amounts of smoke. In Southeast Asia, smoke from agricultural fires can be thick enough to reduce visibility to a few meters, even far from where the burning occurs.

CONCEPT CHECK   **STOP**

1. **What** is the difference between primary and secondary air pollutants?

2. **What** are the seven main classes of air pollutants, and what are some of their effects?

Effects of Air Pollution

LEARNING OBJECTIVES

1. **Relate**, in general terms, the adverse health effects of air pollutants.
2. **Describe** industrial smog, photochemical smog, temperature inversions, urban heat islands, and dust domes.

Air pollution injures organisms, reduces visibility, and attacks and corrodes materials such as metals, plastics, rubber, and fabrics. Air pollutants harm the respiratory tracts of animals, including humans, and can worsen existing medical conditions, such as chronic lung disease, pneumonia, and cardiovascular problems. Most forms of air pollution reduce the overall productivity of crop plants. Air pollution is involved in acid deposition, global warming, and stratospheric ozone depletion (all discussed in Chapter 9).

Air Pollution and Human Health

Generally speaking, exposure to low levels of pollutants leads to irritation of the eyes and inflammation of the respiratory tract (**Table 8.1**). Many air pollutants also suppress the immune system, increasing susceptibility to infection. In addition, exposure to air pollution during respiratory illnesses may result in the development later in life of chronic respiratory diseases, such as emphysema and chronic bronchitis. In *emphysema*, the air sacs (alveoli) in the lungs become irreversibly distended, causing breathlessness and wheezy breathing. *Chronic bronchitis* is a disease in which the air passages (bronchi) of the lungs become permanently inflamed, causing breathlessness and chronic coughing.

Urban Air Pollution

Air pollution in urban areas is often called **smog**. The term *smog* was coined at the beginning of the 20th century for the smoky fog prevalent in London because of coal combustion. Traditional London-type smog—that is, smoke pollution—is sometimes called **industrial smog**. The principal pollutants in industrial smog are sulfur oxides and particulate matter. The worst episodes of industrial smog typically occur during winter months, when combustion of household fuel such as heating oil or coal is high. Because of air quality laws and pollution-control devices, industrial smog is generally not a significant problem in highly developed countries today, but it can be severe in developing countries.

Another important type of smog is **photochemical smog**. First noted in Los Angeles in the 1940s, photochemical smog is

> **photochemical smog** A brownish-orange haze formed by chemical reactions involving sunlight, nitrogen oxides, and hydrocarbons.

Health effects of several major air pollutants • Table 8.1

Pollutant	Source	Effects
Particulate matter	Industries, motor vehicles	Aggravates respiratory illnesses; long-term exposure may cause chronic conditions such as bronchitis
Sulfur oxides	Electric power plants, industries	Irritate respiratory tract; same effects as particulates
Nitrogen oxides	Motor vehicles, industries, heavily fertilized farmland	Irritate respiratory tract; aggravate respiratory conditions such as asthma and chronic bronchitis
Carbon monoxide	Motor vehicles, industries	Reduces blood's ability to transport oxygen; causes headache and fatigue at low levels; mental impairment or death at high levels
Ozone	Formed in atmosphere (secondary air pollutant)	Irritates eyes; irritates respiratory tract; produces chest discomfort; aggravates respiratory conditions such as asthma and chronic bronchitis

generally worst during the summer months. Both nitrogen oxides and hydrocarbons are involved in its formation. A photochemical reaction occurs among nitrogen oxides, largely from automobile exhaust; volatile hydrocarbons; and oxygen in the atmosphere to produce ozone. This reaction requires solar energy (**Figure 8.9**). Ozone reacts with other air pollutants, including hydrocarbons, to form more than 100 different secondary air pollutants (peroxyacyl nitrates [PANs], for example) that injure plant tissues, irritate eyes, and aggravate respiratory illnesses in humans.

The main human source of the ingredients for photochemical smog is automobiles, but bakeries and dry cleaners are also significant contributors. When bread is baked, yeast byproducts that are volatile hydrocarbons are released to the atmosphere, where solar energy powers their interactions with other gases to form ozone. The volatile fumes from dry cleaners also contribute to photochemical smog.

How Weather and Topography Affect Air Pollution

Changes in temperature throughout the day produce air circulation patterns that dilute and disperse air pollutants. During a **temperature inversion**, however, polluting gases and particulate matter remain trapped in high concentrations close to the ground, where people live and breathe. Temperature inversions usually persist for only a few hours before

> **temperature inversion** A layer of cold air temporarily trapped near the ground by a warmer upper layer.

Composition of photochemical smog • Figure 8.9

Photochemical smog is a mixture that includes ozone, peroxyacyl nitrates (PANs), nitric acid, and organic compounds such as formaldehyde.

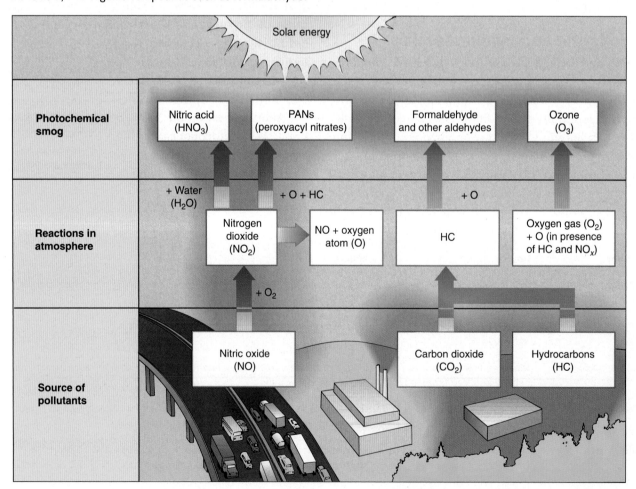

EnviroDiscovery

Air Pollution May Affect Precipitation

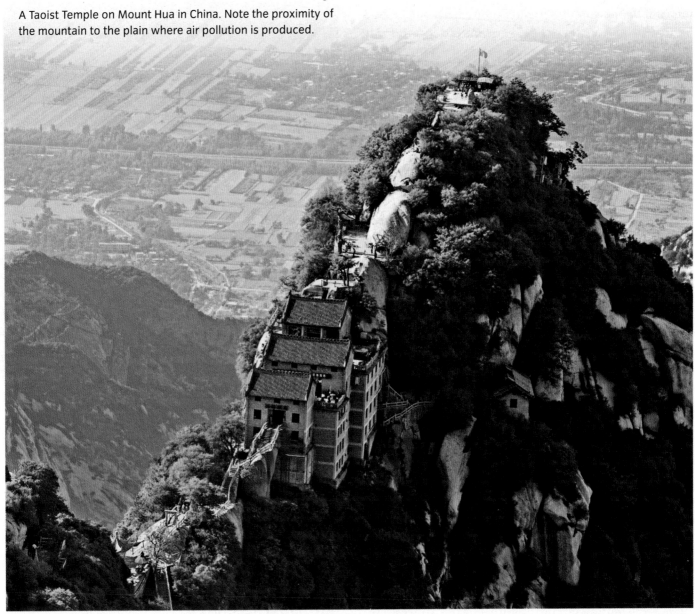

A Taoist Temple on Mount Hua in China. Note the proximity of the mountain to the plain where air pollution is produced.

For several years, climate scientists have noticed that mountainous areas in the western United States are receiving less precipitation than usual. The effect has been particularly pronounced in mountains located downwind from cities, leading scientists to speculate that air pollution may be altering precipitation patterns. However, long-term data to support this hypothesis were not available until recently.

In 2007 climate scientists evaluated weather data taken atop Mount Hua, a sacred mountain in China that overlooks a plain where several cities (which are a source of air pollution) are located. The data, which include precipitation, visibility, and humidity data, have been measured since 1954. By subtracting the effect of humidity on visibility, scientists have been able to estimate the amount of air pollution suspended in the air. The scientists have correlated high visibility—that is, low air pollution—with substantially more precipitation than when air pollution levels were high. They caution, however, that the link between air pollution and precipitation patterns is still tentative and will require more research.

solar energy warms the air near the ground. Sometimes a stalled high-pressure air mass allows a temperature inversion to persist for several days, causing atmospheric stagnation.

Certain types of topography (surface features) increase the likelihood of temperature inversions. Cities located in valleys, near a coast, or on the leeward side of mountains (the side toward which the wind blows) are prime candidates for temperature inversions. The Los Angeles Basin, for example, lies between the Pacific Ocean on the west and mountains to the north and east. During the summer the sunny climate produces a layer of warm dry air at upper elevations. A region of upwelling occurs just off the Pacific coast, bringing cold ocean water to the surface and cooling the ocean air. As this cool air blows inland over the basin, the mountains block its movement further. Thus, a layer of warm, dry air overlies cool air at the surface, producing a temperature inversion.

Air Pollution in Los Angeles

Los Angeles, California, has some of the worst smog in the world. Its location, in combination with its sunny climate, is conducive to the formation of stable temperature inversions that trap photochemical smog near the ground, sometimes for long periods (**Figure 8.10**). Passenger vehicles, heavy-duty trucks, and buses are the source of more than half of the smog-producing emissions in Los Angeles.

In 1969 California became the first state to enforce emissions standards on motor vehicles, largely because of the air pollution problems in Los Angeles. Today Los Angeles has stringent smog controls that regulate everything from low-emission alternative fuels (such as compressed natural gas) for buses to lawn mower emissions to paint vapors. Using the cleanest emission-reduction equipment available significantly reduces emissions from large industrial and manufacturing sources, including oil refineries and power plants. California has no coal-fired power plants; most of its power plants burn natural gas. Future pollution reductions will come in part from requiring auto manufacturers to sell ultra-low-emission cars.

After several decades devoted to improving its air quality, Los Angeles now has the cleanest skies it has had since the 1950s. Despite the impressive progress, Los Angeles still exceeds federal air quality standards on more

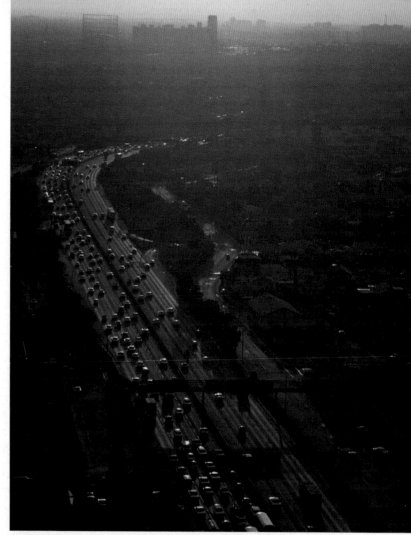

Photochemical smog • Figure 8.10

Photographed in Los Angeles, California, on a day when air pollution exceeded federal air quality standards.

days than almost any other metropolitan region in the United States. Los Angeles experienced 120 days above the federal ozone standard in 2008, down from 203 days in 1977.

Urban Heat Islands and Dust Domes

Streets, rooftops, and parking lots in areas of high population density absorb solar radiation during the day and radiate heat into the atmosphere at night. Heat from human activities such as fuel combustion is also highly concentrated in cities. The air in urban areas therefore forms **urban heat islands** in the surrounding suburban and rural areas (**Figure 8.11**).

urban heat island
Local heat buildup in an area of high population.

Urban heat island • Figure 8.11

This figure shows how temperatures might vary on a summer afternoon. The city stands out as a heat island against the surrounding rural areas.

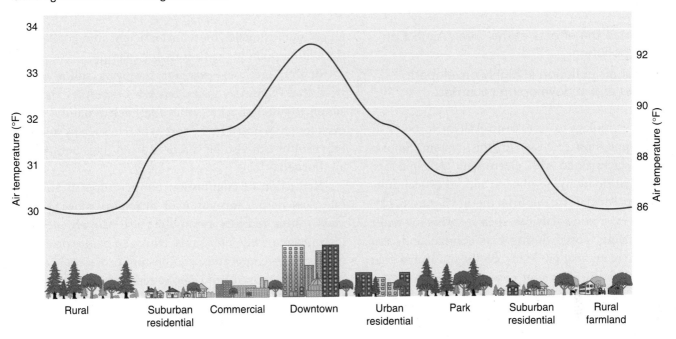

Urban heat islands also contribute to the buildup of pollutants, especially particulate matter, in the form of **dust domes** over cities (**Figure 8.12a**). Pollutants concentrate in a

> **dust dome** A dome of heated air that surrounds an urban area and contains a lot of air pollution.

dust dome because convection (the vertical motion of warmer air) lifts pollutants into the air, where they remain because of the somewhat stable air masses the urban heat island produces. If wind speeds increase, the dust dome moves downwind from the city, and the polluted air spreads over rural areas (**Figure 8.12b**).

Urban heat islands affect local air currents and weather conditions. For example, urban heat islands may increase the number of thunderstorms over the city during summer months. The uplift of warm air over the city produces a low-pressure cell that draws in cooler air from the surroundings. As the heated air rises, it cools, causing water vapor to condense into clouds and producing thunderstorms.

Dust dome • Figure 8.12

a. A dust dome of pollutants forms over a city when the air is somewhat calm and stable.

b. When wind speeds increase, the pollutants move downwind from the city.

CONCEPT CHECK **STOP**

1. **What** are some of the health effects of exposure to air pollution?

2. **What** are urban heat islands? What are dust domes?

Effects of Air Pollution **205**

Controlling Air Pollutants

LEARNING OBJECTIVES

1. **Summarize** the effects of the Clean Air Act on U.S. air pollution.
2. **Contrast** air pollution in highly developed countries and in developing countries.

Technology exists to control all the forms of air pollution discussed in this chapter. Smokestacks fitted with electrostatic precipitators, fabric filters, scrubbers, and other technologies remove particulate matter from the air (**Figure 8.13**). Careful land-excavating activities, such as sprinkling water on dry soil being moved during road construction, also reduce particulate matter. Many of the measures that increase energy efficiency and conservation also reduce air pollution. Smaller, more fuel-efficient automobiles produce fewer polluting emissions, for example.

Several methods exist for removing sulfur oxides from flue (chimney) gases, but it is often less expensive simply to switch to a low-sulfur fuel such as natural gas or even to a non–fossil fuel energy source, such as solar energy. Sulfur can also be removed from fuels before they are burned.

Reduction of combustion temperatures in automobiles lessens the formation of nitrogen oxides. Use of mass transit reduces automobile use, thereby decreasing nitrogen oxide emissions. Nitrogen oxides produced during high-temperature combustion processes in industry can be removed from smokestack exhausts.

Electrostatic precipitator • Figure 8.13

Clean gas out

Electrode (negative charge)

Precipitator wall (positive charge)

Dirty gas (smoke) in

Dust falls off wall into collector

a. In an electrostatic precipitator, the electrode imparts a negative charge to particulates in the dirty gas. These particles are attracted to the positively charged precipitator wall and then fall off into the collector.

b.

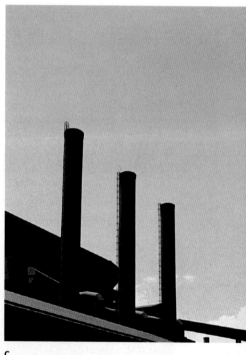

c.

A comparison of emissions from a Delaware Valley steel mill with the electrostatic precipitator turned off (**b**) and on (**c**). The toxic dust must be safely disposed, or it becomes a pollution problem.

Modification of furnaces and engines to provide more complete combustion helps control the production of both carbon monoxide and hydrocarbons. Catalytic afterburners, used immediately following combustion, oxidize most unburned gases. The use of catalytic converters to treat auto exhaust reduces carbon monoxide and volatile hydrocarbon emissions about 85 percent over the life of the car. Careful handling of petroleum and hydrocarbons, such as benzene, reduces air pollution from spills and evaporation.

The Clean Air Act

There is good news and bad news about air pollution in the United States. The bad news is that many locations throughout the country still have unacceptably high levels of one or more air pollutants. Moreover, health experts estimate that air pollution causes the premature deaths of thousands of people in the United States each year. The good news is that overall air quality has improved since 1970.

This improvement is largely due to the U.S. **Clean Air Act (CAA)**, first passed in 1970 and updated and amended in 1977 and 1990. The CAA authorizes the Environmental Protection Agency (EPA) to apply and enforce the CAA by establishing limits on the amount of specific air pollutants permitted everywhere in the United States. Individual states must meet deadlines to reduce air pollution to acceptable levels. States may pass more stringent pollution controls than the EPA authorizes, but they can't mandate weaker limits than those stipulated in the CAA.

The EPA has focused on six air pollutants—lead, particulate matter, sulfur dioxide, carbon monoxide, nitrogen oxides, and ozone—and established maximum acceptable concentrations for each. The most dramatic improvement so far has been in the amount of lead in the atmosphere, which showed a 98 percent decrease between 1970 and 2000, primarily because of the switch from leaded to unleaded gasoline. Atmospheric levels of the other pollutants, with the exception of particulate matter, have also declined (**Figure 8.14**). For example, between 1970 and 2006, sulfur dioxide emissions declined 49 percent. During this same time, U.S. gross domestic product increased more than 160 percent, energy consumption increased 48 percent, and vehicle miles increased about 150 percent.

Emissions in the United States, 1970 and 2006 • Figure 8.14

Carbon monoxide, sulfur dioxide, volatile organic compounds (many of which are hydrocarbons), and nitrogen oxides showed decreases; only particulate matter did not decline. "PM = 10" applies to particles less than or equal to 10 μm (10 micrometers). Since 1990 the EPA has also monitored PM = 2.5, which are very small particles less than or equal to 2.5 μm.

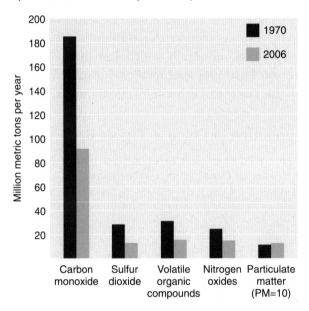

The Clean Air Act of 1970 and its amendments in 1977 and 1990 required progressively stricter controls of motor vehicle emissions. The provisions of the Clean Air Act Amendments of 1990 include the development of "superclean" cars, which emit lower amounts of nitrogen oxides and hydrocarbons, and the use of cleaner-burning gasoline in the most polluted cities in the United States. More recent automobile models do not produce as many pollutants as older models. Yet despite the increasing percentage of newer automobile models on the road, air quality has not improved in some areas of the United States because of the large increase in the number of cars being driven.

The Clean Air Act Amendments of 1990 focus on industrial airborne toxic chemicals in addition to motor vehicle emissions. The Clean Air Act Amendments of 1990 required a 90 percent reduction in the atmospheric emissions of 189 toxic chemicals. To comply with these requirements, both small businesses (such as dry cleaners) and large manufacturers (such as chemical companies) installed pollution control equipment if they had not already done so.

Federal courts also play a role in interpreting the CAA. In 2008 the U.S. Supreme Court ruled that the EPA must regulate carbon dioxide under the CAA, a ruling the EPA is working to implement.

Air Pollution in Developing Countries

As developing nations become more industrialized, they also produce more air pollution. The leaders of most developing countries believe they must industrialize rapidly to compete economically with more highly developed countries. Environmental quality is usually a low priority in the race for economic development. Outdated technologies are adopted because they are less expensive, and air pollution laws, where they exist, are not enforced. Thus, air quality is deteriorating rapidly in many developing nations.

Many cities and towns in China burn so much low-quality coal for heating and industry that residents see the sun only a few weeks of the year (**Figure 8.15**). The rest of the time residents are choked in a haze of orange-colored coal dust. In other developing countries, such as India and Nepal, wood or animal dung is burned indoors, often in poorly designed stoves with little or no outside ventilation, thereby exposing residents to serious indoor air pollution (discussed in the next section).

Air pollution in China • Figure 8.15

A coal-powered steel mill releases pollution in Liaoning Province, China. All forms of pollution are increasing as China becomes industrialized.

Global Locator

CHINA

NATIONAL GEOGRAPHIC

GLOBAL LOCAL

What are the major sources of air pollution where you live?

The growing number of automobiles in developing countries is also contributing to air pollution, particularly in urban areas. Many vehicles in these countries are 10 or more years old and have no pollution control devices. Motor vehicles produce about 60 to 70 percent of the air pollutants in urban areas of Central America, and they produce 50 to 60 percent in urban areas of India. The most rapid proliferation of motor vehicles worldwide is currently occurring in Latin America, Asia, and eastern Europe.

Lead pollution from heavily leaded gasoline is an especially serious problem in developing nations. The gasoline refineries in these countries are generally not equipped to remove lead from gasoline. (The United States was in the same situation until federal law mandated that U.S. refineries upgrade their equipment by 1986.) In Cairo, Egypt, for example, many children have blood lead levels more than two times higher than the level considered at-risk in the United States. Lead can retard children's growth and cause brain damage.

According to the World Health Organization, the five worst cities in the world in terms of exposing children to air pollution are Mexico City, Mexico; Beijing, China; Shanghai, China; Tehran, Iran; and Kolkata, India. Respiratory disease is now the leading cause of death for children worldwide. More than 80 percent of these deaths occur in children under age 5 who live in cities in developing countries.

CONCEPT CHECK

1. **What** is the U.S. Clean Air Act, and how has it reduced outdoor air pollution?
2. **Where** is air pollution worse: in highly developed nations or in developing countries? Why?

Indoor Air Pollution

LEARNING OBJECTIVES

1. **Summarize** at least four sources of indoor air pollution and explain their role in sick building syndrome.
2. **Describe** the effects of indoor air pollution in developing countries.
3. **Explain** why radon gas is an indoors health hazard.

People around the world spend much of their time indoors, and contaminated indoor air can lead to substantial health problems. In rural areas, and particularly in developing countries, cooking with solid fuels (wood, coal, peat, and dung) can have serious health impacts. This has led the World Health Organization to determine that burning solid fuels is among the 10 greatest threats to human health.

The most common contaminants of indoor air in highly developed countries are radon, cigarette smoke, carbon monoxide, nitrogen dioxide (from gas stoves), formaldehyde (from carpet, fabrics, and furniture), household pesticides, cleaning solvents, ozone (from photocopiers), and asbestos. In addition, viruses, bacteria, fungi (yeasts, molds, and mildews), dust mites, pollen, and other organisms are often found in heating, air-conditioning, and ventilation ducts. Because illnesses from indoor air pollution usually resemble common ailments such as colds, influenza, or upset stomachs, they are often not recognized.

Health officials are paying increasing attention to **sick building syndrome**. The Labor Department estimates that more than 20 million employees are exposed to health risks from indoor air pollution. The EPA estimates that annual medical costs for treating the health effects of indoor air pollution in the United States exceed $1 billion. When lost work time and diminished productivity are added to health care costs, the

sick building syndrome Eye irritations, nausea, headaches, respiratory infections, depression, and fatigue caused by indoor air pollution.

Indoor air pollution • Figure 8.16

Cooking indoors with open fires or traditional cooking stoves results in dangerous levels of indoor air pollution.

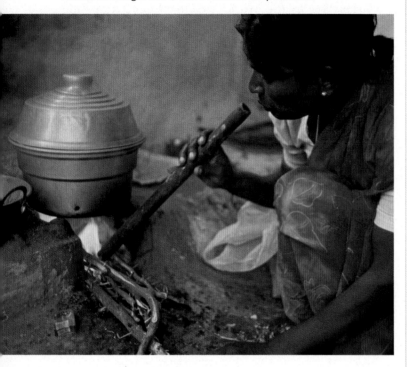

total annual cost to the economy may be as much as $50 billion. Fortunately, most building problems are relatively inexpensive to alleviate.

Indoor air pollution is a particularly serious health hazard in developing countries, where many people burn fuels such as firewood or animal dung indoors to cook and heat water (**Figure 8.16**). Smoke from indoor cooking contains carbon monoxide, particulates, hydrocarbons, and other hazardous air pollutants such as formaldehyde and benzene. Women and children are harmed the most by indoor cooking, which can cause acute lower respiratory infections, pneumonia, eye infections, and lung cancer. The World Health Organization estimates that smoke from indoor cooking kills 1.6 million people each year.

Radon

Radon is a colorless, odorless, tasteless radioactive gas produced naturally as a result of the radioactive decay of uranium in Earth's crust. In the United States, radon has become an increasingly important indoor air contaminant, especially as laws and changes in habits have reduced secondhand cigarette smoke exposure. Radon seeps through the ground and enters buildings, where it sometimes accumulates to dangerous levels (**Figure 8.17**). Radon emitted into the atmosphere gets diluted and dispersed and is of little consequence outdoors.

Only ingested or inhaled radon harms the body. The National Research Council of the National Academy of Sciences estimates that residential exposure to radon causes 12 percent of all lung cancers—between 15,000 and 22,000 lung cancers annually. Cigarette smoking exacerbates the risk from radon exposure; about 90 percent of radon-related cancers occur among current or former smokers.

According to the EPA, about 6 percent of U.S. homes have high enough levels of radon to warrant corrective action—a radon level above 4 picocuries per liter of air. (As a standard of reference, outdoor radon concentrations range from 0.1 to 0.15 picocuries per liter of air worldwide.) The highest radon levels in the United States are found in homes across southeastern Pennsylvania into northern New Jersey and New York.

How radon infiltrates a house • Figure 8.17

Cracks in basement walls or floors, openings around pipes, and pores in concrete blocks provide some of the entries for radon.

Radon escapes outdoors

Wall and foundation cracks

Drain pipes

Floor cracks

Homes may contain higher levels of air pollutants than outside air, even near polluted industrial sites.

Methylene chloride
From: paint strippers and thinners

Gasoline
From: auto, lawn mower

Tobacco smoke
From: cigarettes and pipes

Formaldehyde
From: furniture and carpeting made from synthetic polymers, particle board, foam insulation

Pesticides
From: gardening products

Radon
From: uranium-containing rocks

Tetrachloroethylene
From: dry-cleaning fluid

Carbon monoxide
From: faulty furnace, auto left running

Fungi and bacteria
From: dirty heating and air conditioning ducts

Asbestos
From: pipe insulation, vinyl tiles

Ammonia
From: household cleaners

Nitrogen oxides
From: unvented gas stoves, wood stoves, kerosene heaters

Chloroform
From: chlorine-treated water in hot showers

Para-dichlorobenzene
From: mothball crystals, air fresheners

Ironically, efforts to make our homes more energy efficient have increased the hazard of indoor air pollutants, including radon. Drafty homes waste energy but allow radon to escape outdoors so it does not build up inside. Every home should be tested for radon because levels vary widely from home to home, even in the same neighborhood. Generally, testing and corrective actions are reasonably priced. However, corrective actions can be expensive, with some costing thousands of dollars.

Figure 8.18 summarizes many possible sources of air pollution in homes.

CONCEPT CHECK   STOP

1. **What** are some common indoor air contaminants?

2. **Why** is indoor air pollution such a serious health hazard in developing countries?

3. **How** does radon gas enter buildings?

Curbing Air Pollution in Chattanooga

During the 1960s, the federal government gave Chattanooga, Tennessee, the dubious distinction of having the worst air pollution in the United States. The air was so dirty in this manufacturing city that sometimes people driving downtown had to turn on their headlights in the middle of the day. The orange air soiled their white shirts so quickly that many businesspeople brought extra ones to work. To compound the problem, the mountains surrounding the city kept the pollutants produced by its inhabitants from dispersing.

Today the air in this scenic midsized city of 200,000 people is clean, and Chattanooga ranks high among U.S. cities in terms of air quality (*see* photograph). City and business leaders are credited with transforming Chattanooga's air. Soon after the passage of the federal Clean Air Act of 1970, the city established an air pollution control board to enforce regulations controlling air pollution. New local regulations allowed open burning by permit only, placed limits on industrial odors and particulate matter, outlawed visible automotive emissions, and set a cap on sulfur content in fuel, which controlled the production of sulfur oxides. Businesses installed expensive air pollution control devices. The city started an emissions-free electric bus system. Chattanooga also decided to recycle its solid waste rather than build an emissions-producing incinerator.

In 1984 the EPA declared Chattanooga in attainment for particulate matter; this designation meant particulate levels had been below the federal health limit for one year. The city reached attainment status for ozone in 1989. Since then, the city's levels for all seven EPA-regulated air pollutants have been lower than federal standards require.

In the early 2000s, Chattanoogans continued to move their city toward *environmental sustainability*. The city plans to convert a run-down business district into a community in which people live near their places of work. Businesses located in this district will form an **industrial ecosystem** in which the wastes of one business are raw materials for another business.

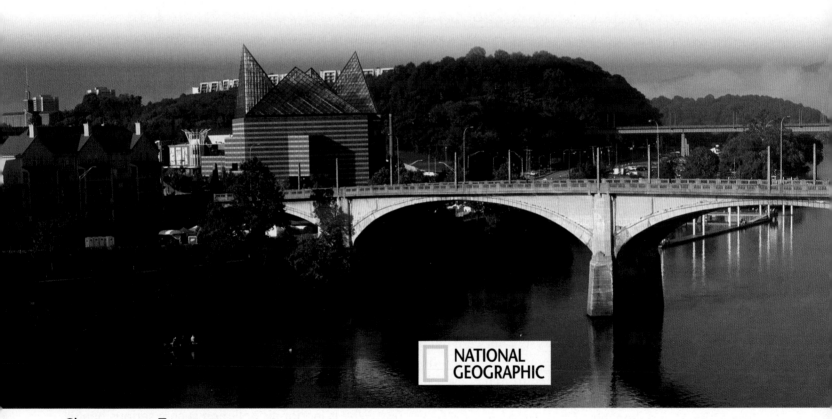

Chattanooga, Tennessee

Chattanooga's air quality has improved dramatically during the past several decades.

Summary

1 The Atmosphere 192

1. Oxygen (21 percent) and nitrogen (78 percent) are the main gases in the **atmosphere**, the gaseous envelope surrounding Earth. Argon, carbon dioxide, other gases, water vapor, and trace amounts of various air pollutants are also present.

2. The **troposphere**, the layer of atmosphere closest to Earth's surface, extends to a height of approximately 12 km (7.5 miles). Temperature decreases with increasing altitude, and weather occurs in the troposphere. In the **stratosphere**, there is a steady wind but no turbulence. The stratosphere contains an ozone layer that absorbs much of the Sun's UV radiation. The **mesosphere**, directly above the stratosphere, has the lowest temperatures in the atmosphere. The **thermosphere** has steadily rising temperatures and gases that absorb x-rays and short-wave UV radiation. The thermosphere reflects outgoing radio waves back toward Earth without the aid of satellites.

3. The **Coriolis effect** is the tendency of moving air or water to be deflected from its path and swerve to the right in the Northern Hemisphere and to the left in the Southern Hemisphere.

associated with photochemical smog and global warming; some are dangerous to human health. **Ozone** is a secondary air pollutant in the lower atmosphere (troposphere) but an essential part of the stratosphere. Tropospheric ozone reduces visibility, causes health problems, stresses plants, and is associated with global warming. Some air pollutants are called **air toxics**, or **hazardous air pollutants**, because they are potentially harmful and may pose long-term health risks to people who are exposed to them; chlorine, lead, hydrochloric acid, formaldehyde, radioactive substances, and fluorides are examples.

2 Types and Sources of Air Pollution 196

1. **Air pollution** consists of various chemicals (gases, liquids, or solids) present in the atmosphere in high enough levels to harm humans, other organisms, or materials. **Primary air pollutants** are harmful chemicals that enter the atmosphere directly due to either human activities or natural processes; examples include carbon oxides, nitrogen oxides, sulfur dioxide, particulate matter, and hydrocarbons. **Secondary air pollutants** are harmful chemicals that form in the atmosphere when primary air pollutants react chemically with each other or with natural components of the atmosphere; ozone and sulfur trioxide are examples.

2. **Particulate matter**—solid particles and liquid droplets suspended in the atmosphere—corrodes metals, erodes buildings, soils fabrics, and can damage the lungs. **Nitrogen oxides** are gases associated with photochemical smog, acid deposition, global warming, and stratospheric ozone depletion; they also corrode metals and fade textiles. **Sulfur oxides** are gases associated with acid deposition; they corrode metals and damage stone and other materials. **Carbon oxides** include the gases carbon monoxide, which is poisonous, and carbon dioxide, which is linked to global warming. **Hydrocarbons** are solids, liquids, or gases

3 Effects of Air Pollution 201

1. Exposure to low levels of air pollutants irritates the eyes and causes inflammation of the respiratory tract. Many air pollutants suppress the immune system, increasing susceptibility to infection. Exposure to air pollution during respiratory illnesses may result in the development of chronic respiratory diseases, such as emphysema and chronic bronchitis.

2. **Industrial smog** refers to smoke pollution. **Photochemical smog** is a brownish-orange haze formed by chemical reactions involving sunlight, nitrogen oxides, and hydrocarbons. A **temperature inversion** is a layer of cold air temporarily trapped near the ground by a warmer upper layer; during a temperature inversion, polluting gases and particulate matter remain trapped in high concentrations close to the ground. An **urban heat island** is local heat buildup in an area of high population. Urban heat islands affect local air currents and weather conditions and contribute to the buildup of pollutants, especially particulate matter, in the form of a **dust dome**, a dome of heated air that surrounds an urban area and contains a lot of air pollution.

4 Controlling Air Pollutants 206

1. Improvements in U.S. air quality since 1970 are largely due to the **Clean Air Act**, which authorizes the EPA to set limits on specific air pollutants. Individual states must meet deadlines to reduce air pollution to acceptable levels and can't mandate weaker limits than those stipulated in the Clean Air Act.

2. Air quality in the United States has slowly improved since passage of the Clean Air Act. The most dramatic improvement is the decline of lead in the air, although levels of sulfur oxides, ozone, carbon monoxide, volatile compounds, and nitrogen oxides have also declined. Air quality is deteriorating in developing nations as a result of rapid industrialization, growing numbers of automobiles, and a lack of emissions standards.

5 Indoor Air Pollution 209

1. Indoor air pollution includes radon, cigarette smoke, nitrogen dioxide (from gas stoves), and formaldehyde (from carpet, fabrics, and furniture). These contribute to a variety of symptoms referred to as **sick building syndrome**.

2. Burning solid fuels indoors in developing countries leads to diseases including respiratory and eye infections, particularly among women and children.

3. Radon, a colorless, odorless, tasteless radioactive gas enters buildings from the ground. In some locations, indoor radon can pose a significant health threat.

Key Terms

- air pollution 196
- atmosphere 192
- Coriolis effect 195
- dust dome 205

- photochemical smog 201
- primary air pollutants 196
- secondary air pollutants 196
- sick building syndrome 209

- temperature inversion 202
- urban heat island 204

What is happening in this picture?

- This Nepalese woman is preparing a meal inside a poorly ventilated room. Cooking meals can take up many hours each day. In this picture, where is the smoke most dense? What does this imply for the health of women, who do much of the cooking in developing countries?

- Young children in developing countries tend to spend much of their time with their mothers; in fact, an infant may be strapped to the mother while she cooks. Explain what sorts of health effects you might expect these children to suffer as a result.

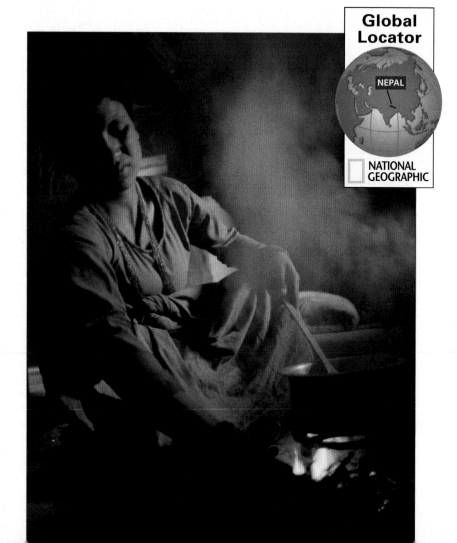

Global Locator

NEPAL

NATIONAL GEOGRAPHIC

Critical and Creative Thinking Questions

1. What two gases comprise most of the atmosphere?

2. The atmosphere of Earth has been compared to the peel covering an apple. Explain the comparison.

3. What basic forces determine the circulation of the atmosphere? Describe the general directions of atmospheric circulation.

4. Distinguish between primary and secondary air pollutants. Give examples of each that you are likely to encounter.

5. Distinguish between mobile and stationary sources of air pollution.

6. What is the relationship between agriculture and air pollution?

7. The graphs below represent air pollutant measurements taken at two different locations. Which location is indoors, and which is outdoors? Explain your answer.

Formaldehyde ■   Radon ▬   Sulfuric acid ■   Methane ■

8. One of the most effective ways to reduce the threat of radon-induced lung cancer is to quit smoking. Explain.

9. What air pollutants do the 1990 amendments to the Clean Air Act target?

10. During a formal debate on the hazards of air pollution, one team argues that ozone is helpful to the atmosphere, and the other team argues that it is destructive. Explain why teams are both correct.

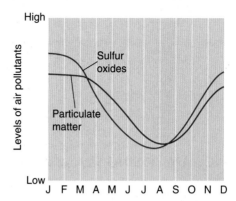

11. The graph above shows air pollutant levels in a city in the Northern Hemisphere, measured throughout a year. Is this city likely to be found in a developing country or in a developed country? Why?

12. This figure shows phase I vapor recovery from an underground gasoline storage tank. Before phase I vapor recovery was developed, gasoline vapors were vented directly into the air. Now the vapor is vented through Hose B into the truck and returned to the gasoline depot, where it is condensed or burned. Which of the following does gasoline vapor recovery control: photochemical smog, urban heat islands, or dust domes? Explain.

THE PLANNER

Global Atmospheric Changes

MELTING ICE AND RISING SEA LEVELS

Powerful evidence that Earth is warming comes from the melting of continental and polar ice. In 2002, an iceberg roughly twice the size of Rhode Island broke off from the Antarctic Peninsula. The Antarctic ice pack has retreated and thinned, losing 40 percent of its volume since 1980.

The Muir Glacier in Alaska was once enormous, with a huge vertical front from which icebergs calved into Glacier Bay. Today, the Muir Glacier has shrunk to a fraction of its former size (see photograph, taken in 2004; the inset shows approximately the same location in 1903).

Human-caused climate change is an established phenomenon. Within the scientific community, the question is no longer whether climate change will occur but at what rate and with what effects, and what can be done about it. The biggest culprit in climate change is an increase in atmospheric carbon dioxide (CO_2), which is generated primarily through the burning of fossil fuels.

In this chapter we examine the challenges of global atmospheric changes: climate change, ozone depletion, and acid deposition. Changes in economics, politics, energy use, agriculture, and human behavior will be necessary to address these issues.

NATIONAL GEOGRAPHIC

CHAPTER OUTLINE

CHAPTER PLANNER ✔

- ❑ Study the picture and read the opening story.
- ❑ Scan the Learning Objectives in each section:
  p. 218 ❑   p. 222 ❑   p. 230 ❑   p. 233 ❑
- ❑ Read the text and study all figures and visuals. Answer any questions.

Analyze key features:

- ❑ Environmental InSight, p. 227 ❑   p. 231 ❑   p. 235 ❑
- ❑ EnviroDiscovery, p. 232 ❑
- ❑ What a Scientist Sees, p. 221 ❑
- ❑ Process Diagram, p. 219 ❑   p. 224 ❑
- ❑ Case Study, p. 237 ❑
- ❑ Stop: Answer the Concept Checks before you go on:
  p. 221 ❑   p. 229 ❑   p. 232 ❑   p. 236 ❑

End of Chapter:

- ❑ Review the Summary and Key Terms.
- ❑ Answer What is happening in this picture?
- ❑ Answer the Critical and Creative Thinking Questions.

The Atmosphere and Climate

LEARNING OBJECTIVES

1. **Distinguish** between weather and climate.
2. **Summarize** the effects of solar energy on Earth's temperature.
3. **Explain** several reasons for regional precipitation differences.

Weather refers to the conditions in the atmosphere at a given place and time; it includes temperature, atmospheric pressure, precipitation, cloudiness, humidity, and wind. Weather changes from one hour to the next and from one day to the next.

The two most important factors that determine an area's overall **climate** are *temperature*—both average temperature and temperature variability—and both average and seasonal *precipitation*. Latitude, elevation, topography, quantity and types of vegetation, distance from the ocean, and geographic location all influence temperature, precipitation, and other aspects of climate. Other climate factors include weather conditions such as wind, humidity, fog, cloud cover, and, in some areas, lightning. Unlike weather, which changes rapidly, climate generally changes slowly, over hundreds or thousands of years.

Earth has many different climates, and because each is relatively constant for many years, organisms have adapted to them. The many kinds of organisms on Earth are here in part because of the large number of different climates—from cold, snow-covered polar climates to tropical climates that are hot and have rain almost every day (**Figure 9.1**).

> **climate** The typical weather patterns that occur in a place over a period of years.

Tropical climate • Figure 9.1

Tropical climates occur in a region that spans the equator, from 15 to 25 degrees latitude north to 15 to 25 degrees latitude south. Photographed at the Na Pali Coast, Kauai, Hawaiian Islands.

Solar Radiation and Climate

The Sun makes life on Earth possible. Sunshine, or **insolation**, warms the planet, including the atmosphere, to habitable temperatures. Without the Sun's energy, all water on Earth would be frozen, including the ocean. Photosynthetic organisms capture the Sun's energy and use it to make the food molecules almost all forms of life require. Most of our fuels, such as wood and **fossil fuels** (oil, coal, and natural gas), originated as solar energy captured by photosynthetic organisms.

> **infrared radiation**
> Electromagnetic radiation with wavelengths longer than those of visible light but shorter than microwaves; perceived as invisible waves of heat.

The amount of insolation a region receives is a major determinant of climate.

Clouds, and to a lesser extent snow, ice, and the ocean, reflect away about 31 percent of the solar radiation that falls on Earth (**Figure 9.2**). The remaining 69 percent is absorbed and runs the hydrologic cycle, carbon cycle, and other biogeochemical cycles; drives winds and ocean currents; powers photosynthesis; and warms the planet. Ultimately, all this energy returns to space as long-wave infrared radiation (heat).

Fate of solar radiation that reaches Earth • Figure 9.2

THE PLANNER

Most of the energy that the Sun produces never reaches Earth. The solar energy that reaches Earth warms the planet's surface, drives the hydrologic cycle and other biogeochemical cycles, produces our climate, and powers almost all life through photosynthesis. On average, about 1366 watts per square meter (W/m^2) enter Earth's atmosphere, resulting in about 1000 W/m^2 at Earth's surface.

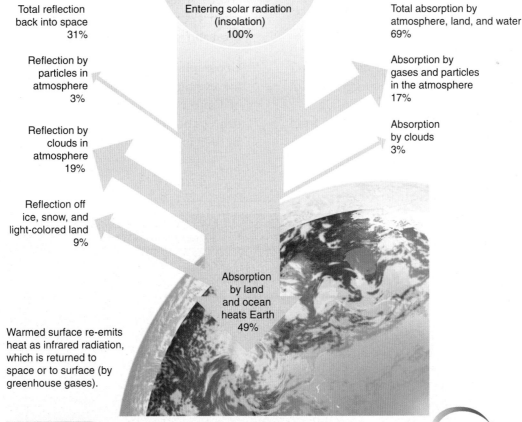

Total reflection back into space
31%

Reflection by particles in atmosphere
3%

Reflection by clouds in atmosphere
19%

Reflection off ice, snow, and light-colored land
9%

Entering solar radiation (insolation)
100%

Total absorption by atmosphere, land, and water
69%

Absorption by gases and particles in the atmosphere
17%

Absorption by clouds
3%

Absorption by land and ocean heats Earth
49%

Warmed surface re-emits heat as infrared radiation, which is returned to space or to surface (by greenhouse gases).

Think Critically Ice is highly reflective compared to water and land. How will this diagram change as ice caps and glaciers melt?

WILEY
RESOURCE KIT

Solar intensity and latitude • Figure 9.3 ___

Light strikes vertically

Light strikes at an angle

Larger area of illumination

Small area of illumination

Temperature Changes with Latitude and Season Earth's roughly spherical shape and the tilt of its axis produce a great deal of variation in the exposure of the surface to solar energy (**Figure 9.3**). Sunlight that shines vertically near the equator (represented by the desk lamp on the left) is concentrated on Earth's surface. As one moves toward the poles, the light hits the surface more and more obliquely (represented by the lamp on the right), spreading the same amount of radiation over larger and larger areas. Because the Sun's energy does not reach all places uniformly, temperature varies locally.

Earth's inclination on its axis (23.5 degrees from a line drawn perpendicular to the orbital plane)

determines the seasons. During half the year (March 21 to September 21) the Northern Hemisphere tilts toward the Sun, and during the other half (September 22 to March 20) it tilts away from the Sun (**Figure 9.4**). The Southern Hemisphere tilts the opposite way, so that summer in the Northern Hemisphere corresponds to winter in the Southern Hemisphere.

Precipitation

Precipitation refers to any form of water, such as rain, snow, sleet, or hail, that falls from the atmosphere. Differences in precipitation depend on three factors:

1. **The amount of water vapor in the atmosphere.** *Equatorial uplift* of warm, moisture-laden air produces heavy rainfall in some areas of the tropics. High surface water temperatures cause vast quantities of water to evaporate from tropical parts of the ocean. Prevailing winds blow the resulting moist air over landmasses. The moist air continues to rise as it is heated by the Sun-warmed land surface. As the air rises, it cools, which decreases its moisture-holding ability. When the air reaches its saturation point—when it can't hold any additional water vapor—clouds form and water is released as precipitation.

2. **Geographic location.** The rising air from the equator eventually descends to Earth near the Tropic of Cancer and Tropic of Capricorn (latitudes 23.5 degrees north and 23.5 degrees south, respectively). By then most of its moisture has precipitated, and the

Progression of seasons • Figure 9.4 ___

Earth's inclination on its axis remains the same as it travels around the Sun. Thus, the Sun's rays hit the Northern Hemisphere obliquely during its winter months and more directly during its summer. In the Southern Hemisphere, the Sun's rays are oblique during its winter, which corresponds to the Northern Hemisphere's summer. At the equator, the Sun's rays are approximately vertical on March 20 and September 22.

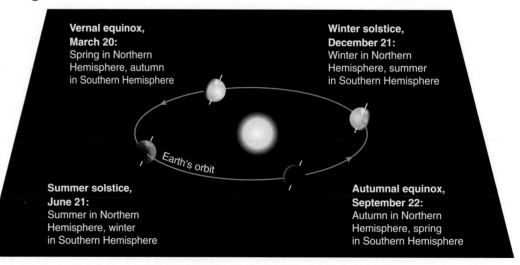

Vernal equinox, March 20: Spring in Northern Hemisphere, autumn in Southern Hemisphere

Winter solstice, December 21: Winter in Northern Hemisphere, summer in Southern Hemisphere

Earth's orbit

Summer solstice, June 21: Summer in Northern Hemisphere, winter in Southern Hemisphere

Autumnal equinox, September 22: Autumn in Northern Hemisphere, spring in Southern Hemisphere

WHAT A SCIENTIST SEES

Rain Shadow

Windward side

Leeward side

Moist air

Dry air

Rain shadow desert

a. A rain shadow refers to arid or semiarid land that occurs on the far side (leeward side) of a mountain. Prevailing winds blow warm, moist air from the windward side. Air temperature cools as it rises, releasing precipitation, so dry air descends on the leeward side. Such a rain shadow exists east of the Cascades.

b. Proxy Falls is in the Cascade Range, which divides the states of Washington and Oregon into a moist western region and an arid region east of the mountains.

dry air returns to the equator. Over land, this dry air produces some of the great tropical deserts, such as the Sahara Desert. Air also dries out as it travels long distances over landmasses. Near the windward coasts of continents (the side from which the wind blows), rainfall may be heavy. However, in the temperate zones (the areas between the tropics and the polar zones), continental interiors are usually dry because they are far from the ocean, which replenishes water in the air passing over it.

3. **Topographic features.** When flowing air encounters mountains, it flows up and over them, cooling as it gains altitude. Because cold air holds less moisture than does warm air, clouds form and precipitation occurs, primarily on the mountains' windward slopes. The air mass is warmed as it moves down on the other side of the mountain, reducing the chance of

precipitation of the remaining moisture. This situation exists on the West Coast of North America, where precipitation falls on the western slopes of mountains close to the coast (see *What a Scientist Sees*). The dry land on the side of the mountains away from the prevailing wind—in this case, east of the mountain range—is called a **rain shadow**. The dry conditions of a rain shadow often occur on a regional scale.

CONCEPT CHECK	STOP

1. **How** do you distinguish between weather and climate?

2. **What** effect does solar energy have on Earth's temperature?

3. **What** are some of the environmental factors that produce areas of precipitation extremes, such as rain forests and deserts?

Global Climate Change

LEARNING OBJECTIVES

1. **List** the five main greenhouse gases and describe the enhanced greenhouse effect.

2. **Discuss** some of the potential effects of global climate change.

3. **Give** examples of strategies to mitigate or adapt to global climate change.

Earth's average temperature is based on daily measurements taken at several thousand land-based meteorological stations around the world, as well as data from weather balloons, orbiting satellites, transoceanic ships, and hundred of sea-surface buoys with temperature sensors. These data show that the years 1998 and 2001 through 2009 were the hottest since records began in the 1880s. According to the National Oceanic and Atmospheric Administration (NOAA), global temperatures in those years may have been the highest in the last millennium. (Although widespread thermometer records have been assembled only since the mid-19th century, scientists reconstruct earlier temperatures using indirect climate evidence in tree rings, lake and ocean sediments, small air bubbles in ancient ice, and coral reefs.) The last two decades of the 20th century were its warmest (**Figure 9.5**).

Other evidence also suggests an increase in global temperature. Several studies indicate that spring in the Northern Hemisphere now comes about six days earlier than it did in 1959, and autumn comes five days later. Since 1949, the United States has experienced an increased frequency of heat waves, resulting in increased heat-related deaths among elderly and other vulnerable people. In the past few decades, the sea level has risen, glaciers worldwide have retreated, and while hurricanes may not have become more frequent, higher ocean surface temperatures have made them increasingly severe.

Scientists around the world have been researching global climate change for the past 50 years. As the evidence has accumulated, those most qualified to address the issue have reached a strong consensus that the 21st century will experience significant climate change for which human activities are largely responsible.

In response to this growing consensus, governments around the world organized the United Nations Intergovernmental Panel on Climate Change (IPCC). With input from hundreds of climate experts, the IPCC provides a definitive scientific assessment of global climate change. In its most recent (2007) report, the IPCC

Mean annual global temperature, 1960 to present • Figure 9.5

Data are presented as surface temperatures (°C) for each year since 1960. The measurements, which naturally fluctuate, show the warming trend of the past several decades. The dip in global temperatures in the early 1990s was caused by the eruption of Mount Pinatubo in 1991.

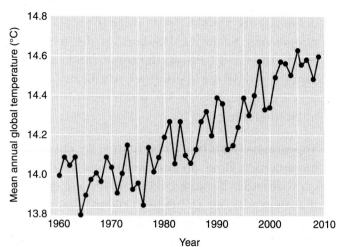

projected a 1.8° to 4.0°C (3.2° to 7.2°F) increase in global temperature by the year 2100. The IPCC predicts that we will observe higher maximum temperatures and more hot days over nearly all land areas, higher minimum temperatures, fewer frost days, fewer cold days, and an increase in the heat index. We may also experience more intense precipitation events over many areas, an increased risk of drought in the continental interiors in the mid-latitudes, and stronger hurricanes in some coastal areas.

Causes of Global Climate Change

Carbon dioxide (CO_2) and certain trace gases, including methane (CH_4), nitrous oxide (N_2O), chlorofluorocarbons (CFCs), and tropospheric ozone (O_3), accumulate in the atmosphere as a result of human activities. The concentration of atmospheric carbon dioxide has increased from about 288 parts per million (ppm) approximately 200 years ago (before the Industrial Revolution began) to 390 ppm in 2009 (**Figure 9.6**). According to the U.N. Food and Agriculture Organization, burning carbon-containing fossil fuels accounts for about 70 to 75 percent of human-generated CO_2 increase. The remaining 25 to 30 percent is released through deforestation, particularly when people fell or burn tropical rain forests. By 2050 the concentration of atmospheric CO_2 may be double what it was in the 1700s.

The combustion of gasoline in your car's engine releases not only CO_2 but also N_2O which triggers the production of tropospheric ozone. Various industrial processes, land-use conversion, and the use of fertilizers also produce nitrous oxide. CFCs (discussed later in the chapter as they relate to depletion of the stratospheric ozone layer) are chemicals released into the atmosphere from old, leaking refrigerators and air conditioners. Methane is produced by the decomposition of carbon-containing organic material by anaerobic bacteria in moist places as varied as rice paddies, sanitary landfills, and the intestinal tracts of large animals. Tundra that is thawing due to warmer global temperatures also contributes methane—an example of an unfortunate feedback.

Global climate change occurs because these gases absorb infrared radiation—that is, heat—given off by Earth's surface. This absorption slows the natural flux of heat into space, warming the lower atmosphere.

Carbon dioxide (CO_2) in the atmosphere, 1958 to present • Figure 9.6

Note the steady increase in the concentration of atmospheric CO_2 since 1958, when measurements began at the Mauna Loa Observatory in Hawaii. This location was selected because it is far from urban areas where factories, power plants, and motor vehicles emit CO_2. The seasonal fluctuations correspond to winter (a high level of CO_2), when plants are not actively growing and absorbing CO_2, and summer (a low level of CO_2), when plants are growing and absorbing CO_2.

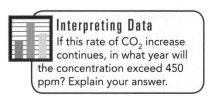

Interpreting Data
If this rate of CO_2 increase continues, in what year will the concentration exceed 450 ppm? Explain your answer.

Because CO_2 and other gases trap the Sun's infrared radiation somewhat like glass does in a greenhouse, they are called **greenhouse gases**. Greenhouse gases accumulating in the atmosphere as a result of human activities

greenhouse gases
Gases—including water vapor, carbon dioxide, methane, and certain other gases—that absorb infrared radiation.

Enhanced greenhouse effect • Figure 9.7

✓ THE PLANNER

The buildup of carbon dioxide (CO_2) and other greenhouse gases warms the atmosphere by absorbing some of the outgoing infrared (heat) radiation. Some of the heat in the warmed atmosphere is transferred back to Earth's surface, warming the land and ocean.

1 Sunlight (radiant energy) is absorbed at surface.

2 Some heat radiated from Earth escapes directly to space.

3 Some heat radiated from Earth is absorbed by greenhouse gases.

4 Some of this heat is transferred back to Earth's surface.

WILEY
RESOURCE KIT

Year	Temperature (°C)	CO_2 concentration (ppm)
1880	13.75	~ 280
1957	13.97	315
1967	14.00	322
1977	14.13	335
1987	14.26	350
1997	14.39	365
2007	14.44	385

are causing an **enhanced greenhouse effect** (**Figure 9.7**).

Some air pollutants, known as atmospheric aerosols, tend to cool the atmosphere. **Aerosols**, which come from both natural and human sources, are particles so small they remain suspended in the atmosphere for days, weeks, or even months. Sulfur haze is an aerosol that reflects sunlight back into space, reducing the amount of solar energy reaching Earth's surface, and thereby cooling the atmosphere. Sulfur haze significantly moderates warming in industrialized parts of the world. Sulfur emissions come from the same smokestacks that emit CO_2. Volcanic eruptions also eject sulfur particles into the atmosphere (**Figure 9.8**).

This cooling effect, however, is much weaker than the enhanced greenhouse effect. Human-produced sulfur emissions typically remain in the atmosphere for

> **enhanced greenhouse effect**
> Additional atmospheric warming produced as human activities increase atmospheric concentrations of greenhouse gases.

less than a year and rarely disperse globally. Greenhouse gases can remain in the atmosphere for hundreds of years. And carbon dioxide and other greenhouse gases help warm the planet 24 hours a day, whereas sulfur haze cools the planet only during the daytime. In addition, sulfur emissions are a respiratory irritant and cause acid deposition (discussed later in this chapter). Most nations are trying to reduce their sulfur emissions, not maintain or increase them.

Effects of Global Climate Change

In addition to melting land and ocean ice, already considered in the chapter introduction, some of the observed and potential effects of global climate change include sea-level rise, changes in precipitation patterns, and impacts on agriculture, human health, and other organisms.

Volcanic eruption • Figure 9.8

The eruption of Mount Pinatubo in the Philippines in 1991 injected massive amounts of sulfur into the atmosphere. Because sulfur haze reduces the amount of sunlight reaching the surface, this eruption caused Earth to cool temporarily. Compared to temperatures during the rest of the 1990s, global temperatures in 1992 and 1993 were relatively cool.

Two factors contribute to sea-level rise. First, as the Antarctic ice cap and continental glaciers melt, the amount of water in the ocean increases. Second, as water heats up, it expands. During the 20th century, the sea level rose 10 to 20 cm (4 to 8 in.). Climate scientists estimate an additional 48 cm (19 in.) rise by 2100. As sea level rises, small island nations such as the Maldives, a low-lying chain of islands in the Indian Ocean, will be increasingly vulnerable to saltwater intrusion and storm surges (**Figure 9.9**).

Computer models of weather changes caused by global climate change indicate that precipitation patterns will change, causing some areas to have more frequent droughts. At the same time, heavier snow and rainstorms may cause more frequent flooding in other areas. These changes could lower the availability and quality of fresh water in many locations, particularly in areas that are currently arid or semiarid, such as the Sahel region just south of the Sahara Desert.

Low-lying island • Figure 9.9

Residents of this low-lying island, which is part of the Maldives (in the Indian Ocean), may need to relocate as sea level continues to rise.

In a climate-warmed world, the mosquito that spreads malaria could expand into temperate areas. Photographed in Ariquemes, Brazil, where most of the town's inhabitants suffer from malaria.

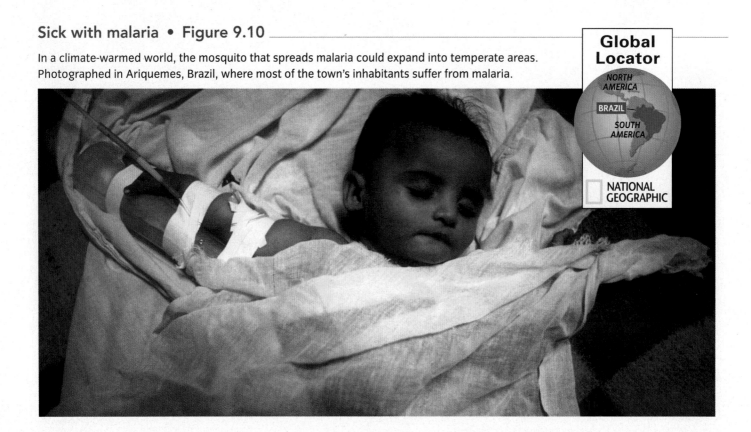

Global Locator

NORTH AMERICA

BRAZIL

SOUTH AMERICA

NATIONAL GEOGRAPHIC

Global climate change will have mixed effects on agriculture. The rise in sea level will inundate some river deltas, which are fertile agricultural lands. Certain agricultural pests and disease-causing organisms will probably proliferate and reduce crop yields. Increased frequency and duration of droughts will be a particularly serious problem, and lack of water for drinking and agriculture may force millions of people to relocate. However, agricultural productivity may increase in some areas.

Currently, most evidence linking climate warming to disease outbreaks is circumstantial. Nonetheless, data linking climate warming and human health problems are accumulating. More frequent and more severe heat waves during summer have increased the number of heat-related illnesses and deaths. Mosquitoes and other disease carriers are expected to expand their range into the newly warm areas and spread dengue fever, schistosomiasis, yellow fever and malaria (**Figure 9.10**). According to the World Health Organization, during 1998, the fourth warmest year on record, the incidence of malaria, Rift Valley fever, and cholera surged in developing countries.

An increasing number of studies report measurable changes in the biology of plant and animal species as a result of climate warming. Climate change also affects populations, communities, and ecosystems. In *Environmental*

InSight: The Effects of Global Climate Change (on facing page), we report on the results of several of the hundreds of studies conducted thus far.

Rising temperatures in the waters around Antarctica have led to a decline in the populations of shrimplike krill and Antarctic silverfish, major food sources for Adélie penguins, reducing Adélie penguin populations (**Figure 9.11a**). Warmer temperatures also cause higher rates of reproductive failure in these penguins by producing puddles of melted snow that kill developing chick embryos at egg-laying sites.

Worldwide, many frog populations have plummeted; these include Puerto Rico's national symbol, the tiny tree frog known as coqui (**Figure 9.11b**). Warmer temperatures and more frequent dry periods have stressed the coqui, making them more vulnerable to infection by a lethal fungus.

Ecosystems considered at greatest risk of climate-change loss are polar seas, coral reefs, mountain ecosystems, coastal wetlands, and tundra. Water temperature increases of 1° to 2°C (1.8° to 3.6°F) cause coral bleaching, which contributes to the destruction of coral reefs (**Figure 9.11c**). In 1998, when tropical waters were some of the warmest ever recorded, about 10 percent of the world's corals died.

As atmospheric CO_2 increases, some of it dissolves in the ocean, producing carbonic acid (**Figure 9.11d**).

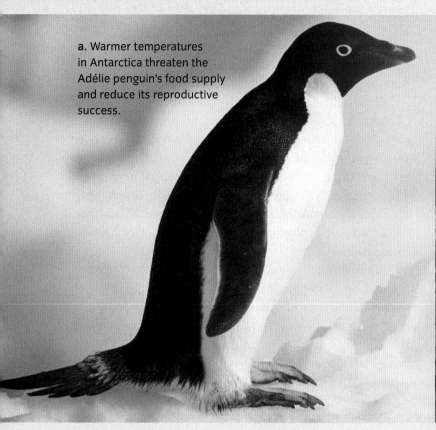

a. Warmer temperatures in Antarctica threaten the Adélie penguin's food supply and reduce its reproductive success.

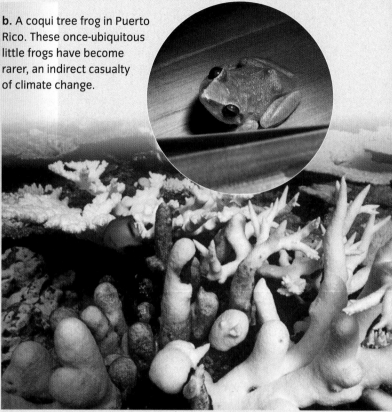

b. A coqui tree frog in Puerto Rico. These once-ubiquitous little frogs have become rarer, an indirect casualty of climate change.

c. Ocean warming and acidification stress corals, causing them to become bleached. Photographed near the Maldives in the Indian Ocean.

1995

2100

TROPIC OF CANCER

EQUATOR

TROPIC OF CAPRICORN

Not corrosive

Acid levels

More corrosive

 No data

d. Scientific models project that ocean water will become increasingly acidic if human-produced CO_2 levels continue to rise. Shown are computer models for 1995 and 2100 from the NOAA Pacific Marine Environmental Laboratory.

 How might plants and animals where you live be affected by a 2°C increase in the lowest winter temperature each year? The same increase in the highest summer temperature?

Interpreting Data
In which parts of the ocean will organisms that are sensitive to acidification be most affected? least affected?

 WILEY
RESOURCE KIT

The consequent acidification could be disastrous for shelled sea animals, particularly zooplankton at the base of the marine food web; the acid would attack and dissolve away their shells. Increased acidity also exacerbates coral bleaching. It could take centuries for the atmosphere–ocean CO_2 balance to stabilize, so even if emissions cease, the ocean will continue to acidify.

Dealing with Global Climate Change: Mitigation and Adaptation

There are basically two ways to manage global climate change: mitigation and adaptation. *Mitigation* is the moderation or postponement of global climate change through measures that reduce greenhouse gas emissions. *Adaptation* is a planned response to changes caused by global climate change. Because the impacts of climate change have already begun, some combination of mitigation and adaptation is necessary to avoid severe or disruptive effects. The earlier we begin implementing strategies for mitigation and adaptation, the more effective they are likely to be.

We will have to deal with all greenhouse gases as we develop strategies to address global climate change, but we focus on CO_2 because it is produced in the greatest quantity and has the largest total effect. Carbon dioxide has an atmospheric lifetime of more than a century, so emissions produced today will still be around in the 22nd century. The extent and severity of global climate change will depend on the amount of additional greenhouse gas emissions we add to the atmosphere. Many studies make the assumption that atmospheric CO_2 will stabilize at 550 ppm, which is roughly twice the concentration of atmospheric CO_2 in the preindustrial world and almost 50 percent higher than the CO_2 currently in the atmosphere.

Mitigation of Global Climate Change
Because most of the CO_2 that human activities result from burning coal, oil, and natural gas, climate change is essentially an energy issue. Developing alternatives to fossil fuels offers a solution to warming caused by CO_2 emissions. We address alternatives to fossil fuels, including solar, hydroelectric, wind, and nuclear power, in Chapters 17 and 18.

Reducing energy use (for example, by driving less) and increasing efficiency (for example, by switching to hybrid cars) will reduce our output of CO_2, and will help mitigate global climate change. Energy-pricing strategies, such as carbon taxes and the elimination of energy subsidies, are other policies that could mitigate global climate change. Most experts think using current technologies and developing such policies could significantly reduce greenhouse gas emissions with little cost to society.

Planting and maintaining forests also mitigates global climate change. Like other green plants, trees remove carbon dioxide from the air and incorporate the carbon into organic matter through photosynthesis. Reasonable estimates suggest that trees could remove 10 percent to 15 percent of the excess CO_2 in the atmosphere, but only through enormous plantings, so such efforts should not be considered a substitute for cutting emissions of greenhouse gases.

Many countries are investigating **carbon management**. Several experimental power plants currently capture CO_2 from their flue gases, but the technology is new (**Figure 9.12**). Technological innovations that more efficiently trap CO_2 from smokestacks would help mitigate global climate change and allow us to continue using fossil fuels (while they last) for energy. The carbon could be sequestered in geologic formations or in depleted oil or natural gas wells on land.

> **carbon management**
> Ways to separate and capture the CO_2 produced during the combustion of fossil fuels and then sequester (store) it.

Additional strategies to mitigate climate change include the following:

- Planting trees on degraded land
- Increasing efficiency of coal-fired power plants
- Replacing coal-fired power plants with nuclear power, hydropower, wind power, or even natural gas
- Increasing fuel economy of motor vehicles
- Redesigning cities to reduce reliance on single-occupant vehicles
- Insulating buildings to reduce the need for heating in the winter and cooling in the summer

Adaptation to Global Climate Change
Because the overwhelming majority of climate experts think human-induced global climate change will continue, government planners and social scientists are developing strategies to help various regions and sectors of society

Vattenhall carbon capture and storage pilot project • Figure 9.12

In the Vattenhall carbon capture and storage pilot project, coal is burned in a gasifier **1**. Particulates, sulfur, and other contaminants are removed **2**. After that, CO_2 is absorbed and removed **3**; the hot, CO_2-free gas is then used to generate electricity, first in a gas turbine **4** and then in a heat recovery steam generator **5**. Meanwhile, the CO_2 is piped away for storage underground **6**.

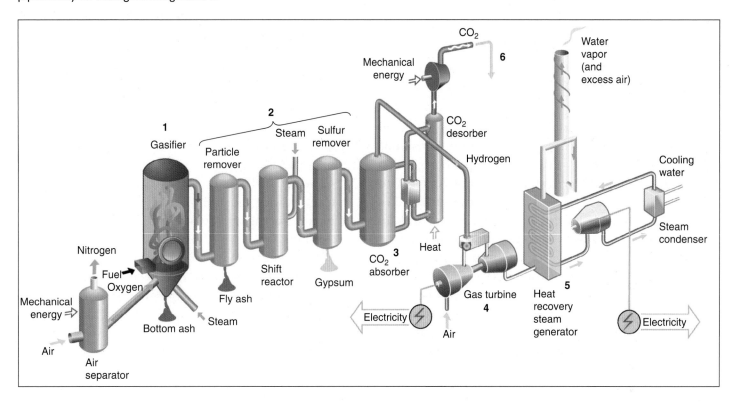

adapt to climate warming. One of the most pressing issues is rising sea level. People living in coastal areas could be moved inland, away from the dangers of storm surges, although the societal and economic costs would be great. Another extremely expensive alternative is the construction of massive sea walls to protect coastal land. Rivers and canals that spill into the ocean could be channeled to prevent *saltwater intrusion* into fresh water and agricultural land.

We must also adapt to shifting agricultural zones. Countries with temperate climates are evaluating semitropical crops to determine the best substitutes for traditional crops as the climate warms. Large lumber companies are developing heat- and drought-resistant strains of trees that will be harvested when global climate change may be well advanced. Evaluating such problems and finding and implementing solutions now will ease future stresses of climate warming.

Adaptation to global climate change is under study at several locations around the United States. One of the

problems identified in a New York City study involves its sewer system. The waterways for storm runoff normally close during high tides. As the sea level rises in response to global climate change, the waterways will have to be shut during many low tides, which will increase the risk of flooding during storms (because excess water will not drain away). City planners will have to rebuild the storm runoff system or find some other way to prevent flooding. Evaluating such problems and implementing solutions now will ease future stresses of climate warming.

CONCEPT CHECK   STOP

1. **What** are greenhouse gases?
2. **How** will climate change affect agriculture? wildlife?
3. **What** are two examples of each of the approaches to manage global climate change: mitigation and adaptation?

Ozone Depletion in the Stratosphere

LEARNING OBJECTIVES

1. **Describe** the importance of the stratospheric ozone layer.

2. **Explain** how ozone thinning takes place and relate some of its harmful effects.

3. **Relate** how the international community is working to protect the ozone layer.

Although ozone (O_3) is a human-made pollutant in the troposphere, it is a naturally produced, essential component in the stratosphere, which encircles our planet some 10 to 45 km (6 to 28 mi) above the surface. The ozone layer shields Earth's surface from much of the high-energy **ultraviolet (UV) radiation** coming from the Sun (**Figure 9.13a** and **b** in Environmental Insight: The Ozone Layer). If ozone disappeared from the stratosphere, Earth would become uninhabitable for most forms of life, including humans.

A slight **ozone thinning** occurs naturally over Antarctica for a few months each year. In 1985, however, the thinning was first observed to be greater than it should be if natural causes were the only factor inducing it. This increased thinning, which occurs each September, is commonly referred to as the "ozone hole" (**Figure 9.13c**). There, ozone levels decrease as much as 70 percent each year.

During the subsequent two decades the ozone-thinned area continued to grow, and by 2006 it had reached the record size of 29.5 million km^2 (11.4 million mi^2), which is larger than the North American continent. A smaller thinning was also detected in the stratospheric ozone layer over the Arctic. In addition, world levels of stratospheric ozone have been decreasing for several decades (**Figure 9.13d**). According to the National Center for Atmospheric Research, ozone levels over Europe and North America have dropped almost 10 percent since the 1970s.

ultraviolet (UV) radiation Radiation from the part of the electromagnetic spectrum with wavelengths just shorter than visible light; can be lethal to organisms at high levels of exposure.

ozone thinning The removal of ozone from the stratosphere by human-produced chemicals or natural processes.

Causes of Ozone Depletion

The primary chemicals responsible for ozone loss in the stratosphere are a group of industrial and commercial compounds called **chlorofluorocarbons (CFCs)**. Scientists first discovered that CFCs can deplete stratospheric ozone in the mid-1970s. Chlorofluorocarbons such as Freon were used as propellants for aerosol cans and coolants in air conditioners and refrigerators. Other CFCs were used as solvents and as foam-blowing agents for insulation and packaging (Styrofoam, for example).

Other compounds that destroy ozone include *halons*, used as fire retardants; *methyl bromide*, a pesticide; *methyl chloroform* and *carbon tetrachloride*, industrial solvents; and *nitrous oxide*, released from the burning of fossil fuels (particularly coal) and from the breakdown of nitrogen fertilizers in the soil.

chlorofluorocarbons (CFCs) Human-made organic compounds that contain chlorine and fluorine; now banned because they attack the stratospheric ozone layer.

Effects of Ozone Depletion

With depletion of the ozone layer, more UV radiation reaches the Earth's surface. Increased levels of UV radiation may disrupt ecosystems. For example, the productivity of Antarctic phytoplankton, the microscopic drifting algae that are the base of the Antarctic food web, has declined due to increased exposure to UV radiation. (The UV radiation inhibits photosynthesis.) Biologists have documented direct UV damage to natural populations of Antarctic fish. Widespread decline of amphibian populations may be linked to increased UV radiation. Because organisms are interdependent, the negative effect on one species has ramifications throughout the ecosystem.

Excessive exposure to UV radiation is linked to several health problems in humans, including eye cataracts, skin cancer, and weakened immunity. Malignant melanoma, the most dangerous type of skin cancer, is increasing faster than any other type of cancer.

30 km (19 mi)

Ozone layer

15 km (9 mi)

Ozone layer

a. Stratospheric ozone absorbs about 99 percent of incoming solar ultraviolet (UV) radiation, effectively shielding the surface.

b. When stratospheric ozone is present at reduced levels, more high-energy UV radiation penetrates the atmosphere to the surface, where its presence harms organisms.

c. Ozone depletion. A computer-generated image of part of the Southern Hemisphere, taken in December 2007, reveals ozone thinning (the purple area over Antarctica). The ozone-thin area is not stationary but moves about as a result of air currents.

d. Average yearly ozone column over New Zealand, 1980 to 2006. Located in the Southern Hemisphere, New Zealand is particularly vulnerable to increasing UV radiation due to ozone thinning.

Total ozone (Dobson units)

110   220   330   440   550

WILEY
RESOURCE KIT

Interpreting Data
In what year did the average yearly column of ozone above New Zealand first drop below 300 Dobson units?

EnviroDiscovery

Links Between Climate and Atmospheric Change

Most environmental studies examine a single issue, such as acid deposition, global climate change, or ozone depletion. In the past few years, however, some researchers have been exploring the interactions of all three problems simultaneously. One study of such interactions found that North American lakes may be more susceptible to damage from UV radiation than the thinning of the ozone hole would indicate. The reason: Organic matter in the lakes, which absorbs some UV radiation and protects the lakes' plant and fish life, is affected by acid deposition and global climate change. Acid deposition reacts with organic matter in lakes, causing it to settle to the lake floor, where it does not absorb as much of the UV radiation as it once did. And a warmer climate increases evaporation, which reduces the amount of organic matter washed into lakes by streams.

Several studies report a link between human-caused climate warming and polar ozone depletion. Greenhouse gases that warm the troposphere also contribute to stratospheric cooling, presumably because heat trapped in the troposphere is not available to warm the stratosphere. The stratospheric temperature has been dropping for the past several years, and these lower temperatures provide better conditions for ozone-depleting chemicals to attack stratospheric ozone. Record ozone holes over Antarctica are attributed to cooler stratospheric temperatures. Some scientists speculate that if the cooling trend in the stratosphere continues, recovery of the ozone layer may be delayed. This means climate warming could prolong ozone depletion in the stratosphere despite the success of the Montreal Protocol.

Scientists now know environmental problems can't be studied as separate issues because they often interact in surprisingly subtle ways. As global climate change, ozone depletion, and acid deposition are studied further, it is likely that other interactions will be discovered.

Helping the Ozone Layer Recover

In 1978 the United States, the world's largest user of CFCs, banned the use of CFC propellants in products such as antiperspirants and hair sprays. Although this ban was a step in the right direction, it did not solve the problem. Most nations did not follow suit, and propellants represented only a small portion of all CFC use.

In 1987 representatives from many countries met in Montreal to sign the **Montreal Protocol**, an agreement that originally stipulated a 50 percent reduction of CFC production by 1998. Despite this effort, stratospheric ozone continued to thin over the heavily populated mid-latitudes of the Northern Hemisphere, and the Montreal Protocol was modified to include even stricter limits on CFC production.

Industrial companies that manufacture CFCs quickly developed substitutes, such as hydrofluorocarbons (HFCs) and hydrochlorofluorocarbons (HCFCs). HFCs do not attack ozone, although they are potent greenhouse gases. HCFCs attack ozone but are less destructive than the chemicals they are replacing.

CFC, carbon tetrachloride, and methyl chloroform production was almost completely phased out in the United States and other highly developed countries in 1996, except for a relatively small amount exported to developing countries. Developing countries were on a different timetable and phased out CFC use in 2005. Methyl bromide was phased out in highly developed countries, which were responsible for 80 percent of its global use, in 2005. HCFCs will be phased out in 2030.

Unfortunately, CFCs are extremely stable and will probably continue to deplete stratospheric ozone for several decades. Human-exacerbated ozone thinning will reappear over Antarctica each year, although the area and degree of thinning will gradually decline over time, until full recovery takes place sometime after 2050.

CONCEPT CHECK

1. **What** is the stratospheric ozone layer and how does it protect life on Earth?
2. **What** is stratospheric ozone thinning? What role do CFCs play in ozone thinning?
3. **How** have governments responded to ozone thinning?

Acid Deposition

LEARNING OBJECTIVES

1. **Define** *acid deposition* and explain how acid deposition develops.
2. **Relate** examples of the effects of acid deposition.

What do fishless lakes in the Adirondack Mountains, recently damaged Mayan ruins in southern Mexico, and dead trees in the Czech Republic have in common? All these problems are the result of acid precipitation or, more properly, **acid deposition** (**Figure 9.14a**). Acid deposition has been around since the Industrial Revolution began. Robert Angus Smith, a British chemist, coined the term *acid rain* in 1872 after he noticed that buildings in areas with heavy industrial activity were being worn away by rain.

> **acid deposition** A type of air pollution that includes sulfuric and nitric acids in precipitation, as well as dry acid particles that settle out of the air.

Acid precipitation, including acid rain, sleet, snow, and fog, poses a serious threat to the environment. Until recently, industrialized countries in the Northern Hemisphere had been hurt the most, especially the Scandinavian countries, central Europe, Russia, and North America. In the United States alone, the annual damage from acid deposition is estimated at $10 billion. Acid deposition is now recognized as a global problem because it also occurs in developing countries as they become industrialized. For example, Chinese scientists have reported that acid deposition affects 40 percent of their country. In the two decades from 1990 to 2010, the amount of sulfur dioxide released from burning high-sulfur coal tripled in China (**Figure 9.14b**). Sulfur dioxide becomes sulfuric acid in the atmosphere.

Sources and effects of acid deposition • Figure 9.14

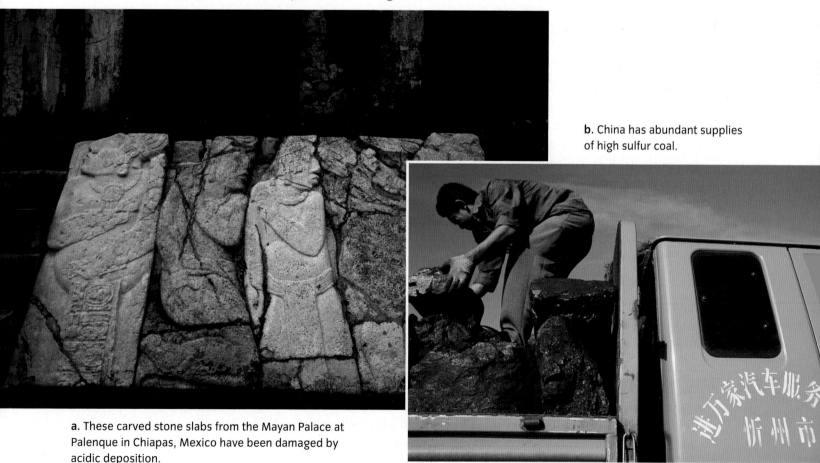

b. China has abundant supplies of high sulfur coal.

a. These carved stone slabs from the Mayan Palace at Palenque in Chiapas, Mexico have been damaged by acidic deposition.

How Acid Deposition Develops

The processes that lead to acid deposition begin when sulfur dioxide and nitrogen oxides are released into the atmosphere (**Figure 9.15a**). Motor vehicles are a major source of nitrogen oxides. Coal-burning power plants, large smelters, and industrial boilers are the main sources of sulfur dioxide emissions and produce substantial amounts of nitrogen oxides as well. Wind carries sulfur dioxide and nitrogen oxides, released into the air from tall smokestacks, for long distances. Tall smokestacks allow England to "export" its acid deposition problem to the Scandinavian countries and the midwestern United States to "export" its acid emissions to New England and Canada.

In the atmosphere, sulfur dioxide and nitrogen oxides react with water to produce dilute solutions of sulfuric acid (H_2SO_4), nitric acid (HNO_3), and nitrous acid (HNO_2). Acid deposition returns these acids to Earth's surface in the form of precipitation or particulates.

Effects of Acid Deposition

The link between acid deposition and declining aquatic animal populations, particularly fish, is well established, but other animals are also adversely affected. Birds living in areas with pronounced acid deposition are at increased risk of laying eggs with thin, fragile shells that break or dry out before the chicks hatch. The inability to produce strong eggshells is attributed to reduced calcium in the birds' diets. Calcium is less available to the food chain because in acidic soils calcium becomes soluble and is washed away, with little left for plant roots to absorb.

Acid deposition also has a serious effect on forest ecosystems. In the Black Forest of Germany, for example, up to 50 percent of trees surveyed are dead or severely damaged. This **forest decline** appears to result from a combination of stressors, including tropospheric ozone, UV radiation (which is more intense at higher altitudes), insect attack, drought, and acid deposition. When one or more stressors weaken a tree, an additional stressor, such as air pollution, may be decisive in causing the tree's death (**Figure 9.15b** and **c**).

forest decline A gradual deterioration and eventual death of many trees in a forest.

Acid deposition can also damage agriculture, and it corrodes metals, building materials, and statues (**Figure 9.15d**). It eats away at important monuments, such as the Washington Monument in Washington, DC, and ancient Mayan ruins in southern Mexico.

The Politics of Acid Deposition

Acid deposition is hard to combat because it does not occur only in the locations where acidic gases are emitted. It is entirely possible for sulfur and nitrogen oxides released in one spot to return to Earth's surface hundreds of kilometers from their source.

The United States has wrestled with this issue. Several states in the Midwest and East—Illinois, Indiana, Missouri, Ohio, Pennsylvania, Tennessee, and West Virginia—produce between 50 percent and 75 percent of the acid deposition that contaminates New England and southeastern Canada. Legislation formulated to deal with acid deposition has led to arguments about who should pay for the installation of expensive devices to reduce emissions of sulfur and nitrogen oxides.

In international disputes, these issues are magnified even more. For example, gases from coal-burning power plants in England move eastward with prevailing winds and return to the surface as acid deposition in Sweden and Norway. Similarly, emissions from mainland China produce acid deposition in Japan, Taiwan, North Korea, and South Korea.

Facilitating Recovery from Acid Deposition

Although the science and the politics surrounding acid deposition are complex, the basic concept of control is straightforward: Reducing emissions of sulfur and nitrogen oxides curbs acid deposition. Simply stated, if sulfur and nitrogen oxides are not released into the atmosphere, they cannot come down as acid deposition. Installing scrubbers in the smokestacks of coal-fired power plants and using clean-coal technologies to burn coal without excessive emissions effectively diminish acid deposition. In turn, a decrease in acid deposition prevents surface waters and soil from becoming more acidic than they already are.

a. Acid deposition.
Sulfur dioxide and nitrogen oxide emissions react with water vapor in the atmosphere to form acids that return to the surface as either dry or wet deposition.

Forest Decline.
Acid deposition is one of several stressors that may interact, contributing to the decline and death of trees.

b. Healthy Sitka spruce branch.

Wind

$SO_2 + NO_x$

Conversion to acids:
sulfuric acid (H_2SO_4)
nitric acid (HNO_3)
nitrous acid (HNO_2)

Wet acid deposition
(droplets of H_2SO_4, HNO_3, and HNO_2 dissolved in rain and snow)

(NO)

Power plant and industrial plumes

Mobile emissions

Dry acid deposition

Increasingly acidic lakes

Surface runoff

c. Sitka spruce branch exhibiting the effects of forest decline. Photographed in Black Forest, Germany.

d. Acid rain damage. Acid rain has destroyed much of the detail of this stone angel's face. Photographed in London.

WILEY
RESOURCE KIT

Rainfall in parts of the Midwest, Northeast, and Mid-Atlantic regions is less acidic today than it was two decades ago, as a result of cleaner-burning power plants and the use of reformulated gasoline. Many power plants in the Ohio Valley switched from high-sulfur to low-sulfur coal. However, solving one environmental problem often creates others. While the move to low-sulfur coal reduced sulfur emissions, it contributed to the problem of global climate change. Because low-sulfur coal has a lower heat value than high-sulfur coal, more of it must be burned—and more CO_2 emitted—to generate a given amount of electricity. Low-sulfur coal also contains higher levels of mercury and other trace metals, so burning it adds more of these hazardous pollutants to the air.

However, a 1991 agreement between the United States and Canada, along with legislation within each country, has reduced acid deposition by as much as 35 percent in some areas.

Despite the fact that the United States, Canada, and many European countries have reduced sulfur emissions, acid precipitation remains a serious problem. Acidified forests and bodies of water have not recovered as quickly as hoped. Trees in the U.S. Forest Service's Hubbard Brook Experimental Forest in New Hampshire, an area damaged by acid deposition, have grown little, even following two decades of declining emissions. Many northeastern streams and lakes, such as those in New York's Adirondack Mountains, remain acidic (**Figure 9.16**). A likely reason for the slow recovery is that the past 30 or more years of acid rain have profoundly altered soil chemistry in many areas. Essential plant minerals such as calcium and magnesium have washed away from forest and lake soils. Because soils take hundreds or even thousands of years to develop, it may take that long for them to recover from the effects of acid rain.

Many scientists are convinced that ecosystems will not recover from acid rain damage until substantial reductions in nitrogen oxide emissions occur. Nitrogen oxide emissions are harder to control than sulfur dioxide emissions because motor vehicles produce a substantial portion of nitrogen oxides. Engine improvements may help reduce nitrogen oxide emissions, but as the human population continues to grow, the increasing number of motor vehicles will probably offset any engineering gains. Dramatic cuts

Acidic stream • Figure 9.16

This mountain stream in the Adirondack Mountains of New York may be acidic because years of acid precipitation have altered soil chemistry.

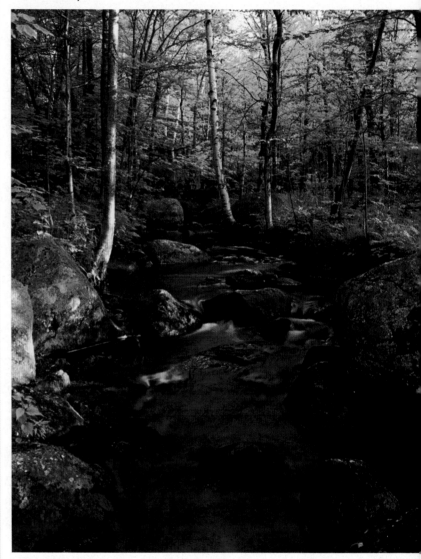

in nitrogen oxide emissions will require a reduction in high-temperature energy generation, especially in gasoline and diesel engines.

CONCEPT CHECK

1. **What** is acid deposition and what are the main sources of atmospheric acid?

2. **What** are the harmful effects of acid deposition on materials, aquatic organisms, and soils?

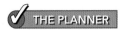
International Implications of Global Climate Change

Various social, economic, and political factors complicate international efforts to deal with global climate change. Although highly developed countries have historically been the major producers of greenhouse gases, many developing countries are rapidly increasing production as they industrialize. But because developing countries have less technical expertise and fewer economic resources, they are often less able to respond to the challenges of global climate change.

The difference between total emissions from a country and the per-person emissions from that country creates tensions among nations, especially between highly developed and developing countries. Most developing countries view fossil fuels as their route to industrial development and resist pressure from highly developed nations to decrease fossil fuel consumption.

Developing countries argue that it would be most fair to limit CO_2 on a per-person basis, since highly developed countries such as the United States, France, and Japan emit several times as much CO_2 per person than do developing countries such as China, India, and Kenya (see figure). However, as both population and per-person energy use increase in developing countries, their total CO_2 emissions are increasing rapidly. The average person in the United States is responsible for more than five times as much CO_2 as the average person in China, but China has surpassed the United States as the largest total emitter.

The international community recognizes that it must stabilize and decrease CO_2 emissions, but progress is slow. At least 174 nations, including the United States, signed the U.N. Framework Convention on Climate Change developed at the 1992 Earth Summit, which established goals for future international policies. In 1997 representatives from 160 countries determined timetables for reductions at a meeting in Kyoto, Japan. By 2005 enough countries had ratified the Kyoto Protocol for it to come into force. Political and economic concerns prevented the United States from joining the Kyoto Protocol, and those countries that have signed on have had limited success in meeting its provisions. Current international negotiations acknowledge that stopping climate change is not an option. Instead, conversation focuses on limiting the amount of change. It appears that the global community may set a maximum global temperature increase of 2°C (3.8°F) between now and 2100 as an achievable target.

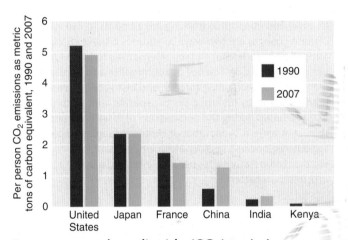

Per-person carbon dioxide (CO_2) emission estimates for selected countries, 1990 and 2007

Currently, industrialized nations produce a disproportionate share of CO_2 emissions. As developing nations such as China and India industrialize, however, their per-person CO_2 emissions increase.

Summary

1 The Atmosphere and Climate 218

1. **Weather** is the conditions in the atmosphere at a given place and time; it includes temperature, atmospheric pressure, precipitation, cloudiness, humidity, and wind. **Climate** is the typical weather patterns that occur in a place over a period of years. The two most important factors that determine an area's climate are temperature and precipitation.

2. Sunlight, or **insolation**, is the primary (almost sole) source of energy available in the biosphere. The Sun's energy runs the hydrologic cycle, drives winds and ocean currents, powers photosynthesis, and warms the planet. Of the solar energy that reaches Earth, 31 percent is immediately reflected away, and the remaining 69 percent is absorbed. Ultimately, all absorbed solar energy is radiated into space as **infrared radiation**, electromagnetic radiation with wavelengths longer than those of visible light but shorter than microwaves.

3. **Precipitation** is greatest where warm air passes over the ocean, absorbing moisture, and is then cooled, such as when mountains force humid air upward. Deserts develop in the rain shadows of mountain ranges or in continental interiors.

2 Global Climate Change 222

1. **Greenhouse gases** are gases that absorb infrared radiation; they include carbon dioxide, methane, nitrous oxide, chlorofluorocarbons, and tropospheric ozone. The **enhanced greenhouse effect** is the additional warming produced as human activities increase the amount of gases that absorb infrared radiation.

2. Global climate change will continue to cause sea level to rise, precipitation patterns to alter, extinction of many species, and problems for agriculture. It could result in the displacement of millions of people, thereby increasing international tensions.

3. Mitigation (slowing down the rate of global climate change) and adaptation (making adjustments to live with climate change) are two ways to address climate change. Mitigation includes developing alternatives to fossil fuels; increasing energy efficiency of automobiles and appliances; planting and maintaining forests; and instigating **carbon management**, by finding ways to separate and capture the CO_2 produced during the combustion of fossil fuels and then sequester it. Adaptation includes strategies to help various regions and sectors of society prepare for warmer temperatures, higher sea level, and altered precipitation patterns.

3 Ozone Depletion in the Stratosphere 230

1. Ozone (O_3) is a human-made pollutant in the troposphere but a naturally produced, essential component in the stratosphere. The stratosphere contains a layer of ozone that shields the surface from much of the Sun's **ultraviolet (UV) radiation**, that part of the electromagnetic spectrum with wavelengths just shorter than those of visible light; UV radiation is a high-energy form of radiation that can be lethal to organisms at high levels of exposure.

2. **Ozone thinning** is the natural and human-caused removal of ozone from the stratosphere. The primary chemicals responsible for ozone thinning in the stratosphere are **chlorofluorocarbons (CFCs)**, human-made organic aerosol compounds that contain chlorine and fluorine. CFCs are now banned because they attack the stratospheric ozone layer. Ozone thinning causes excessive exposure to UV radiation, which can increase cataracts, weaken immunity, and cause skin cancer in humans. Increased levels of UV radiation may also disrupt ecosystems.

3. The **Montreal Protocol** is an international agreement that has phased out much CFC production worldwide.

4 Acid Deposition 233

1. **Acid deposition** is a type of air pollution that includes sulfuric and nitric acids in precipitation as well as dry acid particles that settle out of the air. Acid deposition develops when sulfur and nitrogen oxides are released into the air, where they react to form acids and then return to surface waters and soil.

2. Acid deposition kills aquatic organisms, changes soil chemistry, and may contribute to **forest decline**, a gradual deterioration and eventual death of many trees in a forest.

Key Terms

What is happening in this picture?

- This scientist is drilling into the Antarctic ice sheet to remove an ice core. Do you think the ice deep within the sheet is old or relatively young? Explain your answer.

- Some of the deeper samples were laid down thousands of years ago, when the climate was much cooler. The ice contains bubbles of air. Based on what you have learned in this chapter, do you think the level of carbon dioxide in the air bubbles in the oldest ice is higher or lower than the level in today's atmosphere? Explain your answer.

- If the scientists compared CFCs in the air bubbles in the oldest ice with today's levels of CFCs, what do you think they would find?

Critical and Creative Thinking Questions

1. How does the Sun affect temperature at different latitudes? Why?

2. On the basis of what you know about the nature of science, can we say with absolute certainty that the increased production of greenhouse gases is causing global climate change? Why or why not?

3. Biologists who study plants growing high in the Alps found that plants adapted to cold-mountain conditions migrated up the peaks as fast as 3.7 m (12.1 ft) per decade during the 20th century, apparently in response to climate warming. Assuming that warming continues during the 21st century, what will happen to the plants if they reach the tops of the mountains?

4. Will it be easier for societies to mitigate climate change or to adapt to a changed climate? Explain your answer.

5. Distinguish between the benefits of the ozone layer in the stratosphere and the harmful effects of ozone at ground level.

6. What is the Montreal Protocol, and what environmental problem is it designed to correct?

7. Discuss some of the possible causes of forest decline. How might these factors interact to speed the rate of decline?

8–10. This map shows one model of how warmer global temperatures might alter precipitation in the United States in the next 100 years. Colors indicate the percent change in annual precipitation per century.

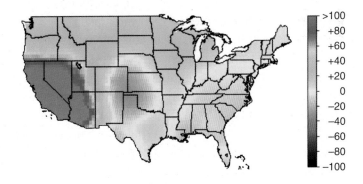

8. Name three states with climates that may become significantly wetter. Name three states with climates that may become significantly drier. Explain your answer.

9. Locate the state where you live. Will it be wetter, drier, or about the same?

10. In what ways might the projected changes in precipitation in the next 100 years impact U.S. agriculture? the U.S. economy?

✓ THE PLANNER

Freshwater Resources and Water Pollution

DROUGHT IN AUSTRALIA

A complex of two major rivers and their various waterways, the Murray-Darling Basin occupies a vast area in southeast Australia (see inset). The basin supports the city of Adelaide and provides 65 percent of the water used by Australia's agriculture. Now the country's dependence on the Murray-Darling is threatened, as in 2009 Australia entered its eighth year of drought, the worst in its recorded history.

About 40 percent of Australia's crops are raised in this semiarid region. Each farmer pays for an annual water allocation that has now been severely cut or eliminated. As crops have succumbed to heat and drought, many farms have collapsed (see photograph).

Farmers compete with other water users, including urban residents. The conflicting demands have altered the environment: Freshwater lakes grow salty as river flow slows, killing off native species, and wetlands become increasingly toxic because they are no longer flushed regularly. The region's Aboriginal people have lost plants and animals critical to their history and customs. Environmental and economic damage has been exacerbated by years of intensive agriculture and alterations to natural river flow.

Environmentalists, other scientists, and politicians acknowledge the likely role of climate change in the drought's severity. Australia's crisis serves as an example of the many water resource conflicts occurring worldwide.

CHAPTER OUTLINE

CHAPTER PLANNER ✓

- ☐ Study the picture and read the opening story.
- ☐ Scan the Learning Objectives in each section:
  p. 242 ☐   p. 244 ☐   p. 252 ☐   p. 256 ☐   p. 262 ☐
- ☐ Read the text and study all figures and visuals. Answer any questions.

Analyze key features

- ☐ What a Scientist Sees, p. 258 ☐
- ☐ Environmental InSight, p. 255 ☐
- ☐ National Geographic Map, pp. 250–251 ☐
- ☐ Process Diagram, p. 262 ☐   p. 263 ☐
- ☐ Case Study, p. 267 ☐
- ☐ Stop: Answer the Concept Checks before you go on:
  p. 244 ☐   p. 252 ☐   p. 254 ☐   p. 261 ☐   p. 266 ☐

End of Chapter

- ☐ Review the Summary and Key Terms.
- ☐ Answer What is happening in this picture?
- ☐ Answer the Critical and Creative Thinking Questions.

15°S

150°E

AUSTRALIA

30°S

Murray-
Darling Basin

INDIAN OCEAN

120°E

135°E

The Importance of Water

LEARNING OBJECTIVES

1. **Explain** how processes of the hydrologic cycle allow water to circulate through the abiotic environment.

2. **Describe** the structure of a water molecule and explain how hydrogen bonds form between adjacent water molecules.

3. **List** the unique properties of water.

Life on planet Earth would be impossible without water. All life forms, from unicellular bacteria to multicellular plants and animals, contain water. Humans are composed of approximately 60 percent water by body weight. We depend on water for our survival as well as for our convenience: We drink it, cook with it, wash with it (**Figure 10.1**), travel on it, and use an enormous amount of it for agriculture, manufacturing, mining, energy production, and waste disposal.

Although Earth has plenty of water, about 97 percent of it is salty and not consumable by most terrestrial organisms. Fresh water is distributed unevenly, resulting in serious regional water supply problems. Water experts predict that by 2025, more than one-third of the human population will live in areas where there isn't enough fresh water for drinking and irrigation.

The Hydrologic Cycle and Our Supply of Fresh Water

In the **hydrologic cycle**, water continuously circulates through the environment, from the ocean to the atmosphere to the land and back to the ocean (see **Figure 10.2**; also see Figure 5.9). The result is a balance of the water resources in the ocean, on the land, and in the atmosphere. The hydrologic cycle provides a continual renewal of the supply of fresh water on land, which is essential to terrestrial organisms.

Young brick workers in India bathe with water from an irrigation pipe • Figure 10.1

Two important components of the hydrologic cycle • Figure 10.2

a. Liquid and solid precipitation continuously falls from the atmosphere to the land and ocean.

b. Evaporation continuously moves water vapor from the land and ocean into the atmosphere.

Surface water is water found in streams, rivers, lakes, ponds, reservoirs, and **wetlands** (areas of land covered with water for at least part of the year). The **runoff** of precipitation from the land replenishes surface waters and is considered a renewable, although finite, resource. A **drainage basin**, or **watershed**, is the area of land drained by a single river or stream. Watersheds range in size from less than 1 km² for a small stream to a huge portion of the continent for a major river system such as the Mississippi River.

> **surface water**
> Precipitation that remains on the surface of the land and does not seep down through the soil.
>
> **runoff** The movement of fresh water from precipitation and snowmelt to rivers, lakes, wetlands, and the ocean.

Earth contains underground formations that collect and store water. This water originates as rain or melting snow that slowly seeps into the soil. It works its way down through cracks and spaces in sand, gravel, or rock until an impenetrable layer stops it; there it accumulates as **groundwater**. Groundwater is eventually discharged into rivers, wetlands, springs, or the ocean. Thus, surface water and groundwater are interrelated parts of the hydrologic cycle. **Aquifers** are underground reservoirs in which groundwater is stored (**Figure 10.3**).

> **groundwater** The supply of fresh water under Earth's surface that is stored in underground aquifers.

Most groundwater is considered a nonrenewable resource because it has taken hundreds or even thousands

Groundwater • Figure 10.3

Excess surface water seeps downward through soil and porous rock layers until it reaches impermeable rock or clay. An unconfined aquifer has groundwater recharged by surface water directly above it. In a confined aquifer, groundwater is stored between two impermeable layers.

WILEY
RESOURCE KIT

of years to accumulate, and usually only a small portion of it is replaced each year by seepage of surface water.

Properties of Water

Water is composed of molecules of H_2O, each consisting of two atoms of hydrogen and one atom of oxygen. Water molecules are **polar**—that is, one end of the molecule has a positive electrical charge, and the other end has a negative charge (**Figure 10.4**). The negative (oxygen) end of one water molecule is attracted to the positive (hydrogen) end of another water molecule, forming a **hydrogen bond** between the two molecules. Hydrogen bonds are the basis for many of water's physical properties, including its high melting/freezing point (0°C, 32°F) and high boiling point (100°C, 212°F). Because most of Earth has a temperature between 0°C and 100°C, most water exists in the liquid form organisms need.

Water absorbs a great deal of solar heat without substantially increasing in temperature. This high heat capacity allows the ocean to have a moderating influence on climate, particularly along coastal areas.

Water is a *solvent*, meaning that it can dissolve many materials. In nature, water is never completely pure because it contains dissolved gases from the atmosphere and dissolved mineral salts from the land. Water's ablities as a solvent have a major drawback: Many of the substances that dissolve in water cause water pollution.

Chemical properties of water • Figure 10.4

a. Each water molecule consists of two hydrogen atoms and one oxygen atom. Water molecules are polar, with positively and negatively charged areas.

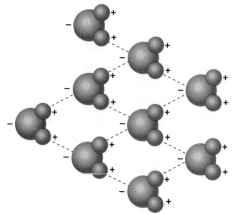

b. The polarity causes hydrogen bonds (represented by dashed lines) to form between the positive areas of one water molecule and the negative areas of others. Each water molecule forms up to four hydrogen bonds with other water molecules.

CONCEPT CHECK

1. **How** do processes in the hydrologic cycle affect the accumulation of groundwater?

2. **How** do hydrogen bonds form between adjacent water molecules?

3. **What** are two unique properties of water?

Water Resource Problems

LEARNING OBJECTIVES

1. **Relate** some of the problems caused by aquifer depletion, overdrawing of surface waters, and salinization of irrigated soil.

2. **Relate** the background behind the water problems of the Ogallala Aquifer and the Colorado River Basin.

3. **Describe** the role of international cooperation in managing shared water resources.

Water resource problems fall into three categories: too much water, too little water, and poor-quality water. Flooding occurs when a river's discharge cannot be contained within its normal channel. Today's floods are more disastrous in terms of property loss than those of the past because humans often remove water-absorbing plant cover from the soil and construct buildings on flood plains. (A **flood plain** is the area bordering a river channel that has the potential to flood.) These activities increase the likelihood of both floods and flood damage.

When a natural area—that is, an area undisturbed by humans—is inundated with heavy precipitation, the plant-protected soil absorbs much of the excess water. What the soil cannot absorb runs off into the river, which may then spill over its banks onto the flood plain. Because rivers meander, the flow is slowed, and the swollen waters rarely cause significant damage to the surrounding area. (See Figure 6.13 for a diagram of a typical river, including its flood plain.)

When an area is developed for human use, construction projects replace much of this protective plant cover. Buildings and paved roads don't absorb water, so runoff, usually in the form of storm sewer runoff, is significantly greater in developed areas (**Figure 10.5**). People who build homes or businesses on the flood plain of a river will most likely experience flooding at some point (**Figure 10.6**).

Arid lands, or deserts, are fragile ecosystems in which plant growth is limited by lack of precipitation. **Semiarid lands** receive more precipitation than deserts but are subject to frequent and prolonged droughts.

Farmers increase the agricultural productivity of arid and semiarid lands with irrigation. Irrigation of these

Flooding in Georgia • Figure 10.6

The swollen Chattahoochee River floods an Atlanta-area neighborhood on September 21, 2009, following a north Georgia storm. The flooding killed nine people and caused property damage estimated at $500 million.

How development changes the natural flow of water • Figure 10.5

Pre-urban

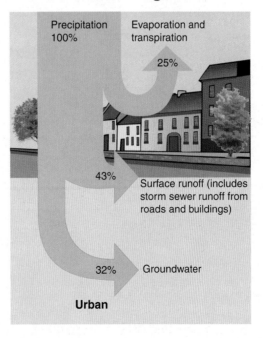

Urban

Shown is the fate of precipitation in Ontario, Canada, before (a) and after (b) urbanization. After Ontario was developed, surface runoff increased substantially, from 10 percent to 43 percent.

lands has become increasingly important worldwide in efforts to produce enough food for burgeoning populations (**Figure 10.7**). Since 1955, the amount of irrigated land has more than tripled; Asia has more agricultural land under irrigation than do other continents, primarily in China, India, and Pakistan. Water use for irrigation will probably continue to increase in the 21st century, but at a slower rate than in the last half of the 20th century.

Aquifer Depletion

Aquifer depletion from excessive removal of groundwater lowers the **water table**, the upper surface of the saturated zone of groundwater. Prolonged aquifer depletion drains an aquifer dry, effectively eliminating

> **aquifer depletion** The removal of groundwater faster than it can be recharged by precipitation or melting snow.

it as a water resource. Even areas with high rainfall can experience aquifer depletion if humans remove more groundwater than can be recharged. In addition, aquifer depletion from porous sediments causes **subsidence**, or sinking, of the land above it. **Saltwater intrusion** occurs along coastal areas when groundwater is depleted faster than it recharges. Saltwater intrusion is also occurring in low-lying areas due to sea-level rise associated with global climate change. Well water in such areas eventually becomes too salty for human consumption or other freshwater uses.

> **saltwater intrusion** The movement of seawater into a freshwater aquifer near the coast.

Agricultural use of water • Figure 10.7

These fields in Kansas use central-pivot irrigation, which minimizes evaporative water loss and gives fields a distinctive circular shape. Each circle is the result of a long irrigation pipe that extends along the radius from the circle's center to its edge and slowly rotates, spraying the crops. This satellite photo, taken in June, shows wheat fields (bright yellow), corn fields (dark green), and newly emerging sorghum (light green).

The Ogallala Aquifer The High Plains cover 6 percent of U.S. land but produce more than 15 percent of the nation's wheat, corn, sorghum, and cotton and almost 40 percent of its livestock. This productivity requires approximately 30 percent of the irrigation water used in the United States. Farmers on the High Plains rely on water from the **Ogallala Aquifer**, the largest groundwater deposit in the world (**Figure 10.8**).

Ogallala Aquifer • Figure 10.8

This massive deposit of groundwater lies under eight states, with extensive portions in Texas, Kansas, and Nebraska. Water in the Ogallala Aquifer takes hundreds or even thousands of years to renew after it is withdrawn to grow crops and raise cattle.

Global Locator

OGALLALA AQUIFER

NATIONAL GEOGRAPHIC

SOUTH DAKOTA

WYOMING

IOWA

NEBRASKA

Denver

COLORADO

KANSAS

MISSOURI

Wichita

Amarillo

NEW MEXICO

OKLAHOMA

ARKANSAS

TEXAS

0   100   200   300 miles
0   100   200   300 kilometers

Thickness of water layer
- 0.0 – 30.0 m
- 30.1 – 120.0 m
- 120.1 – 350.0 m

In some areas farmers are drawing water from the Ogallala Aquifer as much as 40 times faster than nature replaces it. This rapid depletion has lowered the water table more than 30 m (100 ft) in some places. Most hydrologists (scientists who study water supplies) predict that groundwater will eventually drop in all areas of the Ogallala to a level uneconomical to pump. Their goal is to postpone that day through water conservation, including the use of water-saving irrigation systems.

Overdrawing of Surface Waters

Removing too much fresh water from a river or lake can have disastrous consequences in local ecosystems. Humans can remove perhaps 30 percent of a river's flow without greatly affecting the natural environment. In some places, considerably more is withdrawn for human use. In the arid American Southwest, it is not unusual for 70 percent or more of surface water to be removed.

When surface waters are overdrawn, wetlands dry up. **Estuaries**, where rivers empty into seawater, become saltier when surface waters are overdrawn, which reduces their productivity. Wetlands and estuaries, which serve as breeding grounds for many species of birds and other animals, also play a vital role in the hydrologic cycle. When these resources are depleted, the ensuing water shortages and reduced productivity have economic as well as ecological ramifications.

The increased use of U.S. surface water for agriculture, industry, and personal consumption has caused many water supply and quality problems. Some regions that have grown in population during this period—for example, California, Nevada, Arizona, Georgia (metropolitan Atlanta), and Florida—have placed correspondingly greater burdens on their water supplies. If water consumption in these and other areas continues to increase, the availability of surface waters could become a serious regional problem, even in places that have never experienced water shortages.

Nowhere in the country are water problems as severe as they are in the West and Southwest. Much of this large region is arid or semiarid. With the rapid expansion of the population there during the past 25 years, municipal, commercial, and industrial uses now compete heavily with irrigation for available water.

The Colorado River Basin One of the most serious water supply problems in the United States is in the Colorado River Basin. The Colorado River provides water for more than 30 million people, including those in the cities of Denver, Las Vegas, Salt Lake City, Albuquerque, Phoenix, Los Angeles, and San Diego. It supplies irrigation water for 1.4 million hectares (3.5 million acres) of fruit, vegetable, and field crops worth $1.5 billion per year. The Colorado River has 49 dams, 11 of which produce electricity by hydropower. The river produces $1.25 billion per year in revenues from the recreation industry.

An international agreement with Mexico, along with federal and state laws, severely restricts the use of the Colorado's waters. An important state treaty, the 1922 **Colorado River Compact**, stipulates an annual allotment of 7.5 million acre-feet of water each to the upper Colorado (Colorado, Utah, and Wyoming) and the lower Colorado (California, Nevada, Arizona, and New Mexico). Each acre-foot equals 326,000 gal (1.2 million liters), enough for about eight people for 1 year. However, the Colorado River Compact overestimated the average annual flow of the Colorado River, and it locked that estimate into the multistate agreement.

Population growth in the upper Colorado region threatens the lower Colorado region's water supply. Further, people in the states through which the lower Colorado flows take so much water that the remainder is insufficient to meet Mexico's needs, as had been stipulated by international treaty (**Figure 10.9**). To compound the problem, as more and more water is used, the lower Colorado becomes increasingly salty—in some places saltier than the ocean—as it flows toward Mexico.

Some positive steps have been taken. In 2003 California agreed to limit its water withdrawals from the Colorado River to quantities specified in the Colorado River Compact. Also, some California farmers agreed to sell some water they would normally use for irrigation and use the money earned to update their irrigation systems so they make more efficient use of their water.

Salinization of Irrigated Soil

salinization The gradual accumulation of salt in soil, often as a result of improper irrigation methods.

Although irrigation improves the agricultural productivity of arid and semiarid lands, it often causes salt to accumulate in the soil, a phenomenon called **salinization**. Irrigation water contains small amounts of dissolved salts. Normally, through precipitation runoff, rivers carry away salt. Irrigation water, however, normally soaks into the soil and does not run off into rivers. The continued application of such water, season after season, year after year, leads to the gradual accumulation of salt in the soil. Given enough time, the salt concentration can rise to such a high level that

Colorado River Delta, border of Baja, California and Sonora, Mexico • Figure 10.9

As a result of diversion for irrigation and other uses in the United States, the Colorado River often dries up before reaching the Gulf of California in Mexico. Arrows indicate some of the now-dry tributaries that once carried large amounts of water.

The satellite images show the Aral Sea in 1976 and 2008. As water was diverted for irrigation, the sea level subsided.

Global Locator

ARAL SEA

ASIA

NATIONAL GEOGRAPHIC

How do water resource conflicts associated with the Aral Sea compare to those involving the Colorado River Basin? to water resource issues in your state or region?

plants are poisoned or their roots become dehydrated. Thus, salt hurts soil productivity and, in extreme cases, renders soil unfit for crop production.

Global Water Issues

As the world's population continues to increase, global water problems are becoming more serious. In India, where approximately 20 percent of the world's population has access to 4 percent of the world's fresh water, approximately 8000 villages have no local water supply. Water supplies are precarious in much of China, due to population pressures. In many parts of the country, water table levels are dropping, and wells have gone dry. Much of the water in the Yellow River, one of China's main water basins, is diverted for irrigation, depriving downstream areas of water. Mexico is facing the most serious water shortages of any country in the Western Hemisphere. The main aquifer supplying Mexico City is dropping rapidly, and the water table is falling fast in Guanajuato, an agricultural state.

As the needs of the growing human population deplete freshwater supplies, less water will be available for crops. Local famines arising from water shortages are a possibility.

Sharing Water Resources Among Countries In the 1950s, the then Soviet Union began diverting water that feeds into the Aral Sea to irrigate nearby desert areas. Since then the Aral Sea has all but disappeared (**Figure 10.10**). Its total volume has gone down 80 percent, and much of its biological diversity has disappeared. Millions of people living in the Aral Sea's watershed have developed serious health problems, probably due in part to storms lifting into the air toxic salts from the receding shoreline.

Following the breakup of the Soviet Union in 1991, responsibility for saving the Aral Sea shifted to the five Asian countries that share the Aral basin—Uzbekistan, Kazakstan, Kyrgyzstan, Turkmenistan, and Tajikistan. Despite recent cooperative restoration efforts made by these nations and backed by the World Bank and the U.N. Environment Program, the Aral Sea will probably never return to its former size and economic importance. Like the Aral Sea, many of Earth's other watersheds cross political boundaries and face management issues associated with their shared use (**Figure 10.11** on pages 250–251).

Three-fourths of the world's 200 or so major watersheds are shared between at least two nations. International cooperation is required to manage rivers

Earth's primary watersheds • Figure 10.11

Earth's bodies of fresh water can cover enormous areas and cross many political boundaries.

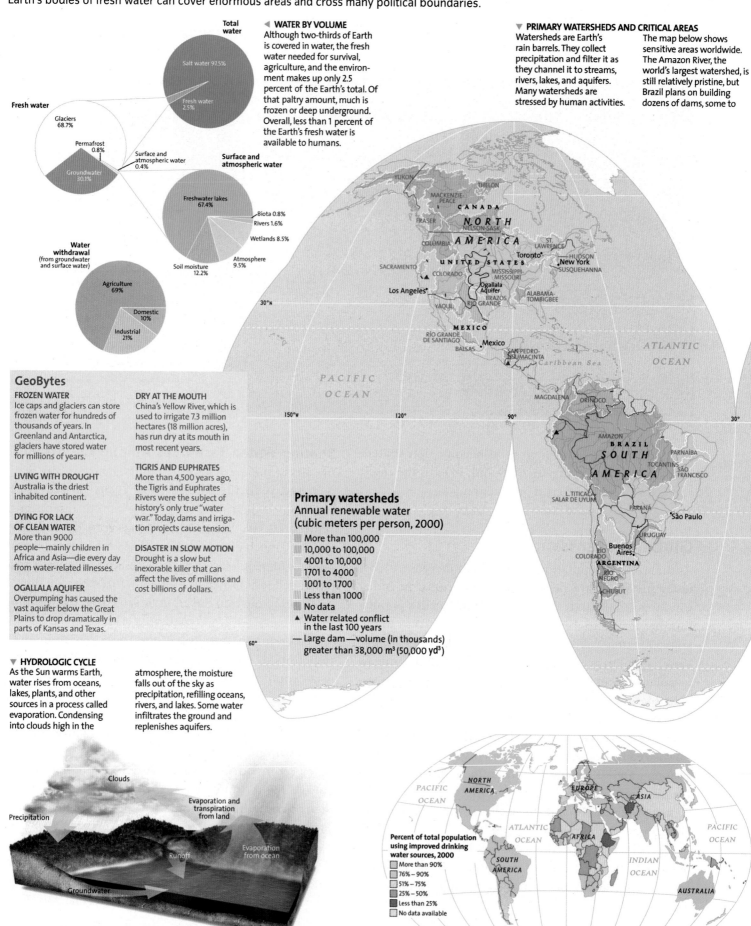

Total water

Salt water 97.5%

Fresh water 2.5%

Fresh water

Glaciers 68.7%

Permafrost 0.8%

Groundwater 30.1%

Surface and atmospheric water 0.4%

Surface and atmospheric water

Freshwater lakes 67.4%

Biota 0.8%

Rivers 1.6%

Wetlands 8.5%

Soil moisture 12.2%

Atmosphere 9.5%

Water withdrawal
(from groundwater and surface water)

Agriculture 69%

Domestic 10%

Industrial 21%

◄ **WATER BY VOLUME**
Although two-thirds of Earth is covered in water, the fresh water needed for survival, agriculture, and the environment makes up only 2.5 percent of the Earth's total. Of that paltry amount, much is frozen or deep underground. Overall, less than 1 percent of the Earth's fresh water is available to humans.

▼ **PRIMARY WATERSHEDS AND CRITICAL AREAS**
Watersheds are Earth's rain barrels. They collect precipitation and filter it as they channel it to streams, rivers, lakes, and aquifers. Many watersheds are stressed by human activities.

The map below shows sensitive areas worldwide. The Amazon River, the world's largest watershed, is still relatively pristine, but Brazil plans on building dozens of dams, some to

GeoBytes

FROZEN WATER
Ice caps and glaciers can store frozen water for hundreds of thousands of years. In Greenland and Antarctica, glaciers have stored water for millions of years.

LIVING WITH DROUGHT
Australia is the driest inhabited continent.

DYING FOR LACK OF CLEAN WATER
More than 9000 people—mainly children in Africa and Asia—die every day from water-related illnesses.

OGALLALA AQUIFER
Overpumping has caused the vast aquifer below the Great Plains to drop dramatically in parts of Kansas and Texas.

DRY AT THE MOUTH
China's Yellow River, which is used to irrigate 7.3 million hectares (18 million acres), has run dry at its mouth in most recent years.

TIGRIS AND EUPHRATES
More than 4,500 years ago, the Tigris and Euphrates Rivers were the subject of history's only true "water war." Today, dams and irrigation projects cause tension.

DISASTER IN SLOW MOTION
Drought is a slow but inexorable killer that can affect the lives of millions and cost billions of dollars.

Primary watersheds
Annual renewable water
(cubic meters per person, 2000)

- More than 100,000
- 10,000 to 100,000
- 4001 to 10,000
- 1701 to 4000
- 1001 to 1700
- Less than 1000
- No data
- ▲ Water related conflict in the last 100 years
- — Large dam—volume (in thousands) greater than 38,000 m³ (50,000 yd³)

▼ **HYDROLOGIC CYCLE**
As the Sun warms Earth, water rises from oceans, lakes, plants, and other sources in a process called evaporation. Condensing into clouds high in the atmosphere, the moisture falls out of the sky as precipitation, refilling oceans, rivers, and lakes. Some water infiltrates the ground and replenishes aquifers.

Clouds

Evaporation and transpiration from land

Precipitation

Runoff

Evaporation from ocean

Groundwater

Percent of total population using improved drinking water sources, 2000
- More than 90%
- 76% – 90%
- 51% – 75%
- 25% – 50%
- Less than 25%
- No data available

ver aluminum smelters.
frica and Asia, lack
ccess to water and
ter-related diseases are
main problems. In
ope and the Middle East,
ruse, pollution, and

disagreement over
diverting water are the
major challenges. Hope
rests in better planning
and community-scale
projects.

WATER

It's as vital to life as air. Yet fresh water is one of the rarest resources on Earth. Only 2.5 percent of Earth's water is fresh, and of that, the usable portion for humans is less than 1 percent of all fresh water, or 0.01 percent of all water on Earth. Water is constantly recycling through Earth's hydrologic cycle. But population growth and pollution are combining to make less and less available per person per year, while global climate change adds new uncertainty.

Efficiency, conservation, and technology can help ensure that the water you use today will still be usable and clean hundreds of years from now.

ACCESS TO FRESH WATER

ccess to clean fresh water
critical for human health.
t in many regions, potable
ater is becoming scarce
cause of heavy demands
d pollution. Especially

worrisome is the poisoning
of aquifers—a primary
source of water for nearly
one-third of the world—by
sewage, pesticides, and
heavy metals.

▼ **GLOBAL IRRIGATED AREAS AND WATER WITHDRAWALS**
Since 1970 global water
withdrawals have correlated
with the rise in irrigated
area. Some 70% of with-

drawals are for agriculture,
mostly for irrigation that
helps produce 40 percent
of the world's food.

NATIONAL GEOGRAPHIC

Freshwater withdrawal as a percentage of total water utilization, 2000

AFRICA
Domestic
Industrial
8.6%
7.3%
84.1%
Agricultural

ASIA
7.2%
11.4%
81.4%

OCEANIA
17.6%
10%
72.4%

SOUTH AMERICA
19.3%
12.5%
68.2%

NORTH AMERICA
14.1%
44.5%
41.4%

EUROPE
15.2%
32.4%
52.4%

Rhine River Basin • Figure 10.12

The Rhine River drains five European countries. (The green area represents the drainage basin.) Water management of such a river requires international cooperation.

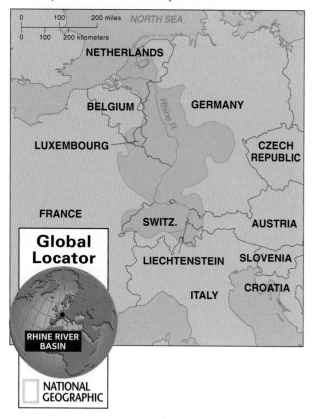

0   100   200 miles
0   100   200 kilometers
NORTH SEA

NETHERLANDS
BELGIUM
Rhine R.
GERMANY
LUXEMBOURG
CZECH REPUBLIC
FRANCE
SWITZ.
AUSTRIA
LIECHTENSTEIN
SLOVENIA
ITALY
CROATIA

Global Locator

RHINE RIVER BASIN

NATIONAL GEOGRAPHIC

that cross international borders. The heavily populated drainage basin for the Rhine River in Europe spans five countries—Switzerland, Germany, France, Luxembourg, and the Netherlands (**Figure 10.12**). All five nations recognize that international cooperation is essential to conserve and protect the supply and quality of the Rhine River. Together they formed the International Commission for Protection of the Rhine, which in 1987 initiated a 15-year Rhine Action Programme. Their efforts have paid off: The main sources of pollution have been eliminated, and water in the Rhine River today is almost as pure as drinking water; long-absent fishes have returned; and projects are under way to restore riverbanks, control flooding, and clean up remaining pollutants.

CONCEPT CHECK STOP

1. **What** problems are associated with overdrawing surface water? with aquifer depletion?

2. **What** issues surround water problems of the Ogallala Aquifer? the Colorado River basin?

3. **How** does international cooperation affect shared water resources?

Water Management

LEARNING OBJECTIVES

1. **Define** sustainable water use.
2. **Contrast** the benefits and drawbacks of dams and reservoirs.
3. **Give** examples of water conservation in agriculture, industry, and individual homes and buildings.

 WILEY

RESOURCE KIT

The main goal of water management is to provide a sustainable supply of high-quality water. **Sustainable water use** means careful human use of water resources so water is available for future generations and for existing nonhuman needs.

Water supplies are obtained by building dams, diverting water, or removing salt from seawater or salty groundwater. Conservation, which includes reusing water, recycling water, and improving water-use efficiency, augments water supplies and is an important aspect of sustainable water use. Economic policies are also important in managing water sustainably: When water is inexpensive, it tends to be wasted.

> **sustainable water use** The wise use of water resources, without harming the essential functioning of the hydrologic cycle or the ecosystems on which present and future humans depend.

Dams and Reservoirs: Managing the Columbia River

Dams generate electricity and ensure a year-round supply of water in areas with seasonal precipitation or snowmelt, but many people think their costs outweigh their benefits. The Columbia River, the fourth-largest river in North America, illustrates the impact of dams on natural fish communities. There are more than 100 dams in the Columbia River system, 19 of which are major generators of hydroelectric power (**Figure 10.13**). The Columbia River system supplies municipal and industrial water to several major urban areas in the northwestern United States and irrigation water for more than 1.2 million hectares (3 million acres) of agricultural land.

As is often the case in natural resource management, one particular use of the Columbia River system may have a negative impact on other uses. The dam impoundments along the Columbia River generate electricity and control floods but have adversely affected fish populations. The salmon population in the Columbia River system is only a fraction of what it was before the watershed was developed. The many dams that impede salmon migrations are widely considered the most significant factor in salmon decline.

Various efforts to assist migrating salmon have not proved particularly successful (**Figure 10.14**). Conservationists and biologists support using a natural approach of releasing water to flush young salmon downstream; they also support adopting a controversial proposal to tear down several dams on the lower Snake River, a tributary of the Columbia River. Farmers and the hydroelectric companies strongly oppose these plans, which they fear would threaten their water supplies.

Grand Coulee Dam on the Columbia River • Figure 10.13 _____

Shown are the dam and part of its reservoir, Lake Roosevelt. Dams provide electricity generation, flood control, and water recreation opportunities, but they disrupt or destroy natural river habitats and are expensive to build.

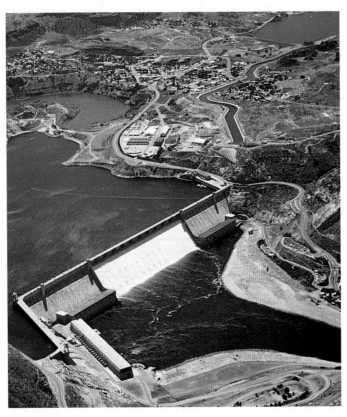

Fish ladder • Figure 10.14 _____

This ladder is located at the Bonneville Dam on the Oregon side of the Columbia River. Fish ladders help migratory fishes to bypass dams in their migration upstream. Despite the installation of fish ladders, the salmon population remains low.

Water Conservation

Today there is more competition than ever before among water users with different priorities, and water conservation measures are necessary to guarantee sufficient water supplies.

Reducing Agricultural Water Waste

Irrigation generally makes inefficient use of water. Traditional irrigation methods involve flooding the land or diverting water to fields through open channels. Plants absorb about 40 percent of the water that flood irrigation applies to the soil; the rest of the water usually evaporates into the atmosphere or seeps into the ground.

One of the most important innovations in agricultural water conservation is **microirrigation**, also called **drip** or

microirrigation
A type of irrigation that conserves water by piping it to crops through sealed systems.

trickle irrigation, in which pipes with tiny holes bored in them convey water directly to individual plants (**Figure 10.15a**). Microirrigation substantially reduces the water needed to irrigate crops—usually by 40 percent to 60 percent compared to traditional irrigation—and also reduces the amount of salt that irrigation water leaves in the soil.

Other measures that can save irrigation water include using lasers to level fields, which allows more even water distribution, and making greater use of recycled wastewater. A drawback of such techniques is their cost, which makes them unaffordable for most farmers in highly developed countries, let alone subsistence farmers in developing nations.

Reducing Water Waste in Industry

Electric power generators and many industrial processes require water. In the United States, five major industries—chemical products, paper and pulp, petroleum and coal, primary metals, and food processing—consume almost 90 percent of industrial water.

Stricter pollution-control laws provide some incentive for industries to conserve water. Industries usually recapture, purify, and reuse water to reduce their water use and their water treatment costs. The U.S. Steel Corporation plant in Granite City, Illinois, for example, recycles approximately two-thirds of the water it uses daily. The Ghirardelli Chocolate Company in San Leandro,

California, installed a recycling system to cool large tanks of its chocolate. The potential for industries to conserve water by recycling is enormous.

International companies also have to consider water issues where they locate plants. As a beverage producer, The Coca-Cola Company uses vast amounts of water. In its efforts to return or recycle the water it consumes, Coca-Cola in China partners in environmental education, river conservation, and water quality protection programs (**Figure 10.15b**). The company's Chinese bottling plants reduced their water usage by 27 percent between 2004 and 2007.

Reducing Municipal Water Waste

Like industries, regions and cities—and the households within them—recycle or reuse water to reduce consumption (**Figure 10.15c**). For example, homes and other buildings can be modified to collect and store gray water. *Gray water* is water that has already been used in sinks, showers, washing machines, and dishwashers. Gray water is recycled to flush toilets, wash cars, or sprinkle lawns. In contrast to water *recycling*, wastewater *reuse* occurs when water is collected and treated before being redistributed. The reclaimed water is generally used for irrigation.

Cities also decrease water consumption by providing consumer education, requiring water-saving household fixtures, developing economic incentives to save water, and repairing leaky water supply systems. Also, increasing the price of water to reflect its true cost promotes water conservation.

The average person in the United States uses 295 L (78 gal) of water per day at home. As a water user, you have a responsibility to use water carefully and wisely. The cumulative effect of many people practicing personal water conservation measures has a significant impact on overall water consumption.

CONCEPT CHECK

1. **What** is *sustainable water use*?
2. **What** are the benefits of dams on the Columbia River? the drawbacks?
3. **How** can individuals conserve and manage water resources?

a. Microirrigation. Close-up of a drip irrigation pipe system releasing water directly between young plants, eliminating much of the waste associated with traditional methods of irrigation. Photographed on an experimental farm in the Negev Desert, Israel.

b. Industrial Water Conservation. The deputy group president of the Pacific Group of the Coca-Cola Company China Division is joined by representatives from the United Nations Development Programme and China's Ministry of Water Resources and Ministry of Commerce at the launch of a program aimed at improving water resources management and drinking water safety in rural China.

Bathroom
Install water-saving shower and faucets and low-flush toilets. Or use a water-displacement device in the tank of a conventional toilet. Fix leaky fixtures.
Modify personal habits: Avoid leaving the faucet running while shaving or brushing teeth. Take shorter showers.

Kitchen
Use a dishwasher, with a full load. It requires less water than washing dishes by hand.

Laundry room
Choose a high-efficiency washing machine to use less water and spin more water out of the clothes.

Gray water storage tank

c. Conserving Water at Home.
In your bathroom, kitchen, and laundry room, you can take many steps to limit water use. Also, to reuse water, individual homes and buildings can be modified to collect and store "gray water," water already used in sinks, showers, washing machines, and dishwashers. Gray water is used when clean water is not required—for example, in flushing toilets, washing the car, and sprinkling the lawn, or, especially in the Southwest, irrigating golf courses. Permits to install gray water systems vary from state to state. Arizona and other states with severe water shortages are more flexible than other states about allowing gray water systems.

NATIONAL GEOGRAPHIC

WILEY RESOURCE KIT

Water Pollution

LEARNING OBJECTIVES

1. **Define** *water pollution*.

2. **Discuss** how sewage is related to eutrophication, biochemical oxygen demand (BOD), and dissolved oxygen.

3. **Distinguish** between the two types of pollution sources and give examples of each.

4. **Describe** sources of groundwater pollution.

ater pollution is a global problem that varies in magnitude and type of pollutant from one region to another. In many locations, particularly in developing countries, the main water pollution issue is providing individuals with disease-free drinking water.

> **water pollution** A physical or chemical change in water that adversely affects the health of humans and other organisms.

Sewage and other organic wastes are measured in terms of their **biochemical oxygen demand (BOD)**. A large amount of sewage in water generates a high BOD, which robs the water of dissolved oxygen (**Figure 10.16**). BOD measures how fast microorganisms remove oxygen from a body of water. When dissolved oxygen levels are low, anaerobic (without oxygen) microorganisms produce compounds with unpleasant odors, further deteriorating water quality.

> **biochemical oxygen demand (BOD)** The amount of oxygen that microorganisms need to decompose biological wastes into carbon dioxide, water, and minerals.

Types of Water Pollution

As discussed earlier in the chapter, water's chemical properties enable it to dissolve many substances, including pollutants. Water pollutants are divided into eight categories: sewage, disease-causing agents, sediment pollution, inorganic plant and algal nutrients, organic compounds, inorganic chemicals, radioactive substances, and thermal pollution. Causes and examples of each type of water pollution are summarized in **Table 10.1** (on facing page). Here we explore pollution threats associated with sewage.

Sewage The release of sewage into water causes several pollution problems. First, because sewage may carry disease-causing agents, water polluted with sewage poses a threat to public health (see Chapter 4). Sewage also generates two serious environmental problems: enrichment and oxygen demand. **Enrichment** of a body of water is due to the presence of high levels of plant and algal nutrients such as nitrogen and phosphorus, both of which are sewage products. When an aquatic ecosystem contains high levels of sewage or other organic material, decomposing microorganisms use up most of the dissolved oxygen, leaving little available for fishes or other aquatic animals.

> **sewage** Wastewater from drains or sewers (from toilets, washing machines, and showers); includes human wastes, soaps, and detergents.

Effect of sewage on dissolved oxygen and biochemical oxygen demand (BOD) • Figure 10.16

Note the initial oxygen depletion (green line) and increasing BOD (brown line) close to the sewage spill (at distance 0). The stream gradually recovers as the sewage is diluted and degraded. As indicated by the dashed line, fishes can't live in water that contains less than 4 mg of dissolved oxygen per liter of water.

Interpreting Data
Are fish located 60 km downstream of the spill likely to survive? What about those located 5 km from the spill? Explain why dissolved oxygen is lower slightly farther from the spill than right next to it.

Types of water pollution • Table 10.1

Type of pollution	Source	Examples	Effects
Sewage	Wastewater from drains or sewers	Human wastes, soaps, detergents	Threatens public health; causes enrichment and high biochemical oxygen demand (BOD)
Disease-causing agents	Wastes of infected individuals	Bacteria, viruses, protozoa, parasitic worms	Spread infectious diseases, including cholera, dysentery, typhoid, infectious hepatitis, and poliomyelitis.
Sediment pollution	Erosion of agricultural lands, forest soils exposed by logging, degraded stream banks, overgrazed rangelands, strip mines, construction	Clay, silt, sand, and gravel, suspended in water and eventually settling out	Reduces light penetration, limiting photosynthesis and disrupting food chain; clogs gills and feeding structures of aquatic animals; carries and deposits disease-causing agents and toxic chemicals
Inorganic plant and algal nutrients	Human and animal wastes, plant residues, atmospheric deposition, fertilizer runoff from agricultural and residential land	Nitrogen and phosphorus	Stimulate growth of excess plants and algae, which disrupts natural balance between producers and consumers and causes enrichment, bad odors, and high BOD; suspected of causing red tides, explosive blooms of toxic pigmented algae that threaten the health of humans and aquatic animals in coastal areas
Organic compounds	Landfills, agricultural runoff, industrial wastes	Synthetic chemicals: pesticides, cleaning solvents, industrial chemicals, plastics	Contaminate groundwater and surface water; threaten drinking water supply; found in some bottled water; some are suspected endocrine disrupters
Inorganic chemicals	Industries, mines, irrigation runoff, oil drilling, urban runoff from storm sewers	Acids, salts, heavy metals such as lead, mercury, and arsenic	Contaminate groundwater and surface water; threaten drinking water supply; found in some bottled water; don't easily degrade or break down
Radioactive substances	Nuclear power plants, nuclear weapons industry, medical and scientific research facilities	Unstable isotopes of radioactive minerals such as uranium and thorium	Contaminate groundwater and surface water; threaten drinking water supply
Thermal pollution	Industrial runoff	Heated water produced during industrial processes, then released into waterways	Depletes water of oxygen and reduces amount of oxygen that water can hold; reduced oxygen threatens fishes

WILEY
RESOURCE KIT

WHAT A SCIENTIST SEES

Oligotrophic and Eutrophic Lakes

a and **b** The average person looking at these two photographs would notice the dramatic differences between them but wouldn't understand the environmental conditions responsible for the differences. **a** Shows Crater Lake, an oligotrophic lake in Oregon; **b** shows a small eutrophic lake in western New York.

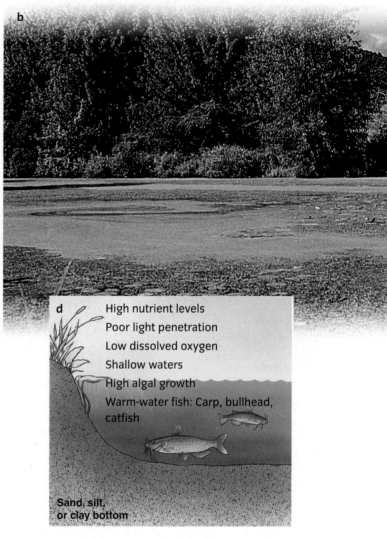

c and d Aquatic ecologists understand the characteristics of oligotrophic and eutrophic lakes. **c** An oligotrophic lake has a low level of inorganic plant and algal nutrients; its fish species require cool, oxygen-rich water. **d** A eutrophic lake has a high level of these nutrients; its fish species tolerate warm, low-oxygen water.

Eutrophication: An Enrichment Problem Lakes, estuaries, and slow-flowing streams that have minimal levels of nutrients are considered unenriched, or oligotrophic. An **oligotrophic** lake has clear water and supports small populations of aquatic organisms (see *What a Scientist Sees*, parts **a** and **c**). **Eutrophication** is the enrichment of a lake, an estuary, or a slow-flowing stream by inorganic plant and algal nutrients such as phosphorus; an enriched body of water is said to be **eutrophic**. The enrichment of water results in an increased photosynthetic productivity. The water in a eutrophic lake is cloudy and usually resembles pea soup because of the presence of vast numbers of algae and cyanobacteria (see parts **b** and **d**).

Over vast periods, oligotrophic lakes, estuaries, and slow-moving streams become eutrophic naturally. The bodies of water are slowly enriched and grow shallower from the immense number of dead organisms that have settled in the sediments. Gradually, plants take root, slowly forming marshes. Some human activities, however, greatly accelerate eutrophication. This fast, human-induced process is usually called **artificial eutrophication** to distinguish it from natural eutrophication. Artificial eutrophication results from enrichment of aquatic ecosystems by nutrients found predominantly in fertilizer runoff and sewage.

> **artificial eutrophication** Overnourishment of an aquatic ecosystem by nutrients such as nitrates and phosphates due to human activities such as agriculture and discharge from sewage treatment plants.

Sources of Water Pollution

Water pollutants come from both natural sources and human activities. Natural sources of pollution such as mercury and arsenic tend to be local concerns, but human-generated pollution is generally more widespread.

The sources of water pollution are classified into two types: point source pollution and nonpoint source pollution. **Point source pollution** is discharged into the environment through pipes, sewers, or ditches from specific sites such as factories or sewage treatment plants (**Figure 10.17**). Point source pollution is relatively easy to control legislatively, but accidents still occur.

Pollutants that enter bodies of water over large areas rather than at a single point cause **nonpoint source pollution**, also called *polluted runoff*. Nonpoint source pollution occurs when precipitation moves over and through the soil, picking up and carrying away pollutants that are eventually deposited in lakes, rivers, wetlands, groundwater, estuaries, and the ocean. Nonpoint source pollution includes agricultural runoff (such as fertilizers, pesticides, livestock wastes, and salt from irrigation), mining wastes (such as acid mine drainage), municipal wastes (such as inorganic plant and algal nutrients), construction sediments, and soil erosion (from fields, logging operations, and eroding stream banks). Although

> **point source pollution** Water pollution that can be traced to a specific spot.
>
> **nonpoint source pollution** Pollution that enter bodies of water over large areas rather than being concentrated at a single point of entry.

nonpoint sources cover more than one site and can be hard to identify, their combined effect can be huge.

According to the Environmental Protection Agency (EPA), agriculture is the leading source of water quality impairment of surface waters nationwide and is responsible for 72 percent of the water pollution in U.S. rivers. Agricultural practices produce several types of pollutants that contribute to nonpoint source pollution. Fertilizer runoff causes water enrichment. Animal wastes and plant residues in waterways produce high BODs and high levels of suspended solids as well as water enrichment. Highly toxic chemical pesticides

Point source pollution • Figure 10.17

Landfill leachate flows out of a pipe in Vashon, Washington.

may leach into the soil and from there into water or may find their way into waterways by adhering to sediment particles. Soil erosion from fields and rangelands causes sediment pollution in waterways. To address the problem of runoff from animal wastes, the EPA launched a largely voluntary program in 1998 that asks U.S. livestock operations to develop Comprehensive Nutrient Management Plans to prevent manure from becoming polluted runoff.

Although sewage is the main pollutant produced by cities and towns, municipal water pollution also has a nonpoint source: urban runoff that carries a variety of contaminants (**Figure 10.18**). The water quality of urban runoff from the storm sewers of city streets is often worse than that of sewage. Urban runoff carries salt from roadways, untreated garbage, animal wastes, construction sediments, and traffic emissions (via rain carrying air pollutants). When Hurricane Katrina breeched New Orleans's levees in 2005, floodwaters picked up contaminants from houses—such as pesticides, cleaners, paints, and food—and from businesses—such as dry-cleaning solvents, lubricants, and medical wastes.

Different industries generate different types of pollutants. Food-processing industries produce organic wastes that decompose quickly but have a high BOD. Pulp and paper mills also release wastes with a high BOD and produce toxic compounds and sludge. The paper industry, however, has begun to adopt new

Urban runoff • Figure 10.18

Many pollutants may be carried from storm drains on streets to streams and rivers. The largest single pollutant in urban runoff is organic waste, which removes dissolved oxygen from water as it decays. Fertilizers cause excessive algal growth, further depleting oxygen levels. Other everyday pollutants include used motor oil, which is often illegally poured into storm drains, and heavy metals.

Sources of groundwater contamination • Figure 10.19

Agricultural practices, sewage (both treated and untreated), landfills, industrial activities, and septic systems are some of the sources of groundwater pollution. Natural processes can only remove groundwater contamination very slowly.

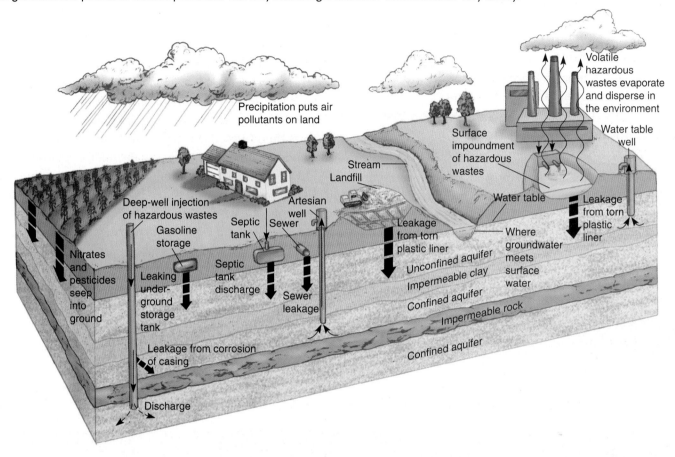

manufacturing methods, such as eliminating chlorine as a bleaching agent, that produce significantly less toxic effluents.

Groundwater Pollution

Roughly half the people in the United States obtain their drinking water from groundwater, which is also withdrawn for irrigation and industry. In recent years, the quality of the nation's groundwater has become a concern. The most common pollutants, such as pesticides, fertilizers, and organic compounds, seep into groundwater from municipal sanitary landfills, underground storage tanks, backyards, golf courses, and intensively cultivated agricultural lands (**Figure 10.19**).

Currently, most of the groundwater supplies in the United States are of good quality and don't violate standards established to protect human health. However, areas that do experience local groundwater contamination face quite a challenge: Cleanup of polluted groundwater is costly, takes years, and in some cases is not technically feasible.

CONCEPT CHECK

1. **What** is water pollution?

2. **What** is biochemical oxygen demand? How is BOD related to sewage?

3. **How** does point source pollution differ from nonpoint source pollution? What are some examples of each?

4. **What** are some common sources of groundwater pollutants?

Improving Water Quality

LEARNING OBJECTIVES

1. **Describe** how most drinking water is purified in the United States.
2. **Distinguish** among primary, secondary, and tertiary treatments for wastewater.
3. **Compare** the goals of the Safe Drinking Water Act and the Clean Water Act.

Water quality is improved by removing contaminants from the water supply before and after it is used. Technology assists in both processes.

Purification of Drinking Water

Most U.S. municipal water supplies are treated before the water is used so it is safe to drink (**Figure 10.20**). Turbid water is treated with a chemical coagulant that causes the suspended particles to clump together and settle out. The water is then filtered through sand to remove remaining suspended materials as well as many microorganisms.

In the final purification step before distribution in the water system, the water is disinfected to kill any remaining disease-causing agents. The most common way to disinfect water is to add chlorine. A small amount

PROCESS DIAGRAM

Treatment of water for municipal use • Figure 10.20

 THE PLANNER

Reservoir
Dam

1 The water supply for a town may be stored in a reservoir, as shown, or obtained from groundwater.

Water supply system

2 The water is treated before use so it is safe to drink.

Treated effluent
Wastewater

4 The quality of the wastewater is fully or partially restored by sewage treatment before the treated effluent is dispersed into a nearby body of water.

Input to sewage treatment system
Sewer lines

3 After use, municipal sewer lines collect the wastewater.

of chlorine is left in the water to provide protection during its distribution through many kilometers of pipes. Other disinfection systems use ozone or ultraviolet (UV) radiation in place of chlorine.

Municipal Sewage Treatment

Wastewater, including sewage, usually undergoes several treatments at a sewage treatment plant to prevent environmental and public health problems. The treated wastewater is then discharged into rivers, lakes, or the ocean.

primary treatment
Treatment of wastewater that involves removing suspended and floating particles through mechanical processes.

Primary treatment removes suspended and floating particles, such as sand and silt, through mechanical processes such as screening and gravitational settling (**Figure 10.21**, left side). The solid material that settles out at this stage is called **primary sludge**. **Secondary treatment** uses microorganisms (aerobic bacteria) to decompose the suspended organic material in wastewater (Figure 10.21, right side).

secondary treatment
Biological treatment of wastewater to decompose suspended organic material; secondary treatment reduces the water's biochemical oxygen demand.

Primary and secondary sewage treatment • Figure 10.21

✓ THE PLANNER

WILEY
RESOURCE KIT

Secondary sewage treatment

Primary sewage treatment

① Raw sewage enters from municipal sewage system.

② Large debris removed; sand settles to bottom.

③ Primary sedimentation tank: Suspended solids sink to bottom.

④ Aeration tank: Wastewater mixes with air (oxygen) to support bacteria that consume suspended organic wastes.

⑤ Secondary sedimentation tank: Cleanest water taken from surface of aeration tank; remaining particles settle.

⑥ Chlorinator (disinfection): Cleanest water taken from surface of secondary sedimentation tank and disinfected by chlorination or ultraviolet light.

⑦ Treated water discharged to a river or other natural water source.

Screen

Chlorinator

Primary sedimentation tank

Aeration tank

Secondary sedimentation tank

Sand

Primary sludge

Activated sludge

Secondary sludge

Sludge digester

Think Critically During what type of treatment, primary or secondary, would you expect items to be removed that had been accidentally flushed, such as coins or jewelry?

⑧ Sludge digester: Sludge from primary and secondary sedimentation tanks pumped to a digester, where bacteria consume organic wastes.

⑨ Digested sewage sludge disposed of in a sanitary landfill, incinerated, or converted into fertilizer.

PROCESS DIAGRAM

After several hours of processing, the particles and microorganisms are allowed to settle out, forming **secondary sludge**, a slimy mixture of bacteria-laden solids. Water that has undergone primary and secondary treatment is clear and free of organic wastes such as sewage. About 11 percent of U.S. wastewater treatment facilities have primary treatment only; about 62 percent have both primary and secondary treatments.

Even after primary and secondary treatments, wastewater still contains pollutants, such as dissolved minerals, heavy metals, viruses, and organic compounds. Advanced wastewater treatment methods, or **tertiary treatment**, include a variety of biological, chemical, and physical processes. Tertiary treatment reduces phosphorus and nitrogen, the nutrients most commonly associated with enrichment, and purifies wastewater for reuse in communities where water is scarce.

> **tertiary treatment** Advanced wastewater treatment methods that are sometimes employed after primary and secondary treatments.

Disposal of primary and secondary sludge is a major problem associated with wastewater treatment. Sludge is generally handled by application to soil as fertilizer, incineration, disposal in a sanitary landfill, or anaerobic digestion. (In anaerobic digestion, bacteria break down the organic material in sludge in the absence of oxygen.)

Some communities have adopted an environmentally innovative and economical approach to wastewater treatment. In the mid-1980s the city of Orlando, Florida, constructed artificial wetlands to treat reclaimed water from a municipal wastewater treatment plant. The Orlando Wetlands Park treats 61 million liters (16 million gallons) each day of treated effluent, across three separate wetland communities established on 494 hectares (1220 acres) of former pastureland (**Figure 10.22**). In what could be called an "advanced tertiary" process, the wetlands absorb and assimilate contaminants normally removed through more expensive treatment methods. The park also provides wildlife habitat for many organisms, and opportunities for human recreation.

Controlling Water Pollution

Many governments have passed legislation to control water pollution. Point source pollutants lend themselves to effective control more readily than do nonpoint source pollutants.

Wastewater treatment in constructed wetlands, Orlando, Florida • Figure 10.22

Influent is pumped in at the Orlando Wetlands Park, where more than 2 million aquatic plants and 200,000 trees planted remove excess nutrients from reclaimed wastewater influent.

The two U.S. laws that have the most impact on water quality today are the Safe Drinking Water Act and the Clean Water Act. The **Safe Drinking Water Act**, passed in 1974, set uniform federal standards for drinking water, to guarantee safe public water supplies throughout the United States. This law required the EPA to determine the **maximum contaminant level**, which is the maximum permissible amount of any water pollutant that might adversely affect human health. The EPA oversees the states to ensure that they adhere to the maximum contaminant levels for specific water pollutants. A 1996 amendment to the Safe Drinking Water Act requires municipal water suppliers to tell consumers what contaminants are present in their city's water and whether these contaminants pose a health risk.

The **Clean Water Act** affects the quality of rivers, lakes, aquifers, estuaries, and coastal waters in the United States. Originally passed as the Water Pollution Control Act of 1972, it was amended and renamed the Clean Water Act of 1977; additional amendments were made in 1981 and 1987. The Clean Water Act has two basic goals: to eliminate the discharge of pollutants in U.S. waterways and to attain water quality levels that make these waterways safe for fishing and swimming. Under the provisions of this act, the EPA is required to set up and monitor **national emission limitations**, the maximum permissible amounts of water pollutants that can be discharged from a sewage treatment plant, factory, or other point source.

Overall, the Clean Water Act has effectively improved the quality of water from point sources. According to the EPA, nonpoint source pollution is a major cause of water pollution yet is much more difficult and expensive to control than point source pollution. The 1987 amendments to the Clean Water Act expanded regulations on nonpoint sources.

The United States has improved its water quality in the past several decades and demonstrated that the environment recovers once pollutants are eliminated. Much remains to be done, however. The EPA's 2004 *National Water Quality Inventory* indicated that water pollution has increased in U.S. rivers, lakes, estuaries, and coastal areas in recent years. According to the report, 44 percent of the nation's rivers, 64 percent of its lakes, and 30 percent of its estuaries were too polluted to support one or more designated uses, including recreation, fishing, or providing drinking water.

Preventing Water Pollution at Home Although individuals produce little water pollution, the collective effect of municipal water pollution, even in a small neighborhood, can be quite large. There are many things you can do to protect surface waters and groundwater from water pollution (see **Table 10.2**); many municipalities have specific regulations or requirements that cover these measures.

Water Pollution in Developing Countries According to the World Health Organization, an estimated 1.4 billion people don't have access to safe drinking water, and about 2.9 billion people don't have access to adequate sanitation systems; most of these people live in rural areas of developing countries. Worldwide, at least 250 million cases of water-related illnesses occur each year, with 5 million or more of these resulting in death—1.8 million from diarrheal illnesses alone.

Preventing water pollution at home • Table 10.2

Location	What you can do
Bathroom	Never throw unwanted medicines down the toilet.
Kitchen	Use the smallest effective amount of toxic household chemicals such as oven cleaners, mothballs, drain cleaners, and paint thinners. Substitute less hazardous chemicals wherever possible. Dispose of unwanted hazardous household chemicals at hazardous waste collection centers.
	Avoid disposing of cooking wastes and uneaten food in the sink drain. Most foods increase BOD levels in sewage. Grease and oils can be hard on wastewater treatment plants. Consider composting fruit and vegetable wastes.
Driveway/car	Never pour used motor oil or antifreeze down storm drains or on the ground. Recycle these chemicals at service stations or local hazardous waste collection centers.
	Clean up spilled oil, brake fluid, and antifreeze, and sweep sidewalks and driveways instead of hosing them off. Dispose of dirt properly; don't sweep it into gutters or storm drains.
	Drive less: Air pollution emissions from automobiles eventually get into surface water and groundwater. Toxic metals and oil byproducts deposited on roads by vehicles are washed into surface waters by precipitation.
Lawn and garden	Pick up pet waste and dispose of it in garbage or toilet. If left on ground, it eventually washes into waterways, where it can contaminate shellfish and enrich water. Replace some grass lawn areas with trees, shrubs, and ground covers, which absorb up to 14 times more precipitation and require little or no fertilizer. To reduce erosion, use mulch to cover bare ground.
	Use fertilizer sparingly; excess fertilizer leaches into groundwater or waterways. Never apply fertilizer near surface water.
	Make sure that gutters and downspouts drain onto water-absorbing grass or graveled areas instead of onto paved surfaces.

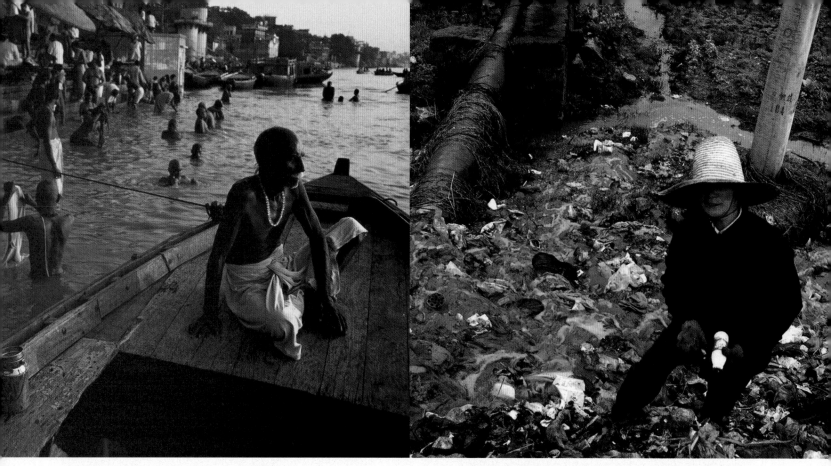

Ganges River • Figure 10.23

Bathing and washing clothes in the Ganges River are common practices in India. The river is contaminated by raw sewage discharged directly into the river at many different locations.

Contaminated water in China • Figure 10.24

A woman in east China's Jiangxi Province collects bottles from a river polluted with red dye and wastewater released from a nearby paper factory. Widespread water pollution limits supplies of safe drinking water in China's cities.

Municipal water pollution from sewage is a greater problem in developing countries, many of which lack water treatment facilities, than in highly developed nations. Sewage from many densely populated cities in Asia, Latin America, and Africa is dumped directly into rivers or coastal harbors. Other major sources of water pollution in developing countries include industrial wastes, agricultural chemicals, and even human remains.

The Ganges River is a holy river that symbolizes the spirituality and culture of the Indian people. Widely used for bathing and washing clothes (**Figure 10.23**), the river is highly polluted, largely with the mostly untreated sewage and industrial waste produced by the 400 million people who live in the Ganges River Basin. Another major source of contamination is the ashes of 35,000 human bodies cremated annually in accordance with Hindu tradition. Government efforts to clean up the Ganges have yet to produce noticeable improvements.

China's rapidly developing economy has resulted in the production of such severe water pollution that many of its densely crowded cities face water safety issues (**Figure 10.24**). Providing safe drinking water for China's 1.3 billion people is a formidable task, even without taking into account the current increase in water pollution.

CONCEPT CHECK STOP

1. **How** is most drinking water purified in the United States?

2. **What** are the stages in municipal sewage treatment? What happens in each stage?

3. **How** has the Safe Drinking Water Act affected U.S. water supplies? How has the Clean Water Act affected them?

CASE STUDY

China's Three Gorges Dam

The Three Gorges Dam on China's Yangtze River is a testament to human alteration of nature on a massive scale (see map). As the world's largest dam on the world's longest river, it is an engineering wonder that provides huge benefits, but its environmental impacts are also likely to be vast and profound.

Completed in late 2009 after 16 years of construction, the Three Gorges Dam project was initiated to control the often devastating flooding that occurred during China's rainy seasons and to make water more available to drought-prone areas during dry seasons. The dam has also improved navigation, enhancing shipping through a system of locks, and generates clean electricity: Its 26 hydropowered generators can produce nearly 85 billion kilowatt-hours each day, an equivalent of the electricity produced by burning 50 million tons of coal.

With the dam height of 185 m (607 ft), the Three Gorges project has raised water levels from a low-water mark of 62 m (203 ft) to a maximum height of 175 m (574 ft) in the reservoir behind the dam (**Figure a**). The rising waters flooded entire communities and required nearly 2 million people to relocate. This transformation in the river ecosystem also fragmented habitat, turning dozens of hilltops into islands.

Many fish species in the Yangtze Basin are now declining. Spawning conditions have been altered for many, including

valuable carp species, and migratory routes have been blocked for others, including sturgeon. Rare freshwater mammals—including the river dolphin (baiji) and finless porpoise (jiangzhu)—appear to be in serious danger (**Figure b**).

Other large-scale environmental impacts of the massive Three Gorges project include the unknown effects of silt buildup behind the dam, apparent declines in water quality as nutrient levels rise in the slowly moving water, potential bank destabilization and erosion that could trigger landslides, and enormous water pollution issues associated with industry and shipping.

The Chinese Academy of Sciences is monitoring these environmental issues. Its findings have generated a government plan to be implemented between 2010 and 2020, to address these varied and serious ecological impacts of the Three Gorges Dam.

Global Locator

HUBEI PROV., CHINA

SICHUAN PROV., CHINA

NATIONAL GEOGRAPHIC

a. The Three Gorges Dam spans the Yangtze River in China's Hubei Province.

b. The finless porpoise has been threatened by the construction of the Three Gorges Dam.

Summary

1 The Importance of Water 242

1. In the **hydrologic cycle**, water continuously circulates through the abiotic environment. **Surface water** is precipitation that remains on the surface. **Runoff** is the movement of fresh water from precipitation and snowmelt to rivers, lakes, wetlands, and the ocean. **Groundwater** is the supply of fresh water that is stored in **aquifers**, underground reservoirs.

2. Water molecules are **polar**: The negatively charged (oxygen) end of one molecule is attracted to the positively charged (hydrogen) end of another molecule, forming a **hydrogen bond**.

3. Hydrogen bonds are the basis for many of water's properties, including its high melting point, high boiling point, high heat capacity, and dissolving ability.

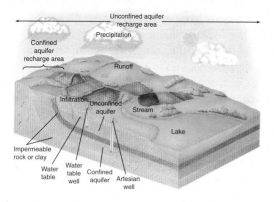

2 Water Resource Problems 244

1. **Aquifer depletion** is the removal of groundwater faster than it can be recharged. **Saltwater intrusion** is the movement of seawater into a freshwater aquifer near the coast. Overdrawing surface water causes **wetlands** to dry up and **estuaries** to become saltier. **Salinization** is the gradual accumulation of salt in soil, often due to improper irrigation.

2. Farmers on the U.S. High Plains are depleting water from the **Ogallala Aquifer** much faster than nature replaces it. In the Colorado River Basin, rapid population growth upstream threatens the water supply of users downstream.

3. Most of the world's major watersheds are shared between at least two nations. International cooperation is often required to manage shared water use.

3 Water Management 252

1. **Sustainable water use** is the wise use of water resources, without harming the hydrologic cycle or the ecosystems on which humans depend.

2. Dams and reservoirs allow rivers to be tapped for hydroelectric power and are used to supply municipal and industrial water, but they are expensive to build and alter the natural environment.

3. **Microirrigation** is an innovative type of irrigation that conserves water by piping it to crops through sealed systems. Industries and cities can employ measures to recapture, purify, and reuse water in homes and buildings.

4 Water Pollution 256

1. **Water pollution** is a physical or chemical change in water that adversely affects the health of humans and other organisms.

2. **Sewage** is wastewater from drains or sewers. It carries disease-causing agents and causes **enrichment**, the fertilization of a body of water due to high levels of nutrients. **Artificial eutrophication** is overnourishment of an aquatic ecosystem due to human activities. Sewage in water also raises the **biochemical oxygen demand (BOD)**, the amount of oxygen that microorganisms need to decompose biological wastes. A high BOD decreases water quality.

3. **Point source pollution** is water pollution that can be traced to a specific spot, such as wastewater released from a factory or sewage treatment plant. **Nonpoint source pollution** includes pollutants that enter bodies of water over large areas, such as agricultural runoff or municipal wastes.

4. Most of the nation's groundwater supplies are of good quality but are threatened by pollutants such as pesticides, fertilizers, and organic compounds.

5 Improving Water Quality 262

1. Most U.S. municipal water supplies are treated so the water is safe to drink. A chemical coagulant traps suspended particles, filtration removes suspended materials and microorganisms, and disinfection kills disease-causing agents.

2. Wastewater usually undergoes several treatments at a sewage treatment plant. **Primary treatment** removes suspended and floating particles from wastewater by mechanical processes. **Secondary treatment**, which reduces water's biochemical oxygen demand, treats wastewater biologically to decompose suspended organic material. **Tertiary treatment** reduces pollutants such as phosphorus and nitrogen.

3. The **Safe Drinking Water Act** protects the safety of the nation's drinking water. The **Clean Water Act** affects the quality of U.S. rivers, lakes, aquifers, estuaries, and coastal waters.

Key Terms

- aquifer depletion 246
- artificial eutrophication 259
- biochemical oxygen demand (BOD) 256
- groundwater 243
- microirrigation 254
- nonpoint source pollution 259

- point source pollution 259
- primary treatment 263
- runoff 243
- salinization 248
- saltwater intrusion 246
- secondary treatment 263

- sewage 256
- surface water 243
- sustainable water use 252
- tertiary treatment 264
- water pollution 256

What is happening in this picture?

- Rain soaks the streets of New Orleans' French Quarter.

- How might the hydrologic cycle be linked to potential groundwater pollution in this type of urban setting?

- What unique property of water allows it to carry pollutants?

- What about the structure of water molecules determines why water on Earth is most often found in this liquid form?

Critical and Creative Thinking Questions

1. Which water resource problems have played a role in Australia's drought crisis? How does that situation compare to conflicts over the Ogallala Aquifer or Colorado River Basin?

2. Briefly describe the complexity of international water use, using the Rhine River or the Aral Sea as examples.

3. Outline a brief water conservation plan for your own daily use. How could you use water more sustainably?

4. Explain why untreated sewage may kill fishes when it is added directly to a body of water.

5. Tell whether each of the following represents point or nonpoint source pollution: fertilizer runoff, thermal pollution from a power plant, urban runoff, sewage from a ship, and erosion sediments from deforestation. Why is nonpoint pollution more difficult to control than point source pollution?

6. What steps are taken in the purification of drinking water to kill disease-causing agents?

7. What roles do bacteria play in primary and secondary treatment of wastewater?

8. Is the Clean Water Act related to the quality of U.S. public drinking water? Explain your answer.

9. Compare the benefits and drawbacks of China's Three Gorges Dam. Do you think the dam will do more harm or more good for China? Explain your answer.

10–11. The graph reflects the monitoring of dissolved oxygen concentrations at six stations along a river. The stations are located 20 m (65.6 ft) apart, with A the farthest upstream and F the farthest downstream.

10. Where along the river did a sewage spill occur?

11. At which station would you most likely discover dead fish?

The Ocean and Fisheries

DEPLETING BLUEFIN TUNA STOCKS

Stocks of the giant, or Atlantic, bluefin tuna, highly prized for sushi, are classified as depleted in the Mediterranean Sea by the U.N. Food and Agriculture Organization. Once harvested sustainably through traditional trapping, Mediterranean bluefins are now fished—often illegally—at approximately four times the sustainable rate. Spotter aircraft locate fish stocks and alert huge fishing fleets, whose ships (see inset) cast purse seines to net schools. Captured bluefins are fattened in offshore pens (see large photograph) before being butchered for market. The enormous economic value of the huge bluefin places it at great risk. Only recently have Mediterranean nations begun implementing conservation measures to protect the species. The European Union in 2008 banned several nations from the purse seine harvest of bluefins in Mediterranean and east Atlantic waters and in 2009 supported a proposed 2010 ban on trade in the species.

Overfishing, the harvesting of fishes faster than they can reproduce, is not limited to the Mediterranean. Worldwide, about 30 percent of fish species have been overfished, including 80 percent of commercial fish stocks in the United States, as demand for fish has grown and harvesting methods have become more sophisticated. Ecologists and economists estimate that if overfishing and ocean pollution aren't curbed, populations of virtually all harvested seafood species could collapse by 2048.

CHAPTER OUTLINE

CHAPTER PLANNER ✓

- ❏ Study the picture and read the opening story.
- ❏ Scan the Learning Objectives in each section:
  p. 272 ❏   p. 276 ❏   p. 282 ❏   p. 289 ❏
- ❏ Read the text and study all figures and visuals. Answer any questions.

Analyze key features

- ❏ Process Diagram, p. 274 ❏
- ❏ Environmental InSight, p. 273 ❏   p. 283 ❏
- ❏ EnviroDiscovery, p. 280 ❏
- ❏ What a Scientist Sees, p. 284 ❏   p. 287 ❏
- ❏ Case Study, p. 291 ❏
- ❏ Stop: Answer the Concept Checks before you go on:
  p. 276 ❏   p. 281 ❏   p. 288 ❏   p. 290 ❏

End of Chapter

- ❏ Review the Summary and Key Terms.
- ❏ Answer What is happening in this picture?
- ❏ Answer the Critical and Creative Thinking Questions.

The Global Ocean

LEARNING OBJECTIVES

1. **Describe** the global ocean and its significance to life on Earth.

2. **Discuss** the roles of winds and the Coriolis effect in producing global water flow patterns, including gyres.

3. **Define** *El Niño–Southern Oscillation (ENSO)* and *La Niña* and describe some of their effects.

The ocean is a vast wilderness, much of it unknown. It teems with life—from warm-blooded mammals such as whales to soft-bodied invertebrates such as jellyfish. The ocean is essential to the hydrologic cycle that provides us with water. It affects cycles of matter on land, influences our climate and weather, and provides foods that enable millions of people to survive. The ocean dominates Earth, and its condition determines the future of life on our planet. If the ocean dies, then we do as well.

The global ocean is a huge body of salt water that surrounds the continents and covers almost three-fourths of Earth's surface. It is a single, continuous body of water, but geographers divide it into four sections separated by the continents: the Pacific, Atlantic, Indian, and Arctic Oceans. The Pacific is the largest: It covers one-third of Earth's surface and contains more than half of Earth's water.

Patterns of Circulation in the Ocean

The persistent prevailing winds blowing over the ocean produce currents, mass movements of surface–ocean water (**Figure 11.1a**). The prevailing winds generate **gyres**, circular ocean currents. In the North Atlantic Ocean, the tropical trade winds tend to blow toward the west, whereas the westerlies in the mid-latitudes blow toward the east. This helps establish a clockwise gyre in the North Atlantic.

> **gyres** Large, circular ocean current systems that often encompass an entire ocean basin.

The **Coriolis effect** influences the paths of surface, or shallow, ocean currents just as it does the winds (see Figure 8.5 on page 195). Earth's rotation from west to east causes surface ocean currents to swerve to the right in the Northern Hemisphere, helping establish the circular, clockwise pattern of water currents. In the Southern Hemisphere, ocean currents swerve to the left, thereby moving in a circular, counterclockwise pattern.

Vertical Mixing of Ocean Water Variations in the **density** (mass per unit volume) of seawater affect deep-ocean currents. Cold, salty water is denser than warmer, less salty water. Colder, salty ocean water sinks and flows under warmer, less salty water, generating currents far below the surface. Deep-ocean currents often travel in different directions and at different speeds than do surface currents, in part because the Coriolis effect is more pronounced at greater depths. **Figure 11.1b** shows the present circulation of shallow and deep currents—the **ocean conveyor belt**—that moves cold, salty deep-sea water from higher to lower latitudes, where it warms up. Note that the Atlantic Ocean gets its cold deep water from the Arctic Ocean, whereas the Pacific Ocean and Indian Ocean get theirs from the water surrounding Antarctica.

The ocean conveyer belt affects regional and possibly global climate. As the Gulf Stream and North Atlantic Drift push into the North Atlantic, they deliver an immense amount of heat from the tropics to Europe (**Figure 11.1c**). As this shallow current transfers its heat to the atmosphere, the water becomes denser and sinks. The deep current flowing southward in the North Atlantic is, on average, 8°C (14.4°F) cooler than the shallow current flowing northward.

Scientific evidence indicates that the ocean conveyor belt shifts from one equilibrium state to another. Historically, these shifts are linked to major changes in global climate.

Ocean–Atmosphere Interaction

The ocean and the atmosphere are strongly linked, with wind from the atmosphere affecting the ocean currents and heat from the ocean affecting atmospheric

a. Surface Ocean Currents. Winds largely cause the basic pattern of ocean currents. The main ocean current flow—clockwise in the Northern Hemisphere and counterclockwise in the Southern Hemisphere—results partly from the Coriolis effect.

b. The Ocean Conveyer Belt. This loop consists of deep-ocean currents that flow in the opposite direction from surface currents. Vertical motions drive the conveyer: Cold, salty water near Antarctica and the Arctic Ocean sinks and eventually flows northward into the Pacific Ocean, where it wells up, eventually becoming warmer and fresher. Cold, salty, deep water in the Atlantic Ocean comes from the Arctic Ocean. The ocean conveyer belt affects regional and global climate.

c. The Gulf Stream. The Gulf Stream is a well-known regional link in the ocean conveyor belt. In this satellite image, the colors represent the water's surface temperature: red = warmest, and blue = coolest. The Gulf Stream flows northeast along the North Carolina coast and then out to sea, toward Europe.

El Niño–Southern Oscillation (ENSO) • Figure 11.2

② Warm, moist air rises, causes low pressure and heavy rainfall.

③ Cool, dry air descends with high pressure and little rainfall.

Southeast trade winds

Equator

South Equatorial Current

① Accumulation of warm water.

④ Upwelling of nutrient-rich water (Humboldt Current).

a. Normal climate conditions

ENSO events depend on the relationship of atmospheric circulation to surface water flow in the Pacific. Normal conditions occur when strong easterly flow pushes warm water into the western Pacific.

Think Critically What are the differences in air circulation patterns across the Arctic Ocean and over North America and Asia during normal climate conditions and ENSO conditions?

② Descending air causes high pressure and warm, dry weather.

③ Warm air rises, causing low pressure and heavy rainfall.

Weak trade winds

Equator

① Warm water flows eastward to South America.

④ Upwelling blocked by warm water.

b. ENSO conditions

An ENSO event occurs when easterly flow weakens, allowing warm water to collect along the South American coast. Note the relationship between precipitation and the location of pressure systems. During an ENSO event, northern areas of the contiguous United States are typically warmer during winter, whereas southern areas are cooler and wetter.

WILEY
RESOURCE KIT

---

circulation. One of the best examples of the interaction between ocean and atmosphere is the **El Niño–Southern Oscillation (ENSO)** event, which is responsible for much of Earth's interannual (from one year to the next) climate variability. As a result of ENSO, some areas are drier, some wetter, some cooler, and some warmer than usual. Normally, westward-blowing trade winds restrict the warmest waters to the western Pacific near Australia (**Figure 11.2a**). Every 3 to 7 years, however, the trade winds weaken, and the warm mass of water expands eastward to South America, increasing surface temperatures in the usually cooler east Pacific (**Figure 11.2b**).

> **El Niño–Southern Oscillation (ENSO)**
> A periodic, large-scale warming of surface waters of the tropical eastern Pacific Ocean that temporarily alters both ocean and atmospheric circulation patterns.

Ocean currents, which normally flow westward in this area, slow down, stop altogether, or even reverse and go eastward. The name for this phenomenon, El Niño (in Spanish, "the boy child"), refers to the Christ child: The warming usually reaches the fishing grounds off Peru just before Christmas. Most ENSOs last between 1 and 2 years.

ENSO can devastate the fisheries off South America. Normally, the colder, nutrient-rich deep water is about 40 m (130 ft) below the surface and **upwells** (comes to the surface) along the coast, partly in response to strong trade winds (**Figure 11.3a**). During an ENSO event, however, the colder, nutrient-rich deep water is about 150 m (490 ft) below the surface, and the warmer surface temperatures and weak trade winds prevent upwelling (**Figure 11.3b**). The lack of nutrients in the

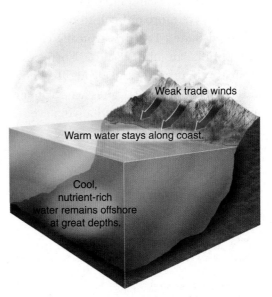

a. Coastal upwelling, where deeper waters come to the surface, occurs in the Pacific Ocean along the South American coast. Upwelling provides nutrients for microscopic algae, which in turn support a complex food web.

b. Coastal upwelling weakens considerably during years with El Niño–Southern Oscillation (ENSO) events, temporarily reducing fish populations.

water results in a severe decrease in the populations of anchovies and many other marine fishes. During the 1982–1983 El Niño, one of the worst ever recorded, the anchovy population decreased 99%. Other species, such as shrimp and scallops, thrive during an ENSO event.

ENSO alters global air currents, directing unusual, sometimes dangerous, weather to areas far from the tropical Pacific where it originates. By one estimate, the 1997–1998 ENSO, the strongest on record, caused more than 20,000 deaths and $33 billion in property damage worldwide. It resulted in heavy snows in parts of the western United States; ice storms in eastern Canada; torrential rains that flooded Peru, Ecuador, California, Arizona, and western Europe; and droughts in Texas, Australia, and Indonesia. An ENSO-caused drought—the worst in 50 years—particularly hurt Indonesia. Fires, many deliberately set to clear land for agriculture, got out of control and burned an area in Indonesia the size of New Jersey.

Climate scientists observe and monitor sea surface temperatures and winds to better understand and predict the timing and severity of ENSO events. The **TAO/TRITON array** consists of 70 moored buoys in the tropical Pacific Ocean. These instruments collect oceanic and weather data during normal conditions and El Niño events. The data are transmitted to scientists onshore by satellite.

Scientists at the National Oceanic and Atmospheric Administration's Climate Prediction Center forecasted the 1997–1998 ENSO 6 months in advance, using data from TAO/TRITON. Such forecasts give governments time to prepare for the extreme weather changes associated with ENSO.

La Niña El Niño isn't the only periodic ocean temperature event to affect the tropical Pacific Ocean. **La Niña** (in Spanish, "the girl child") occurs when the surface water temperature in the eastern Pacific Ocean becomes unusually cool and westbound trade winds become unusually strong. La Niña often occurs after an El Niño event and is considered part of the natural oscillation of ocean temperature.

During the spring of 1998, the surface water of the eastern Pacific cooled 6.7°C (12°F) in just 20 days. Like ENSO, La Niña affects weather patterns around the world, but its effects are more difficult to predict. In the contiguous United States, La Niña typically causes wetter-than-usual winters in the Pacific Northwest, warmer weather in the Southeast, and drought conditions in the Southwest. Atlantic hurricanes are stronger and more numerous than usual during a La Niña event.

CONCEPT CHECK   STOP

1. **What** is the global ocean, and how does it affect Earth's environment?

2. **How** are the Coriolis effect, prevailing winds, and surface ocean currents related?

3. **What** is the El Niño–Southern Oscillation (ENSO)? What are some of its global effects?

Major Ocean Life Zones

LEARNING OBJECTIVE

1. **Describe** and distinguish among the four main ocean life zones.

The immense marine environment is subdivided into several zones (**Figure 11.4**):

- The intertidal zone
- The benthic (ocean floor) environment
- The two provinces—neritic and oceanic—of the pelagic (ocean water) environment

The *neritic province* is that part of the pelagic environment from the shore to where the water reaches a depth of 200 m (650 ft). It overlies the continental shelf. The *oceanic province* is that part of the pelagic environment where the water depth is greater than 200 m, beyond the continental shelf.

The Intertidal Zone: Transition Between Land and Ocean

Although high levels of light, nutrients, and oxygen make the **intertidal zone** a biologically productive habitat, it is a stressful one. On sandy intertidal beaches, inhabitants must contend with a constantly shifting environment that threatens to engulf them and gives them little protection against wave action.

intertidal zone The area of shoreline between low and high tides.

benthic environment The ocean floor, which extends from the intertidal zone to the deep-ocean trenches.

Rocky shores provide fine anchorage for seaweeds and marine animals, but these organisms are exposed to wave action when submerged during high tides and exposed to temperature changes and drying out when in contact with the air during low tides (**Figure 11.5**).

A rocky-shore inhabitant generally has some way of sealing in moisture, perhaps by closing its shell (if it has one), and a means of anchoring itself to the rocks. For example, mussels have tough, threadlike anchors secreted by a gland in the foot, and barnacles secrete a tightly bonding glue that hardens underwater. Some organisms hide in burrows or under rocks or crevices at low tide. Some small crabs run about the splash line, following it up and down the beach.

The Benthic Environment

Most of the **benthic environment** consists of sediments (mainly sand and mud) where many bottom-dwelling animals, such as worms and clams, burrow. Bacteria are common in marine sediments, found even at depths more than 500 m (1625 ft) below the ocean floor. The deeper parts of the benthic environment are divided into three zones, from shallowest to deepest: the bathyal, abyssal, and hadal zones. The communities in the relatively shallow benthic zone that are particularly productive include coral reefs, sea grass beds, and kelp forests.

Zonation in the ocean • Figure 11.4

The intertidal zone, the benthic environment, and the pelagic environment make up the ocean. The pelagic environment consists of the neritic and oceanic provinces. (The slopes of the ocean floor aren't as steep as shown; they are exaggerated here to save space.)

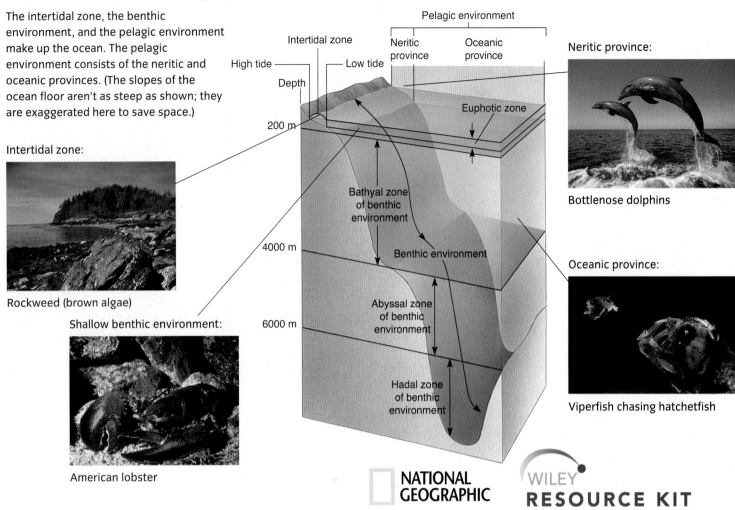

Intertidal zone:

Rockweed (brown algae)

Shallow benthic environment:

American lobster

Pelagic environment

Intertidal zone

Neritic province / Oceanic province

High tide — Low tide

Depth

200 m

4000 m

6000 m

Euphotic zone

Bathyal zone of benthic environment

Benthic environment

Abyssal zone of benthic environment

Hadal zone of benthic environment

Neritic province:

Bottlenose dolphins

Oceanic province:

Viperfish chasing hatchetfish

NATIONAL GEOGRAPHIC

WILEY RESOURCE KIT

Tide zones	Community zonation pattern of rocky shores
Supratidal (splash) zone	Anemones, tube worms, hermit crabs
Level of highest tide	Cyanobacteria, sea hair (*Ulothrix*), rough periwinkles
Intertidal zone	Acorn barnacles, rock barnacles, mussels, limpets, periwinkles, oysters, brown algae
Level of lowest tide	Tide pool
Subtidal zone	Brown algae, mussels, sea stars, brittle stars, sea urchins, spider crabs

Zonation along a rocky shore • Figure 11.5

Three zones are shown: the supratidal, or "splash" zone, which is never fully submerged; the intertidal zone, which is fully submerged at high tide; and the subtidal zone (part of the benthic environment), which is always submerged. Representative organisms are listed for each of these zones.

a. This map shows the distribution of coral reefs around the world. There are more than 6000 of them worldwide.

b. A coral reef in Fiji has a variety of soft corals as well as several fish species.

Corals are small, soft-bodied animals similar to jellyfish and sea anemones. Corals live in hard cups, or shells, of limestone (calcium carbonate) that they produce using the minerals dissolved in ocean water. When the coral animals die, the tiny cups remain, and a new generation of coral animals grows on top of these. Over thousands of generations, a **coral reef** forms from the accumulated layers of limestone.

Coral reefs are found in warm (usually greater than 21°C [70°F]), shallow seawater (**Figure 11.6a**). The living portions of coral reefs grow in shallow waters where light penetrates. The tiny coral animals require light for **zooxanthellae** (symbiotic algae) that live and photosynthesize in their tissues. In addition to obtaining food from the zooxanthellae that live inside them, coral animals capture food at night with stinging tentacles that paralyze **plankton** (small or microscopic organisms carried by currents and waves) and small animals that drift nearby. The waters where coral reefs grow are often poor in nutrients, but other factors are favorable for high productivity, including the presence of zooxanthellae, appropriate temperatures, and year-round sunlight.

Coral reef ecosystems are the most diverse of all marine environments (**Figure 11.6b**). They contain hundreds of species of fishes and invertebrates, such as giant clams, snails, sea urchins, sea stars, sponges, flatworms, brittle stars, sea fans, shrimp, and spiny lobsters. Australia's Great Barrier Reef occupies only 0.1 percent of the ocean's surface, but 8 percent of the world's fish species live there. The multitude of relationships and interactions that occur at coral reefs is comparable only to those of the tropical rain forest. As in the rain forest, competition is intense, particularly for light and space to grow.

Coral reefs are ecologically important because they both provide habitat for many kinds of marine organisms and protect coastlines from shoreline erosion. They provide humans with seafood, pharmaceuticals, and recreation and tourism dollars.

Sea grasses are flowering plants adapted to complete submersion in salty ocean water. They occur only in

Sea grass bed • Figure 11.7

Turtle grasses form underwater meadows that are ecologically important for shelter and food for many organisms. Photographed in the Caribbean Sea, the Cayman Islands.

Kelp forest • Figure 11.8

Underwater kelp forests are ecologically important because they support many kinds of aquatic organisms. Photographed off the coast of California.

shallow water (to depths of 10 m, or 33 ft) where they receive enough light to photosynthesize efficiently. Extensive beds of sea grasses occur in quiet temperate, subtropical, and tropical waters. Eelgrass is the most widely distributed sea grass along the coasts of North America; the world's largest eelgrass bed is in Izembek Lagoon on the Alaska Peninsula. The most common sea grasses in the Caribbean Sea are manatee grass and turtle grass (**Figure 11.7**). Sea grasses have a high primary productivity and are ecologically important: Their roots and rhizomes help stabilize sediments, reducing erosion, and they provide food and habitat for many marine organisms.

In temperate waters, ducks and geese eat sea grasses, and in tropical waters, manatees, green turtles, parrot fish, sturgeon fish, and sea urchins eat them. These herbivores consume only about 5 percent of sea grasses. The remaining 95 percent eventually enters the detritus food web and is decomposed when the sea grasses die. The decomposing bacteria are in turn consumed by animals such as mud shrimp, lugworms, and mullet (a type of fish).

Kelps, known to reach lengths of 60 m (200 ft), are the largest and most complex of all algae commonly called seaweeds (**Figure 11.8**). Kelps, which are brown algae, are common in cooler temperate marine waters of both the Northern and Southern Hemispheres. They

are especially abundant in relatively shallow waters (depths of about 25 m, or 82 ft) along rocky coastlines. Kelps are photosynthetic and are the primary food producers for the kelp "forest" ecosystem. Kelp forests provide habitats for many marine animals, such as tubeworms, sponges, sea cucumbers, clams, crabs, fishes, and sea otters.

Some animals eat kelp fronds, but kelps are mainly consumed in the detritus food web. Bacteria that decompose kelp provide food for sponges, tunicates, worms, clams, and snails. The diversity of life supported by kelp beds almost rivals that of coral reefs.

The Neritic Province: From the Shore to 200 Meters

Organisms that live in the pelagic environment's **neritic province** are all floaters or swimmers. The upper level of the pelagic environment is the **euphotic zone**, which extends from the surface to a maximum depth of 150 m (490 ft) in the clearest open ocean water. Sufficient light penetrates the euphotic zone to support photosynthesis.

> **neritic province**
> The part of the pelagic environment that overlies the ocean floor from the shoreline to a depth of 200 m (650 ft).

EnviroDiscovery

Otters in Trouble

Sea otters play an important role in their environment. They feed on sea urchins, thereby preventing the urchins from eating kelp, which allows kelp forests to thrive. Now scientists have uncovered an alarming decline in sea otter populations in western Alaska's Aleutian Islands—a stunning 90 percent crash since 1990—that in turn poses wide-ranging threats to the coastal ecosystem there. The population of sea urchins in these areas is exploding, and kelp forests are being devastated. Strong evidence identifies killer whales, or orcas, as the culprits. Orcas generally feed on sea lions, seals, and fishes of all sizes. Sea otters, the smallest marine mammal species, are more like a snack to the orca than a desirable meal. So why are the orcas now choosing sea otters? Biologists suggest that it is because seal and sea lion populations have collapsed across the north Pacific.

In a scenario that is partly documented and partly speculative, the starting point of this disastrous chain of events is a drop in fish stocks, possibly caused by overfishing or climate change. With their food fish in decline, seal and sea lion populations have suffered, and orcas have looked elsewhere for food. Even the terrestrial food chain has been affected, as

bald eagles shift away from fish and baby otter prey where otters are scarce. The change in the orcas' feeding behavior has transformed the food chain of kelp forests, putting orcas rather than otters at the top.

Otters in Alaskan waters

Large numbers of **phytoplankton** (microscopic algae) produce food by photosynthesis and are the base of food webs. **Zooplankton**, including tiny crustaceans, jellyfish, comb jellies, and the larvae of barnacles, sea urchins, worms, and crabs, feed on phytoplankton. Zooplankton are in turn consumed by plankton-eating **nekton** (any marine organism that swims freely), such as herring, sardines, squid, baleen whales, and manta rays (**Figure 11.9**). These in turn become prey for carnivorous nekton such as sharks, tuna, porpoises, and toothed whales. Nekton are mostly confined to the shallower neritic waters (less than 60 m, or 195 ft, deep), near their food.

oceanic province
The part of the pelagic environment that overlies the ocean floor at depths greater than 200 m (650 ft).

The Oceanic Province: Most of the Ocean

The oceanic province is the largest marine environment, representing about 75 percent of the ocean's water. Most of the oceanic

Neritic province • Figure 11.9

A manta ray swims slowly through the water, swallowing vast quantities of microscopic plankton as it swims. The wingspan of a mature manta ray can reach about 6 m (20 ft). Note the remoras that are hitching a ride.

Oceanic province • Figure 11.10

a. Found at dark depths of 700 to 3,000 m (2,300 to 9,840 ft), the spiky fanfin anglerfish attracts prey with its glowing lure. Its fin rays allow it to sense movement in the dark water. Photographed in Monterey Bay Canyon, California.

b. Unlike other fish species in the ocean's depths, the dragonfish can see red light. The pockets of red light shining beneath each of its eyes allow it to detect other organisms without being seen.

province is loosely described as the "deep sea." (The average depth of the ocean is 4000 m, more than 2 mi.) All but the shallowest waters of the oceanic province have cold temperatures, high pressure, and an absence of sunlight. These environmental conditions are uniform throughout the year.

Fishes of the deep waters of the oceanic province are strikingly adapted to darkness and scarcity of food (**Figure 11.10**). Adapted to drifting or slow swimming, animals of the oceanic province often have reduced bone and muscle mass. Many of these animals have light-producing organs to locate one another for mating or food capture.

Most organisms of the deep waters of the oceanic province depend on **marine snow**, organic debris that drifts down into their habitat from the upper, lighted regions of the oceanic province. Organisms of this little-known realm are filter feeders, scavengers, and predators. Many are invertebrates, some of which attain great sizes. The giant squid measures up to 18 m (59 ft) in length, including its tentacles.

CONCEPT CHECK STOP

1. **What** are the four main life zones in the ocean, and how do they differ from one another?

Human Impacts on the Ocean

LEARNING OBJECTIVES

1. **Contrast** fishing and aquaculture and relate the environmental challenges of each activity.

2. **Identify** the human activities that contribute to marine pollution and describe their effects.

3. **Explain** how global climate change could potentially alter the ocean conveyor belt.

The ocean is so vast, it's hard to imagine that human activities could harm it. Such is the case, however. Fisheries and aquaculture, marine shipping, marine pollution, coastal development, offshore mining, and global climate change all contribute to the degradation of marine environments. Scientists estimate that in 2008, less than 4 percent of the ocean remained unaffected by human activities, and 41 percent had experienced serious harm (**Figure 11.11a**).

Marine Pollution and Deteriorating Habitat

One of the great paradoxes of human civilization is that the same ocean that provides food to a hungry world is used as a dumping ground. Coastal and marine ecosystems receive pollution from land, from rivers emptying into the ocean, and from atmospheric contaminants that enter the ocean via precipitation. Offshore mining and oil drilling pollute the neritic province with oil and other contaminants. Pollution increasingly threatens the world's fisheries. Events such as accidental oil spills and the deliberate dumping of litter pollute the water. The World Resources Institute estimates that about 80 percent of global ocean pollution comes from human activities on land. In 2003 the Pew Oceans Commission, composed of scientists, economists, fishermen, and other experts, verified the seriousness of ocean problems in a series of studies. Some of their findings are shown in **Figure 11.11b**.

World Fisheries

The ocean contains valuable food resources. About 90 percent of the world's total marine catch is fishes, with clams, oysters, squid, octopus, and other molluscs representing 6 percent of the total catch. Crustaceans, including lobsters, shrimp, and crabs, make up about 3 percent, and marine algae constitute the remaining 1 percent.

Fleets of deep-sea fishing vessels obtain most of the world's marine harvest. Numerous fishes are also captured in shallow coastal waters and inland waters. According to the U.N. Food and Agricultural Organization (FAO), the world annual fish harvest increased substantially, from 19 million tons in 1950 to a high of nearly 95 million tons in 2000, and 91 million tons in 2007, the latest year for which data are available.

Problems and Challenges for the Fishing Industry No nation lays legal claim to the open ocean. Consequently, resources in the ocean are more susceptible to overuse and degradation than land resources, which individual nations own and for which they feel responsible (see a description of this type of dilemma in the Chapter 2 opener, "The Global Commons").

The most serious problem for marine fisheries is that many species, particularly large predatory fish, have been harvested to the point that their numbers are severely depleted. This generally causes a fishery to become unusable for commercial or sport fishermen, as well as for the other marine species that rely on it as part of the food web. Scientists have found that dramatically depleted fish populations recover only slowly. Some show no real increase in population size up to 15 years after the fishery has collapsed.

According to the FAO, at least 75 percent of the world's fish stocks are considered fully exploited, overexploited, or depleted. Fisheries have experienced such pressure for two reasons. First, the growing human population requires protein in its diets, leading to a greater demand for fish. Second, technological advances allow us to fish so efficiently that every single fish is often removed from an area (see *What a Scientist Sees* on page 284).

Fishermen tend to concentrate on a few fish species with high commercial value, such as salmon, tuna,

Impact
- Very low
- Low
- Medium
- Medium high
- High
- Very high

a. Mapping Human Impacts. In 2008, an international team of marine scientists mapped effects of 17 human activities on the ocean. Almost no location remains unaffected, and 41 percent of the ocean has been seriously altered by multiple activities.

Interpreting Data
Which regions exhibit the greatest impacts? Which are least affected? How is the location of these sites related to their status?

Climate change
Coral reefs and polar seas are particularly vulnerable to increasing temperatures.

Aquaculture
Fish farms produce wastes that can pollute ocean water and harm marine organisms.

b. Major Threats to the Ocean.

Invasive species
Organisms are transported and released from ships in ballast water, which contains foreign crabs, mussels, worms, and fishes.

Overfishing
Populations of many commercial fish species are severely depleted.

Nonpoint source pollution (runoff from land)
Agricultural runoff (fertilizers, pesticides, and livestock wastes) pollutes water.

Point source pollution
Passenger cruise ships dump sewage, shower and sink water, and oily bilge water.

Bycatch
Fishermen unintentionally kill dolphins, sea turtles, and sea birds.

Habitat destruction
Trawl nets (fishing equipment pulled along the ocean floor) destroy habitat.

Coastal development
Developers destroy important coastal habitat, such as salt marshes and mangrove swamps.

WILEY
RESOURCE KIT

WHAT A SCIENTIST SEES

Modern Commercial Fishing Methods

a. A full fish net is pulled on board a fishing vessel off the coast of Alaska.

b. Scientific evidence indicates that modern methods of harvesting fish are so effective that many fish species have become rare. Sea turtles, dolphins, seals, whales, and other aquatic organisms are accidentally caught and killed in addition to the target fish. The depth of longlines is adjusted to catch open-water fishes such as sharks and tuna or bottom fishes such as cod and halibut. Purse seines catch anchovies, herring, mackerel, tuna, and other fishes that swim near the water's surface. Trawls catch cod, flounder, red snapper, scallops, shrimp, and other fishes and shellfish that live on or near the ocean floor. Drift nets catch salmon, tuna, and other fishes that swim in ocean waters.

WILEY
RESOURCE KIT

Spotter airplane

Satellite tracking

Sonar

Aquaculture

Drift net

Long lines

Floats

Lines with hooks

Purse seine net

Trawl bag

---

and flounder, and other species, collectively called **bycatch**, are unintentionally caught and then discarded. The FAO reports that about 25 percent of all marine organisms caught—some 27 million metric tons (30 million tons)—are dumped back into the ocean. Most of these unwanted animals are dead or soon die because they are crushed by the fishing gear or are out of the water too long. The United States and other

> **bycatch** The fishes, marine mammals, sea turtles, seabirds, and other animals caught unintentionally in a commercial fishing catch.

countries are trying to significantly reduce the amount of bycatch and develop uses for the bycatch that remains.

In response to harvesting, many nations extended their limits of jurisdiction to 320 km (200 mi) offshore. This action removed most fisheries from international use because more than 90 percent of the world's fisheries are harvested in relatively shallow waters close to land. This policy was supposed to prevent overharvesting by allowing nations to regulate the amounts of fishes and other seafood harvested from their waters. However, many countries also have a policy of **open management**, in which all fishing boats of that country are given unrestricted access to fishes in national waters.

Aquaculture: Fish Farming
Aquaculture is more closely related to agriculture on land than it is to the fishing

aquaculture
The growing of aquatic organisms (fishes, shellfish, and seaweeds) for human consumption.

industry. Aquaculture is carried out both in fresh and marine water; the cultivation of marine organisms is sometimes called **mariculture**. According to the FAO, world aquaculture production has increased substantially, from 544,000 metric tons (600,000 tons) in 1950 to 59.1 million metric tons (65.2 million tons) in 2007.

Aquaculture differs from fishing in several respects. For one thing, although highly developed nations harvest more fishes from the ocean, developing nations produce much more seafood by aquaculture. Developing nations have an abundant supply of cheap labor, which is a requirement of aquaculture because it is labor intensive, like land-based agriculture. Another difference

between fishing and aquaculture is that the limit on the size of a catch in fishing is the size of the natural population, whereas the limit on aquacultural production is primarily the size of the area in which organisms can be grown.

In aquacultural "fish farms," fish populations are concentrated in a relatively small area and produce higher than normal concentrations of waste that pollute the adjacent water and harm other organisms. Aquaculture also causes a net loss of wild fish because many of the fishes farmed are carnivorous. Sea bass and salmon, for example, eat up to 5 kg (11 lb) of wild fish to gain 1 kg (2.2 lb) of weight.

Deep-water, offshore aquacultural facilities, sometimes called "ocean ranches," are becoming more common (**Figure 11.12**). Ocean ranches, which increasingly use cutting-edge technologies such as submersible cages with robotic surveillance, may avoid damaging coastlines but often lack the pollution-restricting oversight associated with other aquaculture operations. Also, caged populations are more genetically homogenous than wild ones; if the two groups interbreed, genetic diversity of wild populations could be diminished. The introduced organisms may also outcompete wild species.

An underwater fish farm • Figure 11.12

Cobia are raised in deep-water cages off Puerto Rico. The open-ocean circulation reduces the waste problems common in shallow-water aquaculture.

Global Locator

PUERTO RICO

NATIONAL GEOGRAPHIC

Shipping, Ocean Dumping, and Plastic Debris

Millions of ships dump oily ballast and other wastes overboard in the neritic and oceanic provinces. The U.N. International Maritime Organization's International Convention for the Prevention of Pollution from Ships (MARPOL) bans marine pollution arising from the shipping industry. MARPOL regulations specifically address six types of marine pollution caused by shipping: oil, noxious liquids, harmful packaged substances, sewage, garbage, and air pollution released by ships. The 2004–2006 revisions to MARPOL regulations included stricter controls on oil tankers and added certain marine sites to the list of special protected areas. Unfortunately, MARPOL is not well enforced in the open ocean.

In the past, U.S. coastal cities such as New York dumped their sewage sludge into the ocean. Disease-causing viruses and bacteria from human sewage contaminated shellfish and other seafood and posed an increasing threat to public health. The **Ocean Dumping Ban Act** barred ocean dumping of sewage and industrial waste, beginning in 1991.

Huge quantities of trash containing plastics are released into the ocean from coastal communities or, sometimes accidentally, from cargo ships. Plastics don't biodegrade; they photodegrade, which means that exposure to light breaks them down into smaller and smaller pieces that exist for an indefinite period. This trash collects in certain areas of the open ocean defined by atmospheric pressure systems. For example, in the north Pacific gyre—halfway between Hawaii and the U.S. mainland—researchers are monitoring a continuous, expanding array of floating plastics dubbed the "Pacific garbage patch," which covered an area roughly twice the size of Texas in late 2009 (**Figure 11.13**).

Not only are marine mammals and birds susceptible to being entangled in and strangled by larger pieces of plastic, but the many filter-feeding organisms near the bottom of the ocean food chain constantly ingest the smaller degraded pieces. These plastic pieces may absorb and transport hazardous chemicals such as PCBs. Scientists have yet to determine whether these substances are incorporated into marine food webs when organisms ingest the plastic.

Coastal Development

Development of resorts, cities, industries, and agriculture along coasts alters or destroys many coastal ecosystems, including mangrove forests, salt marshes, sea grass

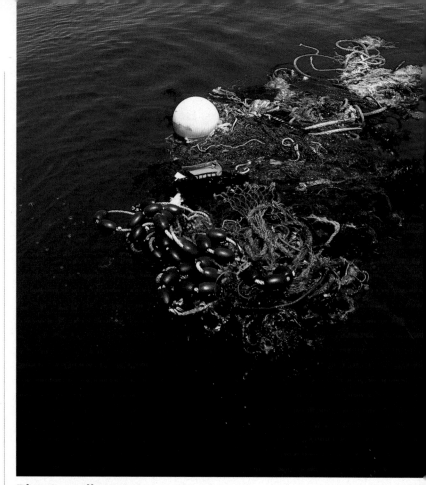

Plastic pollution in the ocean • Figure 11.13

The Pacific garbage patch, in the remote Pacific, doubles in size every decade. Scientists believe it to be one of five trash sites trapped in giant gyres in the ocean.

beds, and coral reefs. Many coastal areas are overdeveloped, highly polluted, and overfished. Although more than 50 countries have coastal management strategies, their goals are narrow and usually deal only with the economic development of the thin strips of land that directly border the oceans. Coastal management plans generally don't integrate the management of both land and water, nor do they take into account the main cause of coastal degradation—sheer human numbers.

Perhaps as many as 3.8 billion people—about 60 percent of the world's population—live within 150 km (93 mi) of a coastline. Demographers project that three-fourths of all humans—perhaps as many as 6.0 billion—will live in that area by 2025. To prevent the world's natural coastal areas from becoming urban sprawl or continuous strips of tourist resorts during the 21st century, coastal management strategies must be developed that take into account projections of human population growth and distribution.

Human Impacts on Coral Reefs

Coral formations, which are important ecosystems, are being degraded and destroyed. Approximately one-fourth of the world's coral reefs are at high risk. In some areas, silt washing downstream from clear-cut inland forests has smothered reefs. High salinity resulting from the diversion of fresh water to supply the growing human population may be killing Florida reefs. Overfishing (particularly the removal of top predators), damage by scuba divers and snorkelers, pollution from ocean dumping and coastal runoff, oil spills, boat groundings, fishing with dynamite or cyanide, hurricane damage, disease, coral bleaching, land reclamation, tourism, and the mining of corals for building material take a heavy toll.

Since the late 1980s, corals in the tropical Atlantic and Pacific have suffered extensive **bleaching** (see *What a Scientist Sees*), in which stressed corals expel their zooxanthellae. Scientists suspect that several environmental stressors, particularly warmer seawater temperatures, contribute to coral bleaching. Many scientists attribute recent record sea temperatures and large die-offs of corals (more than 70 percent losses in some areas) to El Niño effects, global climate change, or a combination of the two. Impaired coral growth in Australia was recently linked to increased ocean acidification associated with warmer water temperatures. Other potential stressors are pollution and coral diseases.

WHAT A SCIENTIST SEES

Ocean Warming and Coral Bleaching

WILEY
RESOURCE KIT

a. Bleached coral off the coast of Indonesia. Scientists have linked coral bleaching to ocean warming. Warmer than usual temperatures stress the coral animals, causing them to lose their zooxanthellae. Without their algae, the corals can't get enough food, and they die.

b. This 1920–2009 time series of annual global mean temperature changes of the ocean surface indicates that the ocean has warmed, particularly during the past 25 years. Most warming has occurred in shallow waters where corals live. Mean temperature anomaly data relative to 1901–2000.

Interpreting Data
If this warming trend continues, what will the mean temperature change be by 2050?

Offshore Extraction of Mineral and Energy Resources

Large deposits of minerals, including **manganese nodules**, lie on or below the ocean floor, most abundantly in the Pacific (**Figure 11.14**). Dredging manganese nodules from the ocean floor would adversely affect sea life, and the current market value for these minerals wouldn't cover the expense of obtaining them using existing technology. Furthermore, it isn't clear which countries have legal rights to minerals in international waters. Despite these concerns, many experts think that deep-sea mining will be technologically feasible in a few decades, and several industrialized nations such as the United States have staked claims in a region of the Pacific known for its large number of nodules. To date, none have been mined.

Offshore reserves of oil have long been tapped as a major source of energy. However, obtaining oil and gas resources from the seafloor generally poses a threat to fishing. Fishermen and conservationists worry that Congress may allow oil and gas wells to threaten fisheries such as the Georges Bank fishery, which is already suffering due to decades of overfishing. The environmental concerns associated with extracting offshore energy resources are discussed in Chapter 17.

Manganese nodules on the ocean floor • Figure 11.14

Manganese nodules—potato-sized rocks that contain manganese and other minerals, such as copper, cobalt, and nickel—are widespread on the ocean floor, particularly in the Pacific Ocean, where this deposit was photographed.

What natural resources are extracted in your region? How are issues related to these resources similar to or different from those involved in offshore mineral extraction?

Climate Change, Sea-Level Rise, and Warmer Temperatures

Our understanding of global climate is so incomplete that unanticipated effects from a globally warmed world will undoubtedly occur. For example, there could be a disruption of the ocean conveyor belt, which transports heat around the globe (see Figure 11.1b). Evidence from seafloor sediments and Greenland ice indicates that the ocean conveyor belt shifts from one equilibrium state to another in a relatively short period (a few years to a few decades). Scientists are concerned that human activities may affect this equilibrium.

Models based on the past behavior of the ocean conveyor belt suggest that climate warming, with its associated freshwater melting off the Greenland ice sheet, could weaken or even shut down the ocean conveyor belt in as short a period as a decade. Such changes in the ocean conveyor belt could cause major cooling in Europe, while greater climate warming occurs elsewhere. In addition, a weakened ocean conveyer belt would not sequester as much carbon in the ocean, leading to a positive feedback loop: Less CO_2 in the ocean would mean more CO_2 in the atmosphere, which would cause additional atmospheric warming and in turn further weaken the ocean conveyer belt.

Until recently, climate scientists couldn't predict whether human-induced global climate change would affect El Niño and La Niña events in the tropical Pacific Ocean. Recent computer models indicate greater extremes of drying and heavy rainfall during El Niño events. Scientists are still uncertain whether El Niño events will occur more frequently with global climate change.

CONCEPT CHECK STOP

1. **What** are some of the harmful environmental effects associated with the fishing industry? with aquaculture?

2. **How** does the widespread use of plastics contribute to ocean pollution?

3. **How** might the effect of global climate change on the ocean alter the ocean conveyer belt?

Addressing Ocean Problems

LEARNING OBJECTIVES

1. **Describe** international initiatives that address problems in the global ocean.
2. **Explain** strategies proposed to correct ocean problems in the future.

The many different threats to the world's ocean are attributed to a range of local, regional, national, and global sources. Problems in the ocean are complex and therefore require complicated solutions.

Industrialized countries' interest in removing manganese nodules from the ocean floor, first expressed in the 1960s, triggered the formation of an international treaty, the **U.N. Convention on the Law of the Sea (UNCLOS)**. UNCLOS, which became effective in 1994, is generally considered a "constitution for the ocean," and its focus is the protection of ocean resources. As of 2010, 159 countries had joined the treaty and are bound to its requirements. (The United States had not yet ratified UNCLOS but voluntarily observes its provisions.) The provisions of UNCLOS are binding only for international waters, not for territorial waters, so seabed mining is not prohibited in territorial waters. For example, hydrothermal vent systems in deep territorial waters off Papua New Guinea contain gold, zinc, copper, and silver, and exploration efforts are currently under way to determine methods for extracting these resources.

In 1995 the United Nations approved the **U.N. Fish Stocks Agreement**, the first international treaty to regulate marine fishing. The treaty went into effect in 2001. Because the overfishing problem continues to escalate, the United Nations has sponsored other fishery protection pacts.

In the United States, the **Magnuson–Stevens Fishery Conservation and Management Act**, which went into effect in 1977 and was reauthorized in 2007, requires that the National Marine Fisheries Service and eight regional councils regulate fishing under U.S. jurisdiction. These groups are required to protect essential fish habitat for more than 600 fish species, reduce overfishing, rebuild the populations of overfished species, and minimize bycatch. Fishing quotas, restrictions of certain types of fishing gear, limits on the number of fishing boats, and closure of fisheries during spawning periods are some of the management tools used to reduce overfishing. The Magnuson-Stevens Act strengthened controls on illegal and unreported fishing in U.S. waters.

Future Actions

A 2004 report by the U.S. Commission on Ocean Policy, the first comprehensive review of federal ocean policy in 35 years, recommended three primary strategies for improving the ocean and coasts:

- **Create a new ocean policy to improve decision making**. Currently, a number of agencies and committees manage U.S. waters, and their respective goals often conflict. The commission recommends strengthening and reorganizing the National Oceanic and Atmospheric Administration (NOAA) and consolidating other federal ocean programs under it.

- **Strengthen science and generate information for decision makers**. There is a critical need for high-quality research on how marine ecosystems function and how human activities affect them.

- **Enhance ocean education to instill in citizens a stewardship ethic**. Environmental education should be part of the curriculum at all levels and should include a strong marine component.

Ensuring the recovery of depleted fisheries may require the establishment of networks of "no-take" reserves and a substantial reduction of fishing fleets. Governments will also have to reduce or remove subsidies that help support the fishing industry. (A **subsidy** is a form of government support given to a business or an institution to promote the activity performed by that business or institution.) Government subsidies encourage modernization and expansion of fishing fleets.

Many scientists think the best way to halt and reverse destruction of the ocean is to adopt an *ecosystem-based approach* to manage ocean environments. This means that rather than focus on a single, narrow goal such as reviving a specific fish population, ocean management should focus on preserving the health and function of the entire marine ecosystem.

Marine reserves • Figure 11.15

a. Fully protected marine reserves can limit destruction to critical habitats such as the Hawaiian breeding grounds of these humpback whales.

Olympic Coast
Cordell Bank
Gulf of the Farallones
Monterey Bay
Channel Islands
Papahānaumokuākea
Marine National Monument
Hawaiian Islands
Humpback Whale
Fagatele Bay
(American Samoa)

Thunder Bay
Stellwagen Bank
Monitor
Gray's Reef
Florida Keys
Flower Garden Banks

b. Map of the National Oceanic and Atmospheric Administration (NOAA) national marine sanctuary system, which includes 13 national marine sanctuaries and one national monument.

One proposed approach that would enhance ecosystem-based management would be to establish networks of fully protected marine reserves, within which no habitat destruction or resource extraction would be allowed. Currently less than 5 percent of U.S. marine environments are set aside as fully protected marine reserves, yet these areas have successfully preserved threatened habitats and increased populations of exploited organisms (**Figure 11.15**).

In 2006, President George W. Bush established the world's largest protected marine area when he designated the northwestern Hawaiian Islands and surrounding waters—an area almost as large as California—as a national monument. This designation provides permanent funding to manage and preserve the area, now named the Papahānaumokuākea Marine National Monument. This protected area is home to more than 7000 species, including seabirds, fishes, marine mammals, coral reef colonies, and other organisms, approximately one-quarter of which are found only there.

Initiatives such as the Papahānaumokuākea Marine National Monument require billions of dollars and years of effort to be fully realized. Like most other countries, the United States recognizes the importance of the ocean to life on this planet. However, it remains to be seen if the United States and other countries will make a strong commitment to protecting and managing the global ocean.

CONCEPT CHECK STOP

1. **Which** international treaties aim to protect ocean resources?

2. **What** three strategies does the U.S. Commission on Ocean Policy recommend?

CASE **STUDY**

The Dead Zone in the Gulf of Mexico

Nitrogen and phosphorus from the Mississippi River—products of fertilizer and manure runoff from midwestern fields and livestock operations—are deemed largely responsible for a huge **dead zone** in the Gulf of Mexico (see Figure a). Except for bacteria that thrive in oxygen-free environments, no life exists in the dead zone because the water there does not contain enough dissolved oxygen to support fishes or other aquatic organisms. Dead zones form seasonally worldwide; more than 405 occur along global coastlines. The Gulf of Mexico dead zone, one of the largest in the ocean, extends from the seafloor up into the water column, sometimes to within a few meters of the surface. In 2008 it covered about 21,000 km^2 (8000 mi^2), an area the size of New Jersey. It generally persists from March to September. In March and April, snowmelt and spring rains flow from the Mississippi River into the Gulf (see Figure b), and the dead zone is most severe during June–August.

The low-oxygen condition in a dead zone, known as **hypoxia**, occurs when algae grow rapidly because of the presence of nutrients in the water. Dead algae sink to the bottom and are decomposed by bacteria, which deplete the water of dissolved oxygen, leaving too little for other sea life.

Scientists are now seeing evidence that ocean warming induced by global climate change may be exacerbating dead zones. Dead zones, including the one in the Gulf of Mexico, are expanding, they are emerging closer to shore than ever before, and they are forming even in areas of the ocean that don't receive agricultural runoff.

Increased frequency and size of dead zones threaten biodiversity and harm coastal fisheries. The EPA has taken some measures to control nitrogen and phosphorus inputs to the Mississippi River but recognizes that the dead zone problem is immense in scope and will take billions of dollars and decades of effort to fix.

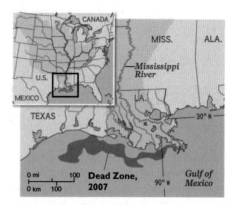

a. The map pictures the location and source of the Gulf of Mexico dead zone.

b. A satellite image shows the Mississippi River depositing sediments—including pollutants—into the Gulf of Mexico.

Southeastern Louisiana

Mouth of the Mississippi River

Gulf of Mexico

Summary

THE PLANNER

1 The Global Ocean 272

1. The global ocean is a huge body of salt water that surrounds the continents. It affects the hydrologic cycle and other cycles of matter, influences climate and weather, and provides food to millions.

2. Prevailing winds over the ocean generate **gyres**, large, circular ocean current systems that often encompass an entire ocean basin. The **Coriolis effect** is a force resulting from Earth's rotation that influences the paths of surface ocean currents, which move in a circular pattern, clockwise in the Northern Hemisphere and counterclockwise in the Southern Hemisphere.

3. The ocean and the atmosphere are strongly linked. The **El Niño–Southern Oscillation (ENSO)** event, which is responsible for much of Earth's interannual climate variability, is a periodic, large-scale warming of surface waters of the

tropical eastern Pacific Ocean that temporarily alters both ocean and atmospheric circulation patterns. A **La Niña** event occurs when surface water in the eastern Pacific Ocean becomes unusually cool. Its effects on weather patterns are less predictable than an ENSO event's.

2 Major Ocean Life Zones 276

1. The vast ocean is subdivided into major life zones. The biologically productive **intertidal zone** is the area of shoreline between low and high tides. The **benthic environment** is the ocean floor, which extends from the intertidal zone to the deep-ocean trenches. Most of the benthic environment consists of sediments where many animals burrow. Shallow benthic habitats include sea grass beds, kelp forests, and coral reefs. The **pelagic environment** is divided into two provinces. The **neritic province** is the part of the pelagic environment from the shore to where the water reaches a depth of 200 m (650 ft). Organisms that live in the neritic province are all floaters or swimmers. The **oceanic province**, "the deep sea," is the part of the pelagic environment where the water depth is greater than 200 m. The oceanic province is the largest marine environment, comprising about 75 percent of the ocean's water.

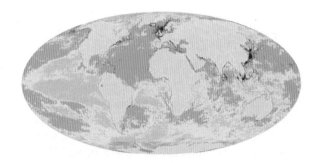

pollute the adjacent water and also causes a net loss of wild fish because many of the fishes farmed are carnivorous.

2. **Marine pollution** is generated by many human activities, including the release of trash and contaminants through commercial shipping, ocean dumping of sludge and industrial wastes, and discarding of plastics that are potentially harmful to marine organisms. Marine environments are also deteriorated by coastal development and the extraction of offshore minerals.

3. The **ocean conveyor belt** moves cold, salty, deep-sea water from higher to lower latitudes, affecting regional and possibly global climate. Global climate change associated with human activities may alter the link between the ocean conveyor belt and global climate.

3 Human Impacts on the Ocean 282

1. The most serious problem for marine fisheries is the overharvesting of many species to the point that their numbers are severely depleted. Fishermen usually concentrate on a few fish species with high commercial value. In doing so, they also catch **bycatch**: fishes, marine mammals, sea turtles, seabirds, and other animals caught unintentionally in a commercial fishing catch and then discarded. **Aquaculture** is the growing of aquatic organisms (fishes, shellfish, and seaweeds) for human consumption. Aquaculture is common in developing nations with abundant cheap labor, and it is limited by the size of the space dedicated to cultivation. Aquaculture produces wastes that

4 Addressing Ocean Problems 289

1. International initiatives aimed at protecting the global ocean include the **U.N. Convention on the Law of the Sea (UNCLOS)**, a "constitution for the ocean" that protects ocean resources, and the **U.N. Fish Stocks Agreement**, the first international treaty to regulate marine fishing.

2. Long-term goals for halting and reversing destruction of the ocean focus on adopting an ecosystem-based approach to management of ocean environments. Consolidating ocean programs, funding research on marine ecosystems, and enhancing ocean education to instill in citizens a stewardship ethic could improve U.S. ocean policy.

Key Terms

- aquaculture 285
- benthic environment 276
- bycatch 284
- El Niño–Southern Oscillation (ENSO) 274
- gyres 272
- intertidal zone 276
- neritic province 279
- oceanic province 280

What is happening in this picture?

- This school of cownose rays is migrating in the Gulf of Mexico.

- In what ocean life zone and province are they found?

- Which human activities are most likely to affect these animals?

- How might Gulf temperatures differ for later generations of this species?

Critical and Creative Thinking Questions

1. How do ocean currents affect climate on land? In particular, describe the role of the ocean conveyor belt.

2. Compare the different global effects of El Niño with those of La Niña. How are the two events similar? How are they different?

3. Identify which of the ocean life zones at right would be home to each of the following organisms: giant squid, kelp, tuna, and mussels. Explain your answers.

4. How might the production of plastic shopping bags contribute to ocean pollution?

5. Describe the global character of the ocean and its importance to life on Earth in terms of the effects of mismanagement of the bluefin tuna fishery in the Mediterranean or the expansion of the Gulf of Mexico dead zone.

6. Imagine that you live in a small Atlantic coast community where a company wants to set up an aquaculture facility in a salt marsh. What are its benefits and its environmental drawbacks? Would you support or oppose this proposal? Explain your answer.

7. If global climate change trends continue, why might Italy grow cooler, and how would impacts on the ocean trigger that change?

Use graph to answer questions 8–9.

8. From 1950 to 1980, did most of the growth in the world seafood harvest occur in the wild catch or in aquaculture?

9. From 1980 to 2006 (most recent data available) did the contribution of aquaculture to the global seafood harvest increase or decrease? In 2006 what percentage (approximately) of the global seafood harvest was from aquaculture?

10. Choose an issue affecting the global ocean and outline a plan to educate other students about this issue, with the goal of instilling a sense of stewardship.

11. Which U.N. treaty might impose limits on the number of fishing vessels allowed to catch tuna in international waters?

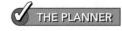

Mineral and Soil Resources

COPPER BASIN, TENNESSEE

Copper Basin, Tennessee, provides an example of environmental degradation caused by smelting, a stage of mineral processing. During the 19th century mining companies in southeastern Tennessee extracted copper ore—rock containing copper—from the ground and dug vast pits to serve as open-air smelters. They cut down the surrounding trees to fire the smelters, producing the high temperatures needed for the separation of copper from other substances in the ore. One of these substances, sulfur, reacted with oxygen in the air to form sulfur dioxide. The sulfur dioxide entered the atmosphere, reacted with water vapor there, and became sulfuric acid that returned to Copper Basin as acid precipitation.

Ecological ruin took only a few short years (see larger photograph). Acid precipitation killed plants.

Without plants to hold the soil in place, erosion cut gullies in the rolling hills. The forest animals disappeared along with the plants, their food and shelter destroyed.

State and federal reclamation efforts were only marginally successful until the 1970s, when specialists began using new replanting techniques. The new plants had a greater survival rate, and as they became established, their roots held the soil in place (see inset). Birds and field mice slowly began to return.

Today, reclamation of Copper Basin continues; the goal is to have the entire area under plant cover by the middle of the 21st century. The return of the original forest ecosystem will take at least a century or two. Plant scientists and land reclamation specialists have learned a lot from Copper Basin, and they will put this knowledge to use in future reclamation projects around the world.

NATIONAL GEOGRAPHIC

CHAPTER OUTLINE

CHAPTER PLANNER ✓

- ❏ Study the picture and read the opening story.
- ❏ Scan the Learning Objectives in each section:
  p. 296 ❏   p. 300 ❏   p. 304 ❏   p. 307 ❏   p. 310 ❏
- ❏ Read the text and study all figures and visuals.
  Answer any questions.

Analyze key features

- ❏ Process Diagram, p. 299 ❏
- ❏ EnviroDiscovery, p. 305 ❏
- ❏ What a Scientist Sees, p. 308 ❏
- ❏ Environmental Insight, p. 313 ❏
- ❏ Case Study, p. 315 ❏
- ❏ Stop: Answer the Concept Checks before
  you go on:
  p. 299 ❏   p. 303 ❏   p. 306 ❏   p. 310 ❏   p. 314 ❏

End of Chapter

- ❏ Review the Summary and Key Terms.
- ❏ Answer What is happening in this picture?
- ❏ Answer the Critical and Creative Thinking Questions.

Plate Tectonics and the Rock Cycle

LEARNING OBJECTIVES

1. **Define** *plate tectonics* and explain its relationship to earthquakes and volcanic eruptions.

2. **Diagram** a simplified version of the rock cycle.

Earth's layers and surface structure • Figure 12.1

a. The Main Layers of Planet Earth.

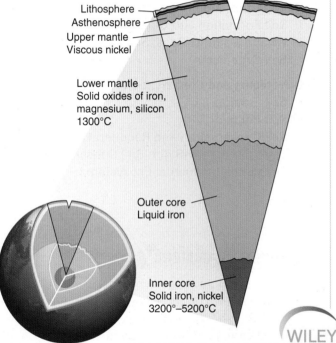

Lithosphere
Asthenosphere
Upper mantle
Viscous nickel

Lower mantle
Solid oxides of iron, magnesium, silicon
1300°C

Outer core
Liquid iron

Inner core
Solid iron, nickel
3200°–5200°C

Geology is an essential part of environmental science. To better understand the environmental effects of humans on mineral and soil resources, you must first know something about the geologic properties of Earth's crust.

Earth's outermost rigid rock layer (the *lithosphere*) is composed of seven large plates, plus a few smaller ones, that float on the *asthenosphere*, the region of the mantle where rocks become hot and soft (**Figure 12.1**). Continents and landmasses are situated on some of these plates. As the plates move across Earth's surface, the continents change their relative positions. **Plate tectonics**, the study of the movement of these plates, explains how most features on Earth's surface originate.

> **plate tectonics**
> The study of the processes by which the lithospheric plates move over the asthenosphere.

An area where two plates meet—a **plate boundary**—is a site of intense geologic activity (**Figure 12.2**). Earthquakes and volcanoes are common in these regions. San Francisco, California (noted for its earthquakes), and the volcano Mount Saint Helens in Washington State are both situated on plate boundaries. Where landmasses meet on the boundary between two plates, mountains may form: When the plate carrying India rammed into the plate carrying

b. Plates and Plate Boundary Locations.
There are seven major independent plates that move horizontally across Earth's surface. Arrows show the directions of plate movements. The three types of plate boundaries are explained in Figure 12.2.

Which plate do you live on? Are you near a plate boundary?

Juan de Fuca Plate
San Andreas Fault
North American Plate
Caribbean Plate
Eurasian Plate
Cocos Plate
Pacific Plate
East Pacific
Nazca Plate
South American Plate
Mid-Atlantic Ridge
African Plate
Indian–Australian Plate
Antarctic Plate

0   1000   2000   3000 miles
0 1000 2000 3000 kilometers

— Divergent boundary
▲▲ Convergent boundary
— Transform boundary

Plate boundaries • Figure 12.2

All three types of plate boundaries occur both in the ocean and on land.

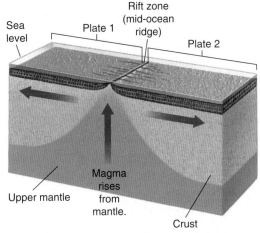

a. Two plates move apart at a divergent plate boundary.

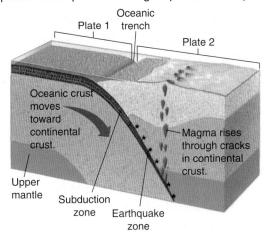

b. When two plates collide at a convergent plate boundary in the seafloor, subduction may occur. Convergent collision can also form a mountain range (not shown).

c. At a transform plate boundary, plates move horizontally in opposite but parallel directions. On land, such a boundary is often evident as a long, thin valley due to erosion along the fault line.

Asia, the resulting pressure pushed up the Himalayas. When two plates collide, one of them is sometimes forced under the other, in the process of **subduction**. When two plates move apart, a ridge of molten rock from the mantle wells up between them, continually expanding as the plates move farther apart. The Atlantic Ocean is growing as a result of the buildup of lava along the Mid-Atlantic Rift Zone, where two plates are separating.

Volcanoes

The movement of tectonic plates on the hot, soft rock of the asthenosphere causes most volcanic activity. In places where the asthenosphere is close to the surface, heat from this part of Earth's mantle melts the surrounding rock, forming pockets of **magma**. When one plate slides under or away from another, this magma may rise to the surface, often forming volcanoes. Magma that reaches the surface is called **lava**.

Volcanoes occur at three kinds of locations: in subduction zones, at spreading centers, and above hot spots. Subduction zones around the Pacific Basin have given rise to hundreds of volcanoes around Asia and the Americas, known as the *Ring of Fire*. Iceland is a volcanic island that formed along the Mid-Atlantic Rift Zone as the adjoining plates there spread apart. The volcanic Hawaiian Islands formed as the Pacific Plate moved over a **hot spot**, a rising plume of magma that flowed from an undersea opening in Earth's crust.

The largest volcanic eruption in the 20th century occurred in 1991, when Mount Pinatubo in the Philippines exploded (see Figure 9.9). Despite the evacuation of more than 200,000 people, several hundred deaths occurred, mostly from the collapse of buildings under the thick layer of wet ash that blanketed the area. The lava and ash ejected into the atmosphere by the eruption blocked much of the sun's warmth and caused a slight cooling of global temperatures for a year or so.

Earthquakes

Forces inside Earth sometimes push and stretch rocks in the lithosphere. The rocks absorb this energy for a time, but eventually, as the energy accumulates, the stress is too great, and the rocks suddenly shift or break. The energy—released as **seismic waves**, vibrations that rapidly spread through rock in all directions—causes one of the most powerful events in nature, an earthquake. Most earthquakes occur along **faults**, fractures in the crust where rock moves forward and backward, up and down, or from side to side. Fault zones are often found at plate boundaries.

The site where an earthquake begins, often far below the surface, is the **focus** (**Figure 12.3**). Directly above the focus, at Earth's surface, is the earthquake's **epicenter**. When seismic waves reach the surface, they cause the ground to shake. Buildings and bridges may collapse, and roads may break. One of the instruments used to measure seismic waves is a seismograph, which helps seismologists (scientists who study earthquakes) determine where an earthquake started, how strong it was, and how long it lasted.

Seismologists record more than 1 million earthquakes each year. Some of these are major, but most are too small for humans to feel, equivalent to readings of about 2 on the *Richter scale*, a measure of the magnitude of energy released by an earthquake. In populated areas, a magnitude 5 earthquake usually causes some property damage, and quakes of 8 or higher cause massive property destruction and kill large numbers of people.

In January 2010, an earthquake with a moment magnitude of 7.0 struck in an area approximately 25 km (16 mi) from the capital of Haiti. (Moment magnitude is a method of measuring earthquakes considered more accurate than the more commonly known Richter

scale.) The earthquake and powerful aftershocks triggered the structural collapse of many buildings. More than 200,000 people were killed, making it one of the deadliest earthquakes on record. About one million people whose homes were destroyed became refugees. The Caribbean region is prone to earthquakes due to movements between the North American, South American, and Caribbean plates. Puerto Rico, Jamaica, Dominican Republic, Martinique, and Guadeloupe have had earthquakes greater than magnitude 7 in the past.

Side effects of earthquakes include landslides and tsunamis. A *landslide* is an avalanche of rock, soil, and other debris that slides swiftly down a mountainside. A 1970 earthquake in Peru resulted in a landslide that buried the town of Yungay and killed 17,000 people.

A *tsunami*, a giant sea wave caused by an underwater earthquake or volcanic eruption, may sweep across the ocean at more than 750 km (450 mi) per hour. Although a tsunami may be only about 1 m (3 ft) high in deep-ocean water, it can build to a wall of water 30 m (about 100 ft)—as high as a 10-story building—when it comes ashore, often far from where the original earthquake triggered it. Tsunamis have caused thousands of deaths, particularly along the Pacific coast. Although the Pacific Tsunami Warning System monitors submarine earthquakes and warns people of approaching tsunamis, deaths still occur because there is little time to respond and not all nations are part of the network.

One of the deadliest natural disasters in modern history was the tsunami generated by an Indian Ocean earthquake in 2004 that killed more than 225,000 people along the coasts of South India, Thailand, Sri Lanka, Indonesia, and eastern Africa. The earthquake was recorded as a magnitude 9 or higher, and the waves it generated reached heights of 30 m (about 100 ft). Unfortunately, no gauges or buoys were in place in the Indian Ocean to detect the tsunami as it developed there.

The Rock Cycle

Rocks, which are aggregates of one or more minerals, fall into three categories, based on how they formed: igneous, metamorphic, and sedimentary. **Igneous rocks** form when magma rises from the mantle and

Earthquakes • Figure 12.3

Earthquakes occur when plates along a fault suddenly move in opposite directions relative to one another. This movement triggers seismic waves that radiate through the crust.

The rock cycle • Figure 12.4

Rocks do not remain in their original form forever. This highly simplified diagram shows how rock cycles from one form to another.

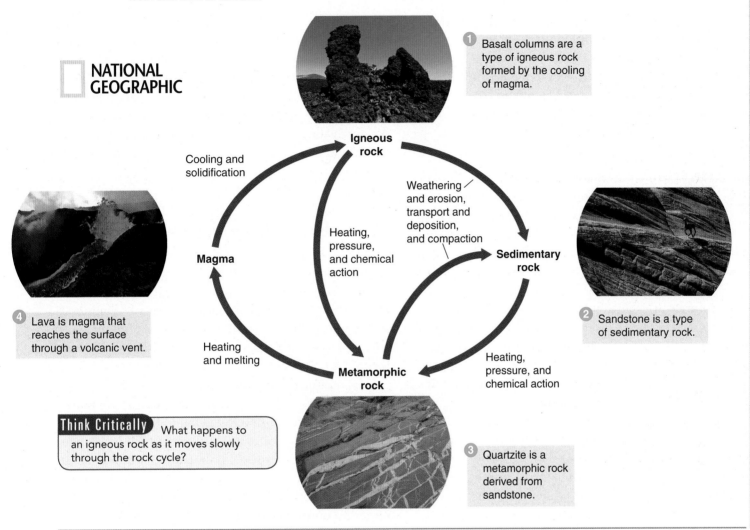

NATIONAL GEOGRAPHIC

❶ Basalt columns are a type of igneous rock formed by the cooling of magma.

Cooling and solidification

Igneous rock

Weathering and erosion, transport and deposition, and compaction

Heating, pressure, and chemical action

Magma

Sedimentary rock

❹ Lava is magma that reaches the surface through a volcanic vent.

❷ Sandstone is a type of sedimentary rock.

Heating and melting

Metamorphic rock

Heating, pressure, and chemical action

Think Critically What happens to an igneous rock as it moves slowly through the rock cycle?

❸ Quartzite is a metamorphic rock derived from sandstone.

cools. **Metamorphic rocks** form when intense heat and pressure alter igneous, sedimentary, or other metamorphic rocks. **Sedimentary rocks** form when small fragments of weathered, eroded rocks (or marine organisms) are deposited, compacted, and cemented together.

Earth's internal structure and the basic geologic processes that we have presented in this chapter result in a **rock cycle**, in which rock moves from one physical state or location to another (**Figure 12.4**). The rock cycle is similar to the other cycles of matter, such as the carbon and hydrologic cycles (see Chapter 5).

However, rocks are formed and move through the environment much more slowly than the elements of the other cycles.

CONCEPT CHECK

1. **What** are tectonic plates and plate boundaries? Where do earthquakes and volcanoes usually occur? Why?

2. **What** are the three types of rock? How are the three types of rock interconnected in the rock cycle?

Economic Geology: Useful Minerals

LEARNING OBJECTIVES

1. **Define** *minerals* and contrast the consumption of minerals by developing and highly developed countries.
2. **Distinguish** between surface mining and subsurface mining, using the terms *overburden* and *spoil bank* in your answer.
3. **Describe** briefly the process of smelting.

Earth's outermost layer, the crust, contains many kinds of minerals that are of economic importance. We now focus on the economic and environmental impacts of extracting and using mineral resources. We then consider soil, the part of the crust where biological and physical processes meet.

Minerals are such an integral part of our lives that we often take them for granted. Steel, an essential building material, is a blend of iron and other metals. Beverage cans, aircraft, automobiles, and buildings all contain aluminum. Copper, which readily conducts electricity, is used for electrical and communications wiring. The concrete used in buildings and roads is made from sand and gravel, as well as cement, which contains crushed limestone. Sulfur, a component of sulfuric acid, is an indispensable industrial mineral. It is used to make plastics and fertilizers and to refine oil. Other important minerals include platinum, mercury, manganese, and titanium.

> **minerals** Elements or compounds of elements that occur naturally in Earth's crust.

Earth's minerals are elements or (usually) compounds of elements and have precise chemical compositions. **Sulfides** are mineral compounds in which certain elements are combined chemically with sulfur, and **oxides** are mineral compounds in which elements are combined chemically with oxygen. Minerals are metallic or nonmetallic (**Figure 12.5**). **Metals** are minerals such as iron, aluminum, and copper, which are malleable, lustrous, and good conductors of heat and electricity. **Nonmetallic minerals**, such as sand, stone, salt, and phosphates, lack these characteristics.

Rocks are naturally formed mixtures of minerals that have varied chemical compositions. **Ore** is rock that contains a large enough concentration of a particular mineral to be profitably mined and extracted. **High-grade ores** contain relatively large amounts of particular minerals, whereas **low-grade ores** contain lesser amounts. Although some minerals are abundant, all minerals are nonrenewable resources that are not replenished by natural processes on a human timescale.

Minerals: An Economic Perspective

At one time, most of the highly developed nations had abundant mineral deposits that enabled them to industrialize. In the process of industrialization, these countries largely depleted their domestic reserves of minerals so that they must increasingly turn to developing countries. This is particularly true for countries in Europe, Japan, and, to a lesser extent, the United States.

As with the consumption of other natural resources, there is a large difference in consumption of minerals between highly developed and developing countries. The United States and Canada, which have about 5.5 percent of the world's population, consume about one-fourth of many of the world's metals. It is too simplistic, however, to divide the world into two groups, the mineral consumers (highly developed countries) and the mineral producers (developing countries). Many of the world's top mineral producers are highly developed countries (for example, the United States, Canada, Australia, and Russia). Many developing countries lack any significant mineral deposits.

Mineral production in China is increasing dramatically, as is China's mineral consumption as the country industrializes. For example, China smelts more than 20 percent of the world's primary aluminum (obtained from ores, not recycling). China also consumes almost all of this aluminum, making it the world's largest producer and largest consumer of primary aluminum. China now accounts for 16 percent of the world's finished goods and services. (In comparison, the United States produces 19 percent of the world's total.)

Some important minerals and their uses • Figure 12.5

Gypsum, silicon, and sulfur are nonmetals. All other minerals shown are metals.

Aluminum
Aircraft, motor vehicles, packaging (cans, foil), water treatment

Chromium
Chrome plate, dyes and paints, steel alloys (cutlery)

Cobalt
Corrosion and wear-resistant alloys, pigments (cobalt blue)

Gold
Jewelry, money, restorative dentistry

Iron
Steel (alloy of iron) buildings and machinery

Magnesium
Beverage cans, electronic devices, firecrackers, flares

Mercury
Industrial chemicals, electric and electronic applications, batteries

Molybdenum
High-temperature alloys for aircraft, industrial motors

Nickel
Coins, metal plating, alloys with various uses

Potassium
Fertilizers, photography

Silver
Jewelry, silverware, photography, electronics

Titanium
Alloy in steel and other industrial alloys, pigment in paints, plastics

Zinc
Galvanizing steel, alloys (brass), anode in alkaline batteries

Gypsum ($CaSO_4-2H_2O$)
Drywall, plaster of Paris, soil conditioner

Silicon
Electronic devices, semiconductors, natural stone, glass, concrete

Sulfur
Industrial chemicals, insecticides, gunpowder, vulcanized tires

Types of mining operations • Figure 12.6

a. This open pit copper mine in Utah is the largest man-made excavation in the world.

b. Strip mining removes overburden along narrow strips to reach the ore beneath.

As China's mineral production and economic growth expand, the natural capital on which its economic growth is based is being degraded. Recall from Chapter 3 that **natural capital** is Earth's resources and processes that sustain humans and other living organisms (see Figure 3.12). Natural capital includes minerals, soils, fresh water, clear air, forests, wildlife, and fisheries.

Because industrialization increases the demand for minerals, developing countries that at one time met their mineral needs with domestic supplies become increasingly reliant on foreign supplies as development occurs.

How Minerals are Extracted and Processed

The process of making mineral deposits available for human consumption occurs in several steps. First, a particular mineral deposit is located. Geologic knowledge of Earth's crust and how minerals are formed is used to estimate locations of possible mineral deposits. Once these sites are identified, geologists drill or tunnel for mineral samples and analyze their composition. Second, mining extracts the mineral from the ground. Third, the mineral is processed, or refined, by concentrating it and removing impurities. Finally, the purified mineral is used to make a product.

Extracting Minerals The depth of a particular mineral deposit determines whether surface or subsurface

mining will be used. In **surface mining**, minerals are extracted near the surface. Surface mining is more common because it is less expensive than subsurface mining. Because even surface mineral deposits occur in rock layers beneath Earth's surface, the overlying soil and rock layers, called **overburden**, must first be removed, along with the vegetation growing in the soil. Then giant power shovels scoop out the minerals.

There are two kinds of surface mining: **open-pit surface mining** and strip mining. Iron, copper, stone, and gravel are usually extracted through **open-pit surface mining**, in which a giant hole, called a quarry, is dug in the ground to extract the minerals (**Figure 12.6a**). In **strip mining**, a trench is dug to extract the minerals (**Figure 12.6b**). Then a new trench is dug parallel to the old one, and the overburden from the new trench is put into the old one, creating a hill of loose rock called a **spoil bank**.

Subsurface mining extracts minerals too deep in the ground to be removed by surface

surface mining The extraction of mineral and energy resources near Earth's surface by first removing the soil, subsoil, and overlying rock strata.

overburden Soil and rock overlying a useful mineral deposit.

spoil bank A hill of loose rock created when the overburden from a new trench is put into the already excavated trench during strip mining.

subsurface mining The extraction of mineral and energy resources from deep underground deposits.

c. In a shaft mine, a hole is dug straight through the overburden to the ore, which is removed up through the shaft in buckets.

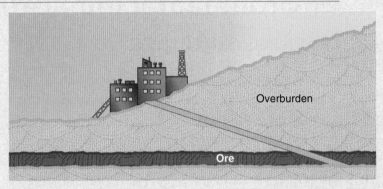

d. In a slope mine, an entry to the ore is dug at an angle so that the ore can be hauled out in carts.

mining. It disturbs the land less than surface mining, but it is more expensive and more hazardous for miners. There is always a risk of death or injury from explosions or collapsing walls, and prolonged breathing of dust in subsurface mines can result in lung disease.

Subsurface mining may be done with underground shaft mines or slope mines. A **shaft mine**, often used for mining coal, is a direct vertical shaft to the vein of ore (**Figure 12.6c**). The ore is broken up underground and then hoisted through the shaft to the surface in buckets. A **slope mine** has a slanting passage that makes it possible to haul the broken ore out of the mine in cars rather than to hoist it up in buckets (**Figure 12.6d**). Sump pumps keep a subsurface mine dry, and a second shaft is usually installed for ventilation.

Processing Minerals Processing minerals often involves smelting. Purified copper, tin, lead, iron, manganese, cobalt, or nickel smelting is done in a blast furnace. **Figure 12.7** shows a blast furnace used to smelt iron. The iron ore reacts with coke (modified coal) to form molten iron and carbon dioxide. The limestone reacts with impurities in the ore to form a molten mixture called **slag**. Note the vent near the top of the iron smelter for exhaust gases. If air pollution control devices are not installed, many dangerous gases are emitted during smelting.

> **smelting** The process in which ore is melted at high temperatures to separate impurities from the molten metal.

Blast furnace • Figure 12.7

Towerlike furnaces separate metal from impurities in the ore. The energy for smelting comes from a blast of heated air.

Iron ore, coke, and limestone

Exhaust gases

230°C

525°C

Hot gases used to preheat air

945°C

1510°C

Preheated air or oxygen

Molten slag

Molten iron

CONCEPT CHECK **STOP**

1. **What** is a mineral? How does mineral consumption differ between highly developed and developing countries?

2. **What** are the steps involved in surface mining? in subsurface mining?

3. **Why** are minerals smelted?

Environmental Implications of Mineral Use

LEARNING OBJECTIVES

1. **Relate** the environmental impacts of mining and refining minerals. Include a brief description of acid mine drainage.

2. **Explain** how mining lands can be restored.

The extraction, processing, and disposal of minerals has **external costs** (see Chapter 3). Mining disturbs and damages the land, and the processing and disposal of minerals pollute the air, soil, and water. Although pollution can be controlled and damaged lands can be restored, these remedies are costly. Historically, the environmental cost of extracting, processing, and disposing of minerals has not been incorporated into the actual price of mineral products to consumers.

Most highly developed countries have regulatory mechanisms in place to minimize environmental damage from mineral consumption, and many developing nations are in the process of putting them in place. Such mechanisms include policies to prevent or reduce pollution, restore mining sites, and exclude certain recreational and wilderness sites from mineral development.

Mining and the Environment

Mining, particularly surface mining, disturbs large areas of land. In the United States, functioning and abandoned metal and coal mines occupy an estimated 9 million hectares (22 million acres). Because mining destroys existing vegetation, mined land is particularly prone to erosion, with wind erosion causing air pollution and water erosion polluting nearby waterways and damaging aquatic habitats.

Open-pit mining of gold and other minerals uses huge quantities of water. As miners dig deeper, they eventually hit the water table and must pump out the water to keep the pit dry. Farmers and ranchers in open-pit mining areas are concerned about depletion of the groundwater they need for irrigation. Environmentalists and others would like the mining operations to reinject the water into the ground after pumping it out.

Mining has contaminated thousands of kilometers of streams and rivers in the United States. Rocks rich in

Acid mine drainage • Figure 12.8

The characteristic orange acid runoff contains sulfuric acid contaminated with heavy metals.

minerals often contain high concentrations of heavy metals such as arsenic and lead. Rainwater seeping through the sulfide minerals in mine waste produces sulfuric acid, which dissolves the heavy metals and other toxic substances in the spoil banks. These acids, called **acid mine drainage**, are highly toxic and are washed into soil and water, including groundwater, by precipitation runoff (**Figure 12.8**). When such acids and toxic compounds make their way into nearby lakes and streams, particularly through "toxic pulses" of thunderstorms or spring snowmelt, they adversely affect aquatic life.

> **acid mine drainage** Pollution caused when sulfuric acid and dangerous dissolved materials such as lead, arsenic, and cadmium wash from mines into nearby lakes and streams.

Environmental Impacts of Refining Minerals

Approximately 80 percent of mined ore consists of impurities that become wastes after processing. These wastes, called **tailings**, are usually left in giant piles on the ground or in ponds near the processing plants (**Figure 12.9**). The tailings contain toxic materials such as cyanide, mercury, and sulfuric acid. When left exposed, they contaminate the air, soil, and water.

EnviroDiscovery

Not-so-Precious Gold

Gold is a precious metal used primarily for jewelry and as a medium of exchange in many countries (see figure). Worldwide demand for gold is increasing, and the environment is suffering from the increased mining. The waste from mining and processing ore is enormous: 6 tons of wastes are produced to yield enough gold to make two wedding rings. One technology, *cyanide heap leaching*, allows profitable mining when minuscule amounts of gold are present, but this process produces up to 3 million pounds of waste for every pound of gold produced. The world's largest gold mine, located in Indonesia but owned by a U.S. company, dumps more than 100,000 tons of cyanide-contaminated waste into the local river each day. The highly toxic cyanide threatens waterfowl and fishes, as well as underground drinking water supplies.

Small-scale miners use other extraction techniques with destructive side effects: soil erosion, production of silt that clogs streams and threatens aquatic organisms, and contamination from mercury used to extract the gold. The environmental hazards of gold mining do not end when the gold is carried away: If not disposed of properly, mining wastes cause long-term problems such as acid mine drainage and heavy-metal contamination.

Jewelry

2,398.7 tons

Industrial

Electronics	Other industrial	Dentistry
310.6	92.7	57.8

Retail investment

Bar hoarding	Official coins	Medals, special coins
235.6	137	72.6

Exchange-traded funds

253.3

Uses of gold

Smelting plants may emit large quantities of air pollutants, particularly sulfur, during mineral processing. Unless expensive pollution control devices are added to smelters, the sulfur escapes into the atmosphere, where it forms sulfuric acid. (The environmental implications of the resulting acid precipitation are discussed in Chapter 9.) Pollution control devices for smelters are the same as the devices used for the burning of sulfur-containing coal—scrubbers and electrostatic precipitators.

Contaminants in ores include the heavy metals lead, cadmium, arsenic, and zinc. These toxic elements may pollute the atmosphere during the smelting process and cause harm to humans. Smelters emit airborne pollutants as well as hazardous liquid and solid wastes that can pollute the soil and water.

One of the most significant environmental impacts of mineral production is the large amount of energy

Copper ore tailings • Figure 12.9

Tailings dumped in mountainous heaps cause air, soil, and water pollution and have serious effects on land use.

required to mine and refine minerals, particularly if they are being refined from low-grade ore. Most of this energy is obtained by burning fossil fuels, which depletes nonrenewable energy reserves and produces carbon dioxide and other air pollutants.

Restoration of Mining Lands

When a mine is no longer profitable to operate, the land can be reclaimed, or restored to a seminatural condition, as has been done to most of the Copper Basin in Tennessee (see the chapter introduction). Reclamation prevents further degradation and erosion of the land, eliminates or neutralizes local sources of toxic pollutants, and makes the land productive for purposes other than mining (**Figure 12.10**). Restoration also makes such areas visually attractive.

Restoring land degraded by mining—called **derelict land**—involves filling in and grading the area to the shape of its natural contours and then planting vegetation to hold the soil in place. The establishment of plant cover is not as simple as throwing a few seeds on the ground. Often the topsoil is completely gone or contains toxic levels of metals, so special types of plants that tolerate such a challenging environment must be used.

The **Surface Mining Control and Reclamation Act** of 1977 requires reclamation of areas that were surface mined for coal. However, no federal law is in place to require restoration of derelict lands produced by other kinds of mines. As a result, restoration of mining lands often does not occur.

CONCEPT CHECK STOP

1. **What** are three harmful environmental effects of mining and processing minerals?

2. **How** are mining lands restored?

Restoration of mining lands • Figure 12.10

Bull elk forage on a reclaimed surface coal mine in Hindman, Kentucky. Restoration of mining lands makes them usable once again, or at least stabilizes them so that further degradation does not occur.

Soil Properties and Processes

LEARNING OBJECTIVES

1. **Define** *soil* and identify the factors involved in soil formation.

2. **Describe** the composition of soil and the organization of soil into horizons.

3. **Relate** at least two ecosystem services performed by soil organisms and briefly discuss nutrient cycling.

The relatively thin surface layer of Earth's crust is **soil**, which consists of mineral and organic matter modified by the natural actions of agents such as weather, wind, water, and organisms. It is easy to take soil for granted. We walk on and over it throughout our lives but rarely stop to think about how important it is to our survival. Vast numbers and kinds of organisms, mainly microorganisms, inhabit soil and depend on it for shelter, food, and water. Plants anchor themselves in soil, and from it they receive essential minerals and water. Terrestrial plants could not survive without soil, and because we depend on plants for our food, humans could not exist without soil either (**Figure 12.11**).

soil The uppermost layer of Earth's crust, which supports terrestrial plants, animals, and microorganisms.

Soil Formation and Composition

Soil is formed from *parent material*, rock that is slowly broken down, or fragmented, into smaller and smaller particles by biological, chemical, and physical **weathering processes**. It takes a long time, sometimes thousands of years, for rock to disintegrate into finer and finer mineral particles. Time is also required for organic material to accumulate in the soil. Soil formation is a continuous process that involves interactions between Earth's solid crust and the biosphere. The weathering of parent material beneath already formed soil continues to add new soil.

Topography, a region's surface features (such as the presence or absence of mountains and valleys), is also involved in soil formation. Steep slopes often have little or no soil on them because soil and rock are continually transported down the slopes by gravity. Runoff from precipitation tends to amplify erosion on steep slopes. Moderate slopes and valleys, on the other hand, may encourage the formation of deep soils.

Soil is composed of four distinct parts: mineral particles, organic matter, water, and air. The mineral portion,

Cut-away view of prairie soil in Kansas • Figure 12.11

Soil is an important natural resource that humans and countless soil organisms rely on.

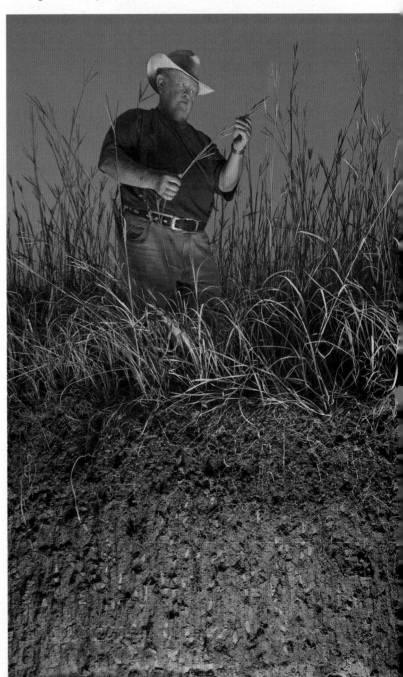

WHAT A SCIENTIST SEES

Soil Profile

O-horizon:
Mostly organic matter and humus; plant litter accumulates and decays.

A-horizon (topsoil):
Dark; high concentration of organic matter.

B-horizon (subsoil):
Light-colored; litter and nutrient minerals leached from A-horizon accumulate here.

C-horizon
(weathered parent material):
Below roots, often saturated with groundwater.

Consolidated bedrock
(parent material).

a. This soil, located on a farm in Virginia, has no O-horizon because it is used for agriculture; the surface litter that would normally compose the O-horizon was plowed into the A-horizon. The shovel gives an idea of the relative depths of each horizon.

b. A "typical" soil profile, as it appears to the trained eye of a soil scientist. Each horizon has its own chemical and physical properties.

WILEY
RESOURCE KIT

---

which comes from parent material, is the main component of soil. It provides anchorage and essential nutrient minerals for plants, as well as pore space for water and air. Litter (dead leaves and branches on the soil's surface), animal dung, and the remains of plants, animals, and microorganisms constitute the organic portion of soil. Organisms such as bacteria and fungi gradually decompose this material.

The black or dark brown organic material that remains after extended decomposition is called **humus** (**Figure 12.12**). Humus, which is a mix of many

Soil rich in humus • Figure 12.12

Humus is partially decomposed organic material, primarily from plant and animal remains. Soil rich in humus has a loose, somewhat spongy structure with several properties, such as increased water-holding capacity, that are beneficial for plants and other organisms living in it.

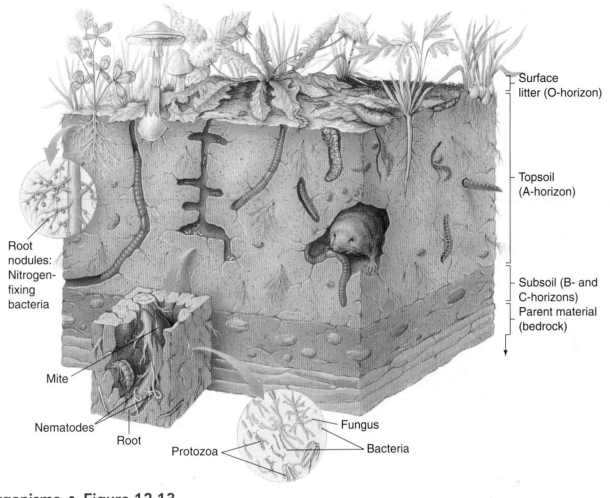

Labels on figure:
- Surface litter (O-horizon)
- Topsoil (A-horizon)
- Subsoil (B- and C-horizons)
- Parent material (bedrock)
- Root nodules: Nitrogen-fixing bacteria
- Mite
- Nematodes
- Root
- Protozoa
- Fungus
- Bacteria

Soil organisms • Figure 12.13

The diversity of life in fertile soil includes plants, algae, fungi, earthworms, flatworms, roundworms, insects, spiders and mites, bacteria, and burrowing animals such as moles and groundhogs. (Soil horizons are not drawn to scale.)

organic compounds, binds to nutrient mineral ions and holds water.

Many soils are organized into distinctive horizontal layers called soil horizons. A **soil profile** is a vertical section from surface to parent material, showing the soil horizons (see *What a Scientist Sees*). The topsoil (or A-horizon) is somewhat nutrient poor due to the leaching of many nutrients into deeper soil layers. **Leaching** is the removal of dissolved materials from the soil by water percolating downward.

> **soil horizons**
> Horizontal layers into which many soils are organized, from the surface to the underlying parent material.

Soil Organisms

Soil organisms, which are usually hidden underground, are remarkably numerous. Organisms that colonize the soil ecosystem include plant roots, insects such as termites and ants, earthworms, moles, snakes, and groundhogs (**Figure 12.13**). Most numerous in soil are bacteria, which number in the hundreds of millions per gram of soil. Other microorganisms that are abundant in soil ecosystems include fungi, algae, microscopic worms such as nematodes, and protozoa.

In a balanced ecosystem, the relationship between soil and the organisms that live in and on it ensure soil fertility. Soil organisms provide several essential **ecosystem**

Nutrient cycling • Figure 12.14

In a balanced ecosystem, nutrient minerals cycle from the soil to organisms and then back to the soil.

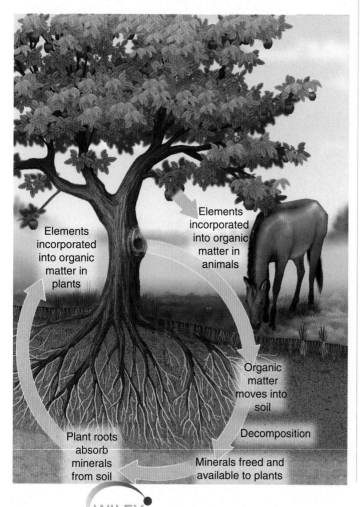

Elements incorporated into organic matter in plants

Elements incorporated into organic matter in animals

Organic matter moves into soil

Decomposition

Plant roots absorb minerals from soil

Minerals freed and available to plants

WILEY RESOURCE KIT

services, such as maintaining soil fertility, preventing soil erosion, breaking down toxic materials, and cleansing water.

Essential nutrient minerals such as nitrogen and phosphorus are cycled from the soil to organisms and back to the soil again. Decomposition, another ecosystem service, is part of **nutrient cycling**. Bacteria and fungi decompose plant and animal detritus and wastes, transforming large organic molecules into small inorganic molecules, including carbon dioxide, water, and nutrient minerals; the nutrient minerals are released into the soil to be reused (**Figure 12.14**; also see Chapter 5). Nonliving processes are also involved in nutrient cycling: The weathering of the parent material replaces some nutrient minerals lost through erosion or agricultural practices.

> **nutrient cycling**
> The pathway of various nutrient minerals or elements from the environment through organisms and back to the environment.

CONCEPT CHECK **STOP**

1. **How** do weathering processes affect soil formation?

2. **What** are soil horizons?

3. **What** role do soil microorganisms play in nutrient cycling?

Soil Problems and Conservation

LEARNING OBJECTIVES

1. **Define** *sustainable soil use.*

2. **Explain** the impacts of soil erosion on plant growth and on other resources, such as water.

3. **Identify** and summarize the major soil conservation methods.

S oil is as important as air and water for human survival. Yet humans disrupt soil systems that would be balanced in nature. We have had a harmful impact on soil resources worldwide, particularly by intensifying agricultural use. These human activities often cause or exacerbate soil problems such as erosion, mineral depletion, and soil pollution, all of which occur worldwide. Such activities do not promote **sustainable soil use**. Soil used in a sustainable way renews itself by natural processes year after year.

> **sustainable soil use** The wise use of soil resources, without a reduction in the amount or fertility of soil, so it is productive for future generations.

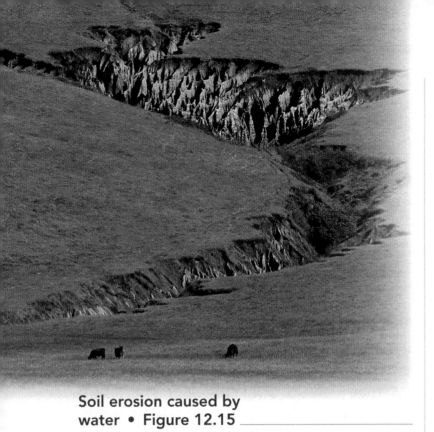

Soil erosion caused by water • Figure 12.15

The branching gullies shown here are the most serious form of erosion and will continue to grow unless checked by some type of erosion control. Photographed on a hillside in San Simeon, California.

Soil Erosion

Water, wind, ice, and other agents promote soil erosion, a natural process often accelerated by human activities. Water and wind are particularly effective in moving soil from one place to another. Rainfall loosens soil particles, which are transported by moving water (**Figure 12.15**). Wind loosens soil and blows it away, particularly if the soil is barren and dry. Erosion reduces the amount of soil in an area and therefore limits the growth of plants.

> **soil erosion** The wearing away or removal of soil from the land.

Humans often accelerate soil erosion with poor soil management. Poor agricultural practices are partly to blame, as are the removal of natural plant communities during road and building construction, and unsound logging practices such as clear-cutting. Soil erosion has an impact on other natural resources as well. Sediment that gets into streams, rivers, and lakes affects water quality and fish habitats (see Chapter 10). If the sediments contain pesticide and fertilizer residues, they further pollute the water.

Sufficient plant cover limits soil erosion. Leaves and stems cushion the impact of rainfall, and roots help to hold soil in place. Although soil erosion is a natural process, the abundant plant cover in many natural ecosystems makes it negligible.

Soil Pollution

Soil pollution is any physical or chemical change in soil that adversely affects the health of plants and other organisms living in or on the soil. Soil pollution is important not only in its own right but because many soil pollutants tend to also pollute surface water, groundwater, and the atmosphere. For example, selenium, an extremely toxic natural element found in many western soils, leaches off irrigated farmlands and poisons nearby lakes, ponds, and rivers. This has caused death and deformity in thousands of migratory birds and other organisms. Most soil pollutants originate as agricultural chemicals such as fertilizers and pesticides (see Chapter 14). Other soil pollutants include salts, petroleum products, and heavy metals.

Irrigation of agricultural fields often results in their becoming increasingly saline, an occurrence known as **salinization** (**Figure 12.16**). In time, salt can become so concentrated in soil that plants are poisoned.

Cut-away view of salinized soil • Figure 12.16

Irrigation water contains small amounts of dissolved salts. Over time, the salt accumulates in the soil. This irrigated soil in Colorado has become too salty for plants to tolerate.

Soil Conservation and Regeneration

Only 11 percent of the world's soil is suitable for agriculture (**Figure 12.17a, b**). We therefore need to protect the soil we use for agriculture. Although agriculture may cause or accelerate soil degradation, good soil conservation practices promote sustainable soil use. Conservation tillage, crop rotation, contour plowing, strip cropping, terracing, and shelterbelts minimize erosion and mineral depletion of the soil. Badly eroded and depleted land can be restored, but restoration is costly and time-consuming.

Conservation Tillage and Crop Rotation Conventional methods of tillage, or working the land, include spring plowing, in which the soil is cut and turned in preparation for planting seeds. Although conventional tillage prepares the land for crops, it greatly increases the likelihood of soil erosion. Conventionally tilled fields contain less organic material and generally hold less water than undisturbed soil.

Conservation tillage is one of the fastest-growing trends in U.S. agriculture (**Figure 12.17c**). More than one-third of U.S. farmland is currently planted using conservation tillage. In addition to reducing soil erosion, conservation tillage increases the organic material in the soil, which improves the soil's water-holding capacity. Decomposing organic matter releases nutrient minerals more gradually than when conventional tillage methods are employed. However, use of conservation tillage requires new equipment, new techniques, and greater use of herbicides to control weeds. Research is under way to develop alternative methods of weed control for use with conservation tillage. (Chapter 14 discusses *sustainable agriculture*, which includes conservation tillage and the other soil conservation practices presented in this chapter.)

Farmers who practice effective soil conservation measures often use a combination of conservation tillage and **crop rotation**. When the same crop is grown over and over in one place, pests for that crop accumulate to destructive levels, and the essential nutrient minerals for that crop are depleted in greater amounts. This makes the soil more prone to erosion, and it makes the crops less productive as well. Crop rotation is effective in decreasing insect damage and disease, reducing soil erosion, and maintaining soil fertility (Figure 12.17c also shows crop rotation).

A typical crop rotation would be corn → soybeans → oats → alfalfa. Soybeans and alfalfa, both members of the legume family, increase soil fertility through their association with bacteria that fix atmospheric nitrogen into the soil. Thus, planting soybeans and alfalfa as part of crop rotation produces higher yields of the grain crops that are also part of the rotation.

Contour Plowing, Strip Cropping, and Terracing Hilly terrain must be cultivated with care because it is more prone than flatland to soil erosion. Contour plowing, strip cropping, and terracing help control erosion of farmland with variable topography. In **contour plowing**, furrows run around hills rather than in straight rows. Strip cropping, a special type of contour plowing, produces alternating strips of different crops along natural contours (see Figure 3.1a on page 50). For example, alternating a row crop such as corn with a closely sown crop such as wheat reduces soil erosion. Even more effective control of soil erosion is achieved when strip cropping is practiced in conjunction with conservation tillage.

Farming is undesirable on steep slopes, but if it must be done, **terracing** can be used to level areas and thereby reduce soil erosion due to gravity or water run-off (**Figure 12.17d**). Nutrient minerals and soil are retained on the horizontal platforms instead of being washed away.

Soil Reclamation Badly eroded land can be reclaimed by (1) stabilizing the land to prevent further erosion and (2) restoring the soil to its former fertility. To prevent further erosion, the bare ground is seeded with plants; they eventually grow to cover the soil, stabilizing it and holding it in place. The plants start to improve the quality of the soil almost immediately,

> **conservation tillage** A method of cultivation in which residues from previous crops are left in the soil, partially covering it and helping to hold it in place until the newly planted seeds are established.

> **crop rotation** The planting of a series of different crops in the same field over a period of years.

> **contour plowing** Plowing that matches the natural contour of the land.

a. Soil and Agriculture. Only 11 percent of the world's total land area has soil that is naturally suitable for agriculture. Soil that is too dry can be irrigated, and soil that is too wet can be drained.

Soil frozen
6%

Soil too wet
10%

Soil suitable for agriculture
11%

Soil too dry
28%

Soil too shallow
22%

Chemical problems (for example, soil too salty)
23%

b. Soil Degradation. This graph shows the degree of soil degradation (eroded, desertified, or salty soil) by continent.

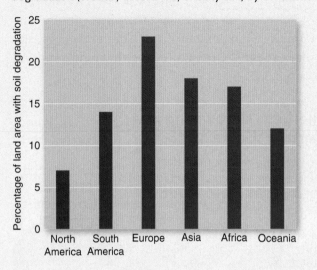

Percentage of land area with soil degradation

North America, South America, Europe, Asia, Africa, Oceania

c. Conservation Tillage. Decaying residues from the previous year's crop (rye) surround young soybean plants in a field in Iowa. Conservation tillage reduces soil erosion as much as 70 percent because plant residues from the previous season's crops are left in the soil.

Interpreting Data
Which continent has the most degraded soils? the least degraded soils? Offer an explanation for why this is the case.

d. Terracing. Terraces are small earthen embankments placed across a steep hillside or mountain. Terracing hilly or mountainous areas, such as these rice terraces on Bali, Indonesia, curbs water flow and reduces the amount of soil erosion. Farmers must maintain these terraced fields.

as dead material decays into humus. The humus holds nutrient minerals in place, releasing them a little at a time; it also improves the water-holding capacity of the soil. One of the best ways to reduce the effects of wind erosion on soil is to plant shelterbelts that reduce the impact of wind (**Figure 12.18**).

Shelterbelts surrounding kiwi orchards • Figure 12.18

Trees protect the delicate fruits from the wind and reduce wind erosion of farmland soil. Photographed on the North Island, New Zealand.

Global Locator

NEW ZEALAND

NATIONAL GEOGRAPHIC

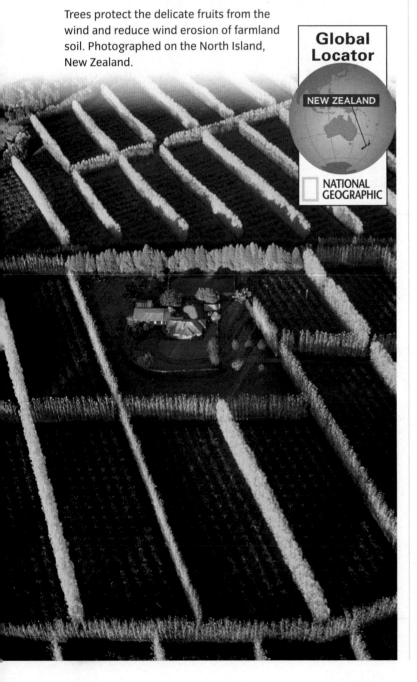

Restoration of soil fertility to its original level is a slow process. The land cannot be farmed or grazed until the soil has completely recovered. Disaster is likely if the land is put back to use before the soil has completely recovered. But the restriction of land use for an indefinite period may be difficult to accomplish. How can a government tell landowners they may not use their own land? How can land use be restricted when people's livelihoods, and maybe even their lives, depend on it?

Soil Conservation Policies in the United States

The Food Security Act (Farm Bill) of 1985 contains provisions for two main soil conservation programs: a conservation compliance program and the Conservation Reserve Program. The conservation compliance program requires farmers with highly erodible land to develop and adopt a 5-year conservation plan for their farms that includes erosion-control measures. If they do not comply, they lose federal agricultural subsidies such as price supports.

The **Conservation Reserve Program (CRP)** is a voluntary subsidy program that pays U.S. farmers to stop producing crops on highly erodible farmland. It requires planting native grasses or trees on such land and then "retiring" it from further use for 10 to 15 years. During that time the land may not be grazed, nor may the grass be harvested for hay. The CRP has benefited the environment. Annual loss of soil on CRP lands planted with grasses or trees has been reduced more than 90 percent. Because the vegetation is not disturbed once it is established, it provides biological habitat. Small and large mammals, birds of prey, and ground-nesting birds such as ducks have increased in number and kind on CRP lands. The reduction in soil erosion has improved water quality and enhanced fish populations in surrounding rivers and streams.

CONCEPT CHECK **STOP**

1. **What** is sustainable soil use?
2. **How** does soil erosion affect plants growing in the soil?
3. **How** do conservation tillage, contour plowing, and shelterbelts contribute to soil conservation?

CASE STUDY

Industrial Ecosystems

Traditional industries operate in a linear fashion: natural resources → products → wastes dumped back into the environment. However, natural resources are finite, and the environment's capacity to absorb waste is limited. The field of **industrial ecology** seeks to use resources efficiently, regarding "wastes" as potential products, and creating **industrial ecosystems** that in many ways mimic natural ecosystems.

One pioneering industrial ecosystem is in Kalundborg, Denmark (see **Figure**). An electric power plant, an oil refinery, a pharmaceutical plant, a wallboard factory, a sulfuric acid producer, a cement manufacturer, a fish farm, greenhouses, and area homes and farms are linked in ways that resemble a food web in a natural ecosystem.

Waste steam from the coal-fired electric power plant is supplied to the oil refinery and pharmaceutical plant, and

the surplus heat warms greenhouses, the fish farm, and area homes. Surplus natural gas from the oil refinery is sold to the power plant and the wallboard factory. Before selling the natural gas, the oil refinery removes excess sulfur from it (as required by law) and sells the sulfur to the sulfuric acid producer.

To meet environmental regulations, the power plant installed pollution control equipment to remove sulfur from its coal smoke. This sulfur, in the form of calcium sulfate, is sold to the wallboard plant and used as a gypsum substitute. The fly ash produced by the power plant goes to the cement manufacturer for use in road building. To fertilize their fields, local farmers use sludge from the fish farm and the pharmaceutical plant.

It took a decade to develop this entire industrial ecosystem. Although initiated for economic reasons, the industrial ecosystem has distinct environmental benefits, from energy conservation to a reduction of pollution.

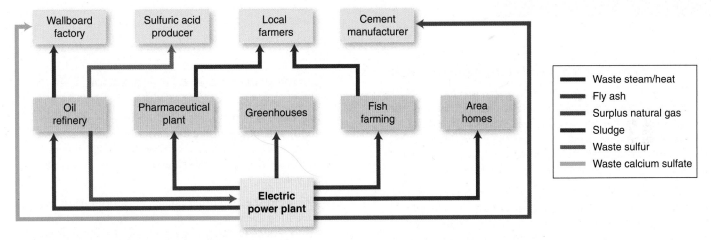

The Kalundborg industrial ecosystem

Kalundborg's industrial ecosystem has significantly reduced resource consumption and waste production.

Think Critically How does Kalundborg's industrial ecosystem mimic a natural ecosystem?

Summary

1 Plate Tectonics and the Rock Cycle 296

1. The lithosphere, Earth's outermost rigid rock layer, is composed of plates that float on the asthenosphere, the region of the mantle where rocks become hot and soft. **Plate tectonics** is the study of the processes by which the lithospheric plates move over the asthenosphere. **Plate boundaries** are often sites of intense geologic activity: earthquakes, volcanoes, and mountain building.

2. The **rock cycle** shows how rock slowly cycles from one form to another. The three categories of rock are **igneous**, **metamorphic**, and **sedimentary rock**.

2 Economic Geology: Useful Minerals 300

1. **Minerals** are metallic or nonmetallic elements or compounds of elements that occur naturally in Earth's crust. Highly developed nations consume a disproportionate share of the world's minerals, but as developing countries become industrialized, their needs for minerals increase.

2. Minerals are extracted through surface or subsurface mining. **Surface mining** removes the **overburden**: the overlying soil, subsoil, and rock strata. Strip mining, a type of surface mining, produces a **spoil bank** when the overburden from a new trench is put into an excavated trench. **Subsurface mining** extracts resources from deep underground deposits.

3. Processing minerals often involves **smelting**, melting the ore in a blast furnace to separate impurities from the metal.

3 Environmental Implications of Mineral Use 304

1. Surface mining destroys vegetation across large areas, increasing erosion. **Open-pit mining** uses huge quantities of water. Mining also affects water quality. **Acid mine drainage** is pollution caused when dissolved toxic materials wash from mines into nearby lakes and streams.

2. **Derelict lands** degraded by mining can be restored by filling in and grading the land to its natural contours and then planting vegetation to hold the soil in place.

4 Soil Properties and Processes 307

1. **Soil** is the uppermost layer of Earth's crust and supports terrestrial plants, animals, and microorganisms. Soil is formed from parent material—rock that is slowly fragmented into small particles through biological, chemical, and physical **weathering processes**.

2. Soil is composed of mineral particles, organic matter, water, and air. **Soil horizons** are the horizontal layers into which many soils are organized, from the surface to the underlying parent material.

3. Soil organisms provide **ecosystem services** such as maintaining soil fertility and preventing soil erosion. Soil organisms carry out **nutrient cycling**, the pathway of nutrient minerals or elements from the environment through organisms and back to the environment.

5 Soil Problems and Conservation 310

1. **Sustainable soil use** is the wise use of soil resources, without a reduction in the amount or fertility of soil, so soil is productive for future generations. Soil used in a sustainable way renews itself by natural processes year after year.

2. Water, wind, ice, and other agents cause **soil erosion**, the wearing away or removal of soil from the land. Soil erosion reduces fertility because essential minerals and organic matter are removed. Erosion causes sediments and pesticide and fertilizer residues to pollute nearby waterways.

3. Good soil conservation practices promote sustainable soil use. In **conservation tillage**, residues from previous crops partially cover the soil to help hold it in place until newly planted seeds are established. **Crop rotation**, the planting of different crops in a field over a period of years, decreases the insect damage, disease, and mineral depletion that occur when one crop is grown continuously. **Contour plowing**, which matches the natural contour of the land, helps control erosion of land with variable topography. **Strip cropping** produces alternating strips of different crops along natural contours. **Terracing** reduces soil erosion on steep slopes. A **shelterbelt** is a row of trees planted as a windbreak to reduce soil erosion.

Key Terms

- acid mine drainage 304
- conservation tillage 312
- contour plowing 312
- crop rotation 312
- minerals 300
- nutrient cycling 310

- overburden 302
- plate tectonics 296
- shelterbelt 314
- smelting 303
- soil 307
- soil erosion 311

- soil horizons 309
- spoil bank 302
- subsurface mining 302
- surface mining 302
- sustainable soil use 310

What is happening in this picture?

- Large mining firms evicted these people from their homes in Ghana. To make a living, they now illegally mine for gold.

- What types of pollution do you see in this picture?

- Do you think this sort of manual mining has greater or lesser effects than mechanized mining? Explain your answer.

- Go online and find the link between illegal gold mining and war in Africa's Great Lakes region.

Critical and Creative Thinking Questions

1. How are plate tectonics and tsunamis related?

2. Through what processes are rocks recycled over geologic time?

3. How many minerals have you come in contact with today? Which were metals; which were nonmetals?

4. What is the difference between surface and subsurface mining? open-pit and strip mines? shaft and slope mines? When is each most likely to be used?

5. Why is pollution control equipment needed for blast furnaces?

6. How does acid mine drainage damage nearby streams and groundwater?

7. How did Copper Basin, Tennessee, become an environmental disaster? Are reclamation efforts making a difference there?

8. What are the roles of weathering, organisms, and topography in soil formation?

9. What are the different soil horizons in a soil profile? Do soil horizons provide any information about a soil's ability to support plant growth? Explain.

10. What is nutrient cycling, and why is it so vital for the plants, animals, and microorganisms that live on land?

11. Why should environmentally sustainable societies protect their soil resources?

12. This graph relates spring wheat production (measured in kg of wheat per hectare of cultivated land) to soil erosion (measured in cm of topsoil lost) in the northern Great Plains. How does soil erosion affect wheat yield?

13. Describe two soil conservation methods that assist in sustainable food production.

Land Resources

U.S. NATIONAL PARKS

In 2003 and 2004 two new U.S. national parks were created, Congaree National Park in South Carolina and Great Sand Dunes National Park in Colorado. One of the smallest national parks, Congaree preserves the largest remaining intact hardwood bottomland (swamp) forest in the United States (see inset). It provides crucial terrestrial and aquatic wildlife habitat for many species.

Designating a national park does not happen overnight. The Sierra Club began campaigning to preserve the Congaree Swamp in 1969, and Congress established the Congaree Swamp National Monument in 1976. More than 25 years elapsed before the Congaree was designated a national park, after its growing recognition as a unique and important land resource.

Great Sand Dunes National Park features the highest sand dunes in North America, some as tall as 230 m (750 ft) (see large photograph, with the Sangre de Christo Mountains in the background), which formed as rain and wind eroded the surrounding mountains. As with Congaree, establishing Great Sand Dunes National Park took several years, and included its designation as a national monument. Land purchases involved the Nature Conservancy, as well as federal, state, and private donors.

National parks represent much more than wildlife sanctuaries. Parks preserve the land for future generations to enjoy the beauty of fast-disappearing natural areas.

NATIONAL GEOGRAPHIC

CHAPTER OUTLINE

CHAPTER PLANNER ✓

- ❏ Study the picture and read the opening story.
- ❏ Scan the Learning Objectives in each section:
  p. 320 ❏ p. 322 ❏ p. 331 ❏ p. 334 ❏ p. 339 ❏
- ❏ Read the text and study all figures and visuals. Answer any questions.

Analyze key features

- ❏ Process Diagram, p. 322 ❏
- ❏ What a Scientist Sees, p. 325 ❏
- ❏ EnviroDiscovery, p. 326 ❏
- ❏ Environmental InSight, p. 329 ❏ p. 335 ❏
- ❏ Case Study, p. 342 ❏
- ❏ National Geographic Map, pp. 340–341 ❏
- ❏ Stop: Answer the Concept Checks before you go on:
  p. 321 ❏ p. 330 ❏ p. 333 ❏ p. 338 ❏ p. 339 ❏

End of Chapter

- ❏ Review the Summary and Key Terms.
- ❏ Answer What is happening in this picture?
- ❏ Answer the Critical and Creative Thinking Questions.

Land Use in the United States

LEARNING OBJECTIVE

1. **Summarize** current land ownership in the United States.

Private citizens, corporations, and nonprofit organizations own more than 60 percent of the land in the United States, and Native American tribes own more than 2 percent. State and local governments own another 9 percent. The federal government owns the rest (nearly 28 percent).

Government-owned land encompasses all types of ecosystems, from tundra to desert, and includes land that contains important resources such as minerals and fossil fuels, land that possesses historical or cultural significance, and land that provides critical biological habitat. Most federally owned land is in Alaska and 11 western states (**Figure 13.1**).

Federal land is managed primarily by four agencies, three in the U.S. Department of the Interior—the Bureau of Land Management (BLM), the Fish and Wildlife Service (FWS), and the National Park Service (NPS)—and one in the Department of Agriculture—the U.S. Forest Service (USFS) (**Table 13.1**).

Government-owned lands provide vital **ecosystem services** that benefit humans living far from public forests,

Selected federal lands • Figure 13.1

Shown are national parks and preserves; national wildlife refuges; national forests, grasslands, and wilderness; and national marine sanctuaries in the United States. Note the preponderance of federal lands in western states and Alaska. Other federal lands, such as military installations and research facilities, aren't shown.

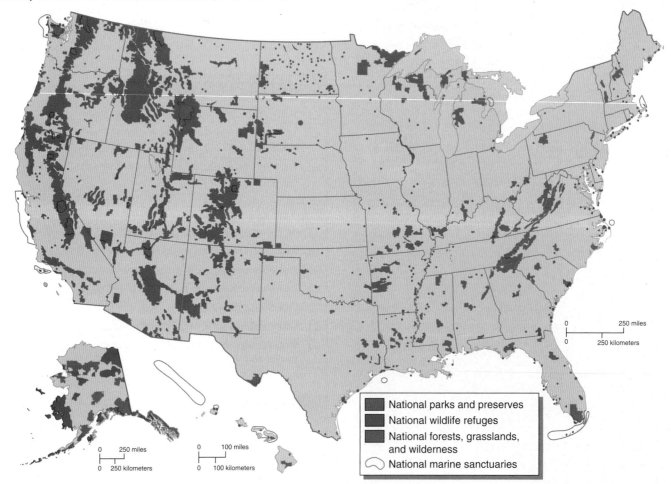

Legend:
- National parks and preserves
- National wildlife refuges
- National forests, grasslands, and wilderness
- National marine sanctuaries

Administration of federal lands • Table 13.1

Agency	Land held	Primary uses	Area in millions of hectares (acres)
Bureau of Land Management (Dept. of the Interior)	National resource lands	Mining, livestock grazing, oil and natural gas extraction	102 (253)
U.S. Forest Service (Dept. of Agriculture)	National forests	Logging, recreation, conservation of watersheds, wildlife habitat, mining, livestock grazing, oil and natural gas extraction	78 (193)
U.S. Fish and Wildlife Service (Dept. of the Interior)	National wildlife refuges	Wildlife habitat; also logging, hunting, fishing, mining, livestock grazing, oil and natural gas extraction	38 (93)
National Park Service (Dept. of the Interior)	National Park Service	Recreation, wildlife habitat	34 (84)
Other—includes Department of Defense, Corps of Engineers (Dept. of the Army), and Bureau of Reclamation (Dept. of the Interior)	Remaining federal lands	Military uses, wildlife habitat	23 (57)
Total federal lands			**275 (680)**

grasslands, deserts, and wetlands. These services include wildlife habitat, flood and erosion control, groundwater recharge, and the breakdown of pollutants.

Undisturbed public lands are ecosystems that scientists use as a benchmark, or point of reference, to determine the impact of human activity. Geologists, zoologists, botanists, ecologists, and soil scientists are some of the scientists who use government-owned lands for scientific inquiry. These areas provide perfect settings for educational experiences not only in science but also in history, because they can be used to demonstrate the condition of the land when humans originally settled it (**Figure 13.2**).

Public lands are important for their recreational value, providing places for hiking, swimming, boating, rafting, sport hunting, and fishing. Wild areas—forest-covered mountains, rolling prairies, barren deserts, and other undeveloped areas—are important to the human spirit. We can escape the tensions of the civilized world by retreating, even temporarily, to the solitude of natural areas.

Not all public lands remain undeveloped. As you will see throughout this chapter, many public lands are developed for uses ranging from logging to cattle grazing to mineral extraction.

Hurricane Ridge in Olympic National Park, Washington • Figure 13.2

Olympic National Park, 1 of 58 national parks and nearly 400 total sites in the National Park Service system, includes rugged mountains as well as rain forests and Pacific beaches.

CONCEPT CHECK STOP

1. **What** percentage of land in the United States is privately owned? What percentage is owned by the federal government?

Forests

LEARNING OBJECTIVES

1. **Define** *sustainable forestry* and explain how monocultures and wildlife corridors are related to it.

2. **Define** *deforestation*, including *clear-cutting*, and list the main causes of tropical deforestation.

3. **Describe** national forests and state which government agencies administer them and current issues of concern.

Forests, important ecosystems that provide many goods and services to support human society, occupy less than one-third of Earth's total land area. Timber harvested from forests is used for fuel, construction materials, and paper products. Forests supply nuts, mushrooms, fruits, and medicines. Forests provide employment for millions of people worldwide and offer recreation and spiritual sustenance in an increasingly crowded world.

Forests also provide a variety of beneficial ecosystem services, such as influencing climate conditions. If you walk into a forest on a hot summer day, you will notice that the air is cooler and moister than it is outside the forest. This is the result of a biological cooling process called *transpiration*, in which water from the soil is absorbed by roots, transported through plants, and then evaporated from their leaves and stems. Transpiration provides moisture for clouds, eventually resulting in precipitation (**Figure 13.3**). Thus, forests help maintain local and regional precipitation.

PROCESS DIAGRAM

Role of forests in the hydrologic cycle • Figure 13.3

THE PLANNER

Forests return most of the water that falls as precipitation to the atmosphere by transpiration. When an area is deforested, almost all precipitation is lost as runoff.

Think Critically

In which part of the hydrologic cycle do trees and other plants play a huge role? How would clearing the trees on a stretch of land disrupt this process? What would happen to water in soil there?

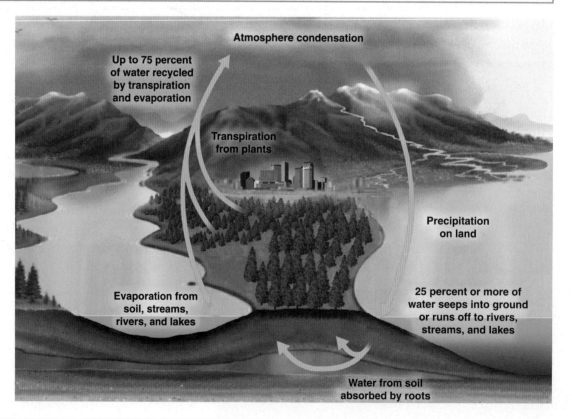

Atmosphere condensation

Up to 75 percent of water recycled by transpiration and evaporation

Transpiration from plants

Precipitation on land

Evaporation from soil, streams, rivers, and lakes

25 percent or more of water seeps into ground or runs off to rivers, streams, and lakes

Water from soil absorbed by roots

Forests play an essential role in regulating global biogeochemical cycles like those for carbon and nitrogen. Photosynthesis by trees removes large quantities of heat-trapping carbon dioxide from the atmosphere and fixes it into carbon compounds, while releasing oxygen back into the atmosphere. Forests thus act as carbon "sinks," which may help mitigate climate warming, and they produce oxygen, which almost all organisms require for cellular respiration.

Tree roots hold vast tracts of soil in place, reducing erosion and mudslides. Forests protect watersheds because they absorb, hold, and slowly release water; this moderation of water flow provides a more regulated flow of water downstream, even during dry periods, and helps control floods and droughts. Forest soils remove impurities from water, improving its quality. In addition, forests provide a variety of essential habitats for many organisms, such as mammals, reptiles, amphibians, fishes, insects, lichens and fungi, mosses, ferns, conifers, and numerous kinds of flowering plants.

Forest Management

Management for timber production disrupts a forest's natural condition and alters its species composition and other characteristics. Specific varieties of commercially important trees are planted, and those trees not as commercially desirable as others are thinned out or removed. Traditional forest management often results in low-diversity forests. In the southeastern United States, many tree plantations of young pine grown for timber and paper production are all the same age and are planted in rows a fixed distance apart (**Figure 13.4**). These "forests" are essentially **monocultures**—areas uniformly covered by one crop, like a field of corn. Herbicides are sprayed to kill shrubs and herbaceous plants between the rows. One of the disadvantages of monocultures is that they are at increased risk of damage from insect pests and disease-causing microorganisms. Consequently, pests and diseases must be controlled in managed forests, usually by applying insecticides and fungicides. Also, because managed forests contain few kinds of food, they can't support the variety of organisms typically found in natural forests.

> **monoculture**
> Ecological simplification in which only one type of plant is cultivated over a large area.

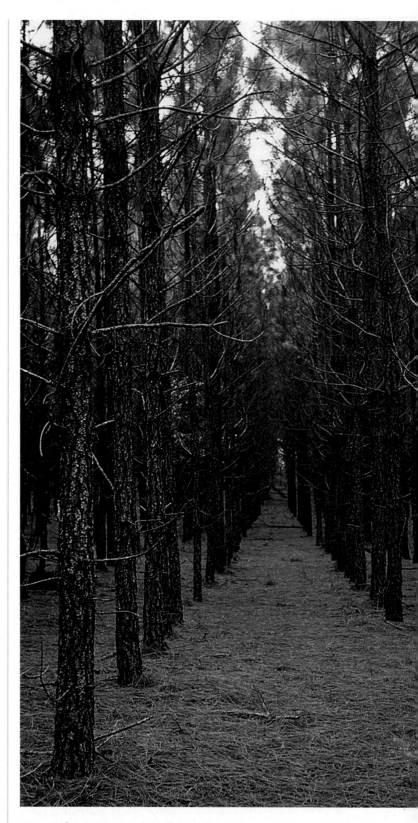

Tree plantation • Figure 13.4 _____

This intensively managed pine plantation in the southern United States is a monoculture, with trees of uniform size and age. Such plantations supplement harvesting of trees in wild forests to provide the United States with the timber it requires.

In recognition of the many ecosystem services performed by natural forests, a newer method of forest management, known as **ecologically sustainable forest management**, or simply sustainable forestry, is evolving. Sustainable forestry maintains a mix of forest trees, by age and species, rather than imposing a monoculture. This broader approach seeks to conserve forests for the long-term commercial harvest of timber and nontimber forest products. Sustainable forestry also attempts to sustain biological diversity by providing improved habitats for a variety of species, to prevent soil erosion and improve soil conditions, and to preserve watersheds that produce clean water. Effective sustainable forest management involves cooperation among environmentalists, loggers, farmers, indigenous peoples, and local, state, and federal governments.

> **sustainable forestry** The use and management of forest ecosystems in an environmentally balanced and enduring way.

When loggers use sustainable forestry principles, they set aside unlogged areas and wildlife corridors as sanctuaries for organisms. The purpose of wildlife corridors is to provide animals with escape routes, should they be needed, and to allow them to migrate so they can interbreed. (Small, isolated, inbred populations may have an increased risk of extinction.) Wildlife corridors may also allow large animals such as the Florida panther to maintain large territories. Some scientists question the effectiveness of wildlife corridors, although recent research in fragmented landscapes suggests that wildlife corridors help certain wildlife populations persist. Additional research is needed to determine the effectiveness of wildlife corridors for all endangered species.

> **wildlife corridor** A protected zone that connects isolated unlogged or undeveloped areas.

Methods for ecologically sustainable forest management are under development. Such practices vary from one forest ecosystem to another, in response to different environmental, cultural, and economic conditions. In Mexico, many sustainable forestry projects involve communities that are economically dependent on forests. Because trees have such long life spans, scientists and forest managers of the future will judge the results of today's efforts.

Harvesting Trees According to the U.N. Food and Agricultural Organization (FAO), about 3.5 million m^3 (120 million ft^3) of wood are harvested annually (for fuelwood, timber, and other products). The five countries with the greatest tree harvests are the United States, Canada, Russia, Brazil, and China; these countries currently produce more than half the world's timber. About 50 percent of harvested wood is burned directly as fuelwood or used to make charcoal. Most fuelwood and charcoal are used in developing countries (see Chapter 18). Highly developed countries consume more than three-fourths of the remaining 50 percent for paper and wood products.

Loggers harvest trees in several ways—selective cutting, shelterwood cutting, seed tree cutting, and clear-cutting (see *What a Scientist Sees* on facing page). **Selective cutting**, in which mature trees are cut individually or in small clusters while the rest of the forest remains intact, allows the forest to regenerate naturally.

The removal of all mature trees in an area over an extended period is **shelterwood cutting**. In the first year of harvest, undesirable tree species and dead or diseased trees are removed. Subsequent harvests occur at intervals of several years, allowing time for remaining trees to grow.

In **seed tree cutting**, almost all trees are harvested from an area; a scattering of desirable trees is left behind to provide seeds for the regeneration of the forest.

Clear-cutting is harvesting timber by removing all trees from an area and then either allowing the area to reseed and regenerate itself naturally or planting the area with one or more specific varieties of trees. Timber companies prefer clear-cutting because it is the most cost-effective way to harvest trees. However, clear-cutting over wide areas is ecologically unsound. It destroys biological habitats and increases soil erosion, particularly on sloping land, sometimes degrading land so much that reforestation doesn't take place. Clear-cut areas at lower elevations are usually regenerated successfully, whereas those at high elevations are difficult to regenerate. Obviously, most recreational benefits of forests are lost when clear-cutting occurs.

> **clear-cutting** A logging practice in which all the trees in a stand of forest are cut, leaving just the stumps.

WHAT A SCIENTIST SEES

Harvesting Trees

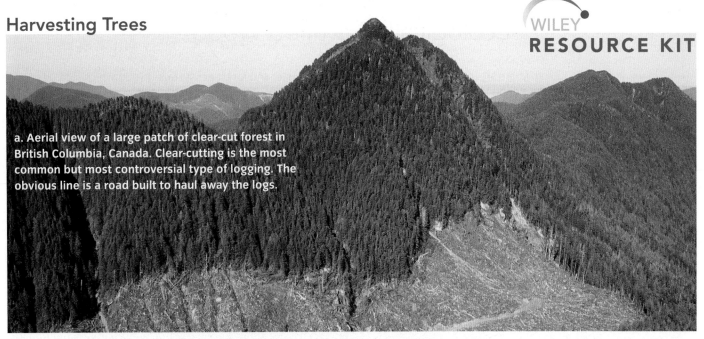

a. Aerial view of a large patch of clear-cut forest in British Columbia, Canada. Clear-cutting is the most common but most controversial type of logging. The obvious line is a road built to haul away the logs.

b. As a forest scientist looks at a clear-cut forest, he or she may think about the various kinds of tree harvesting (1 to 3) that are less environmentally destructive than clear-cutting (4).

(1) In **selective cutting**, the older, mature trees are selectively harvested from time to time, and the forest regenerates itself naturally.

(2) In **shelterwood cutting**, less desirable and dead trees are harvested. As younger trees mature, they produce seedlings, which continue to grow as the now-mature trees are harvested.

(3) **Seed tree cutting** involves the removal of all but a few trees, which are allowed to remain to provide seeds for natural regeneration.

(4) In **clear-cutting**, all trees are removed from a particular site. Clear-cut areas may be reseeded or allowed to regenerate naturally.

EnviroDiscovery

Ecologically Certified Wood

Many homebuilders and homeowners are interested in "green" wood for flooring and other building materials (see photograph). Such wood is ecologically certified by a legitimate third party, such as the Germany-based Forest Stewardship Council (FSC), to have come from a forest managed with environmentally sound and socially responsible practices. Although these areas remain a small percentage of world forests, by late 2009 the FSC had certified as well managed more than 118 million hectares (more than 291 million acres) in more than 82 countries. Certification is based on sustainability of timber resources, socioeconomic benefits provided to local people, and forest ecosystem health, which includes such considerations as preservation of wildlife habitat and watershed stability.

Green forestry has its detractors. Traditional forestry organizations are skeptical about the reliability of FSC investigations and the economic viability of this type of forestry. Trade experts caution that government efforts to specify the purchase of certified timber could violate global free-trade agreements. Still, green timber is gaining market share, pleasing business owners and consumers alike, and offering the promise of better conservation practices in managed forests.

The Forest Stewardship Council ecologically certifies "green" wood. Often, the consumer pays no additional premium, or only slightly more, for ecologically certified wood, which has become so popular that demand threatens to exceed supply.

Deforestation

The most serious problem facing the world's forests is **deforestation**. According to latest FAO estimates, world forests shrank by about 36 million hectares (89 million acres) between 2000 and 2005. This estimate represents a total loss of about 1 percent of global forested area and does not include thinned forests or those degraded by overharvesting, declining biological diversity, and reduced soil fertility. Causes of the decades-long trend of deforestation include fires caused by drought and land clearing practices, expansion of agriculture, construction of roads, tree harvests, insects, disease, and mining.

Most of the world's deforestation is currently taking place in Africa and South America, according to the FAO. Africa lost about 3.2 percent of its forested area from 2000 to 2005, and South America lost about 2.5 percent.

> **deforestation**
> The temporary or permanent clearance of large expanses of forest for agriculture or other uses.

Central America and the Caribbean nations lost about 3.9 percent of their forests during this period. There was a comparatively small loss of forested area in North America, whereas Europe and Asia actually gained forested areas, through either natural regrowth or increasing forest plantations.

Results of Deforestation Deforestation results in decreased soil fertility, as the essential mineral nutrients found in most forest soils leach away rapidly without trees to absorb them. Uncontrolled soil erosion, particularly on steep deforested slopes, affects the production of hydroelectric power as silt builds up behind dams. Increased sedimentation of waterways caused by soil erosion harms downstream fisheries. In drier areas, deforestation contributes to the formation of deserts. Regulation of water flow is disrupted when a forest is removed, so that the affected region experiences alternating periods of flood and drought.

Deforestation contributes to the extinction of many species. (See Chapter 15 for a discussion of the importance of tropical forests as repositories of biological diversity.) Many tropical species, in particular, have limited ranges within a forest, so they are especially vulnerable to habitat modification and destruction. Migratory species, including birds and butterflies, also suffer because of deforestation.

Deforestation causes regional and global climate changes. Trees release substantial amounts of moisture into the air; in the hydrologic cycle, about 97 percent of the water that roots absorb from the soil is evaporated directly into the atmosphere and then falls back to Earth. When a large forested area is removed, local rainfall may decline, droughts may become more common in that region, and temperatures may rise slightly. Deforestation also contributes to an increase in global temperature by releasing carbon originally stored in the trees into the atmosphere as carbon dioxide, which enables the air to retain heat. When an old-growth forest is harvested, researchers estimate that it takes about 200 years for the replacement forest to accumulate the equivalent amount of carbon stored in the original trees.

Boreal Forests and Deforestation Extensive deforestation in boreal forests due to logging began in the late 1980s. **Boreal forests** occur in Alaska, Canada, Scandinavia, and northern Russia and are dominated by coniferous evergreen trees such as spruce, fir, cedar, and hemlock. The boreal forest biome is the world's largest, covering about 11 percent of Earth's land. Harvested primarily through clear-cut logging, boreal forests are the primary source of the world's industrial wood and wood fiber. The annual loss of boreal forests is estimated to encompass an area twice as large as the rain forests of Brazil.

About 1 million hectares (2.5 million acres) of forest in Canada—currently the world's biggest timber exporter—are logged annually (**Figure 13.5**). Most of Canada's forests are subject to logging contracts, known as tenures, between provinces and companies. On the basis of current harvest quotas, logging is unsustainable in Canada. According to the World Resources Institute, the Canadian government promotes sustainable forest policies but does not always implement them.

Extensive tracts of Siberian forests in Russia are harvested, although estimates are unavailable. Alaska's boreal forests are at risk because the U.S. government may increase logging on public lands in the future.

Logging in Canada's boreal forest • Figure 13.5

About 80 percent of Canada's forest products are exported to the United States.

Tropical Forests and Deforestation There are two types of tropical forests: tropical rain forests and tropical dry forests. **Tropical rain forests** prevail in warm areas that receive 200 cm (79 in) or more of precipitation annually. Tropical rain forests are found in Central and South America, Africa, and Southeast Asia, but almost half of them are in just three countries: Brazil, Democratic Republic of the Congo, and Indonesia (**Figure 13.6a**). **Tropical dry forests** occur in other tropical areas where annual precipitation is less but is still enough to support trees. India, Kenya, Zimbabwe, Egypt, and Brazil are a few of the countries that have tropical dry forests.

Most of the remaining undisturbed tropical forests, which lie in the Amazon and Congo river basins of South America and Africa, respectively, are being cleared and burned at a rate unprecedented in human history. Tropical forests are also being destroyed at an extremely rapid rate in southern Asia, Indonesia, Central America, and the Philippines.

Why are tropical rain forests disappearing? Several studies show a strong statistical correlation between population growth and deforestation. More people need more food, and so forests are cleared for agricultural expansion.

However, tropical deforestation can't be attributed simply to population pressures because it is also affected by a variety of interacting economic, social, and government factors that vary from place to place. Government policies sometimes provide incentives that favor the removal of forests. The Brazilian government opened the Amazonian frontier, beginning in the late 1950s, by constructing the Belem–Brasilia Highway, which cut through the Amazon Basin. Such roads open a forest for settlement (**Figure 13.6b**). An example of economic conditions encouraging deforestation is a farmer converting forest to pasture so that he can maintain a larger herd of cattle.

Keeping in mind that the origins of tropical deforestation are complex, three agents—subsistence agriculture, commercial logging, and cattle ranching—are considered its most immediate causes. Subsistence agriculture, in which a family produces just enough food to feed itself, accounts for more than half of tropical deforestation. Subsistence farmers carry out **slash-and-burn agriculture** (discussed further in Chapter 14). Subsistence farmers often follow loggers' access roads until they find a suitable spot. They cut down trees and allow them to dry; then they burn the area and plant crops immediately after burning. The yield from the first crop is often quite high because the nutrients that were in the burned trees are made available in the soil. Soil productivity subsequently declines at a rapid rate, and subsequent crops are poor. In a short time, the people farming the land must move to a new part of the forest and repeat the process. Cattle ranchers often claim the abandoned land for grazing because land not rich enough to support crops can still support livestock.

Slash-and-burn agriculture done on a small scale, with plenty of forest to shift around in so that there are periods of 20 to 100 years between cycles, is sustainable. The forest regrows rapidly after a few years of farming. But when millions of people try to obtain a living in this way, the land is not allowed to lie uncultivated long enough to recover.

Vast tracts of tropical rain forests, particularly in southeast Asia, are harvested by commercial logging operations for export abroad. Most tropical countries allow commercial logging to proceed at a much faster rate than is sustainable. Unmanaged logging does not contribute to economic development; rather, it depletes a valuable natural resource faster than it can regenerate for sustainable use.

Some tropical deforestation, especially in Latin America, is carried out to provide open rangeland for cattle. Other causes of tropical rainforest destruction include the development of hydroelectric power and mining. A considerable portion of forestland is cleared for plantation-style agriculture, which produces export crops such as citrus fruits, bananas, soy, and palm oil (**Figure 13.6c**).

Why are tropical dry forests disappearing? Tropical dry forests are being destroyed at a rapid rate, primarily for fuelwood (**Figure 13.6d**). About half of the wood consumed worldwide is used as heating and cooking fuel by much of the developing world. Often the wood cut for fuel is converted to charcoal, which is then used to power steel, brick, and cement factories. Charcoal production is extremely wasteful: 3.6 metric tons (4 tons) of wood produce only enough charcoal to fuel an average-sized iron smelter for 5 minutes.

a. Distribution of Tropical Rain Forests.
Rain forests (green areas) occur in Central and South America, Africa, and Southeast Asia. Much of the remaining forested area is highly fragmented. The three countries with the largest area of tropical rain forest are outlined.

India

Congo

Indonesia

Brazil

b. Human Settlements Along a Road in Brazil's Tropical Rain Forest.
This satellite photograph shows numerous smaller roads extending perpendicularly from the main roads. As farmers settle along the roads, they clear out more and more forest (dark green) for croplands and pastures (tans and pinks).

d. Deforestation for Fuelwood.
Women in Kabini, India, gather firewood to take back to their village. About 94 percent of wood removed from Indian forests is burned as fuel.

c. Deforestation Due to Agriculture.
Trees are stripped and burned in Malaysian Borneo to make room for an oil palm plantation. Palm oil—used worldwide for cooking, manufacture of processed foods, and biodiesel—is a major export crop for Malaysia and Indonesia.

NATIONAL GEOGRAPHIC

Forests in the United States

In recent years, most temperate forests in the Rocky Mountains, Great Lakes region, and New England and other eastern states have been holding steady or even expanding. In Vermont, the amount of land covered by forests has increased from 35 percent in 1850 to 75 percent in 2010. Expanding forests are the result of *secondary succession* on abandoned farms (see Figure 6.22), the commercial planting of tree plantations on both private and public lands, and government protection. Although these second- and third-growth forests generally don't have the biological diversity of virgin stands, many organisms have successfully become reestablished in the regenerated areas.

Slightly more than one-half of U.S. forests are privately owned (**Figure 13.7**), and three-fourths of these private lands are in the Northeast and Midwest. Many private owners are under economic pressure to subdivide the land and develop tracts for housing or shopping malls, as they seek ways to recoup their high property taxes. Projected conversion of forests to agricultural, urban, and suburban lands over the next 40 years will have the greatest potential impact in the South, where more than 85 percent of forest is privately owned and logging is largely unregulated.

U.S. National Forests According to the USFS, the United States has 155 national forests encompassing 78 million hectares (193 million acres) of land, mostly in Alaska and western states. The USFS manages most national forests, and the BLM oversees the remainder. National forests have been established to provide U.S. citizens with the maximum benefits of natural resources such as fish, wildlife, and timber. Multiple uses include timber harvesting; mining; livestock foraging; hunting, fishing, and other forms of outdoor recreation; water resources and watershed protection; and habitat for fishes and wildlife. Recreation, which increased dramatically in national forests during the 1990s and early 2000s, ranges from camping at designated campsites to backpacking in the wilderness. Visitors to national forests swim, boat, picnic, and observe nature. With so many possible uses of national forests, conflicts inevitably arise, particularly between timber interests and those who wish to preserve the trees for other purposes.

Road building is a particularly contentious issue, in part because the USFS builds taxpayer-funded roads to allow private logging companies access to forests to remove timber. (See the *Case Study* at the end of the chapter for an example.) Road building in national forests is environmentally destructive when improper construction accelerates soil erosion and mudslides (particularly on steep terrain) and causes water pollution in streams. Biologists are concerned that the many roads that are built fragment wildlife habitat and provide entries for disease organisms and invasive species.

Forest ownership in the United States • Figure 13.7

Most forests are privately owned.

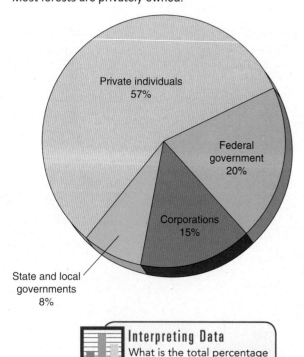

Private individuals 57%

Federal government 20%

Corporations 15%

State and local governments 8%

Interpreting Data
What is the total percentage of forest owned by all types of government?

CONCEPT CHECK

1. **What** is sustainable forestry?
2. **What** is deforestation? What are four important causes of tropical deforestation?
3. **Why** is the fact that U.S. national forests have been created for multiple uses often a contentious issue?

Rangelands

LEARNING OBJECTIVES

1. **Describe** rangelands and their general uses.
2. **Define** *desertification* and explain its relationship to overgrazing.
3. **Describe** how conservation easements help protect privately owned rangelands.
4. **Identify** the government agencies that administer public rangelands and describe current issues of concern.

Rangelands are grasslands, in both temperate and tropical climates, that serve as important areas of food production for humans by providing fodder for livestock such as cattle, sheep, and goats (**Figure 13.8**). Rangelands may be mined for minerals and energy resources, used for recreation, and preserved for biological habitat and for soil and water resources. The predominant vegetation of rangelands includes grasses, *forbs* (small plants other than grasses), and shrubs.

> **rangeland** Land that is not intensively managed and is used for grazing livestock.

Rangeland Degradation and Desertification

Grasses, the predominant vegetation of rangelands, have *fibrous root systems*, in which many roots form diffuse networks in the soil to anchor the plants. Plants with fibrous roots hold the soil in place quite well, thereby reducing soil erosion. Grazing animals eat the leafy shoots of the grass, and the fibrous roots continue to develop, allowing the plants to recover and regrow to their original size.

Carefully managed grazing is beneficial for grasslands. Because rangeland vegetation is naturally adapted to grazing, when grazing animals remove mature vegetation, the activity stimulates rapid regrowth. At the same time, the hooves of grazing animals disturb the soil surface enough to allow rainfall to more effectively reach the root systems of grazing plants. Several studies have reported that moderate levels of grazing encourage greater plant diversity.

The **carrying capacity** of a rangeland is the maximum number of animals the natural vegetation can sustain over an indefinite period without deterioration of

Rangeland • Figure 13.8

Rangeland is considered a renewable resource when its carrying capacity—the number of animals it can sustain without suffering deterioration—is not exceeded. Photographed along the Salmon River in Idaho.

the ecosystem. When the carrying capacity of a rangeland is exceeded, **overgrazing** of grasses and other plants occurs. When plants die, the ground is left barren, and the exposed soil is susceptible to erosion. Sometimes plants that do not naturally grow in a rangeland but that can tolerate the depleted soil invade an overgrazed area.

Most of the world's rangelands lie in semiarid areas that have natural extended droughts. Under normal conditions, native grasses in these dry lands can survive severe drought. The aboveground portion of the plant dies back, but the extensive root system remains alive and holds the soil in place. But when an extended drought occurs in conjunction with overgrazing, once-fertile rangeland may be converted to desert as reduced grass cover allows winds to erode the soil. Even when the rains return, the degradation may be so extensive that the

overgrazing A situation that occurs when too many grazing animals consume the plants in a particular area, leaving the vegetation destroyed and unable to recover.

desertification Degradation of once-fertile rangeland or tropical dry forest into nonproductive desert.

land cannot recover. Water erosion removes the little remaining topsoil, and the sand left behind forms dunes.

Land degradation is both a natural and a human-induced process that decreases the future ability of the land to support crops or livestock. This progressive degradation, which induces unproductive desert-like conditions on formerly productive rangeland (or tropical dry forest), is **desertification** (**Figure 13.9**). It reduces the agricultural productivity of economically valuable land, forces out many organisms, and threatens endangered species. Worldwide, desertification seems to be on the increase. The United Nations estimates that each year since the mid-1990s, 3,560 km^2 (1,374 mi^2)—an area about the size of Rhode Island—has turned into desert.

Desertification in the African Sahel region, Niger • Figure 13.9

As goats consume the remaining grass and shrubs, the dunes of the Sahara Desert will inevitably continue to encroach upon this pastureland in Niger. There is still much that scientists don't understand about desertification, such as the extent to which desertification results from natural fluctuations in climate versus population pressures and human activities.

Global Locator

SAHEL

AFRICA

NATIONAL GEOGRAPHIC

Rangeland Trends in the United States

Rangelands make up approximately 30 percent of the total land area in the United States, mostly in the western states. Of this, approximately one-third is publicly owned and two-thirds is privately owned. Much of the private rangeland is under increasing pressure from developers, who want to subdivide the land into lots for homes and condominiums. To preserve the open land, conservation groups often pay ranchers for conservation easements that prevent future owners from developing the land. An estimated 400,000 hectares (1 million acres) of private rangelands are protected by conservation easements.

Excluding Alaska, there are at least 89 million hectares (220 million acres) of public rangelands in the United States. The BLM manages approximately 69 million hectares (170 million acres) of public rangelands, and the USFS manages an additional 20 million hectares (50 million acres).

Overall, the condition of public rangelands in the United States has slowly improved since the low point of the Dust Bowl in the 1930s, when the combined effects of poor agricultural practices, severe winds, and extended drought led to devastating soil erosion and dramatic declines in soil productivity. Much of this improvement is attributed to fewer livestock being permitted to graze the rangelands after the passage of the **Taylor Grazing Act** of 1934, the **Federal Land Policy and Management Act** of 1976, and the **Public Rangelands Improvement Act** of 1978. Better livestock management practices, such as controlling the distribution of animals on a range through fencing or herding, as well as scientific monitoring, have also contributed to rangeland recovery.

But restoration is slow and costly, and more is needed. Rangeland management includes seeding in places where plant cover is sparse or absent, conducting controlled burns to suppress shrubby plants, constructing fences to allow rotational grazing, controlling invasive weeds, and protecting habitats of endangered species. Most livestock operators use public rangelands in a way that results in their overall improvement.

> **conservation easement** A legal agreement that protects privately owned forest, rangeland, or other property from development for a specified number of years.

Issues Involving Public Rangelands

The federal government distributes permits that allow private livestock operators to use public rangelands for grazing in exchange for a fee that is much lower than the cost of grazing on private land. The permits are held for many years and are not open to free-market bidding by the general public—that is, only ranchers who live in the local area are allowed to obtain grazing permits. Some environmental groups are concerned about the ecological damage caused by overgrazing of public rangelands and want to reduce the number of livestock animals allowed to graze. They want public rangelands managed for other uses, such as biological habitat, recreation, and scenic value, rather than exclusively for livestock grazing. To accomplish this goal, they would like to purchase grazing permits and set aside the land for nongrazing purposes.

Conservative economists have joined environmentalists in criticizing the management of federal rangelands. According to policy analysts at Taxpayers for Common Sense, in 2004 taxpayers contributed at least $115 million more than the grazing fees collected in order to support grazing on public rangelands. This money is used to manage and maintain the rangelands, including installing water tanks and fences, and to repair damage caused by overgrazing. Taxpayers for Common Sense and other free-market groups want grazing fees increased to cover all costs of maintaining herds on publicly owned rangelands.

CONCEPT CHECK   STOP

1. **What** are rangelands?
2. **How** can overgrazing of rangeland lead to desertification?
3. **How** do conservation easements help protect privately owned rangelands?
4. **Which** agencies manage public rangelands? What management issues do they face?

National Parks and Wilderness Areas

LEARNING OBJECTIVES

1. **State** which government agency administers U.S. national parks and describe current issues of concern.

2. **Define** *wilderness* and discuss the administration and goals of the National Wilderness Preservation System and the problems faced by wilderness areas.

3. **Explain** the differences between the wise-use and environmental movements' views on the use of public lands.

Many acres of federal land are set aside either as national park property or as wilderness areas. Both types of land were established to encourage the protection of the natural environment, and both experience conflicts associated with how best to use and manage these protected areas.

National Parks

In 1872 Congress established the world's first national park, Yellowstone National Park, in federal lands in the territories of Montana and Wyoming. Created in 1916 as a federal bureau in the Department of the Interior, the **National Park Service** (NPS) was originally composed of large, scenic areas in the West such as Yellowstone, the Grand Canyon, and Yosemite Valley (**Figure 13.10a**). Today the NPS has more cultural and historical sites—battlefields and historically important buildings and towns—than places of scenic wilderness. The NPS currently administers 388 sites, 58 of which are national parks (see the introduction to this chapter), encompassing 34.1 million hectares (84.3 million acres).

Because the NPS believes that knowledge and understanding increase enjoyment, one of its primary roles is to teach people about the natural environment, management of natural resources, and history of a site by providing nature walks and guided tours of its parks. Exhibits along roads and trails, evening campfire programs, museum displays, and lectures are other common educational tools.

The popularity and success of U.S. national parks (**Figure 13.10b**) have encouraged many other nations to establish national parks. Today the U.N. Environment Programme identifies more than 3500 national parks, as defined by the World Conservation Union (IUCN), in nearly 100 countries. As in the United States, in other countries parks usually have multiple roles, from providing biological habitat to facilitating human recreation.

Threats to U.S. Parks Some national parks are overcrowded (**Figure 13.10c**). Problems plaguing urban areas are also found in popular national parks during peak seasonal use, including crime, vandalism, litter, traffic jams, and pollution of the soil, water, and air. In addition, thousands of resource violations, from cutting live trees and collecting plants, minerals, and fossils, to defacing historical structures with graffiti and setting fires, are investigated in national parks each year. Park managers have had to reduce visitor access to park areas that have become degraded from overuse, and in some cases to restrict vehicle traffic (**Figure 13.10d**).

Some national parks have imbalances in wildlife populations. Populations of many mammal species are in decline, including bears, white-tailed jackrabbits, and red foxes. For example, grizzly bear populations in national parks of the western United States are threatened. Grizzlies are territorial and require large areas of wilderness as habitat, and the presence of humans in national parks may adversely affect them. Most importantly, the parks may be too small to support grizzlies. While grizzly bears have survived in sustainable numbers in Alaska and Canada, the populations native to the more restricted spaces of the western continental United States are in considerable danger.

Other mammal populations—notably elk—have proliferated. Elk in Yellowstone National Park's northern range increased from a population of 3100 in 1968 to a record high of 19,000 in 1994. Ecologists documented that elk reduced the abundance of native vegetation, such as willow and aspen, and seriously eroded stream banks. The reintroduction of gray wolves to

a. Yosemite National Park in California.
This winter view shows the Merced River flowing past the rock formation El Capitan.

c. Heavy Traffic at Cades Cove, Great Smoky Mountains National Park in Tennessee.
Great Smoky Mountains National Park receives the most visitors of any U.S. national park. The popularity of certain national parks threatens to overwhelm them.

b. The 10 Most Popular National Parks.

National park	Number of recreational visitors in 2008 (in millions)
Great Smoky Mountains (North Carolina, Tennessee)	9.0
Grand Canyon (Arizona)	4.4
Yosemite (California)	3.4
Olympic (Washington)	3.1
Yellowstone (Wyoming, Montana, Idaho)	3.1
Cuyahoga Valley (Ohio)	2.8
Rocky Mountain (Colorado)	2.8
Zion (Utah)	2.7
Grand Teton (Wyoming)	2.5
Acadia (Maine)	2.1
Total visitors to the national park system	274.8

d. Shuttle Bus at Zion National Park in Utah.
In operation since 2000, the Zion National Park shuttle system eliminates traffic congestion. During peak months, personal vehicles are banned from popular areas; visitors instead ride the propane-powered shuttle.

WILEY
RESOURCE KIT

Gray wolves prey on elk in Yellowstone National Park • Figure 13.11

Since their reintroduction, gray wolf populations have gained a secure foothold in Yellowstone National Park. Early studies of the effects of these predators support scientists' predictions that wolves will help reduce the burgeoning elk population.

Yellowstone, which began in 1995, has helped reduce the elk population (**Figure 13.11**), which in turn has led to improved aspen and willow growth and growing numbers of herbivores.

National parks are increasingly becoming islands of natural habitat surrounded by human development. Development on the borders of national parks limits the areas in which wild animals may range, forcing them into isolated populations. Ecologists have found that when environmental stressors occur, several small "island" populations are more likely to become threatened than a single large population occupying a sizable range (see Chapter 15).

Wilderness Areas

Wilderness encompasses regions where the land and its community of organisms are not greatly disturbed by human activities, where humans may visit but don't live permanently. The U.S. Congress recognized that increased human population and expansion into wilderness areas might result

> **wilderness** A protected area of land in which no human development is permitted.

in a future where no lands exist in their natural condition. Accordingly, the **Wilderness Act** of 1964 authorized the U.S. government to set aside federally owned land that retains its primeval character and lacks permanent improvements or human habitation, as part of the **National Wilderness Preservation System** (NWPS). These federal lands range in size from tiny islands of uninhabited land to portions of national parks, national forests, and national wildlife refuges (**Figure 13.12**). Although mountains are the most common wilderness areas, portions of other ecosystems have been set aside, including tundra, desert, and wetlands.

Areas designated as wilderness are given the highest protection of any federal land. These areas are to remain natural and unchanged so they will be unimpaired for future generations to enjoy. The same four government agencies that regulate all publicly owned land—the NPS, USFS, FWS, and BLM—oversee 756 wilderness areas comprising 44.1 million hectares (109 million acres) of land. More than one-half of the lands in the NWPS lie in Alaska, and western states are home to much of the remainder.

Millions of people visit U.S. wilderness areas each year, and some areas are overwhelmed by this traffic: Eroded trails, soil and water pollution, litter and trash, and human congestion predominate over quiet, unspoiled land. Government agencies now restrict the number of people allowed into each wilderness area at one time so that human use doesn't seriously affect the wilderness.

Some of the most popular wilderness areas may require more intensive future management, such as the development of trails, outhouses, cabins, and campsites. These amenities are not encountered in true wilderness, posing a dilemma between wilderness preservation and human use and enjoyment of wild lands.

Limiting the number of human guests in a wilderness area doesn't control all the factors that threaten wilderness, however. **Invasive species** have the potential to upset the natural balance among native species. For example, white pine blister rust, a foreign (non-native) fungus that kills white pine trees, has invaded the wilderness in the northern Rocky Mountains. Wilderness managers are concerned that declining white pine populations could affect the population of grizzly bears in the region because pine seeds are a major part of the grizzlies' diet. The Wilderness Act specifies the avoidance of intentional ecological management. In this example, should

a. At more than 3.7 million hectares (9.1 million acres), the Wrangell–Saint Elias Wilderness in Alaska is the nation's largest designated wilderness area.

b. The Cabeza Prieta Wilderness, Arizona's largest wilderness area, includes 325,000 hectares (803,000 acres) of isolated desert landscapes.

Diverse wilderness areas protected in the National Wilderness Preservation System • Figure 13.12 _____

c. The Pelican Island National Wildlife Refuge in Florida is home to the 2.4-hectare (6-acre) Pelican Wilderness, the smallest protected area in the NWPS.

the white pine populations be scientifically manipulated to help preserve the original wilderness?

Large tracts of wilderness, most of it in Alaska, have been added to the NWPS since passage of the Wilderness Act in 1964. People who view wilderness as a nonrenewable resource support the designation of additional wilderness areas, particularly in the lower 48 states. In March 2009 President Barack Obama signed the Omnibus Public Land Management Act of 2009, which designated 52 new U.S. wilderness areas and added almost 1 million hectares (more than 2 million acres) to the NWPS, most of it outside Alaska. Increasing the amount of federal land designated as wilderness in the NWPS is opposed by groups who operate businesses on public lands (such as timber, mining, ranching, and energy companies) and by their political representatives.

A logging site in Gifford Pinchot National Forest, Cascade Range, Washington • Figure 13.13

The wise-use movement favors opening federal lands to logging and other types of economic development.

Management of Federal Lands

How do we best manage the legacy of federal lands? Should federal lands be managed under multiple uses, or should they be preserved so that they benefit U.S. citizens for generations to come? These questions have divided many Americans into two groups, each a coalition of several hundred grassroots organizations. Those who wish to exploit resources on federal lands are known collectively as the *wise-use movement*; this group also includes many corporations. Those who wish to preserve the resources on federally owned lands are known collectively as the *environmental movement*. The descriptions of the two movements that follow are mainstream; any given group may not support all of the listed goals of a movement.

In general, people who support the wise-use movement think that the government overregulates environmental protection and that property owners should have more flexibility to use natural resources. They believe that the primary purpose of federal lands is to enhance economic growth (**Figure 13.13**).

Some of the goals of the wise-use movement include the following:

1. Put all national forests, including old-growth forests, under timber management.

2. Permit mining and commercial development of wilderness areas, wildlife refuges, and national parks, where appropriate.

3. Allow unrestricted development of wetlands.

4. Sell parts of resource-rich federal lands to private interests, such as mining, oil, coal, ranching, and timber groups, for resource extraction.

Many organizations that embrace the wise-use movement have environmentally friendly names. The National Wetlands Coalition, for example, consists primarily of real estate developers and energy companies that want to drain and develop wetlands. Similarly, logging companies support the American Forest Resource Alliance.

In contrast to the wise-use movement, the environmental movement views federal lands as a legacy of U.S. citizens. They think that:

1. The primary purpose of public lands is to protect biological diversity and ecosystem integrity.

2. Those who extract resources from public lands should pay U.S. citizens compensation equal to the fair market value of the resource and not be subsidized by taxpayers.

3. Those who use public lands should be held acountable for any environmental damage they cause.

CONCEPT CHECK **STOP**

1. **What** government agency administers the National Park System? What problems does it face?

2. **What** is wilderness, and how does the U.S. National Wilderness Preservation System seek to protect it? What problems do protected wilderness areas face?

3. **How** do the wise-use and environmental movements differ in their views on the use of public lands?

Conservation of Land Resources

LEARNING OBJECTIVES

1. **Name** at least three of the most endangered ecosystems in the United States.
2. **Describe** several of the criteria used to evaluate whether an ecosystem is endangered.

O ur ancestors considered natural areas an unlimited resource to exploit. They appreciated prairies as valuable agricultural land and forests as immediate sources of lumber and eventual farmland. This outlook was practical as long as there was more land than people needed. But as the population increased and the amount of available land decreased, people began to view land as a limited resource. Thus, exploitation has increasingly shifted to preservation of the remaining natural areas in the United States and elsewhere around the world (**Figure 13.14** on pages 340–341).

Although all types of ecosystems must be conserved, several are in particular need of protection. The U.S. Geological Survey (USGS) and the Defenders of Wildlife commissioned studies that ranked the most endangered ecosystems in the United States. They used four criteria:

1. The area lost or degraded since Europeans colonized North America

2. The number of present examples of a particular ecosystem, or the total area

3. An estimate of the likelihood that a given ecosystem will lose a significant area or be degraded during the next 10 years

4. The number of threatened and endangered species living in that ecosystem

Table 13.2 lists the 15 most endangered U.S. ecosystems based on these criteria. Examples include the South Florida landscape, southern Appalachian spruce-fir forests, and longleaf pine forests and savannas. As these ecosystems are lost and degraded, the organisms that compose them decline in number and in genetic

The 15 most endangered ecosystems in the United States (in order of priority) • Table 13.2
South Florida landscape
Southern Appalachian spruce-fir forests
Longleaf pine forests and savannas
Eastern grasslands, savannas, and barrens
Northwestern grasslands and savannas
California native grasslands
Coastal communities in the lower 48 states and Hawaii
Southwestern riparian communities
Southern California coastal sage scrub
Hawaiian dry forest
Large streams and rivers in the lower 48 states and Hawaii
Cave systems
Tallgrass prairie
California river- and stream-bank communities and wetlands
Florida scrub

diversity. Implementing conservation strategies that set aside ecosystems is the best way to preserve an area's biodiversity.

As you have seen in this chapter, government agencies, private conservation groups, and private citizens have begun to set aside natural areas for permanent preservation. Such activities ensure that our children and grandchildren will inherit a world with wild places and other natural ecosystems.

CONCEPT CHECK

1. **What** are three U.S. ecosystems that need protection?

2. **What** are three criteria used to evaluate whether an ecosystem is endangered?

Protected ecosystems around the world • Figure 13.14

GeoBytes

LARGEST NATIONAL PARK
North East Greenland National Park, Greenland, 972,000 km² (375,000 mi²)

LARGEST MARINE PARK
Northwestern Hawaiian Islands Marine National Monument, U.S., 360,000 km² (140,000 mi²)

LARGEST TROPICAL FOREST PARK
Tumucumaque National Park in the Brazilian Amazon 24,135 km² (9319 mi²)

BIODIVERSITY HOTSPOTS
Conservation International identifies world regions suffering from a severe loss of biodiversity.

WORLD HERITAGE SITES
The United Nations Educational, Scientific, and Cultural Organization (UNESCO) recognizes natural and cultural sites of "universal value."

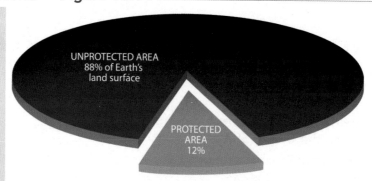

UNPROTECTED AREA
88% of Earth's land surface

PROTECTED AREA
12%

Protected areas worldwide represent 12 percent of the Earth's land surface, according to the U.N Environmental Programme World Conservation Monitoring Centre. Only 0.5 percent of the marine environment is within protected areas—an amount considered inadequate by conservationists because of the increasing threats of overfishing and coral reef loss worldwide.

HAWAII VOLCANOES NATIONAL PARK, HAWAII
The park includes two of the world's most active volcanoes, Kilauea and Mauna Loa. The landscape shows the results of 70 million years of volcanism, including calderas, lava flows, and black sand beaches. Lava spreads out to build the island, and seawater vaporizes as lava hits the ocean at 1149°C (2100°F). The national park, created in 1916, covers 10 percent of the island of Hawaii and is a refuge for endangered species such as the hawksbill turtle and Hawaiian goose. It was made a World Heritage site in 1987.

GALÁPAGOS NATIONAL PARK, ECUADOR
Galápago means "tortoise" in Spanish, and at one time 250,000 giant tortoises roamed the islands. Today about 15,000 remain, and 3 of the original 14 subspecies are extinct—and the Pinta Island tortoise may be extinct soon. In 1959, Ecuador made the volcanic Galápagos Islands a national park, protecting the giant tortoises and other endemic species. The archipelago became a World Heritage site in 1978, and a marine reserve surrounding the islands was added in 2001.

WESTERN UNITED STATES
An intricate public lands pattern—including national forests, wilderness areas, wildlife refuges, and national parks such as Arches (above)—embraces nearly half the surface area of 11 western states. Ten out of 19 World Heritage sites in the United States are found here. It was in the West that the modern national park movement was born in the 19th century, with the establishment of Yellowstone and Yosemite National Parks.

MADIDI NATIONAL PARK, BOLIVIA
Macaws may outnumber humans in Madidi, Bolivia's second largest national park, established in 1995. A complex community of plants, animals, and native Indian groups share this 18,900-km² (7300-mi²) reserve, part of the Tropical Andes biodiversity hotspot. Indigenous communities benefit from ecotourism.

AMAZON BASIN, BRAZIL
Indigenous peoples help manage reserves in Brazil that are linked with Jaú National Park. The park and reserves are part of the Central Amazon Conservation Complex, a World Heritage site covering more than 60,000 km² (23,000 mi²). It is the largest protected area in the Amazon Basin and one of the most biologically rich regions on the planet.

ARCTIC REGIONS
Polar bears find safe havens in Canadian parks, such as on Ellesmere Island, and in Greenland's huge protected area—Earth's largest—that preserves the island's frigid northeast. In 1996 countries with Arctic lands adopted the Circumpolar Protected Areas Network Strategy and Action Plan to help conserve ecosystems. Today 15 percent of Arctic land area is protected.

What do protected areas around the world have in common? How are their concerns shared by state and local governments and organizations?

PROTECTED AREAS WORLDWIDE
What are protected areas? Most people agree that such territories are dedicated to protecting and maintaining biodiversity and are often managed through legal means. Yellowstone National Park, established in 1872, is often cited as the start of the modern era of protected areas. From a mere handful in 1900, the number of protected areas worldwide now exceeds 104,790, covering more than 20 million km² (7.7 million mi²). North America claims the most protected land of any region (in terms of total area), amounting to almost 18 percent of its territory. South Asia, at about 7 percent, has the least amount of land under some form of protection. Not all protected areas are created or managed equally, and management categories developed by the International Union for Conservation of Nature (IUCN) range from strict nature reserve to areas for sustainable use. Management effectiveness varies widely and can be affected by such factors as conservation budgets, and political stability. Throughout the world—but especially in tropical areas—protected areas are threatened by illegal hunting, overfishing, pollution, and the removal of native vegetation. Countries and international organizations no longer choose between conservation and development; rather the goal for societies is to balance the two for equitable and sustainable resource use.

Wildest biomes
- Wildlands
- Ice or snow cover

WILDEST AREAS

Although generally far from cities, the world's remaining wild places play a vital role in a healthy global ecosystem. The boreal forests of Canada and Russia, for instance, help cleanse the air we breathe by absorbing carbon dioxide and providing oxygen. With the human population increasing by an estimated 1 billion over the next 15 years, many wild places could fall within reach of the plow or under a cloud of smog.

F or millennia, lands have been set aside as sacred ground or as hunting reserves for the powerful. Today, great swaths are protected for recreation, habitat conservation, biodiversity preservation, and resource management. Some groups may oppose protected spaces because they want access to resources now. Yet local inhabitants and governments are beginning to see the benefits of conservation efforts and sustainable use for human health and future generations.

SAREKS NATIONAL PARK, SWEDEN
This remote 1970-km^2 (760-mi^2) park, established in 1909 to protect the alpine landscape, is a favorite of backcountry hikers. It boasts some 200 mountains more than 1800 m (5900 ft) high, narrow valleys, and about 100 glaciers. Sareks forms part of the Laponian Area World Heritage site and has been a home to the Saami (or Lapp) people since prehistoric times.

AFRICAN RESERVES
Some 120,000 elephants roam Chobe National Park in northern Botswana. Africa has more than 7500 national parks, wildlife reserves, and other protected areas, covering about 9 percent of the continent. Protected areas are under enormous pressure from expanding populations, civil unrest and war, and environmental disasters.

WOLONG NATURE RESERVE, CHINA
Giant pandas freely chomp bamboo in this 2000-km^2 (772-mi^2) reserve in Sichuan Province, near the city of Chengdu. Misty bamboo forests host a number of endangered species, but the critically endangered giant panda—among the rarest mammals in the world—is the most famous resident. Only about 1600 giant pandas exist in the wild.

KAMCHATKA, RUSSIA
Crater lakes, ash-capped cones, and diverse plant and animal species mark the Kamchatka Peninsula—a World Heritage site—located between the icy Bering Sea and Sea of Okhotsk. The active volcanoes and glaciers form a dynamic landscape of great beauty, known as "The Land of Fire and Ice." Kamchatka's remoteness and rugged landscape help fauna flourish, producing record numbers of salmon species and half of the Steller's sea-eagles on Earth.

GUNUNG PALUNG NATIONAL PARK, INDONESIA
A tree frog's perch could be precarious in this 900-km^2 (347-mi^2) park on the island of Borneo, in the heart of the Sundaland biodiversity hotspot. The biggest threat to trees and animals in the park and region is illegal logging. Gunung Palung contains a wider range of habitats than any other protected area on Borneo, from mangroves to lowland and cloud forests. A number of endangered species, such as orangutans and sun bears, depend on the dense forests.

AUSTRALIA & NEW ZEALAND
Uluru, a red sandstone monolith (formerly known as Ayers Rock), and the vast Great Barrier Reef, one of the largest marine parks in the world, are outstanding examples of Australia's protected areas—which make up more than 10 percent of the country's area and conserve a diverse range of unique ecosystems. About a third of New Zealand is protected, and it is a biodiversity hotspot because of threats to flightless native birds, such as the kakapo and kiwi. Cats, stoats, and other predators, introduced to New Zealand by settlers, kill thousands of birds each year.

ANTARCTICA
The Antarctic Treaty, signed in 1959, regulates the continent and the marine environment (south of 60° south latitude) as a "natural reserve devoted to peace and science." The treaty is recognized as one of the most successful international agreements. The annual Antarctic Treaty Consultative Meeting brings national representatives together to discuss topics such as scientific cooperation and environmental protection.

NATIONAL GEOGRAPHIC

The Tongass Debate Over Clear-Cutting

Despite its northern location along Alaska's southeastern coast, the Tongass National Forest is one of the world's few temperate rain forests (**Figures a** and **b**; also see Chapter 6 for a description of the temperate rainforest biome). It is one of the wettest places in the United States. This moisture supports old-growth forest of giant Sitka spruce, yellow cedar, and western hemlock, some of which are 700 years old. This 6.9-million-hectare (17-million-acre) forest, the largest in the National Forest System, provides habitat for a wealth of wildlife, such as grizzly bears and bald eagles.

The Tongass is a prime logging area because a single large Sitka spruce may yield as much as 23.6 m² (10,000 board ft) of high-quality timber. The logging industry forms the basis of much of the local economy but conflicts with environmental interests seeking to avoid overharvesting. Regeneration of mature forest after it is clear-cut can take several centuries.

As in most other national forests, it is expensive to log in the Tongass. To cover high operating costs, timber interests such as pulp mills rely on obtaining the timber from the federal government at below-market prices. This right was granted in 1954 by a contract that expired in the 1990s. In 1990, congressional efforts to pass the Tongass Timber Reform Act, which would force timber interests to pay market prices, were bitterly opposed. The compromise agreement, reached in 1997, provided timber to the mills at market prices. As a result of this legislation, clear-cut logging continued in the Tongass, but at lower rates than in the past.

In the closing months of the Clinton administration, the USFS officially adopted the *Roadless Area Conservation Rule* to protect roadless national forests from road building and forest harvest. A federal judge under the Bush administration blocked the roadless rule in 2001, opening to logging and development forest that had formerly been off-limits (**Figure c**). A 2006 court decision overturned the repeal of the roadless rule, but the Tongass remains exempt from the rule's protection. As of late 2009, ongoing litigation involving native tribe members and environmentalists challenged continued road construction in the Tongass.

The take-home message is that the USFS, like other government agencies, takes its lead from current presidential policies. Changes in administrations can shift support for established rulings.

Alaska's Tongass National Forest

a. This temperate rain forest (light green area) is in southeastern Alaska along the Pacific Ocean.

b. An aerial view of the Tongass reveals the vast size of its old-growth forest.

c. USFS logging roads in the Tongass open the forest to logging and other kinds of development.

Summary

1 Land Use in the United States 320

1. More than one-half of U.S. land is privately owned. Approximately one-third—including many types of ecosystems and land uses—is owned by the federal government. Nine percent belongs to state and local governments, and more than 2 percent to Native American tribes.

2 Forests 322

1. **Sustainable forestry** is the use and management of forest ecosystems in an environmentally balanced and enduring way. Sustainable forestry maintains a mix of forest trees, by age and species, rather than a **monoculture**, in which only one type of plant is cultivated over a large area. Adopting sustainable forestry principles requires setting aside sanctuaries and **wildlife corridors**, protected zones that connect isolated unlogged or undeveloped areas.

2. **Deforestation** is the temporary or permanent clearing of large expanses of forest for agriculture or other uses. **Clear-cutting** is a logging practice in which all the trees in a stand of forest are cut, leaving just the stumps; clear-cutting over a wide area is ecologically unsound. The major causes of tropical deforestation are subsistence farming, commercial logging, and cattle ranching, all accelerated by growing human populations. Increased needs for fuelwood drive deforestation of tropical dry forests.

3. Most U.S. national forests are managed by the U.S. Forest Service (USFS); the rest are overseen by the Bureau of Land Management (BLM). National forests face conflicts associated with supporting multiple uses: timber harvest; livestock forage; water resources and watershed protection; mining; hunting, fishing, and other forms of recreation; and habitat for fishes and wildlife.

3 Rangelands 331

1. **Rangelands** are grasslands that aren't intensively managed and are used for grazing livestock. Rangelands are also mined for mineral and energy resources, used for recreation, and preserved for biological habitat and for soil and water resources.

2. **Overgrazing** is the destruction of vegetation caused by too many grazing animals consuming the plants in a particular area, leaving them unable to recover. Overgrazing accelerates **land degradation**, which decreases the future ability of the land to support crops or livestock. **Desertification** is the degradation of once-fertile rangeland or tropical dry forest into nonproductive desert.

3. **A conservation easement** is a legal agreement that protects privately owned forest or other property from development for a specified number of years. Conservation groups often pay for conservation easements to preserve open rangeland.

4. The BLM manages more than three-fourths of U.S. public rangelands, excluding Alaska; the USFS manages the remainder. Current issues on public rangelands include conflicts between environmental groups and ranchers over the number of livestock allowed to graze and the potential to manage the areas for uses such as biological habitat, recreation, and scenic value. Conflicts also arise over whether grazing fees paid by livestock operators on public lands should be high enough to cover all costs of maintaining herds, removing taxpayer burden.

Seed tree cutting

Clear-cutting

4 National Parks and Wilderness Areas 334

1. **The National Park Service** administers 388 sites in the United States, including 58 national parks. The problems the sites encounter include overcrowding, pollution, crime, resource violations, and imbalanced wildlife populations.

2. **Wilderness** is a protected area of land in which no human development is permitted. The **National Wilderness Preservation System** consists of four U.S. government agencies—the NPS, USFS, FWS, and BLM—that oversee 756 wilderness areas. The problems these areas face include overuse and overcrowding by visitors, pollution, erosion, and the introduction of invasive species.

3. Those who support the **wise-use movement** believe a primary purpose of federal lands is to enhance economic growth. They think that the government overregulates environmental protection and that property owners should have more flexibility to use natural resources. Those who support the **environmental movement** view federal lands as a legacy of U.S. citizens and thus want to preserve resources on federally owned lands.

5 Conservation of Land Resources 339

1. Endangered U.S. ecosystems include the south Florida landscape, southern Appalachian spruce-fir forests, and longleaf pine forests and savannas.

2. Criteria used to evaluate whether an ecosystem is endangered and to what degree it is threatened include its history of land loss and degradation, its prospects for future loss or degradation, the area the ecosystem occupies, and the number of threatened and endangered species living in that ecosystem.

Key Terms

- clear-cutting 324
- conservation easement 333
- deforestation 326
- desertification 332
- monoculture 323
- overgrazing 332
- rangeland 331
- sustainable forestry 324
- wilderness 336
- wildlife corridor 324

What is happening in this picture?

Julia "Butterfly" Hill lived in this 600–1,000-year-old, 180-foot-tall California redwood for more than 2 years in the late 1990s, to keep a lumber company from cutting down the tree.

- Would Hill's perspective on wilderness better fit the wise-use movement or the environmental movement?

- Explain the likely differences in the perspectives of Hill and the lumber company, especially given that the tree is on the company's land.

- Based on issues faced in Tongass National Forest, why do ecologists and environmentalists think that the logging of old-growth trees causes particular damage?

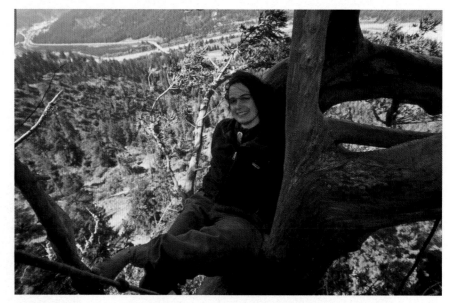

Critical and Creative Thinking Questions

1. Why is deforestation a serious global environmental problem?

2. What are the environmental effects of clear-cutting on steep mountain slopes? on tropical rainforest land?

3. Distinguish between rangeland degradation and desertification. Why is moderate grazing beneficial to rangelands, yet overgrazing leads to erosion?

4. Explain the various uses that must be considered in the management of a national park. Which federal agency administers U.S. parks?

5. Debate conflicts over logging in Tongass National Forest from two points of view: that of the wise-use movement and that of the environmental movement.

6. Do you think additional federal lands should be added to the wilderness system? Why or why not?

7. Consider the message that the cartoon above sends about the popularity of U.S. national parks. What might park managers do to protect parks while encouraging their use?

8. What percentage of U.S. land is privately owned? Should private landowners have control over what they wish to do to their land? How would you as a landowner handle land-use decisions that could affect the public?

9. Explain how economic growth and sustainable use of natural resources can be compatible goals.

10. Why are ecosystems in areas with fast-growing human populations most likely to meet the criteria for endangered ecosystems? Describe the four criteria in your answer.

11. Which federal agencies are responsible for managing public rangelands? What environmental and economic issues do they face?

12. Given the important contributions of forests in providing both timber and ecosystem services, how would you manage U.S. public forests if you were in charge?

13. For three decades, Forest A has been harvested using clear-cutting, and Forest B has been managed through sustainable forestry. Describe five differences you would expect to see between the two forests, including sustainable forest practices in your description.

For questions 14 and 15, examine the following graph of world land use that was assembled by the World Resources Institute and the U.N. Food and Agricultural Organization.

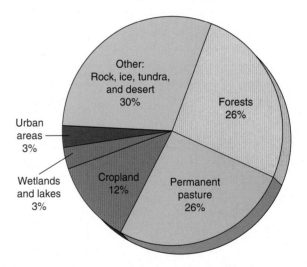

14. How much of the world's total land area is used for agriculture? How might conservation easements protect some of these areas?

15. Based on types of habitats represented, which of the categories shown are most likely to include the most endangered U.S. ecosystems?

✓ THE PLANNER

Agriculture and Food Resources

MAINTAINING GRAIN STOCKS

When people have access at all times to adequate amounts and kinds of food needed for healthy, active lives, they are said to have *food security*. World grain stocks provide a measure of food security (see larger photograph). World grain stocks are the amounts of rice, wheat, corn, and other grains remaining from previous harvests as a cushion against poor harvests and rising costs. Grain stocks have been decreasing since their all-time highs in the mid-1980s and late 1990s. For example, the amount of grain stockpiled in 2007 would have fed the world's people for only 51 days. According to the United Nations, world grain stocks should not fall below a minimum of 70 days' supply in a given year.

World grain stocks have dropped in the past few years for several reasons. Environmental conditions such as rising temperatures, falling water tables, and droughts have caused poor harvests.

World grain stocks have also fallen because consumption of beef, pork, poultry, and eggs has increased in developing countries such as China (see inset), where growing affluence has led some people to diversify their diets. This trend represents a global pattern: In highly developed countries, animal products account for nearly half of the calories people consume, compared to only 5 percent of the calories people in developing countries consume. Increased consumption of meat and meat products has prompted a surge in the amount of grain used to feed the world's billions of livestock animals: More than one-third of the world's grain is now used to feed livestock. Thus, the global trend of eating more meat and other animal products is linked to increased use of grains and other feed crops for livestock.

CHAPTER OUTLINE

CHAPTER PLANNER ✓

- ❏ Study the picture and read the opening story.
- ❏ Scan the Learning Objectives in each section:
  p. 348 ❏   p. 351 ❏   p. 353 ❏   p. 358 ❏   p. 362 ❏
- ❏ Read the text and study all figures and visuals. Answer any questions.

Analyze key features

- ❏ Environmental InSight, p. 349 ❏   p. 357 ❏
- ❏ Process Diagram, p. 351 ❏   p. 360 ❏
- ❏ EnviroDiscovery, p. 359 ❏
- ❏ What a Scientist Sees, p. 364 ❏
- ❏ Case Study, p. 366 ❏
- ❏ Stop: Answer the Concept Checks before you go on:
  p. 350 ❏   p. 352 ❏   p. 356 ❏   p. 361 ❏   p. 365 ❏

End of Chapter

- ❏ Review the Summary and Key Terms.
- ❏ Answer What is happening in this picture?
- ❏ Answer the Critical and Creative Thinking Questions.

World Food Problems

LEARNING OBJECTIVES

1. **Differentiate** between undernutrition and overnutrition.
2. **Define** *food insecurity* and relate it to human population, poverty, and world hunger.

The U.N. Food and Agricultural Organization (FAO) reported in 2008 that about 963 million people—more than three times the population of the United States—lack the food needed for healthy, productive lives. Most of these people live in roadless rural areas of the poorest developing countries.

The average adult human must consume enough food to get approximately 2600 *kilocalories*, or simply *Calories*, per day. People who receive fewer calories than needed are undernourished. Over an extended period of undernourishment, their health and stamina decline, even to the point of death. Worldwide, an estimated 182 million children under age 5 suf-

undernutrition A type of malnutrition in which there is an underconsumption of calories or nutrients that leaves the body weakened and susceptible to disease.

fer from **undernutrition** and are seriously underweight, according to the World Health Organization (WHO).

People might receive enough calories in their diets but still be malnourished because they do not receive enough essential nutrients, such as proteins, vitamin A, iodine, or iron. Adults suffering from malnutrition are more susceptible to disease and have less strength to function productively than those who are well fed. In addition to being more susceptible to disease, malnourished children do not grow or develop normally. Because malnutrition affects cognitive development, malnourished children typically do not perform well in school. Currently, WHO estimates that more than 3 billion people worldwide—the greatest number in history—are malnourished (**Figure 14.1a, b, c**). In addition, more than half the deaths in children younger than 5 years old in

developing countries are associated with malnutrition (**Figure 14.1d**).

People who eat more food than necessary are overnourished. Generally, a person suffering from **overnutrition** has a diet high in saturated (animal) fats, sugar, and salt. Overnutrition, which is most common among people in highly developed nations such as the United States, results in obesity, high blood pressure, and an increased likelihood of

overnutrition A type of malnutrition in which there is an overconsumption of calories that leaves the body susceptible to disease.

disorders such as diabetes, heart disease, and some cancers. Overnutrition is also emerging in some developing countries, particularly in urban areas, where, as people earn more money, their diets shift from consumption of cereal grains to consumption of more livestock products and processed foods high in fat and sugar.

Population and World Hunger

According to the FAO, 86 countries are considered low income and food deficient, which means they cannot produce enough food or afford to import enough food to feed the entire population. South Asia, with an estimated 270 million hungry people, and sub-Saharan Africa, with an estimated 175 million who are hungry, are the regions of the world with the greatest **food insecurity**.

food insecurity The condition in which people live with chronic hunger and malnutrition.

Today, sub-Saharan Africa produces less food per person than it did in 1950. Several factors contribute to the food shortages in Africa; these include civil wars and military actions, HIV/AIDS (which has killed or incapacitated much of the agricultural workforce in some countries), floods, droughts, and soil erosion from hilly, marginal farmlands. People with food insecurity always live under the threat of starvation.

a. Millions of children suffer from *kwashiorkor*, caused by severe protein deficiency. Note the characteristic swollen belly, which results from fluid retention. Photographed in Haiti.

b. *Marasmus* is progressive emaciation caused by a diet low in both total calories and protein. Symptoms include a pronounced slowing of growth and extreme wasting of muscles. Photographed in Somalia.

c. Globally, millions of adult men and women are hungry. This homeless man is suffering from severe malnutrition and starvation. Photographed in New Delhi, India.

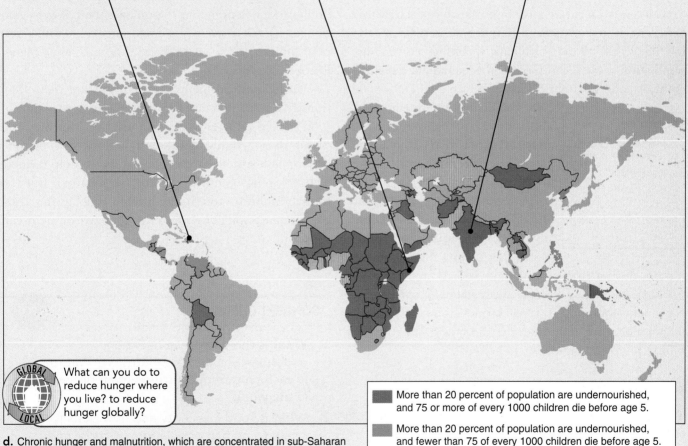

What can you do to reduce hunger where you live? to reduce hunger globally?

More than 20 percent of population are undernourished, and 75 or more of every 1000 children die before age 5.

More than 20 percent of population are undernourished, and fewer than 75 of every 1000 children die before age 5.

Less than 20 percent of population are undernourished, and 75 or more of every 1000 children die before age 5.

d. Chronic hunger and malnutrition, which are concentrated in sub-Saharan Africa and South Asia, are linked to increased mortality for infants and children.

WILEY
RESOURCE KIT

Total world grain production and grain production per person, 1970 to 2007 • Figure 14.2

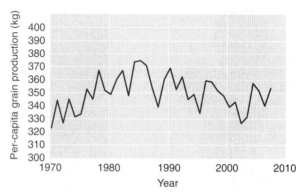

a. Total world grain production increased from 1.1 billion tons in 1970 to 2.3 billion tons in 2007.

b. The amount of grain produced per person has not changed significantly in the past 40 years.

Experts agree that world hunger, population, poverty, and environmental problems are interrelated, but they disagree on the relative importance of each factor. Different groups propose different solutions for resolving the world's food problems, including controlling population growth, promoting the **economic development** of countries that do not produce adequate food, and correcting the inequitable distribution of resources. All experts agree that population pressures exacerbate world food problems.

> **economic development**
> An expansion in a government's economy, viewed by many as the best way to raise the standard of living.

Although annual grain production almost doubled from 1970 to 2007 (**Figure 14.2**), the world population increased so rapidly during that period that the amount of grain *per person* has not changed appreciably. According to National Geographic, by 2030 farmers will have to grow 30 percent more grain than they do now so that the 8.3 billion people living then can be fed.

Poverty and Food

The main cause of undernutrition is poverty. Infants, children, and the elderly are most susceptible to poverty and chronic hunger. The world's poorest people—those living in developing countries in Asia, Africa, and Latin America—do not own land on which to grow food and do not have sufficient money to purchase food. Poverty and hunger are not restricted to developing nations, however; poor hungry people are also found in the United States, Europe, and Australia.

World food problems are many, as are their solutions. We must increase the sustainable production of food (discussed later in the chapter) and improve food distribution. Highly developed nations can provide economic assistance and technical aid to help farmers in developing countries produce more food. But the ultimate solution to chronic hunger is achieving a stable human population at a level the environment can support.

CONCEPT CHECK

1. **What** is the difference between undernutrition and overnutrition? Where is each type of malnutrition most prevalent in the world?

2. **What** is food insecurity and how does it relate to human population, poverty, and world hunger?

The Principal Types of Agriculture

LEARNING OBJECTIVES

1. **Contrast** industrialized agriculture with subsistence agriculture.

2. **Describe** three kinds of subsistence agriculture.

Agriculture can be roughly divided into two types: industrialized agriculture and subsistence agriculture. Most farmers in highly developed countries and some in developing countries practice *high-input agriculture*, or **industrialized agriculture**. It relies on large inputs of capital and energy (in the form of fossil fuels) to

> **industrialized agriculture** Modern agricultural methods that require large capital inputs and less land and labor than traditional methods.

make and run machinery, purchase seed, irrigate crops, and produce agrochemicals such as commercial inorganic fertilizers and pesticides (**Figure 14.3**). Industrialized agriculture produces high **yields** (the amount of food produced per unit of land), which allows forests and other natural areas to remain wild instead of being converted to agricultural land. However, the productivity of industrialized agriculture comes with costs, such as soil degradation and increased pesticide resistance in agricultural pests; we discuss these and other problems later in the chapter (and in Chapters 4 and 12).

PROCESS DIAGRAM

THE PLANNER ✓

Energy inputs in industrialized agriculture • Figure 14.3

WILEY RESOURCE KIT

Think Critically Industrialized agriculture requires a lot of capital. Why is it so expensive?

Nomadic sheep herders in Kenya • Figure 14.4

Most farmers in developing countries practice sub-sistence agriculture, the production of enough food to feed oneself and one's family, with little left over to sell

subsistence agriculture
Traditional agricultural methods that are dependent on labor and a large amount of land to produce enough food to feed oneself and one's family.

or reserve for hard times. Subsistence agriculture also requires large inputs of energy, but from humans and draft animals rather than from fossil fuels.

Some types of subsistence agriculture require large tracts of land. **Shifting cultivation** is a form of subsistence agriculture in which short periods of cultivation are followed by longer periods of fallow (land being left uncultivated), during which the land reverts to forest. Shifting cultivation supports relatively small populations. **Slash-and-burn agriculture** is a type of shifting cultivation that involves clearing small patches of tropical forest to plant crops (see Chapter 13). Because tropical soils lose their productivity quickly when they are cultivated, farmers using slash-and-burn agriculture must move from one area of forest to another every 3 years or so.

Nomadic herding, in which livestock is supported by land too arid for successful crop growth, is a

similarly land-intensive form of subsistence agriculture (**Figure 14.4**). Nomadic herders must continually move their livestock to find adequate food for the animals.

Intercropping is a form of intensive subsistence agriculture that involves growing a variety of plants on the same field simultaneously. When certain crops are grown together, they produce higher yields than when they are grown as **monocultures**. (A monoculture is the cultivation of only one type of plant over a large area.) **Polyculture** is a type of intercropping in which several kinds of plants that mature at different times are planted together. In polyculture practiced in the tropics, fast- and slow-maturing crops are often planted together so that different crops can be harvested throughout the year.

CONCEPT CHECK **STOP**

1. **What** are some differences between industrialized agriculture and subsistence agriculture?

2. **What** are shifting cultivation, nomadic herding, and intercropping?

Challenges of Agriculture

LEARNING OBJECTIVES

1. **Discuss** recent trends in loss of U.S. agricultural land, global declines in domesticated plant and animal varieties, and efforts to increase crop and livestock yields.

2. **Relate** the benefits and problems associated with the green revolution.

3. **Describe** the environmental impacts of industrialized agriculture, including land degradation and habitat fragmentation.

The United States has more than 121 million hectares (300 million acres) of **prime farmland**, land that has the soil type, growing conditions, and available water to produce food, forage, fiber, and oilseed crops. U.S. agriculture faces a decline in prime farmland. Other challenges include coping with declining numbers of domesticated varieties, improving crop and livestock yields, and addressing environmental impacts.

Loss of Agricultural Land

There is considerable concern that much of the nation's prime agricultural land is falling victim to urbanization and suburban sprawl by being converted to parking lots, housing developments, and shopping malls (**Figure 14.5**). In certain areas of the United States, loss of rural land is a significant problem. According to the American Farmland Trust, more than 162,000 hectares (400,000 acres) of prime U.S. farmland are lost each year.

The 1996 Farm Bill included funding for the establishment of a national *Farmland Protection Program*. This voluntary program lets farmers sell **conservation easements** that prevent their farmland from being converted to nonagricultural uses. The easements are in effect from a minimum of 30 years to forever. As with other conservation easements, the farmers retain full rights to use their property—in this case, for agricultural purposes.

Suburban sprawl onto agricultural land • Figure 14.5

Virginia housing developments encroach on farmland.

White leghorn hen • Figure 14.6

If you eat eggs, they are probably laid by white leghorn hens, which produce almost all of the eggs in the United States as well as those in many other countries. This breed is prolific at egg laying, but it is not particularly good for meat production.

Global Decline in Domesticated Plant and Animal Varieties

A global trend is currently under way to replace the many local varieties of a particular crop or domesticated farm animal with just a few kinds (**Figure 14.6**). When farmers abandon traditional varieties in favor of more modern ones, which are bred for uniformity and maximum production, the traditional varieties frequently face extinction. This represents a great loss in genetic diversity because each variety's characteristic combination of genes gives it distinctive nutritional value, size, color, flavor, resistance to disease, and adaptability to different climates and soil types.

To preserve older, more diverse varieties of plants and animals, many countries, including the United States, are collecting **germplasm**: seeds, plants, and plant tissues of traditional crop varieties and the sperm and eggs of traditional livestock breeds.

> **germplasm** Any plant or animal material that may be used in breeding.

Increasing Crop Yields

Until the 1940s, agricultural yields among various countries, whether highly developed or developing, were generally equal. Advances by research scientists since then have dramatically increased food production in highly developed countries (**Figure 14.7**). Greater knowledge of plant nutrition has resulted in production of fertilizers that promote high yields. The use of pesticides to control insects, weeds, and disease-causing organisms has also improved crop yields.

The Green Revolution By the middle of the 20th century, serious food shortages occurred in many developing countries coping with growing populations. The development and introduction during the 1960s of high-yield varieties of wheat and rice to Asian and Latin American countries gave these nations the chance to provide their people with adequate supplies of food (**Figure 14.8**). But the high-yield varieties required intensive industrial cultivation methods, including the use of commercial inorganic fertilizers, pesticides, and mechanized machinery, to realize their potential. These agricultural technologies were passed from highly developed nations to developing nations.

Average U.S. wheat yields, 1950 to 2005 • Figure 14.7

Each year shown is actually an average of 3 years to minimize the effects that poor weather conditions might have in a single year. Similar increases in yield have occurred in other grain crops.

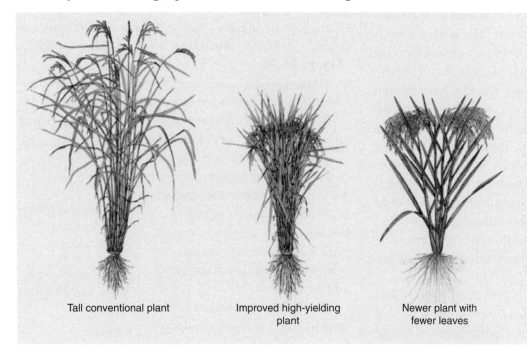

Tall conventional plant

Improved high-yielding plant

Newer plant with fewer leaves

Traditional rice plants on the left are taller and do not yield as much grain (clusters at top of plant) as the more modern varieties shown in the middle and on the right. The rice plant in the middle was developed during the 1960s by crossing a high-yield, disease-resistant variety with a dwarf variety to prevent the grain-heavy plants from falling over. Improvements since the green revolution have been modest, as shown in the rice variety developed during the 1990s (right). Some researchers think rice and certain other genetically improved crops are near their physical limits of productivity.

Using modern cultivation methods and the high-yield varieties of certain staple crops to produce more food per acre of cropland is known as the **green revolution**. Some of the success stories of the green revolution are remarkable. During the 1920s, Mexico produced less than 700 kg (0.77 ton) of wheat per hectare annually. During the green revolution years that began in 1965, Mexico's annual wheat production rose to more than 2400 kg (2.65 tons) per hectare. Indonesia, another green revolution success story, formerly imported more rice than any other country in the world. Today Indonesia produces enough rice to feed its people and export some.

Critics argue that the green revolution has made developing countries dependent on imported technologies, such as agrochemicals and tractors, at the expense of traditional agriculture. The two most important problems associated with higher crop production are the high energy costs built into this type of agriculture and the environmental problems caused by the intensive use of fertilizers and pesticides.

Increasing Crop Yields in the Post–Green Revolution Era

In 1999, the International Food Policy Research Institute projected that the world demand for rice, wheat, and corn will increase 40 percent between 2000 and 2020. This rise in demand will require a corresponding rise in grain production to feed the human population, as well as the livestock needed to satisfy the appetites of the increasing numbers of affluent people who can afford to buy meat.

This challenge cannot be met by increasing the amount of land under cultivation, as the best arable lands are already being cultivated. Projected freshwater shortages, rising costs of agricultural chemicals, and deteriorating soil quality caused by intensive agricultural techniques may further constrain productivity. As Figure 14.8 demonstrates, recent progress in coaxing more grain out of crops genetically improved during the green revolution has resulted in diminishing returns. Grain yields have continued to rise since the 1960s, but in recent years the rates of increase have not been as great as they were previously.

Despite these problems, most plant geneticists think we can produce enough food in the 21st century to meet demand if countries spend more money in support of a concerted scientific effort to improve crops. Many scientists think genetic engineering is one of the keys to breeding more productive varieties. In addition, modern agricultural methods, such as water-efficient irrigation, will need to be introduced to developing countries that do not currently have them if we are to continue increasing crop yields.

Increasing Livestock Yields

The use of hormones and antibiotics, although controversial, increases animal production. **Hormones**, usually administered by ear implants, regulate livestock bodily functions and promote faster growth. Although U.S. and Canadian farmers use hormones, the European Union (EU) currently restricts imports of hormone-treated beef. It cites studies which suggest that these hormones or their breakdown products, both found in trace amounts in meat and meat products, could cause cancer or affect the growth of young children. In 1999 an international scientific committee organized by the FAO and WHO concluded that the trace amounts of hormones found in beef are safe because they are very low compared to the normal hormone concentrations found in the human body.

Modern agriculture has also embraced the addition of low doses of **antibiotics** to feed for pigs, chickens, and cattle. These animals gain 4 to 5 percent more weight than untreated animals, presumably because they expend less energy fighting infections.

Several studies link the indiscriminate use of antibiotics in humans and livestock to the evolution of bacterial strains that are resistant to antibiotics (**Figure 14.9a**). Because there is increasing evidence that the use of antibiotics in agriculture reduces their medical effectiveness for humans, WHO recommended in 2003 that routine use of antibiotics in livestock be eliminated. Many European countries have complied, but the United States and many other countries continue the practice.

Environmental Impacts

Industrialized agriculture has many environmental effects. The agricultural use of fossil fuels and pesticides produces air pollution. Untreated animal wastes and agricultural chemicals such as fertilizers and pesticides cause water pollution, which reduces biological diversity, harms fisheries, and leads to outbreaks of nuisance species. According to the Environmental Protection Agency, agricultural practices are the single largest cause of surface-water pollution in the United States.

Industrialized agriculture has favored the replacement of traditional family farms with large agribusiness conglomerates. In the United States, most cattle, hogs, and poultry are now grown in feedlots and livestock factories (**Figure 14.9b**). The large concentrations of animals in livestock factories create many environmental problems, including air and water pollution (**Figure 14.9c**).

Many insects, weeds, and disease-causing organisms have developed or are developing resistance to **pesticides**. Pesticide resistance forces farmers to apply progressively larger quantities of pesticides (**Figure 14.9d**). Pesticide residues contaminate our food supply and reduce the number and diversity of beneficial microorganisms in the soil. Fishes and other aquatic organisms are sometimes killed by pesticide runoff into lakes, rivers, and estuaries.

Land degradation is a reduction in the potential productivity of land. Soil erosion, which is exacerbated by large-scale mechanized operations, causes a decline in soil fertility, and the eroded sediments damage water quality. Other types of degradation are compaction of soil by heavy farm machinery and waterlogging and **salinization** (salting) of soil from improper irrigation methods.

Clearing grasslands and forests and draining wetlands to grow crops result in **habitat fragmentation** that reduces biological diversity. Many species are endangered or threatened as a result of habitat loss to agriculture. The most dramatic example of habitat loss in North America is tallgrass prairie, more than 90 percent of which has been converted to agriculture.

> **pesticide** A toxic chemical used to kill pests.
>
> **degradation (of land)** Natural or human-induced reduction in the potential ability of the land to support crops or livestock.
>
> **habitat fragmentation** The breakup of large areas of habitat into small, isolated patches.

CONCEPT CHECK

1. **What** is happening to the number of domesticated plant and animal varieties? Why?

2. **What** is the green revolution? What are some of its benefits and problems?

3. **What** are the major environmental problems associated with industrialized agriculture?

a. **Evolution of Antibiotic Resistance.** Shown is the increasing resistance of *E. coli* in blood and cerebrospinal infections to the antibiotic ciprofloxacin.

b. **Hog Factory.** The hogs remain indoors and are fed and watered by machines at timed intervals throughout the day.

c. **Waste Lagoon in an Automated Hog Farm.** Some livestock factories produce as much sewage as small cities, yet the wastes do not currently fall under water quality laws.

RESOURCE KIT

d. **Colorado Potato Beetle on Potato Leaf.** As a result of exposure to heavy pesticides over the years, these beetles are resistant to most insecticides registered for use on potatoes.

Some common problems of industrialized agriculture

Air pollution

• Pesticide sprays

• Soil particles from wind erosion

• Odors from livestock factories

• Greenhouse gases from combustion of fossil fuels

• Other air pollutants from combustion of fossil fuels

Water issues

• Groundwater depletion from irrigation

• Pollution from fertilizer and pesticide runoff

• Sediment pollution from eroding soil particles

• Pollution from animal wastes (livestock factories)

• Enrichment of surface water from fertilizer runoff and livestock wastes

Loss of biological diversity

• Habitat fragmentation (clearing land and draining wetlands)

• Monocultures (lack of diversity in croplands)

• Stress from air and water pollution

• Stress from pesticides

• Replacement of many traditional crop and livestock varieties with just a few

Land degradation

• Soil erosion

• Loss of soil fertility

• Soil salinization

• Soil pollution (pesticide residues)

• Waterlogged soil from improper irrigation

Solutions to Agricultural Problems

LEARNING OBJECTIVES

1. **Define** *sustainable agriculture* and contrast it with industrialized agriculture.

2. **Identify** the potential benefits and problems of genetic engineering.

The green revolution and industrialized agriculture have unquestionably met the food requirements of most of the human population, even as that population has more than doubled since 1960. But we pay for these food gains with serious environmental problems, and we do not know if industrialized agriculture is sustainable for more than a few decades. To compound the issue, we must continue to increase food production to feed the growing human population—but the resulting damage to the environment may reduce our chances of continuing those food increases into the future.

Fortunately, farming practices and techniques exist that ensure a sustainable output at yields comparable to those of industrialized agriculture. Farmers who practice industrialized agriculture can adopt these alternative agricultural methods, which cost less and are less damaging to the environment. Advances are also being made in sustainable subsistence agriculture.

Moving to Sustainable Agriculture

In **sustainable agriculture** (also called **alternative**, or **low-input**, **agriculture**) certain modern agricultural techniques are carefully combined with traditional farming methods. Sustainable agriculture relies on beneficial biological processes and environmentally friendly chemicals that disintegrate quickly and do not persist as residues in the environment. A sustainable farm consists of field crops, trees that bear fruits and nuts, small herds of livestock, and even tracts of forest (**Figure 14.10**).

> **sustainable agriculture** Agricultural methods that maintain soil productivity and a healthy ecological balance while having minimal long-term impacts.

Diversification protects a sustainable farmer against unexpected changes in the marketplace. The breeding of disease-resistant crop plants and the maintenance of animal health rather than the continual use of antibiotics to prevent disease are important parts of sustainable agriculture. Biological diversity is maintained as a way to minimize pest problems. Water and energy conservation are also practiced in sustainable agriculture.

An important goal of sustainable agriculture is to preserve the quality of agricultural soil. Crop rotation,

Some goals of sustainable agriculture • Figure 14.10

Sustainable agriculture protects the soil so that it does not become depleted.

Increase biological diversity in crops and livestock to enhance food security, maximize natural processes (such as pest control and pollination), and minimize pesticide inputs.

Manage wetlands and water resources carefully.

Enhance soil health and minimize erosion through crop rotation, multiple cropping, conservation tillage, and planting tracts of forest.

Improve soil fertility by adding organic matter and managing soil biology (for example, using legume-based nitrogen fixation) to minimize commercial inorganic fertilizer inputs.

Emphasize total agricultural ecosystem rather than single crop (monoculture).

Focus on biological processes (nutrient cycling, energy flow, predator–prey relationships) rather than single actions (effect of application of fertilizer on crop growth).

conservation tillage, and contour plowing help control erosion and maintain soil fertility (see Figures 12.17c on page 313 and 3.1a on page 50). Sloping hills converted to mixed-grass pastures erode less than hills planted with field crops, thereby conserving the soil and supporting livestock.

Animal manure added to soil not only cuts costs but decreases the need for high levels of commercial inorganic fertilizers. Legumes such as soybeans, clover, and alfalfa convert atmospheric nitrogen into a form that plants can use in a process called *biological nitrogen fixation*, which reduces the need for nitrogen fertilizers.

Sustainable agriculture is not a single program but a series of programs adapted for specific soils, climates, and farming requirements. Some sustainable farmers—those who practice **organic agriculture**—use no pesticides; others use a system of *integrated pest management (IPM)*, which incorporates the limited use of pesticides with pest-controlling biological and cultivation practices.

In growing recognition of the environmental problems associated with industrialized agriculture, more and more mainstream farmers are trying some methods of sustainable agriculture. These methods cause fewer environmental problems to the agricultural ecosystem, or **agroecosystem**, than industrialized agriculture. This trend away from using intensive techniques that produce high yields and toward methods that focus on long-term sustainability of the soil is sometimes referred to as the **second green revolution**.

EnviroDiscovery

A New Weapon for Locust Swarms

A non-insect biological control agent targets desert locusts, which periodically increase in number and swarm across the African Sahel, threatening crops across 12 million hectares (30 million acres) located south of the Sahara Desert. The most recent locust outbreak was during the 2004–2005 growing season (see photo). During the last major swarm, in 1988, African farmers sprayed massive quantities of pesticides into the environment to bring the locust population under control. However, many people were concerned about the adverse ecological and health effects of using such large quantities of pesticides. Economists were also concerned about high short-term costs of using pesticides ($300 million for the 1988 swarm) as well as long-term economic costs.

After the 1988 swarm, an international team of scientists worked for 15 years to develop a biological control for locusts that consists of fungal spores. Called Green Muscle, it is environmentally safe and very effective. Unfortunately, Green Muscle wasn't developed in time to prevent the 2004 outbreak, but if used widely in the future, it might prevent the locusts from swarming again.

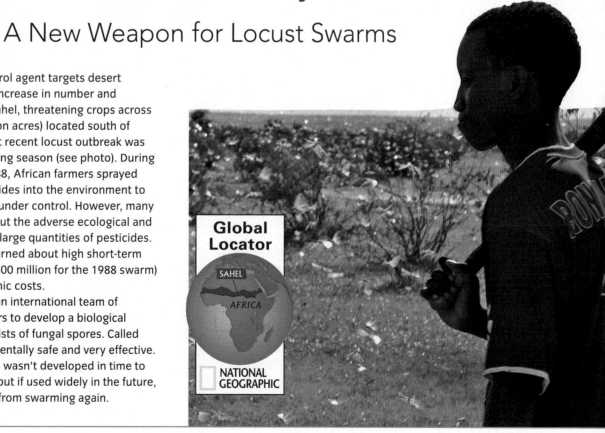

Genetic engineering • Figure 14.11

This example of genetic engineering uses a plasmid, a small circular molecule of DNA (genetic material) found in many bacteria. The plasmid of the bacterium *Agrobacterium* introduces into a plant desirable genes from another organism.

1 The foreign DNA is spliced into a plasmid.

Agrobacterium chromosome

Plasmid

Plasmid

Foreign gene from soybean codes for desirable trait

2 Plasmid is inserted into *Agrobacterium*.

Recombinant plasmid

3 *Agrobacterium* transfers plasmid to plant cell.

Plant cell

Chromosome   Nucleus

4 Foreign gene is inserted into the plant's chromosome.

5 Plant cells divide in tissue culture; each cell contains the foreign gene.

6 Using tissue culture techniques, cells are regenerated into plants.

7 Genetically modified plants are then produced from the cultured plant cells.

Think Critically How does the bacterium *Agrobacterium* help scientists produce genetically modified plants?

WILEY
RESOURCE KIT

Genetic Engineering: A Solution or a Problem?

Genetic engineering is a controversial technology that has begun to revolutionize medicine and agriculture. The agricultural goals of genetic engineering are not new. Using traditional breeding methods, farmers and scientists have developed desirable characteristics in crop plants and agricultural animals for centuries. It takes time—15 years or more—to develop such genetically improved organisms using traditional breeding methods. Genetic engineering has the potential to accomplish the same goal in a fraction of that time.

Genetic engineering differs from traditional breeding methods in that desirable genes from any organism can be used, not just those from the species of the plant or animal being improved. If a gene for disease resistance found in soybeans would be beneficial to tomatoes, a genetic engineer can splice the soybean gene into the tomato plant's DNA (**Figure 14.11**). Traditional breeding methods could not do this because soybeans and tomatoes belong to separate groups of plants and do not interbreed.

Genetic engineering has the potential to produce more nutritious food plants that contain all the essential amino acids (which no single food crop currently does)

genetic engineering The manipulation of genes (for example, taking a specific gene from one species and placing it into an unrelated species) to produce a particular trait.

World production of GM crops • Figure 14.12

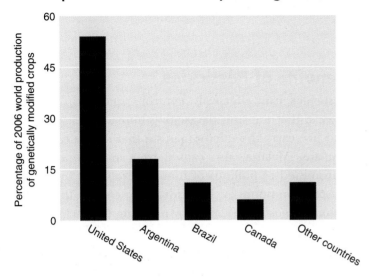

a. The world's top producers of GM crops.

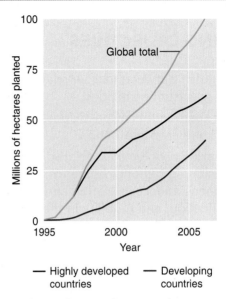

b. The production of GM crops has increased rapidly.

or that would be rich in necessary vitamins. Crop plants resistant to viral diseases, drought, heat, cold, herbicides, salty or acidic soils, and insect pests are also being developed.

The first **genetically modified (GM)** crops were approved for commercial planting in the United States in the early 1990s. The United States is the world's top producer of GM crops (**Figure 14.12a**). Since 2000, however, GM production in developing nations has increased faster than in highly developed countries (**Figure 14.12b**).

Genetic engineering has been used to develop more productive farm animals, including rapidly growing hogs and fishes. Perhaps the greatest potential contribution of animal genetic engineering is the production of vaccines against disease organisms that harm agricultural animals. For example, genetically engineered vaccines have been developed to protect cattle against the deadly viral disease rinderpest, which is economically devastating in parts of Asia and Africa.

Concerns About Genetically Modified Foods

During the late 1990s and early 2000s, opposition to genetically engineered crops increased in many countries in Europe and Africa. In 1999 the EU placed a 5-year moratorium on virtually all approvals of GM crops (the moratorium has since been lifted). The EU refused to buy U.S. corn because it might be genetically modified. (Currently, about 60 percent of the U.S. corn crop

is genetically modified, mostly to protect against insect pests and provide resistance to herbicides [weed killers].) One concern is that the inserted genes could spread from GM crops to weeds or wild relatives of crop plants and possibly harm natural ecosystems in the process. Scientists recognize this concern as legitimate and must take special precautions to avoid this possibility. Critics also worry that some consumers might develop food allergies to GM foods, although scientists routinely screen new GM crops for allergenicity.

The scientific consensus is that the risks associated with consuming food derived from GM varieties are the same as those associated with consuming food derived from varieties produced using traditional breeding techniques. A growing body of evidence, summarized in the FAO *State of Food and Agriculture 2003–2004*, concludes that current genetically modified (GM) crop plants are as safe for human consumption as crops grown using conventional or organic agriculture. However, strict guidelines exist in areas of genetic engineering research in which there are unanswered questions about possible effects on the environment.

CONCEPT CHECK

1. **What** is sustainable agriculture? What are some features of a sustainable farm?

2. **Why** are some people opposed to GM crops?

Controlling Agricultural Pests

LEARNING OBJECTIVES

1. **Distinguish** between narrow-spectrum and broad-spectrum pesticides.
2. **Relate** the benefits of pesticides in disease control and crop protection.
3. **Summarize** problems associated with pesticide use, including genetic resistance, ecosystem imbalances, bioaccumulation and biological magnification, and mobility in the environment.
4. **Describe** alternative ways to control pests.

Any organism that interferes in some way with human welfare or activities is a **pest**. Some weeds, insects, rodents, bacteria, fungi, nematodes (microscopic worms), and other pest organisms compete with humans for food; other pests cause or spread disease. People try to control pests, usually by reducing the size of the pest population. Using *pesticides* is the most common way of doing this, particularly in agriculture. Pesticides can be grouped by their target organisms—that is, by the pests they are supposed to eliminate. **Insecticides** kill insects, **herbicides** kill plants, **rodenticides** kill rodents such as rats and mice, and **fungicides** kill fungi (**Figure 14.13**).

Benefits of Pesticides

Pesticides can effectively control organisms, such as insects, that transmit devastating human diseases. Fleas and lice carry the microorganism that causes typhus in humans. Malaria, also caused by a microorganism, is transmitted to millions of humans each year by female *Anopheles* mosquitoes. Pesticides help control the population of mosquitoes, thereby reducing the incidence of malaria.

Pesticides also protect crops. It is estimated that pests eat or destroy more than one-third of the world's crops. Given our expanding population and world hunger, it is easy to see why control of agricultural pests is desirable. Pesticides reduce the amount of a crop lost through competition with weeds, consumption by insects, and diseases caused by plant **pathogens** (microorganisms, such as fungi and bacteria, that cause disease).

Why are agricultural pests found in such great numbers in our fields? Part of the reason is that agriculture is usually a **monoculture**, in which the field cultivated with a single species represents a simple ecosystem. In contrast, forests, wetlands, and other natural ecosystems are complex and contain many species, including predators and

Oranges being sprayed with fungicide before shipment to grocery stores • Figure 14.13 _____

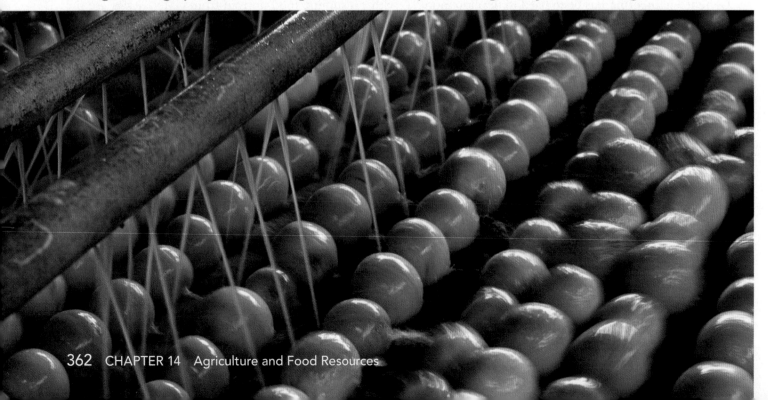

parasites that control pest populations, as well as plant species that pests do not use for food. A monoculture reduces the dangers and accidents that might befall a pest as it searches for food. In the absence of many natural predators and in the presence of plenty of food, a pest population thrives and grows, damaging more of the crop.

Problems with Pesticides

The ideal pesticide is a **narrow-spectrum pesticide** that kills only the intended organism and does not harm other species. The perfect pesticide would readily break down into safe materials such as water, carbon dioxide, and oxygen. The ideal pesticide would stay exactly where it was put and would not move around in the environment. Unfortunately, no pesticide is perfect. Most pesticides are **broad-spectrum pesticides.**

> **broad-spectrum pesticide** A pesticide that kills a variety of organisms, including beneficial organisms, in addition to the target pest.

Some pesticides do not degrade readily, or they break down into compounds as dangerous as—if not more dangerous than—the original pesticide. And most pesticides move around the environment quite a bit.

> **genetic resistance** An inherited characteristic that decreases the effect of a given agent (such as a pesticide) on an organism (such as a pest).

The prolonged use of a particular pesticide can cause a pest population to develop **genetic resistance** to the pesticide. In the 50 years during which pesticides have been widely used, at least 520 species of insects and mites and at least 84 weed species have evolved genetic resistance to certain pesticides (**Figure 14.14**).

An organism exposed to a chemically stable pesticide that takes years to break down may accumulate high concentrations of the toxin, a phenomenon known as **bioaccumulation**. Organisms at higher levels on food webs tend to have greater concentrations of bioaccumulated pesticide stored in their bodies than those lower on food webs, through a process known as **biological magnification** (**Figure 14.15**; also see Figure 4.8 on p. 82).

Genetic resistance • Figure 14.14

The number of pest species evolving genetic resistance to pesticides has increased.

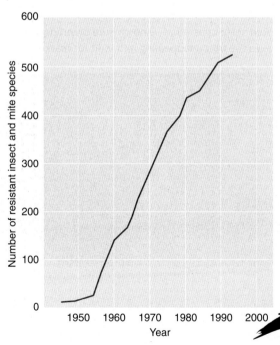

Interpreting Data
How is this figure similar to Figure 14.9a on page 357, which shows the evolution of resistance to antibiotics?

Peregrine falcons make a comeback • Figure 14.15

Peregrine falcons were once an endangered species in the United States. Bioaccumulation of pesticides such as DDT caused the birds to lay eggs with extremely thin shells, causing the chicks' deaths. With the ban of DDT and similar pesticides, the falcons recovered.

One of the worst problems associated with pesticide use is that pesticides affect more species than the pests for which they are intended. Beneficial insects and other organisms are harmed and killed as effectively as pest insects. Quite often the stress of carrying pesticides in their tissues makes organisms more vulnerable to predators, diseases, or other stressors in their environments. Because the natural enemies of pests often starve or migrate in search of food after a pesticide is sprayed in an area, pesticides are indirectly responsible for a large reduction in the populations of these natural enemies. Pesticides also kill natural enemies directly because predators may consume lethal amounts of a pesticide while consuming the pests. After a brief period, the pest population rebounds and gets larger than ever, partly because no natural predators are left to keep its numbers in check. In some instances, the use of a pesticide has resulted in a pest problem that did not exist before (see *What a Scientist Sees*).

Another problem associated with pesticides is that they do not stay where they are applied but tend to move through the soil, water, and air, sometimes traveling long distances (**Figure 14.16**). Pesticides applied to agricultural lands wash into rivers and streams, where they can harm fishes. Pesticide

WHAT A SCIENTIST SEES

Pesticide Use and New Pest Species

An infestation of red scale insects on lemons occurred in California after DDT was sprayed to control a different pest. Prior to DDT treatment, red scale did not cause significant economic injury to citrus crops.

a. Red scale infestation—shown here on oranges—makes the fruit unappealing and unfit for market.

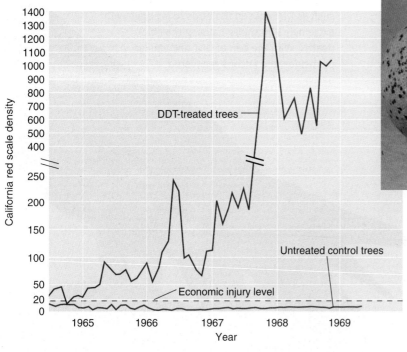

b. Scientists understand that the red scale infestation is a direct result of DDT spraying. This graph compares red scale populations on DDT-treated trees (red line) and untreated trees (blue line).

Mobility of pesticides in the environment • Figure 14.16

A helicopter sprays pesticides on a crop—and everything else in its pathway—in California.

mobility is also a problem for humans. About 14 million U.S. residents drink water containing traces of five widely used herbicides, and some people living where the herbicides are commonly used face a slightly elevated cancer risk because of their exposure.

Alternatives to Pesticides

Given their many problems, pesticides are clearly not the final solution to pest control. Fortunately, pesticides are not the only weapons in our arsenal. Alternative ways to control pests include cultivation methods, **biological controls**, **pheromones** and hormones, reproductive controls, genetic controls, quarantine, and irradiation (**Table 14.1**). A combination of these methods in agriculture, often including a limited use of pesticides as a last resort, is known as **integrated pest management (IPM)**.

CONCEPT CHECK

1. **What** is a broad-spectrum pesticide?
2. **What** are two important benefits of pesticide use?
3. **Discuss** two problems associated with the use of pesticides.
4. **How** can pests be controlled without pesticides?

Alternative methods of controlling agricultural pests • Table 14.1

Pest control method	How it works	Disadvantages
Cultivation methods	Interplanting mixes different plants, as by alternating rows; strip cutting alternates crop harvest by portion—remaining portions protect natural predators and parasites of pests	No appreciable disadvantages; more care must be taken in harvest
Biological controls	Naturally occurring predators, parasites, or disease organisms are used to reduce pest populations	Organism introduced for biological control can unexpectedly affect the environment or other organisms
Pheromones and hormones	Sexual attractants (pheromones) lure pest species to traps; synthetic regulatory chemicals (hormones) disrupt pests' growth and development	Hormones might affect beneficial species
Reproductive controls	Sterilizing some members of pest population reduces population size	Expensive; must be carried out continually
Genetic controls	Selective breeding or genetic engineering develops pest-resistant crops	Plant pathogens evolve rapidly, adapting to disease-resistant host plant; plant breeders forced to constantly develop new strains
Quarantine	Governments restrict importation of foreign pests, diseases	Not foolproof; pests are accidentally introduced
Irradiating foods	Harvested foods are exposed to ionizing radiation that kills potentially harmful microorganisms	Consumers concerned about potential radioactivity (not a true risk); irradiation forms traces of potentially carcinogenic chemicals (free radicals)

Integrated Pest Management

Many pests cannot be controlled effectively with a single technique; a combination of control methods is often more effective. Integrated pest management (IPM) combines the use of a variety of biological, cultivation, and pesticide controls tailored to the conditions and crops of an individual farm (**Figures a, b**). Biological and genetic controls, including GM crops designed to resist pests, are used as much as possible, and conventional pesticides are used sparingly and only when other methods fail. When pesticides are required, the least toxic pesticides are applied in the lowest possible effective quantities. Thus, IPM allows a farmer to control pests with a minimum of environmental disturbance and often at minimal cost.

To be effective, IPM requires a thorough knowledge of the life cycles and feeding habits of the pests as well as all their interactions with their hosts and other organisms. The timing of planting, cultivation, and treatments with biological controls is determined by carefully monitoring the concentration of pests. IPM is an important part of sustainable agriculture.

IPM is based on two fundamental premises. First, IPM is the *management* rather than the eradication of pests. Farmers who adopt IPM allow a low level of pests in their fields and accept a certain amount of economic damage from the pests. These farmers do not spray pesticides at the first sign of a pest. Instead, they periodically sample the pest population in the field to determine when the benefit of using pesticides exceeds the cost of that action.

Second, IPM requires that farmers be educated to understand what strategies will work best in their particular situations. Managing pests is more complex than trying to eradicate them. A farmer must know what pests to expect on each crop and what to do to minimize their effects.

Adoption of IPM by U.S. farmers has steadily increased since the 1960s, but the overall proportion of farmers using IPM is still small. One reason that IPM is not more widespread is that the knowledge required to use pesticides is relatively simple compared to the sophisticated knowledge needed to implement IPM.

a. Tools of Integrated Pest Management (IPM).

b. Pheromone Trap, a Tool of IPM.
This trap, which was placed in the middle of a pea crop, contains a pheromone to attract pea moth males. Note the dead males in the trap.

Summary

1 World Food Problems 348

1. **Undernutrition** is a serious underconsumption of calories or nutrients that leaves the body weakened and susceptible to disease. **Overnutrition** is a serious overconsumption of calories.

2. **Food insecurity**, in which people live with chronic hunger, is exacerbated by population growth, environmental problems, and poverty.

2 The Principal Types of Agriculture 351

1. **Industrialized agriculture** uses modern methods requiring large capital input and less land and labor than traditional methods. **Subsistence agriculture** requires labor and a large amount of land to produce enough food to feed a family.

2. There are three types of subsistence agriculture. In **shifting cultivation**, short periods of cultivation are followed by longer periods of fallow. In **nomadic herding**, carried out on arid land, herders move livestock continually to find food for them. **Intercropping** involves simultaneously growing a variety of plants on the same field.

3 Challenges of Agriculture 353

1. **Prime farmland** in the United States is being lost to urbanization and urban sprawl. Global declines in plant and animal varieties have led many countries to collect **germplasm**, plant and animal material that may be used in breeding. Farmers and ranchers strive to increase yields by administering **hormones** and **antibiotics** to livestock.

2. The **green revolution** introduced modern cultivation methods and high-yield crop varieties to Asia and Latin America. These methods require developing nations to import energy-intensive technologies and to face environmental problems caused by the use of inorganic fertilizers and pesticides.

3. Environmental problems caused by industrialized agriculture include air pollution from the use of fossil fuels and pesticides, water pollution from untreated animal wastes and agricultural chemicals, pesticide-contaminated foods and soils, and increased resistance of pests to pesticides. **Land degradation** decreases the future ability of the land to support crops or livestock. Clearing grasslands and forests and draining wetlands to grow crops have resulted in **habitat fragmentation**.

4 Solutions to Agricultural Problems 358

1. **Sustainable agriculture** uses methods that maintain soil productivity and a healthy ecological balance while minimizing long-term impacts.

2. **Genetic engineering**, the manipulation of genes to produce a particular trait, can produce more nutritious crops or crop plants that are resistant to pests, diseases, or drought. Concerns about genetic engineering include unknown environmental effects.

5 Controlling Agricultural Pests 362

1. A **pesticide** is any toxic chemical used to kill pests. Most pesticides are **broad-spectrum pesticides** that kill other organisms in addition to the intended pest.

2. Pesticides can effectively control disease-carrying organisms and crop pests.

3. Pesticide use leads to several problems: Pests develop **genetic resistance**, an inherited characteristic that decreases the effect of a given agent (such as a pesticide) on an organism; ecosystem imbalances occur when pesticides affect species other than the intended pests; and some pesticides exhibit persistence, degrading very slowly. **Bioaccumulation** is the buildup of a persistent pesticide in an organism's body. **Biological magnification** is the increased concentration of pesticide in organisms at higher levels in food webs. Pesticides also show mobility, moving to places other than where they were applied.

4. Alternatives to pesticides include **biological controls**, which use disease organisms, parasites, or predators to control pests. **Pheromones**, produced by animals to stimulate a response in other members of the same species, attract and trap pest species. **Integrated pest management** is a combination of pest control methods that keep a pest population small enough to prevent substantial economic loss.

Key Terms

- broad-spectrum pesticide 363
- degradation (of land) 356
- economic development 350
- food insecurity 348
- genetic engineering 360

- genetic resistance 363
- germplasm 354
- habitat fragmentation 356
- industrialized agriculture 351
- overnutrition 348

- pesticide 356
- subsistence agriculture 352
- sustainable agriculture 358
- undernutrition 348

What is happening in this picture?

This aerial view of a crop circle shows a center-pivot irrigation system that supplies water to a 99-acre corn field in Zambia. You can see the arc that was most recently watered at the 10 o'clock position on the field. The spots are nonfertile areas where crops do not thrive.

Center-pivot irrigation saves water compared to the traditional gravity-flow method of irrigation, in which water is directed to flow in ditches between crop rows; perhaps only 40 percent of the water reaches the crops using traditional irrigation methods.

- Use the Internet to investigate the low-energy, precision application (LEPA) sprinklers that allow 90 to 95 percent of the water in center-pivot irrigation to reach crops.

- Highly developed countries have most of the world's center-pivot irrigation systems. Why do you think this is the case?

- The wavy lines are small dikes constructed to control runoff because the field is not flat. Why do you think these dikes are constructed the way they are? How is using these dikes similar to contour plowing?

Global Locator

ZAMBIA

NATIONAL GEOGRAPHIC

Critical and Creative Thinking Questions

1. Compare undernutrition and overnutrition. Describe the types of people and the regions of the world most likely to be affected by each.

2. How and why do (a) population growth and (b) the rising consumption of meat affect world grain carryover stocks and world food security?

Use the following graph to answer questions 3 and 4.

School attainment and undernourishment by region, 2000

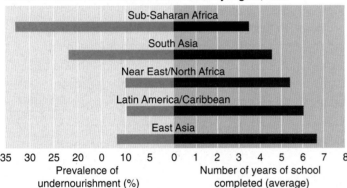

Sub-Saharan Africa
South Asia
Near East/North Africa
Latin America/Caribbean
East Asia

35 30 25 20 0 10 5 0 1 2 3 4 5 6 7 8
Prevalence of undernourishment (%)   Number of years of school completed (average)

3. Based on the graph, what conclusion can you make about how the level of education people attain in a region is related to that region's prevalence of undernutrition? Give examples of data from specific regions to support your conclusion.

4. How do these results fit in with our understanding of the role that poverty plays in the world's hunger problems?

5. Distinguish between shifting cultivation and slash-and-burn agriculture.

6. What are the environmental impacts of urban/suburban sprawl on agriculture land?

7. What was the green revolution? Describe a few of its successes and shortcomings.

8. Name two environmental problems associated with industrialized agriculture and give at least three examples of ways that industrialized agriculture could be made more sustainable.

9. Contrast the use of the following in industrialized and sustainable agriculture: water resources, fertilizers, and chemical pesticides.

10. How is developing new genetically modified crops and livestock promising in feeding the world's people? What are some concerns associated with growing genetically modified crops?

11. These charts compare the 2007 global acreage of genetically modified crops (screened areas) to the total production. How many acres of genetically modified soybeans were planted? cotton? corn?

Soybeans

64% of 225 million acres

Cotton

43% of 86 million acres

Corn

24% of 366 million acres

12. Which type of pesticide has fewer deleterious environmental effects, broad-spectrum or narrow-spectrum? Explain your answer.

13. Overall, do you think the benefits of pesticide use outweigh its disadvantages? Give at least two reasons for your answer.

14. Why does the heavy use of a particular pesticide eventually cause the pesticide to lose its effectiveness?

15. Is a major goal of integrated pest management (IPM) to eradicate a pest species? Explain your answer.

Biological Resources

15

DISAPPEARING FROGS

Since the 1970s, many of the world's frog populations have dwindled or disappeared. Almost 168 amphibian species are thought to have become extinct in the past 2 decades, and about one-third (1856) of all amphibian species are threatened, even in remote, pristine locations, such as the high reaches of California's Sierra Nevada Mountains (see large photograph).

Biologists cannot pinpoint a single cause of the massive die-offs and suspect they reflect a combination of threats. Habitat loss likely causes the greatest harm to these species, but chemical pollutants, infectious diseases, UV-B radiation, and climate change have all been implicated through scientific investigation.

Chytridiomycosis, a rapidly spreading disease caused by a chytrid fungus, had been recorded in 43 countries and in 36 U.S. states. Responsible for the disappearance of more than 100 amphibian species worldwide, and afflicting at least 200 species, this disease is thought to exacerbate the harm caused by other threats.

Rescue efforts around the globe include testing and treating animals affected by chytrids (see inset). Large-scale conservation programs, which include the captive breeding of vulnerable amphibian species, resulted in the establishment of the Amphibian Ark, an international group that seeks to preserve at least 500 species for reintroduction.

CHAPTER PLANNER ✔

- ❑ Study the picture and read the opening story.
- ❑ Scan the Learning Objectives in each section:
  p. 372 ❑   p. 376 ❑   p. 384 ❑   p. 388 ❑
- ❑ Read the text and study all figures and visuals. Answer any questions.

Analyze key features

- ❑ EnviroDiscovery, p. 378 ❑
- ❑ What a Scientist Sees, p. 379 ❑
- ❑ Environmental InSight, p. 381 ❑   p. 387 ❑
- ❑ Case Study, p. 391 ❑
- ❑ Stop: Answer the Concept Checks before you go on:
  p. 375 ❑   p. 384 ❑   p. 388 ❑   p. 390 ❑

End of Chapter

- ❑ Review the Summary and Key Terms.
- ❑ Answer What is happening in this picture?
- ❑ Answer the Critical and Creative Thinking Questions.

Species Richness and Biological Diversity

LEARNING OBJECTIVES

1. **Describe** factors associated with species richness.
2. **Define** *biological diversity* and distinguish among species richness, genetic diversity, and ecosystem diversity.
3. **Relate** several important ecosystem services provided by biological diversity.

A **species** is a group of distinct organisms that are capable of interbreeding with one another in the wild but that do not interbreed with organisms outside their group. We do not know exactly how many species exist. In fact, biologists now realize how little we know about Earth's diverse organisms.

How Many Species Are There?

Scientists estimate that there may be as few as 5 million or as many as 100 million different species inhabiting Earth. To date, about 1.8 million species have been scientifically named and described, including more than 330,000 plant species, 45,000 vertebrate animal species, and some 950,000 insect species. About 10,000 new species are identified each year.

Species richness varies greatly from one community to another. It is related to the abundance of potential ecological niches. A complex community, such as a

> **species richness**
> The number of different species in a community.

tropical rain forest or a coral reef, offers a greater variety of potential ecological niches than does a simple community such as mountain chaparral. A similar effect is seen in comparisons between more structurally complex and less structurally complex areas within the same type of habitat (**Figure 15.1**).

Species richness is inversely related to the geographic isolation of a community. Isolated island communities are much less diverse than communities in similar environments found on continents, for two reasons. Many species have difficulty reaching and successfully colonizing an island, and locally extinct species are not readily replaced in isolated environments such as islands or mountaintops.

Species richness is also inversely related to environmental stress. Only species capable of tolerating extreme environmental conditions can live in an environmentally stressed community, such as a polluted stream or a polar region exposed to a harsh climate. Species richness is also reduced when one species enjoys a decided dominance within a community because that species may appropriate a disproportionate share of resources, thus crowding out other species.

Species richness is usually greater at the edges of adjacent communities than in their centers. This is because an **ecotone**—a transitional zone where communities meet—contains all or most of the ecological niches of the adjacent communities as well as some niches unique to the ecotone. The change in species composition produced at ecotones is known as the **edge effect**.

Effect of community complexity on species richness • Figure 15.1

The structural complexity of chaparral vegetation in California (*x*-axis) is based on vegetation height and density, from low complexity (very dry scrub) to high complexity (woodland). Note that species richness in birds increases as vegetation becomes more structurally complex.

Interpreting Data
How many bird species are supported by the most structurally complex areas?

Geologic history greatly affects species richness. Tropical rain forests are probably old, stable communities that have undergone few climate changes over time, which allowed a large number of species to evolve. In contrast, glaciers have repeatedly altered temperate and arctic regions during Earth's history. An area recently vacated by glaciers will have a low species richness because few species will as yet have had a chance to enter it and become established.

Why We Need Biodiversity

The variation among organisms is referred to as **biological diversity**, or **biodiversity**. Biological diversity occurs at all levels of biological organization, from populations to ecosystems (Chapter 5). It takes into account three components: **genetic diversity**, the

> **biological diversity** The number and variety of Earth's organisms; consists of three components: genetic diversity, species richness, and ecosystem diversity.

genetic variety *within* all populations of that species; species richness; and **ecosystem diversity**, the variety of ecosystems found on Earth and the variety of interactions among organisms in natural communities (**Figure 15.2**). For example, a forest community—with its trees, shrubs, vines, herbs, insects, worms, vertebrate animals, fungi, bacteria, and other microorganisms—has greater ecosystem diversity than a cornfield.

Humans depend on the contributions of thousands of species for their survival. For example, insects are instrumental in several ecological and agricultural processes, including pollination of crops, weed control, and insect pest control. Bacteria and fungi provide us with foods, antibiotics and other medicines, and biological processes such as nitrogen fixation (see Chapter 5). However, relatively few species have been evaluated for their potential usefulness to humans. As many as 280,000 of the approximately 330,000 known plant species have yet to be assessed for industrial, medicinal, or agricultural potential. The same is true for most of the millions of microorganisms, fungi, and animals.

Ecosystem Services and Species Richness The

living world functions much like a complex machine. Each ecosystem is composed of many parts that are organized and integrated to maintain the ecosystem's overall performance. The activities of all organisms are

Levels of biodiversity • Figure 15.2

a. The variation in color patterns is evidence of the genetic diversity in the land snail species *Cepaea nemoralis*, common in western and central Europe.

b. Species richness in a European grassland or woodland is represented by the land snail species, the grasshopper, and the grass.

c. A scenic view in Switzerland reflects ecosystem diversity across central Europe.

Species Richness and Biological Diversity **373**

Alligators in the environment • Figure 15.3

The American alligator plays an integral role in its natural ecosystem. Alligators help maintain populations of smaller fishes by eating gar, a fish that preys on those smaller fishes. Alligators dig underwater holes that other aquatic organisms use during droughts when the water level is low. Their nest mounds eventually form small islands colonized by trees and other plants. The trees on these islands support heron and egret populations. The alligator habitat is maintained in part by underwater "gator trails," which help clear out aquatic vegetation that might eventually form a marsh.

interrelated; we depend on one another and on the physical environment, often in subtle ways (**Figure 15.3**). When one species declines, other species linked to it may either decline or increase in number.

Ecosystems supply human societies with many environmental benefits, or **ecosystem services** (**Table 15.1**). Forests are not just a source of lumber; they provide watersheds from which we obtain fresh water, limit the number and severity of local floods, and reduce soil erosion. Many flowering plant species depend on insects to transfer pollen for reproduction. Soil dwellers, from earthworms to bacteria, develop and maintain soil fertility for plants. Bacteria and fungi perform the crucial task of decomposition, which allows nutrients to cycle in the ecosystem. Conservationists maintain that ecosystems with greater species richness supply ecosystem services better than ecosystems with lower species richness.

> **ecosystem services**
> Environmental benefits, such as clean air, clean water, and fertile soil, provided by an ecosystem.

You might think that the loss of some species from an ecosystem would not endanger the rest of the organisms, but this is far from true. If enough species are removed, the entire ecosystem will change. Species richness within an ecosystem provides the ecosystem with resilience, the ability to recover from environmental changes or disasters.

Importance of Genetic Diversity

The maintenance of a broad genetic base is critical for the long-term health and survival of each species. Consider economically important crop plants. During the 20th century, plant scientists developed genetically uniform, high-yielding varieties of important food crops such as wheat. However, genetic uniformity resulted in increased susceptibility to pests and disease. Crossing the "super strains" with more genetically diverse relatives can

Ecosystem services • Table 15.1

Ecosystem	Services provided
Forests	Purify air and water
	Produce and maintain soil
	Absorb carbon dioxide (carbon storage)
	Provide wildlife habitat
	Provide humans with wood and recreation
Freshwater systems (rivers, lakes, and groundwater)	Moderate water flow and mitigate floods
	Dilute and remove pollutants
	Provide wildlife habitat
	Provide humans with drinking and irrigation water, food, transportation corridors, electricity, and recreation
Grasslands	Purify air and water
	Produce and maintain soil
	Absorb carbon dioxide (carbon storage)
	Provide wildlife habitat
	Provide humans with livestock and recreation
Coasts	Provide a buffer against storms
	Dilute and remove pollutants
	Provide wildlife habitat
	Provide humans with food, harbors, transportation routes, and recreation

 Interpreting Data
Which ecosystem service is provided by all of the listed ecosystems? How does each ecosystem contribute to pollution control?

allow disease and pest resistance to be reintroduced into such plants.

Genetic engineering, the incorporation of genes from one organism into a different species (see Chapter 14), makes it possible to use organisms' genetic resources on a wide scale. Genetic engineering has provided new vaccines, more productive farm animals, and disease-resistant agricultural plants.

Evolution has taken hundreds of millions of years to produce the genetic diversity found on our planet today. This diversity may hold solutions to today's problems and to future problems we have not begun to imagine. It would be unwise to allow such an important part of our heritage to disappear.

Medicinal, Agricultural, and Industrial Importance of Organisms

The genetic resources of organisms are vitally important to the pharmaceutical industry, which incorporates hundreds of chemicals derived from plants and other organisms into its medicines (**Figure 15.4**). Many of the natural products taken directly from marine organisms are promising anticancer or antiviral drugs. The AIDS (acquired immunodeficiency syndrome) drug AZT (azidothymidine), for example, is a synthetic derivative of a compound from a sponge. The 20 best-selling prescription drugs in the United States are either natural products, natural products that are slightly modified chemically, or synthetic drugs whose chemical structures were obtained from organisms.

The agricultural importance of plants and animals is indisputable because we must eat to survive. However, the number of different kinds of foods we eat is limited compared to the total number of edible species available in any given region. Many species are probably nutritionally superior to our common foods.

Modern industrial technology depends on a broad range of products from organisms. Plants supply oils and lubricants, perfumes and fragrances, dyes, paper, lumber, waxes, rubber and other elastic latexes, resins, poisons, cork, and fibers. Animals provide wool, silk, fur, leather, lubricants, waxes, and transportation, and they are important in medical research. The armadillo, for example, is used for research in Hansen's disease (leprosy) because it is the only species besides humans known to be susceptible to that disease. Insects secrete a large assortment of chemicals that represent a wealth of potential products. Certain beetles produce steroids

Medicinal value of the rosy periwinkle • Figure 15.4

The rosy periwinkle produces chemicals that are effective against certain cancers. Drugs from the rosy periwinkle have increased the chance of surviving childhood leukemia from about 5 percent to more than 95 percent. These leaves and flowers are being harvested in the Berenty Private Reserve, Madagascar.

with birth-control potential, and fireflies produce a compound that may be useful in treating viral infections. Biologists estimate that perhaps 90 percent of all insects have yet to be identified, and insects represent an important potential biological resource.

Aesthetic, Ethical, and Spiritual Value of Organisms

Organisms not only contribute to human survival and physical comfort, they provide recreation, inspiration, and spiritual solace. Our natural world is a thing of beauty largely because of the diversity of living forms found in it. Artists have attempted to capture this beauty in drawings, paintings, sculpture, and photography, and poets, writers, architects, and musicians have created works reflecting and celebrating the natural world.

CONCEPT CHECK   STOP

1. **What** are two determinants of species richness? Give an example of each.

2. **What** is biological diversity?

3. **What** are ecosystem services? Describe some ecosystem services that a forest provides.

Endangered and Extinct Species

LEARNING OBJECTIVES

1. **Define** *extinction* and distinguish between background extinction and mass extinction.
2. **Contrast** threatened and endangered species.
3. **Describe** four human causes of species endangerment and extinction.

xtinction, the death of a life form, occurs when the last member of a species dies. When a species is extinct, it will never reappear. Biological extinction is the fate of all species, much as death is the fate of all individuals.

> **extinction** The elimination of a species from Earth.

According to one estimate, for every 2000 species that have ever lived, 1999 of them are extinct today.

During the time in which organisms have occupied Earth, a continuous, low-level extinction of species, or **background extinction**, has occurred. A second kind of extinction, commonly referred to as **mass extinction**, has occurred perhaps five or six times throughout Earth's history. During a mass extinction event, a large number of species disappear during a relatively short period of geologic time.

The causes of past mass extinctions are not well understood, but biological and environmental factors were probably involved. A major climate change or a catastrophe such as a collision between a large asteroid or comet and Earth could have triggered a mass extinction.

Although extinction is a natural biological process, it is greatly accelerated by human activities. The burgeoning human population has spread into almost all areas of Earth. Whenever humans invade an area, the habitats of many organisms are disrupted or destroyed, which contributes to their extinction.

Earth's biological diversity is disappearing at an unprecedented rate (**Figure 15.5**). Conservation biologists estimate that species are now becoming extinct at a rate of 100 to 1000 times the natural rate of background extinctions. More than 34,000 plant species are currently threatened with extinction.

Endangered and Threatened Species

The Endangered Species Act legally defines an **endangered species** as a species in imminent danger of extinction throughout all or a significant portion of its **range**. (The area in which a particular species is found is its range.) A species is endangered when its numbers are so severely reduced that it is in danger of becoming extinct without human intervention. A species is legally defined as **threatened** when extinction is less imminent but its population is quite low and the species is likely to become endangered in the foreseeable future.

> **endangered species** A species that faces threats that may cause it to become extinct within a short period.
>
> **threatened species** A species whose population has declined to the point that it may be at risk of extinction.

Endangered and threatened species represent a decline in biological diversity because as their numbers decrease, their genetic variability is severely diminished. Long-term survival and evolution depend on genetic diversity, so a decline in genetic diversity heightens the risk of extinction for endangered and threatened species compared to species that have greater genetic variability.

Areas of Declining Biological Diversity

Declining biological diversity is a concern throughout the United States but is most serious in the states of Hawaii (where 63 percent of species are at risk) and California (where about 29 percent of species are at risk). At least two-thirds of Hawaii's native forests are gone.

As serious as declining biological diversity is in the United States, it is even more serious abroad, particularly in tropical rain forests. Tropical rain forests are being destroyed faster than almost all other ecosystems; approximately 1 percent of these ecosystems are being cleared or severely degraded each year. The forests are making way for human settlements, banana plantations, oil and mineral explorations, and other human activities.

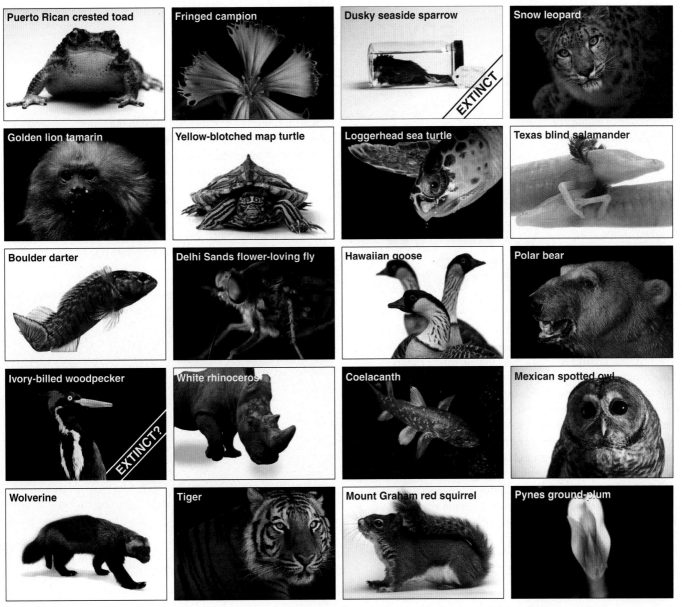

Representative endangered or extinct species from around the world • Figure 15.5

Officials at the U.S. Fish and Wildlife Service estimate that more than 500 U.S. species have become extinct during the past 200 years. Of these, roughly half have become extinct since 1980.

(For further discussion of tropical rain forests, see Chapters 6 and 13.)

Tropical rain forests are home to thousands or even millions of species. Many species in tropical rain forests are **endemic** (that is, they are not found anywhere else in the world), and the clearing of tropical rain forests contributes to their extinction.

endemic species
Organisms that are native to or confined to a particular region.

Perhaps the most unsettling outcome of tropical **deforestation** is its disruptive effect on evolution. In Earth's past, mass extinctions were followed over millions of years by the evolution of new species as replacements for those that died out. In the past, tropical rain forests may have supplied ancestral organisms from which other organisms evolved. Destroying tropical rain forests may reduce nature's ability to replace its species.

EnviroDiscovery

NATIONAL GEOGRAPHIC

Is Your Coffee Bird Friendly®?

Many species of migratory songbirds, favorites among North American bird lovers, are in decline, and Americans' coffee habits may play a role. In the tropics, high-yield farms cultivating coffee in full sunlight—known as *sun plantations*—are rapidly replacing traditional *shade plantations*. This switch is affecting wintering birds common to southern Mexico, the Caribbean, Costa Rica, and Colombia. Shade plantations grow coffee plants in the shade of tropical rainforest trees. These trees support a vast diversity of songbird species that winter in the tropics (one study counted 150 species in 5 hectares [12.4 acres]), as well as large numbers of other vertebrates and insects. In contrast, sun plantations provide poor bird habitat. Sun-grown varieties of coffee, treated with large inputs of chemical pesticides and fertilizers, outproduce the shade-grown varieties but lack the diverse products that come from the shade trees. About half of the region's shade plantations have been converted to sun plantations since the 1970s.

Songbird populations have declined alarmingly during this period. Researchers counted 94 to 97 fewer bird species on

The SMBC allows farmers of certified shade-grown and organic coffee to use this label on their product.

sun plantations in Colombia and Mexico than on shade-grown coffee plantations. Various conservation organizations and development agencies, such as the Smithsonian Migratory Bird Center (SMBC) and the U.S. Agency for International Development (USAID) have initiated programs to certify coffee as "shade grown," which allows consumers the chance to support the preservation of tropical rain forest. Shade-grown coffee typically costs more than sun-grown coffee because it is hand picked, involves more care in selecting only ripe beans, and is often certified organic.

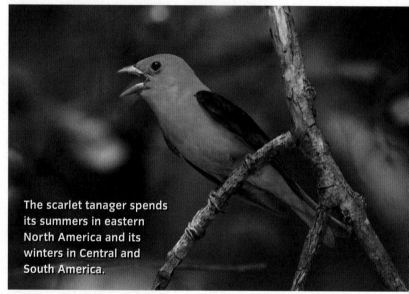

The scarlet tanager spends its summers in eastern North America and its winters in Central and South America.

Earth's Biodiversity Hotspots

In the 1980s ecologist Norman Myers of Oxford University coined the term **biodiversity hotspots**. In 2000, using plants as their criteria, Myers and ecologists at Conservation International identified 25 biological hotspots around the world (see *What a Scientist Sees*).

> **biodiversity hotspots** Relatively small areas of land that contain an exceptional number of endemic species and are at high risk from human activities.

Interestingly, these 25 hotspots for plants contain 29 percent of the world's endemic bird species, 27 percent of endemic mammal species, 38 percent of endemic reptile species, and 53 percent of endemic amphibian species. Many humans—nearly 20 percent of the world's population—live in the hotspots. Of the 25 hotspots, 15 are tropical, and 9 are mostly or solely islands. Many biologists recommend that conservation planners focus on preserving land in these hotspots to reduce the mass extinction of species that is currently under way.

Human Causes of Species Endangerment

Scientists generally agree that the single greatest threat to biological diversity is loss of habitat. Pollution, the spread of invasive species, and overexploitation are also important. Underlying these direct effects are increases

WHAT A SCIENTIST SEES

Where Is Declining Biological Diversity the Most Serious?

a. If you've visited a zoo and seen a black-and-white ruffed lemur, you may know that lemurs are found in nature only on Madagascar and the small neighboring Comoro Islands. However, most people do not realize that the ruffed lemur comes from a biological hotspot. (Photographed at Folsom Childrens' Zoo in Lincoln, Nebraska.)

b. Ecologists have identified Madagascar and neighboring islands as one of the world's 25 biological hotspots. These hotspots, which are rich in endemic species, are at great risk from human activities. All of Madagascar's 33 lemur species are currently in danger of extinction.

in human population, economic activity, and uses of technology, as well as social, political, and cultural factors (**Figure 15.6a**).

Habitat Destruction, Fragmentation, and Degradation

Most species facing extinction today are endangered because of the destruction, fragmentation, or degradation of habitats by human activities (**Figure 15.6b** and **c**; also see Figure 13.6b and c). We demolish or alter habitats when we build roads, parking lots, bridges, and buildings; clear forests to grow crops or graze domestic animals; and log forests for timber. We drain marshes to build on aquatic habitats, thus converting them to terrestrial ones, and we flood terrestrial habitats when we build dams. Exploration for and mining of minerals, including fossil fuels, also destroy habitats. Habitats are altered by outdoor recreation, including using off-road vehicles, hiking off-trail, golfing, skiing, and camping. Most organisms depend utterly on a particular type of environment, and habitat destruction reduces their biological range and ability to survive.

As the human population has grown, the need for increased amounts of food has resulted in a huge conversion of natural lands into croplands and permanent pastures. According to the U.N. Food and Agriculture Organization, total agricultural lands (croplands and pasturelands) currently occupy 38 percent of Earth's land area. Agriculture also has a major impact on aquatic ecosystems because it diverts water for irrigation. Little habitat remains for many endangered species. For example, human population growth and resource extraction have destroyed most of the grizzly bear's U.S. habitat—it now occupies about 2 percent of its original habitat in the lower 48 states.

Habitat fragmentation, the breakup of large areas of habitat into small, isolated patches (that is, islands), is a major threat to the long-term survival of many species. In ecological terms, *island* refers not only to any land mass surrounded by water (Figure 15.6b) but also to any isolated habitat surrounded by an expanse of unsuitable territory. Therefore, a small patch of forest surrounded by agricultural and suburban lands is considered an island (Figure 15.6c). Habitat destruction, fragmentation, and degradation are happening worldwide, causing the extinction of many species and reducing genetic diversity within many surviving species.

Africa provides a vivid example of the conflict over land use between growing human populations and other species. For example, African elephants require a large area of natural landscape in which to forage for the hundreds of kilograms of food that each consumes daily. People in Africa are increasingly pushing into the elephants' territory to grow crops and graze farm animals. The elephants often trample or devour crops, ruining a year's growth of crops in a single night; they have even killed people. Farmers cannot shoot at or kill elephants because they are a protected species. Researchers have found that elephants move out of areas when they become too crowded with people. Unfortunately, the wild areas to which elephants can move are steadily shrinking.

Pollution

Human-produced acid precipitation, stratospheric ozone depletion, and climate warming degrade even wilderness habitats that are considered natural and undisturbed. Acid precipitation is thought to have contributed to the decline of large stands of forest trees and the biological death of many freshwater lakes. Because ozone in the upper atmosphere shields the ground from a large proportion of the sun's harmful ultraviolet (UV) radiation, ozone depletion in the upper atmosphere represents a threat to all terrestrial life. Overwhelming evidence links climate warming, caused in part by an increase in atmospheric carbon dioxide released when fossil fuels are burned, to biodiversity declines. Further climate change is expected to increase the rate of extinction. Such habitat modifications reduce the biological diversity of species with particularly narrow and rigid environmental requirements. Other types of pollutants that affect organisms include industrial and agricultural chemicals, organic pollutants from sewage, acid mine drainage, and thermal pollution from heated industrial wastewater.

Invasive Species

The introduction of a non-native or foreign species into an ecosystem in which it did not evolve often upsets the balance among the organisms living in that area and interferes with the ecosystem's normal functioning. The foreign species may compete with native species for food or habitat or may prey on them. Generally, an introduced competitor or predator has a greater negative effect on local organisms than do native competitors or predators. Foreign species whose introduction causes economic or environmental

Interpreting Data
How might increased use of technology contribute to declining biological diversity?

a. Causes of Declining Biological Diversity.
In this highly simplified diagram, indirect causes (tan) interact with and amplify the effects of one another and of direct causes (gray). All of these factors interact in complex ways.

b. Destruction of the World's Wildlife Habitats.
This tiny island, located in the Panama Canal, was once a hilltop in a forest that was flooded when the Panama Canal was constructed.

c. Isolating Wildlife Habitats.
Roads and agricultural lands effectively isolate the scattered remnants, or "islands," of forest. Photographed in Paraná State, Brazil.

Invasive species • Figure 15.7

Selected examples of the many established foreign species accidentally or deliberately introduced into the United States.

Asian tiger mosquito | Caulerpa | European gypsy moth | Nutria

Sea lamprey | Zebra mussel | Puerto Rican frog | Brazilian pepper tree

Brown tree snake | Water hyacinth | Fire ant | European wild boar | Purple loosestrife

invasive species Foreign species that spread rapidly in a new area if free of predators, parasites, or resource limitations that may have controlled their population in their native habitat.

harm are called **invasive species** (**Figure 15.7**). Although invasive species may be introduced into new areas by natural means, humans are usually responsible for such introductions, either knowingly or unknowingly.

Cargo-carrying ocean vessels carry *ballast water* from their ports of origin to increase their stability in the ocean. When they reach their destinations, they discharge this water into local bays, rivers, or lakes. Ballast water may contain clams, mussels, worms, small fishes, and crabs, along with millions of microscopic aquatic organisms. These organisms, if they establish themselves, may threaten the area's aquatic environment and contribute to the extinction of native organisms.

One of North America's greatest biological threats—and its most costly aquatic invader—is the zebra mussel, a native of the Caspian Sea, probably introduced through

ballast water flushed into the Great Lakes by a foreign ship in 1985 or 1986. Since then, the tiny freshwater mussel has clustered in extraordinary densities on hulls of boats, piers, buoys, and, most damaging of all, water intake systems (**Figure 15.8**).

The zebra mussel's strong appetite for algae and zooplankton reduces the food supply of native fishes, mussels, and clams, threatening their survival. The U.S. Coast Guard estimates that economic losses and control efforts associated with the zebra mussel cost the United States about $5 billion each year.

Estimates of the number of non-native species now in the United States vary widely, but there may be as many as 50,000. Of these, approximately 4300 are considered invasive species. Worldwide, most regions are estimated to contain 10 percent to 30 percent foreign species.

Overexploitation Sometimes species become endangered or extinct as a result of deliberate efforts to eradicate or control their numbers. Many of these species prey on

Zebra mussels • Figure 15.8 _____

Zebra mussels have caused billions of dollars in damage, in addition to displacing native clams and mussels.

game animals or livestock. Ranchers, hunters, and government agents have reduced populations of large predators such as wolves and grizzly bears. Some animals are killed because they cause problems for humans. The Carolina parakeet, a beautiful green, red, and yellow bird endemic to the southern United States, was exterminated as a pest by farmers because it ate fruit and grain crops. It was extinct by 1920.

Prairie dogs and pocket gophers were poisoned and trapped so extensively by ranchers and farmers that between 1900 and 1960 they disappeared from most of their original geographic range. As a result of sharply decreased numbers of prairie dogs, the black-footed ferret, the natural predator of these animals, became endangered. A successful captive-breeding program has allowed black-footed ferrets to be reintroduced into the wild and reproduce successfully, though some populations have been decimated by disease.

Unregulated hunting, or overhunting, was a factor contributing to the extinction of certain species in the past but is now strictly controlled in most countries. The passenger pigeon was one of the most common birds in North America in the early 1800s, but a century of overhunting resulted in its extinction in the early 1900s. Unregulated hunting was one of several factors that caused the near extinction of the American bison.

Illegal commercial hunting, or **poaching**, endangers many larger animals, such as the tiger, cheetah, and snow leopard, whose beautiful furs are quite valuable. Rhinoceroses are slaughtered primarily for their horns, used for ceremonial dagger handles in the Middle East, and for purported medicinal purposes in Asian medicine. Bears are killed for their gallbladders, used in Asian medicine to treat ailments ranging from indigestion to heart problems. Endangered American turtles are captured and exported illegally to China, where they are killed for food. Caimans (reptiles similar to crocodiles) are killed for their skins and made into shoes and handbags. Although all these animals are legally protected, the demand for their products on the black market has led to their being hunted illegally.

In West Africa, poaching has contributed to the decline in lowland gorilla and chimpanzee populations. The meat (called *bushmeat*) of these rare primates and other protected species, such as anteaters, elephants, and mandrill baboons, provides an important source of protein for indigenous people. Bushmeat is also sold to urban restaurants. This demand for a meat source increases the incidence of poaching.

Live organisms collected through **commercial harvest** end up in zoos, aquaria, biomedical research laboratories, circuses, and pet stores. Several million birds are commercially harvested each year for the pet trade, but unfortunately many of them die in transit, and many more die from improper treatment after they are in their owners' homes. Although it is illegal to capture endangered animals from nature, there is a thriving black market, mainly because collectors in the United States, Canada, Europe, and Japan are willing to pay large amounts to obtain a variety of species, particularly rare tropical birds. At least 40 parrot species are

Illegal animal trade • Figure 15.9

These green parrots, captured illegally in the Amazon rain forest, are held for sale at a market in Iquitos, Peru.

now threatened or endangered, in part because of unregulated commercial trade (**Figure 15.9**).

Animals are not the only organisms threatened by excessive commercial harvest. Many unique and rare plants have been collected from nature to the point

that they are endangered. These include carnivorous plants, wildflowers, grasses, and ferns, certain cacti, and orchids. Such unregulated harvests benefit few individuals and contribute to species' decline, whereas carefully monitored and regulated commercial use of animal and plant resources can create an economic incentive to ensure that these resources do not disappear.

CONCEPT CHECK

1. **What** is background extinction? mass extinction?
2. **What** is the difference between a threatened species and an endangered species?
3. **How** do human activities cause species to become endangered or extinct?

Conservation Biology

LEARNING OBJECTIVES

1. **Define** *conservation biology* and compare in situ and ex situ conservation.

2. **Describe** restoration ecology.

> **conservation biology** The scientific study of how humans affect organisms and of the development of ways to protect biological diversity.

Studies in the field of **conservation biology** cover everything from the processes that influence biological diversity to the protection and restoration of endangered species and the preservation of entire ecosystems and landscapes.

Conservation biology includes two problem-solving techniques to save organisms from extinction: in situ and ex situ conservation. **In situ conservation**, which includes the establishment of parks and reserves, concentrates on preserving biological diversity in nature. With increasing demands on land, in situ conservation cannot guarantee the preservation of all types of biological diversity. Sometimes only ex situ conservation can save a species. **Ex situ conservation** involves conserving biological diversity in human-controlled settings. The breeding of captive species in zoos and the seed storage of genetically diverse plant crops are examples of ex situ conservation.

Protecting Habitats

Protecting animal and plant habitats—that is, conserving and managing the ecosystem as a whole—is the single best way to preserve biological diversity. Because human activities adversely affect the sustainability of many ecosystems, direct conservation management of protected areas is often required (**Figure 15.10**).

Currently, more than 3500 national parks, sanctuaries, refuges, forests, and other protected areas exist worldwide (see Figure 13.14). Protected areas are not always effective in preserving biological diversity. Many existing protected areas are too small or too isolated from other protected areas to efficiently conserve species. In developing countries where biological diversity is greatest, there is little money or expertise to manage them. Finally, many of the world's protected areas are in lightly populated mountain areas, tundra, and the driest deserts, places that often have spectacular scenery but relatively few species. In contrast, ecosystems in which biological diversity is greatest often receive little protection. Protected areas are urgently needed in tropical rain forests, deserts, the tropical grasslands and savannas of Brazil and Australia, many islands and temperate river basins, and dry forests all over the world.

Wildlife Refuges The **National Wildlife Refuge System**, established in 1903 by President Theodore Roosevelt, is the most extensive network of lands and waters committed to wildlife habitat in the world. The National Wildlife Refuge System contains more than 545 refuges, with at least 1 in each of the 50 states, and encompasses 38 million hectares (95 million acres) of land (see Figure 13.1). The refuges represent all major U.S. ecosystems, from tundra to temperate rain forest to desert, and are home to some of North America's most endangered species, such as the whooping crane. The mission of the National Wildlife Refuge System, which the U.S. Fish and Wildlife Service (FWS) administers, is to preserve lands and waters for the conservation of fishes, wildlife, and plants of the United States.

Restoring Damaged or Destroyed Habitats

When preserving habitats is not possible, scientists can reclaim disturbed lands and convert them into areas with high biological diversity. In **restoration ecology**, an important part of in situ conservation, ecological principles are used to help return a degraded environment to a more functional and sustainable one.

> **restoration ecology** The study of the historical condition of a human-damaged ecosystem, with the goal of returning it as closely as possible to its former state.

Restoration of disturbed lands creates biological habitats and provides additional benefits, such as the regeneration of soil damaged by agriculture or mining. The disadvantages of restoration include the expense and the time required to restore an area. Even so, restoration is an important aspect of in situ conservation, as restoration may reduce extinction.

Some challenges in conservation management • Figure 15.10

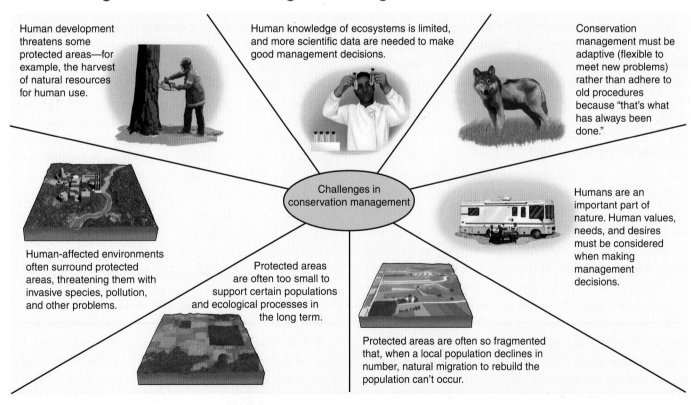

Human development threatens some protected areas—for example, the harvest of natural resources for human use.

Human knowledge of ecosystems is limited, and more scientific data are needed to make good management decisions.

Conservation management must be adaptive (flexible to meet new problems) rather than adhere to old procedures because "that's what has always been done."

Human-affected environments often surround protected areas, threatening them with invasive species, pollution, and other problems.

Challenges in conservation management

Protected areas are often too small to support certain populations and ecological processes in the long term.

Protected areas are often so fragmented that, when a local population declines in number, natural migration to rebuild the population can't occur.

Humans are an important part of nature. Human values, needs, and desires must be considered when making management decisions.

GLOBAL LOCAL What are some specific human activities that have likely harmed wildlife habitats around the world? in environments near you?

a. The University of Wisconsin–Madison Arboretum pioneered restoration ecology. The restoration of the prairie was at an early stage in November 1935.

b. The prairie as it looks today. This picture was taken at approximately the same location as the 1935 photograph.

Since 1934 the University of Wisconsin-Madison Arboretum has carried out one of the world's most famous examples of restoration ecology (**Figure 15.11**). Several distinct natural communities—including prairie and forests—were carefully developed on damaged agricultural land.

Conserving Species

Zoos, aquaria, botanical gardens, and other organizations practicing ex situ conservation often play critical roles in saving species on the brink of extinction. Eggs or seeds may be collected from nature, or the few remaining wild animals may be captured and bred in research environments (**Figure 15.12a**). But attempting to save a species approaching extinction is expensive, and only a small proportion of endangered species can be saved.

Conservation organizations are an essential part of the effort to maintain biological diversity through species and habitat conservation. These groups help educate policymakers and the public about the importance of biological diversity. In certain instances, they galvanize public support for important biodiversity preservation efforts. They provide financial support for conservation projects, from basic research to the purchase of land that is critical habitat for a particular species or group of species (**Figure 15.12b**).

Reintroducing Endangered Species to Nature

The ultimate goal of the captive-breeding programs practiced by zoos, aquaria, and other conservation organizations is to produce offspring in captivity and then release them into nature to restore wild populations. However, only 1 of every 10 reintroductions using animals raised in captivity is successful. Before attempting a reintroduction, conservation biologists conduct a feasibility study. This includes determining what factors originally caused the species to become extinct in nature, whether these factors still exist, and whether any suitable habitat still remains. Captive-breeding programs are sometimes unsuccessful because it is impossible to teach critical survival skills to animals raised in captivity.

Seed Banks More than 100 seed collections, called **seed banks**, or *gene banks*, exist around the world and collectively hold several million samples at low temperatures (**Figure 15.12c**). The Svalbard Global Seed Vault in Norway, opened in 2008, is designed to store as many as 4.5 million distinct seed samples. Seed banks offer the advantage of storing a large amount of plant genetic material in a small space. Seeds stored in seed banks are safe from habitat destruction, climate change and general neglect.

a. Captive Breeding.
Lucky (right) is the first whooping crane born to parents raised in captivity and then released. In late 2009 the captive population totaled 152 cranes, and the total number of wild cranes was 383.

b. Reforestation in Brazil.
These seedlings are to be used in Brazil's Atlantic Forest Restoration Project. The Nature Conservancy is facilitating the project, whose goal is to form a 60-km (37-mi) corridor of tropical forest linking two national parks in the Brazilian state of Bahia. The Atlantic Forest is widely recognized as one of the world's biodiversity hotspots.

WILEY
RESOURCE KIT

c. Seeds from a Seed Bank.
Shown are seeds to be stored at the Wakehurst Place Seed Bank in Sussex, England.

WILEY
RESOURCE KIT

Seed banks have some disadvantages. Seeds of many types of plants, such as avocados and coconuts, cannot be stored because they do not tolerate being dried out, a necessary step in freezing the seeds. Seeds do not remain alive indefinitely and must be germinated periodically so new seeds can be collected. Also, growing, harvesting, and returning seeds to storage is expensive. Perhaps the most important disadvantage of seed banks is that plants stored in this manner remain stagnant in an evolutionary sense. Thus, they may be less fit for survival when they are reintroduced into nature. Despite these shortcomings, seed banks are increasingly viewed as important because they safeguard seeds for future generations.

CONCEPT CHECK

1. **What** is conservation biology? When is in situ conservation used? ex situ conservation?
2. **What** are the goals of restoration ecology?

Conservation Policies and Laws

LEARNING OBJECTIVES

1. **Describe** the benefits and shortcomings of the U.S. Endangered Species Act.
2. **Relate** the purpose of the World Conservation Strategy.

In 1973 the **Endangered Species Act (ESA)** was passed in the United States, authorizing the FWS to protect endangered and threatened species in the United States and abroad. Many other countries now have similar legislation. International laws and policies also seek to conserve Earth's biological resources.

The Endangered Species Act

Currently, about 1300 species in the United States are listed as endangered or threatened (**Table 15.2, Figure 15.13**). The ESA provides legal protection to listed species to reduce their danger of extinction. The ESA requires the FWS to select critical habitats and design a detailed recovery plan for each species listed. The recovery plan includes an estimate of the current population size, an analysis of the factors contributing to its endangerment, and a list of activities to help the population recover.

The ESA was updated in 1982, 1985, and 1988. It is considered one of the strongest pieces of U.S. environmental legislation, in part because species are designated as endangered or threatened entirely on biological

U.S. organisms listed as endangered or threatened, January 2010 • Table 15.2

Type of organism	Number of endangered species	Number of threatened species
Mammals	70	15
Birds	74	16
Reptiles	13	24
Amphibians	14	11
Fishes	74	65
Snails	24	11
Clams	62	8
Crustaceans	19	3
Corals	0	2
Insects	47	10
Spiders	12	0
Flowering plants	573	145
Conifers	2	1
Ferns and other plants	24	2
Lichens	2	0
TOTAL	1010	313

Endangered species • Figure 15.13

The Florida panther is an endangered subspecies of cougar that exists in small pockets of isolated habitat in southern Florida.

on property rights. Another contentious issue is the financial cost of the law.

Some critics view the ESA as an impediment to economic progress, as when the timber industry was blocked from logging old-growth forests in certain parts of the Pacific Northwest to protect the habitat of the northern spotted owl. Those who defend the ESA point out that of 34,000 past cases of endangered species versus development, only 21 cases were not resolved through some sort of a compromise. When the black-footed ferret was reintroduced on the Wyoming prairie, for example, it was classified as an "experimental, nonessential species" so that its reintroduction would not block ranching and mining in the area. Thus, the ferret release program obtained the support of local landowners, support that was deemed necessary to the ferrets' survival in nature.

This type of compromise is crucial to the success of saving endangered species because, according to the U.S. General Accounting Office, more than 90 percent of endangered species live on at least some privately owned land. Some critics of the ESA think the law should be changed so that private landowners are given economic incentives to help save endangered species living on their lands. For example, tax cuts for property owners who are good land stewards could make the presence of endangered species on their properties an asset instead of a liability.

Defenders of the ESA agree that it is not perfect. Few endangered species have recovered enough to be delisted—that is, removed from protection of the ESA. However, the FWS says that hundreds of listed species are stable or improving; it expects as many as several dozen species to be delisted in the next decade or so.

Conservationists would like the ESA to be strengthened in such a way as to manage whole ecosystems and maintain complete biological diversity rather than attempt to save endangered species as isolated entities. This approach offers collective protection to many declining species rather than to single species.

International Conservation Policies and Laws

The **World Conservation Strategy**, a plan designed to conserve biological diversity worldwide, was formulated in 1980 by the International Union for the Conservation of Nature, the World Wildlife Fund, and the U.N.

grounds. Currently, economic considerations cannot influence the designation of endangered or threatened species. Biologists generally agree that fewer species have become extinct than would have if the ESA had not been passed.

The ESA is also one of the most controversial pieces of environmental legislation. The ESA does not provide compensation for private property owners who suffer financial losses because they cannot develop their land if a threatened or endangered species lives there. The ESA has also interfered with some federally funded development projects.

The ESA was scheduled for congressional reauthorization in 1992 but has been entangled since then in political wrangling between conservation advocates and supporters of private property rights. Conservation advocates think the ESA does not do enough to save endangered species, whereas those who own the land on which rare species live think the law goes too far and infringes

Environment Programme. In addition to conserving biological diversity, the World Conservation Strategy seeks to preserve the vital ecosystem services on which all life depends for survival and to develop sustainable uses of organisms and their ecosystems.

The Convention on Biological Diversity produced by the 1992 Earth Summit requires that each signatory nation inventory its own biodiversity and develop a **national conservation strategy**, a detailed plan for managing and preserving the biological diversity of that

Illegal trade in products made from endangered species • Figure 15.14

A merchant in Myanmar (Burma) deals in wildlife products.

specific country. As of 2010, 193 nations were participating in the Convention on Biological Diversity.

The exploitation of endangered species is somewhat controlled through legislation at the international level: 175 countries participate in the **Convention on International Trade in Endangered Species of Wild Flora and Fauna (CITES)**, which went into effect in 1975. Originally drawn up to protect endangered animals and plants considered valuable in the highly lucrative international wildlife trade, CITES bans the hunting, capturing, and selling of endangered or threatened species and regulates trade of organisms listed as potentially threatened. CITES protects at various levels more than 30,000 species of plants and animals. Unfortunately, enforcement of this treaty varies from country to country. Even where enforcement exists, the penalties are not severe. As a result, illegal trade continues in rare, commercially valuable species (**Figure 15.14**).

The goals of CITES often stir up controversy over issues such as who actually owns the world's wildlife and whether global conservation concerns take precedence over competing local interests. These conflicts often highlight socioeconomic differences between wealthy consumers of CITES products and poor people who trade the endangered organisms.

The case of the African elephant, discussed earlier in the chapter, is a good example of these controversies. Listed as an endangered species since 1989 to halt the slaughter of elephants driven by the ivory trade, the species seems to have recovered in southern Africa. However, poaching rebounded in the 2000s. According to wildlife specialists, by 2008 about 8 percent of the African elephant population was being killed for ivory each year. This level of slaughter is greater than the level in 1989, when the ban went into effect.

CONCEPT CHECK

1. **What** are the goals of the Endangered Species Act? Why is the ESA considered controversial?
2. **What** is the World Conservation Strategy?

Fighting the Asian Carp Invasion

Two species of Asian carp, bighead and silver, were introduced into the southern United States in the 1970s to maintain aquaculture ponds. Flooding in the 1990s released the species into the Mississippi River, where they have invaded rapidly and established self-sustaining populations.

The bighead carp is a large fish, reaching a maximum size of about 150 cm (60 in) and 50 kg (110 lb) (**Figure a**); the silver carp, a jumping species, grows to slightly more than half that size (**Figure b**). Both species are fast-growing filter feeders whose populations have doubled annually in portions of the upper Mississippi, Illinois, and Missouri Rivers.

Because the Illinois River connects the Mississippi River to Lake Michigan, scientists, government officials, and environmentalists are highly concerned that the Asian carp will reach the Great Lakes. The voracious appetites and rapid reproduction of these species make them a significant threat to the Great Lakes ecosystem, where the invading carp have no known predators. Asian carp could disrupt the food chain supporting the native species, which in turn could decimate the Great Lakes sport and commercial fisheries, valued at $4.5 billion per year.

The U.S. Army Corps of Engineers, Environmental Protection Agency, U.S. Fish and Wildlife Service, State of Illinois, and other parties have worked together to install electric barriers between the known Asian carp populations and the Chicago-area canal linking the Illinois River to Lake Michigan (**Figure c**). An experimental effort, the barrier may not be completely effective. As of early 2010, fragments of Asian carp DNA had been found in Lake Michigan, though no actual fish had been spotted there.

In December 2009 the State of Michigan filed a lawsuit with the U.S. Supreme Court, seeking the emergency closure of the shipping canal system to prevent the invasion of Asian carp into the Great Lakes. Fought by the State of Illinois and interests in the massive shipping industry in the region, the request was rejected by the Court in April 2010.

c. Waterways connecting Lake Michigan and the Illinois River, showing the electric barrier blocking carp.

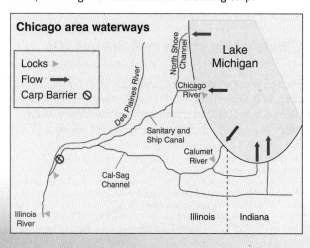

a. This 12-kg (26.5-lb) bighead carp caught in the Illinois River is one-fourth the maximum size of this species.

b. The silver carp is known to jump higher than 3 m (10 ft) when startled by boats, injuring and potentially killing boaters.

Summary

1 Species Richness and Biological Diversity 372

1. **Species richness** is the number of different species in a community. High species richness is associated with communities that are ecologically complex, not isolated, geologically old and stable, and not subject to environmental stress. Species richness is also higher when no one species dominates the community.

2. **Biological diversity** is the number and variety of Earth's organisms; it consists of three components: genetic diversity, species richness, and ecosystem diversity. **Genetic diversity** is the genetic variety within all populations of a given species. **Ecosystem diversity** is the variety of interactions among organisms in natural communities.

3. Ecosystems with greater species richness are better able to supply **ecosystem services**: environmental benefits such as clean air to breathe, clean water to drink, and fertile soil in which to grow crops.

2 Endangered and Extinct Species 376

1. **Extinction** is the elimination of a species from Earth. **Background extinction**, a continuous, low-level extinction of species, has occurred throughout Earth's history. **Mass extinction**, in which many species disappear during a relatively short period of geologic time, has occurred only a few times in Earth's history.

2. An **endangered species** is a species that faces threats that may cause it to become extinct within a short period. A species is defined as **threatened** when extinction is less imminent but its population is quite low.

3. Humans cause species endangerment through habitat destruction, fragmentation, and degradation; pollution; the spread of invasive species; and the overexploitation of biological resources. **Endemic species** are organisms that are native to or confined to a specific place. **Biodiversity hotspots** are areas that contain particularly high numbers of endemic species. **Invasive species** are foreign species, usually introduced by humans, that spread rapidly in a new area where they are free of predators, parasites, or resource limitations that may have controlled their population in their native habitat.

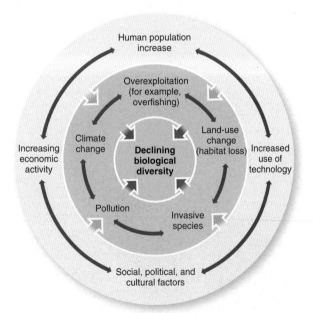

3 Conservation Biology 384

1. **Conservation biology** is the scientific study of how humans affect organisms and of the development of ways to protect biological diversity. **In situ conservation** includes the establishment of parks and reserves to preserve biological diversity in nature; **ex situ conservation** involves conservation of biological diversity in human-controlled settings such as zoos and seed banks.

2. **Restoration ecology** is the study of the historical condition of a human-damaged ecosystem, with the goal of returning it as closely as possible to its former state.

4 Conservation Policies and Laws 388

1. **The Endangered Species Act (ESA)** authorizes the U.S. Fish and Wildlife Service (FWS) to protect endangered and threatened species in the United States and abroad. The ESA requires the FWS to select critical habitats and design a detailed recovery plan for each species listed. Species are designated as endangered or threatened entirely on biological grounds, not economic factors. The act does not compensate private property owners who suffer financial losses related to its enforcement.

2. The **World Conservation Strategy**, formulated by the International Union for the Conservation of Nature, the World Wildlife Fund, and the U.N. Environment Programme, seeks to conserve biological diversity worldwide, to preserve vital ecosystem services, and to develop sustainable uses of organisms and their ecosystems.

Key Terms

- biodiversity hotspots 378
- biological diversity 373
- conservation biology 384
- ecosystem services 374
- endangered species 376
- endemic species 377
- extinction 376
- invasive species 382
- restoration ecology 385
- species richness 372
- threatened species 376

What is happening in this picture?

Habitat destruction, poaching, pollution, and other hazards caused populations of the California condor to decline dramatically. As shown here, researchers use puppets to rear captive condor chicks for release into the wild.

- How do you suppose the Endangered Species Act has assisted in the condor recovery effort?

- Explain what it means that the species is still listed as endangered.

Critical and Creative Thinking Questions

1. Is biological diversity a renewable or nonrenewable resource? Why could it be seen both ways?

2. List at least five important ecosystem services provided by living organisms.

3. Describe factors affecting species richness and explain how and why the species richness of a wheat field might differ from that of a coral reef.

4. What are the four main causes of species endangerment and extinction? Which cause do biologists consider most important?

5. What are invasive species?

6. Incorporate what you have learned about disappearing frogs and other declining species to compare threatened and endangered species and define *extinction*.

7. If you had the assets and authority to take any measure to protect and preserve biological diversity, but could take only one, what would it be?

8. The most recent version of the World Conservation Strategy includes stabilizing the human population. How would stabilizing the human population affect biological diversity?

9. In *A Sand County Almanac* and *Sketches Here and There*, Aldo Leopold wrote, "To keep every cog and wheel is the first precaution of intelligent tinkering." How does his statement relate to this chapter?

10. Why do you suppose the Svalbard Seed Vault in Norway is called the "Doomsday vault"? How is this name reflected in the vault's location, buried in a frozen hillside?

11. What are the goals of restoration ecology? Are there any disadvantages associated with restoring disturbed areas?

The Nature Conservancy evaluated the extent of human-caused habitat disturbance in the world's various biomes. Use the data in the graph below to answer questions 12 and 13.

12. On which global region have humans had the greatest impact—polar, temperate, or tropical? Suggest why human impact has been greatest in this region.

13. Which biome has had the lowest percentage of habitat disturbance? Suggest two possible reasons why human impacts on this biome may increase greatly in the future.

THE PLANNER

Solid and Hazardous Waste

REUSING AND RECYCLING OLD AUTOMOBILES

In the United States, about 35 million motor vehicles leave service each year. Most are exported to developing countries, but about 11 million cars and trucks are discarded (see photo of a Virginia salvage yard). Although 75 percent of a car can be reused as secondhand parts or recycled, the remaining 25 percent—glass, metals, plastics, fabrics, rubber, foam, and leather—usually ends up in landfills. Economics is an important aspect of the problem, because companies must make money as they reuse and recycle auto components.

About 37 percent of the iron and steel scrap reprocessed in the United States comes from old cars. According to the Environmental Protection Agency, recycling scrap iron and steel produces 86 percent less air pollution and 76 percent less water pollution than mining and refining an equivalent amount of iron ore.

Recycling plastic, which automakers use because it is lightweight and improves fuel efficiency, is one of the biggest challenges in auto recycling. No industry standards currently exist for plastic parts, so the kinds and amounts used in cars vary a great deal. As many as 15 plastics comprise some dashboards, and because many of these plastics are chemically incompatible, they cannot be melted together for recycling.

Auto manufacturers around the world have begun to address the challenge of reusing and recycling old cars. Japan and the European Union have mandated that by 2015, 95 percent of each discarded car must be recoverable. Toyota has developed a way to recover urethane foam and other shredded materials to make soundproofing material. Honda, Mercedes-Benz, Peugeot, Toyota, Volkswagen, Volvo, and other auto manufacturers have started to design cars so that each part of an old automobile can be reused or recycled (see inset of recyclable plastic and composite parts on a Toyota vehicle).

CHAPTER PLANNER ✓

- ❑ Study the picture and read the opening story.
- ❑ Scan the Learning Objectives in each section:
  p. 396 ❑   p. 402 ❑   p. 407 ❑   p. 410 ❑
- ❑ Read the text and study all figures and visuals. Answer any questions.

Analyze key features

- ❑ What a Scientist Sees, p. 399 ❑
- ❑ Process Diagram, p. 401 ❑   p. 406 ❑
- ❑ EnviroDiscovery, p. 402 ❑   p. 408 ❑
- ❑ Environmental InSight, p. 405 ❑
- ❑ Case Study, p. 413 ❑
- ❑ Stop: Answer the Concept Checks before you go on:
  p. 402 ❑   p. 406 ❑   p. 409 ❑   p. 412 ❑

End of Chapter

- ❑ Review the Summary and Key Terms.
- ❑ Answer What is happening in this picture?
- ❑ Answer the Critical and Creative Thinking Questions.

Solid Waste

LEARNING OBJECTIVES

1. **Distinguish** between municipal and nonmunicipal solid waste.

2. **Describe** the features of a modern sanitary landfill and relate some of the problems associated with sanitary landfills.

3. **Describe** the features of a mass burn incinerator and relate some of the problems associated with incinerators.

4. **Explain** the composting process.

T he United States generates more solid waste per capita than any other country. (Canada is a close second.) Each person in the United States produces an average of 2.1 kg (4.6 lb) of solid waste per day. This amount corresponded to a total of 257 million metric

> **municipal solid waste** Solid materials discarded by homes, offices, stores, restaurants, schools, hospitals, prisons, libraries, and other commercial and institutional facilities.

tons (283 million tons) in 2007. The problem worsens each year as the U.S. population increases.

Waste generation is an unavoidable consequence of the prosperous, high-technology, industrial economies of the United States and other highly developed nations. Many products that would be repaired, reused, or recycled in less affluent nations are simply thrown away. Nobody likes to think about solid waste, but it is certainly a concern of modern society—we keep producing it, and places to dispose of it safely are dwindling in number (**Figure 16.1**).

Types of Solid Waste

Municipal solid waste consists of the combined residential and commercial waste produced in a municipal area. Municipal solid waste is a heterogeneous mixture composed primarily

Empty plastic bottles • Figure 16.1

Global demand for bottled water is increasing, but most of these empty bottles are not recycled. The average American drinks 104 L (110 qt) of bottled water annually, which translates into 60 million plastic water bottles per day.

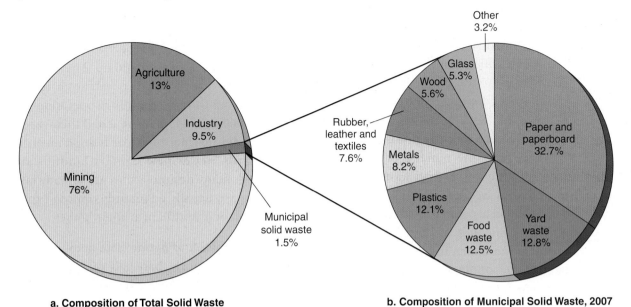

a. Composition of Total Solid Waste

b. Composition of Municipal Solid Waste, 2007

of paper and paperboard; yard waste; plastics; food waste; metals; rubber, leather, and textiles; wood; and glass (**Figure 16.2**). The proportions of the major types of solid waste in this mixture change over time. Today's solid waste contains more paper and plastics than in the past, whereas the amounts of glass and steel have declined.

Municipal solid waste makes up only a small proportion—less than 2 percent—of the total solid waste produced each day. **Nonmunicipal solid waste**, which includes mining, agricultural, and industrial wastes, is produced in substantially larger amounts. Most solid waste generated in the United States is from nonmunicipal sources.

nonmunicipal solid waste Solid waste generated by industry, agriculture, and mining.

Disposal of Solid Waste

Solid waste has traditionally been regarded as material that is no longer useful and should be disposed of. We can get rid of solid waste in four ways: dump it, bury it, burn it, or recycle or compost it (**Figure 16.3**).

Open Dumps The old method of solid waste disposal was dumping. Open dumps, which are now illegal, were unsanitary, malodorous places where disease-carrying vermin such as rats and flies proliferated. Methane gas was released into the surrounding air as microorganisms

decomposed the solid waste, and fires polluted the air with acrid smoke. Liquid oozed and seeped through the heaps of solid waste, often leaching hazardous materials that then contaminated soil, surface water, and groundwater.

U.S. disposal of municipal solid waste • Figure 16.3

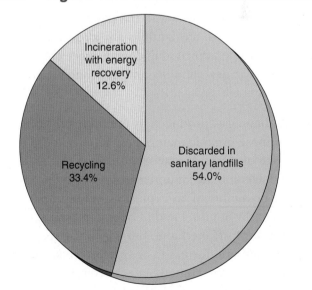

Think Critically In 2050, do you think these percentages will be the same or different? Why?

Sanitary landfills Open dumps have been replaced by **sanitary landfills**, which receive 54 percent of the municipal solid waste generated in the United States today. Sanitary landfills differ from open dumps in that the solid waste is placed in a hole, compacted, and covered with a thin layer of soil every day (see *What a Scientist Sees*). This process reduces the number of rats and other vermin usually associated with solid waste, lessens the danger of fires, and decreases the amount of odor. If a sanitary landfill is operated in accordance with solid waste management–approved guidelines, it does not pollute local surface water and groundwater. Layers of compacted clay and plastic sheets at the bottom of the landfill prevent liquid waste from seeping into groundwater.

> **sanitary landfill**
> The most common disposal site for solid waste, where waste is compacted and buried under a shallow layer of soil.

Newer landfills possess a double liner system (plastic, clay, plastic, clay) and use sophisticated systems to collect **leachate** (liquid that seeps through the solid waste) and gases that form during decomposition.

The choice of where to locate a sanitary landfill is based on a variety of factors, including the geology of the area, soil drainage properties, and the proximity of nearby bodies of water, wetlands, and groundwater. A landfill should be far enough away from centers of dense population so it is inoffensive but close enough so as not to require high transportation costs.

Although the operation of sanitary landfills has improved over the years with the passage of stricter and stricter guidelines, few landfills are ideal. Most sanitary landfills in operation today do not meet current legal standards for new landfills and encounter a variety of problems:

- Methane gas, produced by microorganisms that decompose organic material anaerobically (in the absence of oxygen), may seep through the solid waste and accumulate in underground pockets, creating the possibility of an explosion—even in basements of nearby homes. Landfill operators typically collect and burn off methane, but a growing number use the methane for gas-to-energy projects. About 425 landfills in the United States currently use methane gas to generate electricity.

- Leachate that seeps from unlined landfills or through cracks in the lining of lined landfills can potentially contaminate surface water and groundwater. Because even household trash contains toxic chemicals such as heavy metals, pesticides, and organic compounds, the leachate must be collected and treated to neutralize its negative effects.

- Landfills, by their nature, fill up. They are not a long-term remedy for waste disposal. From 1988 to 2007, the number of U.S. landfills in operation decreased from 8000 to 1754; many reached their capacity, and others did not meet state or federal environmental standards. Fewer new sanitary landfills are being opened; many desirable sites are already taken, and people are usually adamantly opposed to the construction of a landfill near their homes.

The special problem of plastic The amount of plastic in our solid waste, more than half of it from packaging, is growing faster than any other component of municipal solid waste. Most plastics are chemically stable and do not readily decompose. This characteristic, although essential in the packaging of products such as food and medicine, causes long-term problems: Most plastic debris disposed of in sanitary landfills will probably last for centuries. In response to concerns about the volume of plastic waste, some areas have banned the use of certain types of plastic, such as the polyvinyl chloride employed in packaging.

Special plastics that have the ability to degrade or disintegrate have been developed. Some of these are **photodegradable**—that is, they break down after being exposed to sunlight—which means they will not break down if buried in a sanitary landfill. Other plastics are **biodegradable**—they are decomposed by microorganisms such as bacteria. Whether biodegradable plastics actually break down under the conditions found in a sanitary landfill is not yet clear, although preliminary studies indicate that they probably do not. (Other waste management options for plastic are discussed later in this chapter.)

Incineration When solid waste is incinerated, two positive things are accomplished. First, the volume of solid waste is reduced by up to 90 percent: Ash is more compact than unburned solid waste. Second, incineration produces heat that can make steam to warm buildings or generate electricity. In 2007 the United States had 88 waste-to-energy incinerators, which produce substantially less carbon dioxide emissions than power plants that burn fossil fuels. (Recall from Chapter 9 that carbon dioxide is a potent greenhouse gas.)

WHAT A SCIENTIST SEES

Sanitary Landfills

a. A bulldozer compacts trash at a sanitary landfill in California. To the average person, a sanitary landfill is just a "dump."

b. Environmental engineers know that sanitary landfills constructed today have protective liners of compacted clay and high-density plastic and sophisticated leachate collection systems that minimize environmental problems such as groundwater contamination. Solid waste is spread in a thin layer, compacted into small sections called *cells*, and covered with soil.

61 cm (2 ft) final earth cover.

Final layer of soil seeded and planted.

Portable fences used to catch windblown debris.

Compacted trash covered daily with soil to keep out vermin and reduce odor.

Earth cover obtained by excavation.

Pipes sunk through compacted layers collect methane gas and either burn it off or convert it to energy.

Trash unloaded.

Trash pushed into place and compacted by bulldozers.

30 cm (1 ft) intermediate cover

Part of drainage grid: Perforated drains collect leachate for treatment.

Gravel

Original ground

Layers of compacted clay and high-density plastic liners protect groundwater.

Some materials are best removed from solid waste before incineration occurs. Glass does not burn, and when it melts, it is difficult to remove from the incinerator. Although food waste burns, its high moisture content often decreases incineration efficiency, so it is better to remove it before incineration. Removal of batteries, thermostats, and fluorescent lights is desirable because doing so eliminates most mercury emissions produced during combustion.

The best materials for incineration are paper, plastics, and rubber, all of which produce a lot of heat. Paper burns readily, and 1 kg (2.2 lb) of plastic waste yields almost as much heat as 1 kg of fuel oil.

Tires produce as much heat as coal and often generate less pollution. Some electric utilities in the United States and Canada burn tires instead of or in addition to coal (**Figure 16.4**). About 45 percent of all discarded tires are incinerated annually.

Most problems associated with incineration arise from the potential for environmental contamination:

- Incinerators pollute the air with carbon monoxide, particulates, heavy metals such as mercury, and other toxic materials, unless expensive air pollution control devices are used.

- Incinerators produce large quantities of ash, which must be disposed of properly. **Bottom ash**, or slag, is the ash left at the bottom of the incinerator when combustion is completed. **Fly ash** is the ash from the flue (chimney) that is trapped by air pollution control devices. Fly ash usually contains more toxic materials, including heavy metals and possibly dioxins, than bottom ash. Both types of incinerator ash are best disposed of in specially licensed hazardous waste landfills (discussed later in this chapter).

- As with sanitary landfills, site selection for incinerators is controversial. People may recognize the need for an incinerator, but they do not want it near their homes.

- Incinerators are expensive to run. Prices have escalated because costly pollution control devices are now required. Economic factors have also restricted construction of new plants.

The three types of incinerators are mass burn, modular, and refuse-derived fuel incinerators. Most mass burn incinerators are large and designed to recover the energy produced from combustion (**Figure 16.5**). **Modular incinerators** are smaller incinerators that burn all solid waste. Assembled at factories, they are less expensive to build. **Refuse-derived fuel incinerators** burn the combustible portion of

> **mass burn incinerator** A large furnace that burns all solid waste except for unburnable items such as refrigerators.

Tires that will be burned to generate electricity • Figure 16.4

This mountain in Westley, California, contains 4 to 6 million old tires. The power plant that burns them supplies electricity to 3500 homes. (The person wearing red gives a sense of scale.)

Mass burn, waste-to-energy incinerator • Figure 16.5

Modern incinerators have pollution control devices such as scrubbers and electrostatic precipitators to trap dangerous and dirty emissions.

WILEY
RESOURCE KIT

→ Path of smoke

⟹ Wastes to be treated or disposed of

❹ Steam piped to generate electricity or warm buildings.

❺ Pollution control devices trap dangerous and dirty emissions.

❻ Clean smoke

❷ Crane feeds waste into furnace.

❸ Smoke (hot air and fire ash)

Electrostatic precipitator

❶ Enclosed receiving area

Furnace

Scrubber

Boiler

Waste bunker

Ash conveyor

Bottom ash

Polluted water

Fly ash

Think Critically Why should bottom ash and flyash be disposed of in a hazardous waste landfill?

Hazardous waste landfill

Treated and reused

Hazardous waste landfill

---

solid waste. First, noncombustible wastes, such as glass and metals, are removed by machine or by hand. The remaining solid waste, including plastic and paper, is shredded or shaped into pellets and burned.

Composting Yard waste, such as grass clippings, branches, and leaves, is a substantial component of municipal solid waste (see Figure 16.2). As space in sanitary landfills becomes more limited, other ways to dispose of yard waste are being implemented.

One of the best recovery methods for yard waste is to convert it into soil conditioners such as compost or mulch (**Figure 16.6**). Food scraps, sewage sludge, and agricultural manure are other forms of solid waste that can be used to make compost. Compost provides nutrients to the soil and

Home composting of household waste • Figure 16.6

The drum of this composting bin can be rotated to mix the decomposing materials, thereby speeding up decomposition.

EnviroDiscovery

The U.S.–China Recycling Connection

Most of the materials that Americans recycle—from scrap metal to old cardboard boxes to used soda bottles—are redeveloped into products in the United States, but a growing amount are exported abroad. During the 2000s, China became the biggest importer of America's recyclable materials, collectively called *scrap*. When the scrap arrives in China, it becomes the raw materials for Chinese factories, paper mills, and steel mills. According to the Institute of Scrap Recycling Industries, scrap is now the third largest product exported from the United States to China, after airplanes and semiconductors.

China does not have the natural resource base that countries such as the United States are lucky to possess. To fuel its economic growth, China relies on scrap—used paper to replace its dearth of wood pulp and steel scrap to replace its dearth of iron ore. Some of the scrap shipped from the United States to China makes a round-trip, returning to the United States as auto parts, polyester shirts, and toys. Because Chinese workers are paid much less than U.S. workers, products made in China are generally less expensive for U.S. consumers than the equivalent domestic products. The economic downside of the U.S. scrap–Chinese product cycle is that reliance on Chinese imports reduces the number of jobs available in the United States.

---

reduces the need for fertilizers and pesticides. Compost and mulch are used for landscaping in public parks and playgrounds and as part of the daily soil cover at sanitary landfills. Compost and mulch are also sold to gardeners.

Composting as a means of managing solid waste first became popular in Europe. Many municipalities in the United States have composting facilities as part of their comprehensive solid waste management plans, and many states have banned yard waste from sanitary landfills. This trend is likely to continue, making composting even more desirable.

CONCEPT CHECK

1. **How** do municipal and nonmunicipal solid waste differ?
2. **What** are some features of sanitary landfills? What problems are associated with them?
3. **What** are the main features of a mass burn incinerator? What problems are associated with incinerators?
4. **How** do composters work?

Reducing Solid Waste

LEARNING OBJECTIVES

1. **Define** *source reduction.*
2. **Summarize** how source reduction, reuse, and recycling help reduce the volume of solid waste.
3. **Define** *integrated waste management.*

iven the problems associated with sanitary landfills and incinerators, it makes sense to do whatever we can to reduce the wastes we generate. The three goals of waste prevention, in order of priority, are (1) reduce the amount of waste as much as possible, (2) reuse products as much as possible, and (3) recycle materials as much as possible.

Reducing the amount of waste includes purchasing products that have less packaging and that last longer or are repairable (**Figure 16.7**). Consumers can also decrease their consumption of products to reduce waste. Before deciding to purchase a product, a consumer should ask, "Do I really *need* this product, or do I merely *want* it?"

Source Reduction

The most underutilized aspect of waste management is **source reduction**. Source reduction is accomplished in a variety of ways, including using raw materials that introduce less waste during manufacturing and reusing and recycling wastes at the plants where they are generated.

> **source reduction**
> An aspect of waste management in which products are designed and manufactured in ways that decrease the amount of solid and hazardous waste in the solid waste stream.

Innovation and product modifications play a key role. Consider aluminum cans: They are 35 percent lighter now than they were in the 1970s because less material is introduced into their manufacture. Dry-cell batteries are another example: They contain much less mercury today than they did in the early 1980s.

Dematerialization, the progressive decrease in the size and weight of a product as a result of technological improvements, is an example of source reduction only if the new product is at least as durable as the one it replaces. If the smaller, lighter product has a shorter life span and must be replaced more often, source reduction is not achieved.

Reusing Products

One example of reuse is refillable glass beverage bottles. Years ago, refillable beverage bottles were used a great deal in the United States. Today they are rarely used. For a glass bottle to be reused, it must be considerably thicker (and heavier) than a single-use bottle. With increased weight, transportation costs are higher. In the past, reuse of glass bottles made sense because there were many small bottlers scattered across the United States, minimizing transportation costs. Today there are approximately one-tenth as many bottlers. The centralization of bottling facilities makes it economically difficult to go back to the days of refillable bottles. Several other countries still reuse glass extensively, including Japan, Ecuador, Denmark, Finland, Germany, the Netherlands, Norway, Sweden, and Switzerland.

Recycling Materials

Many materials found in solid waste can be collected and reprocessed into new products. Recycling is preferred over landfill disposal because it conserves natural resources and is more environmentally benign. Every ton of recycled paper saves 17 trees, 26,500 L (7,000 gal) of water, 4,100 kilowatt-hours of energy, and 2.3 m³ (3 yd³) of landfill space. Recycling also has a positive effect on the economy by generating jobs and revenues from selling the recycled materials. Recycling does have environmental costs: Like all other human activities, it uses energy and generates pollution. For example, the de-inking process in paper recycling requires energy and produces a toxic sludge that contains heavy metals.

The many different materials in municipal solid waste must be separated before recycling. The separation of materials in items with complex compositions is difficult. Some food containers are composed of thin layers of metal foil, plastic, and paper, and trying to separate these layers is a daunting prospect.

The number of communities with recycling programs increased remarkably during the 1990s but leveled off somewhat in the early 2000s. Recycling programs include curbside collection, drop-off centers, buy-back programs, and deposit systems. In 2007 the annual recycling rate of aluminum and steel cans, plastic bottles, glass containers, newspapers, and cardboard was 170 kg (374 lb) per person.

Recyclables are usually sent to a materials recovery facility, where they are sorted and prepared for remanufacturing. Currently, the United States recycles about 32 percent of its municipal solid waste, including composting of yard trimmings; this percentage is higher than in most other highly developed nations (**Figure 16.8a**). (Recall, however, that the United States also generates more municipal solid waste than any other country.)

Most people think recycling involves merely separating certain materials from the solid waste stream, but that is only the first step. For recycling to work, there must be a market for the recycled goods, and the recycled products must be used in preference to virgin products. Prices paid by processors for recyclable materials vary significantly from one year to the next, depending largely on the demand for recycled products. In some places, recycling—particularly curbside collection—is not economically feasible.

Greater recycling generally occurs when the economy is strong, and the amount of recycling nationwide varies from year to year. Thus, although the amount of municipal solid waste generated is fairly constant, the amount of recycling varies.

Recycling Paper The United States currently recycles about 55 percent of its paper and paperboard. Many highly developed countries have higher recycling rates. Denmark, for example, recycles 97 percent of its paper. Part of the reason paper is not recycled more in the United States is that many older paper mills are not equipped to process waste paper. The number of mills that can process waste paper has increased in recent years, in part because of consumer demand. In addition to a slow increase in paper recycling in the United States, there is a growing demand for U.S. waste paper in other countries. China, Mexico, Taiwan, and Korea import large quantities of waste paper and cardboard from the United States.

Recycling Glass The United States currently recycles about 28 percent of its glass containers. Recycled glass costs less than glass made from virgin materials. Glass food and beverage containers are crushed to form **cullet**,

which glass manufacturers can use to make new products. Cullet is more valuable when glass containers of different colors are separated before being crushed (**Figure 16.8b**).

Recycling Metals The recycling of aluminum is one of the best success stories in U.S. recycling, largely because of economic factors (**Figure 16.8c**). Making a new aluminum can from a recycled one requires a fraction of the energy it would take to make a new can from raw metal (**Figure 16.8d**). According to the EPA, in 2007, about 49 percent of discarded aluminum beverage cans were recycled.

Other recyclable metals include lead, gold, iron and steel, silver, and zinc. For example, according to the Institute of Scrap Recycling Industries, new steel products contain an average of 56 percent recycled scrap steel. One of the obstacles to recycling metal products discarded in municipal solid waste is that their metallic compositions are often unknown. It is also difficult to extract metal from products such as stoves that contain materials besides metal (plastic, rubber, or glass, for instance).

Recycling Plastic About 12 percent of plastic containers and packaging in the United States is recycled. Depending on the economic situation, it is sometimes less expensive to make plastic from raw materials (petroleum and natural gas) than to recycle it. Some local and state governments support or require the recycling of plastic. According to the EPA, 37 percent of soft drink and water bottles are recycled annually to make such diverse products as carpet, automobile parts, tennis ball felt, and polyester cloth. One of the challenges associated with recycling plastic is that there are many different kinds. Forty-six different plastics are common in consumer products, and many products contain multiple kinds of plastic that must be separated or sorted before they can be recycled.

Recycling Tires Recycled tires are used for only a few products: retread tires; playground equipment; trash cans, garden hoses, and other consumer products; and rubberized asphalt for pavement. More recently, rubber from old tires has been used to make carpets, roofing materials, and molded products. Research in product

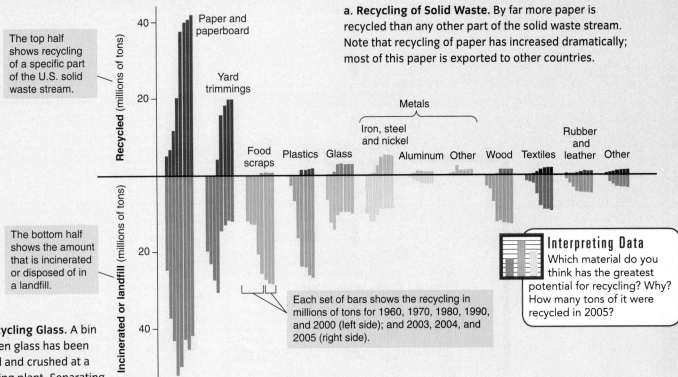

The top half shows recycling of a specific part of the U.S. solid waste stream.

The bottom half shows the amount that is incinerated or disposed of in a landfill.

Paper and paperboard

Yard trimmings

Food scraps

Plastics

Glass

Metals

Iron, steel and nickel

Aluminum

Other

Wood

Textiles

Rubber and leather

Other

Recycled (millions of tons)

Incinerated or landfill (millions of tons)

a. Recycling of Solid Waste. By far more paper is recycled than any other part of the solid waste stream. Note that recycling of paper has increased dramatically; most of this paper is exported to other countries.

Each set of bars shows the recycling in millions of tons for 1960, 1970, 1980, 1990, and 2000 (left side); and 2003, 2004, and 2005 (right side).

Interpreting Data
Which material do you think has the greatest potential for recycling? Why? How many tons of it were recycled in 2005?

b. Recycling Glass. A bin of green glass has been sorted and crushed at a recycling plant. Separating the different colors of glass in a cost-effective manner is challenging. Note in part **a** that the amount of glass recycling has not increased since 1990.

Used once
Recycled

Energy consumption (thousand kilocalories)

Glass bottles

Aluminum cans

c. Recycling Aluminum. Bales of crushed aluminum beverage cans are ready for processing. A typical aluminum can contains about 40 percent recycled aluminum. The amount of aluminum recycling is declining because of the increased use of plastic beverage containers.

d. Energy and Recycling. Comparison of energy required to manufacture glass bottles and aluminum cans that are either used once or recycled.

WILEY PLUS Video

Integrated waste management • Figure 16.9

Source reduction, reuse, recycling, and composting are part of integrated waste management, in addition to incineration and disposal in landfills.

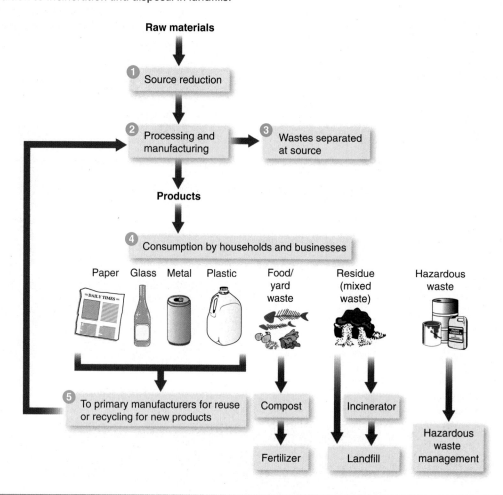

development continues, and almost all states now have tire-recycling programs. As noted earlier in this chapter, tires are also burned in waste-to-energy incinerators to produce electricity.

Integrated Waste Management

integrated waste management
A combination of the best waste management techniques into a consolidated program to deal effectively with solid waste.

The most effective way to deal with solid waste is with a combination of techniques. In **integrated waste management**, a variety of waste minimization methods, including the three Rs of waste prevention (reduce, reuse, and

recycle), are incorporated into an overall waste management plan (**Figure 16.9**). Even on a large scale, recycling and source reduction will not entirely eliminate the need for disposal facilities such as incinerators and landfills. However, recycling and source reduction will substantially reduce the amount of solid waste requiring disposal in incinerators and landfills.

CONCEPT CHECK

1. **What** is source reduction?
2. **How** do source reduction, reuse, and recycling reduce the volume of solid waste?
3. **What** is integrated waste management?

Hazardous Waste

LEARNING OBJECTIVES

1. **Define** *hazardous* waste.
2. **Briefly characterize** two types of hazardous waste: dioxins and PCBs.

Hazardous waste (also called **toxic waste**) accounts for about 1 percent of the solid waste stream in the United States. Hazardous waste includes dangerously reactive, corrosive, ignitable, or toxic chemicals. The chemicals may be solids, liquids, or gases. More than 700,000 different chemicals are known to exist. How many are hazardous is unknown because most have never been tested for toxicity, but without a doubt, there are thousands.

hazardous waste
A discarded chemical that threatens human health or the environment.

Hazardous waste has periodically been in the news since 1977. At that time it was discovered that toxic waste from an abandoned chemical dump had contaminated homes and possibly people in **Love Canal**, a small neighborhood on the edge of Niagara Falls, New York. Love Canal became synonymous with chemical pollution caused by negligent hazardous waste management. In 1978 it became the first location ever declared a national emergency disaster area because of toxic waste; more than 700 families were evacuated (**Figure 16.10**).

From 1942 to 1953 a local industry, Hooker Chemical Company, disposed of about 20,000 metric tons (22,000 tons) of toxic chemical waste in the 914-m-long (3000-ft-long) Love Canal. When the site was filled, Hooker added topsoil and donated the land to the local board of education. A school and houses were built on the site, which began oozing toxic waste several years later. More than 300 chemicals, many of them carcinogenic, have been identified in Love Canal's toxic waste.

In 1990, after almost 10 years of cleanup, the EPA and the New York Department of Health declared the area safe for resettlement. Today, the canal is a 16.2-hectare (40-acre) mound covered by clay and surrounded by a chain-link fence and warning signs.

The Love Canal episode resulted in the passage of the federal Superfund Act, which holds polluters accountable for the cost of cleanups (discussed shortly). Its passage generated immediate and ongoing concern about hazardous waste.

Aerial view of Love Canal toxic waste site in the early 1980s • Figure 16.10

All the homes shown in this photograph were evacuated and demolished.

Global Locator

LOVE CANAL

NORTH AMERICA

NATIONAL GEOGRAPHIC

EnviroDiscovery

NATIONAL GEOGRAPHIC

Handling Nanotechnology Safely

Nanotechnology is in the news a lot these days. **Nanomaterials**, which are unique materials and devices designed on the ultra-small scale of atoms or molecules, have numerous possible applications. For example, nanoparticles of cadmium selenide might be injected into cancerous tissue, where they would accumulate inside cancer cells; when exposed to ultraviolet radiation, these nanoparticles glow, and surgeons could more easily excise the cancerous tissues and leave the healthy tissues intact. Nanocrystals have the potential to be used in thin-film solar panels to convert solar energy to electricity. Silica nanoparticles embedded in glass make a heat-resistant glass capable of withstanding temperatures of up to 1800°C for several hours (see photograph)

Despite the potential of nanotechnology, particles on the nanometer scale (a nanometer is one-billionth of a meter) might pose health, safety, or environmental risks. No one knows for sure. The EPA has adopted a precautionary approach (see the section on the precautionary principle in Chapter 4) and decided to regulate nanomaterials that might adversely affect the environment. This means that the burden of proof about product safety will fall on companies that sell nanotechnology. Similarly, the Food and Drug Administration will have to oversee regulation of nanotechnology that has potential health and safety risks.

Nanotechnology

The glass between the man and the fire has fire-resistant properties because of the addition of silica nanoparticles.

Types of Hazardous Waste

Hazardous chemicals include a variety of acids, dioxins, abandoned explosives, heavy metals, infectious wastes, nerve gas, organic solvents, polychlorinated biphenyls (PCBs), pesticides, and radioactive substances. Many of these chemicals are discussed in other chapters; see Chapters 4, 8, 10, 14, and 17, which examine endocrine disrupters, air pollution, water pollution, pesticides, and radioactive waste. Here we discuss dioxins and PCBs because they are some of the most persistent hazardous compounds that contaminate our environment.

Dioxins **Dioxins** are a group of 75 similar chemical compounds formed as byproducts during the combustion of chlorine compounds. Incineration of medical and municipal wastes accounts for 70 to 95 percent of known human emissions of dioxins. Some other known sources of dioxins are iron ore mills, copper smelters, cement kilns, metal recycling, coal combustion, pulp and paper plants that use chlorine for bleaching (**Figure 16.11**),

Air pollution from a paper mill • Figure 16.11

These emissions may contain dioxins.

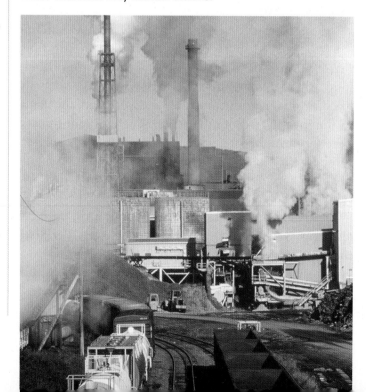

and chemical accidents. Motor vehicles, outdoor grills, and cigarette smoke emit minor amounts of dioxins. Forest fires and volcanic eruptions are natural sources of dioxins. Dioxins also form during the production of some pesticides.

Dioxins are emitted in smoke and then settle on plants, the soil, and bodies of water; from there they are incorporated into the food web. When humans and other animals ingest dioxins—primarily in contaminated meat, dairy products, and fish—they store and accumulate the dioxins in their fatty tissues (see the bioaccumulation and biomagnification discussion in Chapter 4). Because dioxins are so widely distributed in the environment, virtually everyone has dioxins in their body fat.

Dioxins cause several kinds of cancer in laboratory animals, but the data conflict on their cancer-causing ability in humans. A 2001 EPA report suggests that dioxins probably cause several kinds of cancer in humans and likely affect the human reproductive, immune, and nervous systems. Because dioxins are passed through human milk, nursing infants are considered particularly at risk.

PCBs Polychlorinated biphenyls (PCBs) are a group of 209 industrial chemicals composed of carbon, hydrogen, and chlorine. PCBs were manufactured in the United States between 1929 and 1979 for a wide variety of uses: as cooling fluids in electrical transformers, electrical capacitors, vacuum pumps, and gas-transmission turbines; and in hydraulic fluids, fire retardants, adhesives, lubricants, pesticide extenders, inks, and other materials. Prior to the EPA ban in the 1970s, PCBs were dumped in large quantities into landfills, sewers, and fields. Such improper disposal is one of the reasons PCBs are still a threat today.

The dangers of PCBs first became evident in Japan in 1968, where hundreds of people ate rice bran oil accidentally contaminated with PCBs and consequently experienced serious health problems, including liver and kidney damage. A similar mass poisoning tied to PCBs occurred in Taiwan in 1979. Since then, toxicity tests conducted on animals indicate that PCBs harm the skin, eyes, reproductive organs, and gastrointestinal system. PCBs are endocrine disrupters: They interfere with hormones released by the thyroid gland. Several studies have demonstrated that in utero exposure to PCBs can lead to certain intellectual impairments in children. PCBs may be carcinogenic; they are known to cause liver cancer in rats, and studies in Sweden and the United States have shown a correlation between high

Studying bacteria that break down PCBs in contaminated soil • Figure 16.12

A microbiologist adds soil to a "bioreactor" to test the ability of certain bacteria to treat contaminated soil. Note the three bioreactors in the foreground.

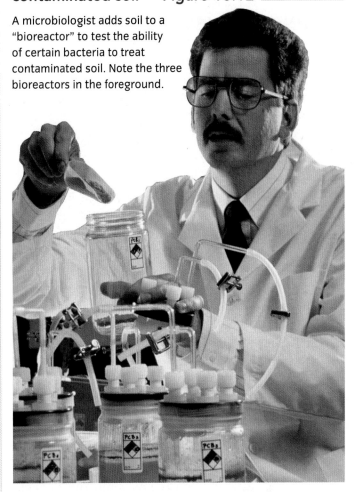

PCB concentrations in the body and incidences of certain cancers.

Although high-temperature incineration is one of the most effective ways to destroy PCBs in most solid waste, it is too costly to be used for the removal of PCBs that have leached into soil and water. One way to remove PCBs from soil and water is to extract them using solvents. This method is undesirable because the solvents themselves are hazardous chemicals, and these extraction methods are also costly.

Recently, researchers have discovered several bacteria that degrade PCBs at a fraction of the cost of incineration. Additional research is needed to make the biological degradation of PCBs practical (**Figure 16.12**).

CONCEPT CHECK

1. **What** is hazardous waste?

2. **What** are two sources of dioxins? of PCBs?

Managing Hazardous Waste

LEARNING OBJECTIVES

1. **Compare** the Resource Conservation and Recovery Act and the Comprehensive Environmental Response, Compensation, and Liability Act (the Superfund Act).
2. **Explain** how green chemistry is related to source reduction.

W e have the technology to manage toxic waste in an environmentally responsible way, but it is extremely expensive. Although great strides have been made in educating the public about the problems of hazardous waste, we have only begun to address many issues of hazardous waste disposal. No country currently has an effective hazardous waste management program, but several European countries have led the way by producing smaller amounts of hazardous waste and by using fewer hazardous substances.

Chemical Accidents

When a chemical accident occurs in the United States, whether at a factory or during the transport of hazardous chemicals, the National Response Center (NRC) is notified. The NRC assigns an on-scene coordinator to determine the size and chemical nature of the accident. Most chemical accidents reported to the NRC involve oil, gasoline, or other petroleum spills. The remaining accidents involve more than 1000 other hazardous chemicals, such as ammonia, sulfuric acid, and chlorine.

Chemical safety programs have traditionally stressed accident mitigation and adding safety systems to existing procedures. More recently, industry and government agencies have stressed accident prevention through the **principle of inherent safety**, in which industrial processes are redesigned to involve less toxic materials so that dangerous accidents are less likely to occur in the first place. The principle of inherent safety is an important aspect of source reduction.

Public Policy and Toxic Waste Cleanup

Currently, two federal laws dictate how hazardous waste should be managed: (1) the Resource Conservation and Recovery Act, which is concerned with managing hazardous waste being produced now, and (2) the Superfund Act, which provides for the cleanup of abandoned and inactive hazardous waste sites.

The **Resource Conservation and Recovery Act (RCRA)** was passed in 1976 and amended in 1984. Among other things, RCRA instructs the EPA to identify which wastes are hazardous and to provide guidelines and standards to states for hazardous waste management programs. RCRA bans hazardous waste from land disposal unless it is treated to meet the EPA's standards of reduced toxicity. In 1992 the EPA initiated a major reform of RCRA to expedite cleanups and streamline the permit system to encourage hazardous waste recycling.

In 1980 the **Comprehensive Environmental Response, Compensation, and Liability Act (CERCLA)**, commonly known as the **Superfund Act**, established a program to tackle the huge challenge of cleaning up abandoned and illegal toxic waste sites across the United States. At many of these sites, hazardous chemicals have migrated deep into the soil and have polluted groundwater. The greatest threat to human health from toxic waste sites comes from drinking water laced with such contaminants.

Cleaning Up Existing Toxic Waste: The Superfund Program The federal government estimates that the United States has more than 400,000 hazardous waste sites with leaking chemical storage tanks and drums (both above and below ground), pesticide dumps, and piles of mining waste. This estimate does not include the hundreds or thousands of toxic waste sites at military bases and nuclear weapons facilities.

By 2007, 10,753 sites were in the CERCLA inventory, which means the EPA had identified them as qualifying for cleanup (**Figure 16.13**). (This count does not include more than 1000 sites that had been cleaned up and removed from the CERCLA inventory since 1980.)

The sites posing the greatest threat to public health and the environment are placed on the **Superfund National Priorities List**, and the federal government will assist in their cleanup. As of 2007, 1556 sites were on the

National Priorities List. The five states that led the list in 2007 were New Jersey (115 sites), California (94 sites), Pennsylvania (94 sites), New York (86 sites), and Michigan (65 sites). The average cost of cleaning up a site is $20 million.

One reason for the urgency over cleaning up the sites on the National Priorities List is their locations. With the growth of cities and their suburbs, residential developments now surround many of the dumps. Because the federal government cannot clean up every old dump in the United States, the current landowner, prior owners, and anyone who has dumped waste on or transported waste to a particular site may be liable for cleanup costs.

Although critics decry the slow pace and high cost of cleaning up Superfund sites, the existence of CERCLA is a deterrent to further polluting. Companies that produce hazardous waste are now fully aware of the costs of liability and cleanup and are more likely to properly dispose of their hazardous wastes.

Managing Toxic Waste Production

The Superfund Act deals only with hazardous waste produced in the past, not the large amount of toxic waste produced today. There are three ways to manage hazardous waste: (1) source reduction, (2) conversion to less hazardous materials, and (3) long-term storage.

As with municipal solid waste, the most effective approach is source reduction—that is, using less hazardous or nonhazardous materials in industrial processes.

Source reduction relies on the increasingly important field of **green chemistry**.

For example, chlorinated solvents are widely used in electronics, dry cleaning, foam insulation, and industrial cleaning. To accomplish source reduction, it is sometimes possible to substitute a less hazardous water-based solvent for the toxic chlorinated one. Substantial source reduction of chlorinated solvents can also be accomplished by reducing solvent emissions. Installing solvent-saving devices benefits the environment and also saves money because smaller amounts of chlorinated solvents must be purchased. No matter how efficient source reduction becomes, however, it will never entirely eliminate hazardous waste.

The second-best way to deal with hazardous waste is to reduce its toxicity by chemical, physical, or biological means, depending on the nature of the waste. High-temperature incineration, for example, reduces dangerous compounds such as pesticides, PCBs, and organic solvents to safe products such as water and carbon dioxide. The resulting ash is hazardous and must be disposed of in a landfill designed for hazardous materials. Incineration using a *plasma torch* produces such high temperatures (up to 10,000°C, five times higher than temperatures of conventional incinerators) that hazardous waste is almost completely converted to nontoxic gases, such as carbon dioxide and nitrogen.

> **green chemistry**
> A subdiscipline of chemistry in which commercially important chemical processes are redesigned to significantly reduce environmental harm.

Cleaning up hazardous waste • Figure 16.13

a. Toxic waste in deteriorating drums at a site near Washington, DC. The metal drums in which much of the waste is stored have corroded and started to leak. Old toxic waste dumps are commonplace around the United States.

b. Cleanup of a hazardous waste site near Minneapolis, Minnesota. Removal and destruction of the wastes are complicated by the fact that usually nobody knows what chemicals are present.

Cutaway view through a hazardous waste landfill • Figure 16.14

The bottom of this hazardous waste landfill has two layers of compacted clay, each covered by a high-density plastic liner. (Some hazardous waste landfills have three layers of compacted clay.) A drain system located above the plastic and clay liners allows liquid leachate to collect in a basin where it can be treated, and a leak detection system is installed between the clay liners. Barrels of hazardous waste are placed above the liners and covered with soil.

Native vegetation

1.2 m (4 ft) of soil

Biota layers (crushed concrete)

Synthetic clay/plastic liner

Soil

Stacked barrels of hazardous waste

Protective soil layer (0.9 to 1.2 m, or 3 to 4 ft)

Collection system for leachate

Detection system for leachate

High-density plastic liner

0.9 m (3 ft) of compacted clay liner

Barrels of hazardous waste are placed above the liners and covered with soil.

A drain system located above the liners allows liquid leachate to collect in a basin where it can be treated.

The bottom of this hazardous waste landfill has two or three layers of compacted clay, each covered by a high-density plastic liner.

GLOBAL LOCAL Would it be more likely to find this type of hazardous waste landfill in the United States or Russia? Why?

Hazardous waste that is not completely detoxified must be placed in long-term storage. Hazardous waste landfills are subject to strict environmental criteria and design features. They are located as far as possible from aquifers, streams, wetlands, and residences. Such a landfill includes several layers of compacted clay and high-density plastic liners at the bottom of the landfill to prevent leaching of hazardous substances into surface water and groundwater (**Figure 16.14**). Leachate is collected and treated to remove contaminants. The entire facility and nearby groundwater deposits are carefully monitored to make sure there is no leakage.

Some toxic liquid waste, such as explosives and pesticides, is disposed of by *deep-well injection*. In this technique, the liquid waste is injected through pipes that extend hundreds of meters into an injection zone located between two impermeable areas. In such geologic formations, the waste is not likely to migrate into aquifers that could be used for drinking or irrigation.

CONCEPT CHECK

1. **How** are the Resource Conservation and Recovery Act and the Comprehensive Environmental Response, Compensation, and Liability Act alike and what is the focus of each act?

2. **How** is green chemistry applied to reducing sources of hazardous waste?

High-Tech Waste

In the United States and other highly developed countries, the average computer is replaced every 18 to 24 months, not because it is broken but because rapid technological developments and new generations of software make it obsolete. Old computers may still be in working order, but they have no resale value and are even difficult to give away. As a result, they sit in warehouses, garages, and basements—or are frequently thrown away with the trash. According to the EPA, more than 300 million electronic devices were discarded in 2006 (latest data available), and 80 percent of these were disposed of in sanitary landfills.

This disposal represents a huge waste of the high-quality plastics and metals (aluminum, copper, tin, nickel, palladium, silver, and gold) that make up computers. Computers also contain the toxic heavy metals lead, cadmium, mercury, and chromium, which could potentially leach from landfills into soil and groundwater. Most computers contain 1.4 to 3.6 kg (3 to 8 lb) of lead, for example. Several states have passed legislation requiring businesses and residents to **e-cycle** consumer electronics—that is, recycle PCs, color monitors, cell phones, and color televisions.

Currently, about 20 percent of discarded U.S. computer and electronic components are e-cycled. Although some companies handle obsolete computers in the United States (see photograph), many U.S. computers are shipped overseas to be recycled in developing countries such as India, Pakistan, and China. There the computers are disassembled, often using methods that are potentially dangerous to the workers taking them apart. For example, circuit boards are often burned to obtain the small amount of gold in them, and burning releases hazardous fumes into the air.

Some highly developed countries have been more progressive than the United States in dealing with their computer waste. The European Union implemented a Waste Electrical and Electronic Equipment plan in 2005 to recover, recycle, and dispose of electronic waste and remove some of the most hazardous chemicals. Japan and other industrialized nations have put in place similar policies.

Obsolete computer equipment

The computer monitors at this Texas electronic recycler are being disassembled. Functional tubes will be exported to Thailand, where they will be used to manufacture inexpensive televisions. Broken tubes will be recycled or disposed of in the United States.

NATIONAL GEOGRAPHIC

Summary

1 Solid Waste 396

1. **Municipal solid waste** consists of solid materials discarded by homes, office buildings, stores, restaurants, schools, hospitals, prisons, libraries, and other facilities. **Nonmunicipal solid waste** consists of solid waste generated by industry, agriculture, and mining.

2. Use of **sanitary landfills** is the most common method of solid waste disposal, involving compacting and burying waste under a shallow layer of soil. Layers of compacted clay and plastic sheets prevent **leachate** (liquid waste) from seeping into groundwater. Problems with sanitary landfills include the potential for methane gas to seep out and cause explosions, the accidental leaking of toxic leachate, a lack of existing landfill space, and resistance to new landfills near homes and businesses.

3. A **mass burn incinerator** is a large furnace that burns all solid waste except for unburnable items such as refrigerators. Problems associated with incineration of solid waste include the potential for air pollution, difficulties in disposing of the toxic ash produced, the high costs of the process, and difficulties in choosing incinerator sites.

4. In composting, yard waste, food scraps, and other organic wastes are transformed by microbial action into a material that, when added to soil, improves its condition.

2 Reducing Solid Waste 402

1. In **source reduction**, products are designed and manufactured in ways that decrease the volume of solid waste and the amount of hazardous waste in the solid waste stream.

2. The volume of solid waste produced can be decreased through source reduction, reuse of products, and recycling of materials. Recycling conserves natural resources and is more environmentally benign than landfill disposal but requires a market for the recycled goods.

3. **Integrated waste management** is a combination of the best waste management techniques into a consolidated program to deal effectively with solid waste.

3 Hazardous Waste 407

1. **Hazardous waste** is a discarded chemical that threatens human health or the environment. Hazardous chemicals may be solids, liquids, or gases and include a variety of acids, dioxins, abandoned explosives, heavy metals, infectious wastes, nerve gas, organic solvents, PCBs, pesticides, and radioactive substances.

2. **Dioxins** are hazardous chemicals formed as unwanted byproducts during the combustion of many chlorine compounds. **Polychlorinated biphenyls (PCBs)** are hazardous, oily, industrial chemicals composed of carbon, hydrogen, and chlorine.

4 Managing Hazardous Waste 410

1. The **Resource Conservation and Recovery Act (RCRA)** instructs the EPA to identify hazardous waste and to provide guidelines and standards for states' hazardous waste management programs. The **Comprehensive Environmental Response, Compensation, and Liability Act (CERCLA)**, or **Superfund Act**, established a program whose goal is to clean up abandoned and illegal toxic waste sites across the United States.

2. The most effective approach to managing hazardous waste is source reduction, reducing the amount and toxicity of hazardous materials used in industrial processes. Source reduction relies on **green chemistry**, a subdiscipline of chemistry in which commercially important chemical processes are redesigned to reduce environmental harm.

Key Terms

- green chemistry 411
- hazardous waste 407
- integrated waste management 406
- mass burn incinerator 400
- municipal solid waste 396
- nonmunicipal solid waste 397
- sanitary landfill 398
- source reduction 403

What is happening in this picture?

This clothing market in Tanzania sells used clothing from Europe and America.

• How does this practice affect the volume of solid waste produced?

• These clothes, which come from nonprofit charities such as Goodwill, are sold in local marketplaces. How does this practice affect African clothing industries?

• Africa now exports about $1.6 billion of its textiles and clothing to the United States each year. However, Africa's main competition for the U.S. market is China and other Asian countries. Explain this global connection.

Critical and Creative Thinking Questions

1. Compare the advantages and disadvantages of disposing of municipal solid waste in sanitary landfills and by incineration.

2. How could source reduction efforts reduce the volume of waste that arises from abandoned automobiles?

3. List what you think are the best ways to treat each of the following types of solid waste and explain the benefits of the processes you recommend: paper, plastic, glass, metals, food waste, and yard waste.

4. What are dioxins, and how are they produced? What harm do they cause?

5. Suppose hazardous chemicals were suspected to be leaking from an old dump near your home. Outline the steps you would take to (1) have the site evaluated to determine whether there is a danger and (2) mobilize the local community to get the site cleaned up.

6. What are the goals, strengths, and weaknesses of the Superfund Act?

7. What is integrated waste management? Why must a sanitary landfill always be included in any integrated waste management plan?

8. What is hazardous waste, and how does green chemistry help address the problem of hazardous waste?

In an effort to reduce municipal solid waste, many communities have required customers to pay for garbage collection according to the amount of garbage they generate, an approach termed "unit pricing," or "pay as you throw." The figure below applies to questions 9 and 10 and illustrates the effects of unit pricing in San Jose, California, on garbage sent to landfills and on wastes diverted through recycling and through separation of yard wastes.

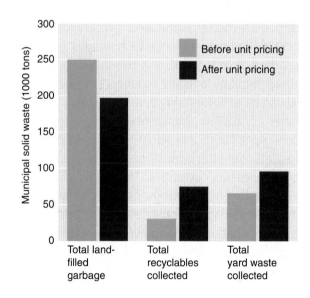

9. How did the implementation of unit pricing in San Jose affect the amount of garbage sent to landfills?

10. How did the implementation of unit pricing affect the quantity of materials recycled or of yard wastes collected?

Nonrenewable Energy Resources

COAL AND THE ENVIRONMENT

About 300 years ago, industrialists discovered that burning coal could heat water to drive steam turbines. That discovery allowed the Industrial Revolution to occur, transforming almost all aspects of life for people, first in Europe, then in the Americas, and now around the word. The downsides of coal—mining dangers, air and water pollution, and solid waste—have always been apparent. Through the centuries, as coal use has increased, people have developed technologies to manage these problems. Nonetheless, as long as coal remains a major energy source, we will constantly struggle with its many environmental consequences.

In many parts of the world, coal is burned using old technology, leading to air, water, and ground contamination. Further, while newer technology can remove particulate matter, sulfur, and mercury from the exhaust of coal-fired power plants, technology to eliminate carbon dioxide is just now being developed. Even where technology is highly advanced, large-scale use of coal can have devastating consequences. On December 22, 2008, a pond containing about 4 million m³ (140 ft³) of fly ash slurry—water mixed with the ash left over after coal is burned—broke open in Kingston, Tennessee, covering 1.2 km² (0.5 mi²) of the surrounding area (see photograph; the bluish-gray material is fly ash slurry). The spill destroyed houses and roads, and it contaminated the Emory and Clinch Rivers, which feed the Tennessee River, the Ohio River, and eventually the Mississippi River before reaching the ocean.

Line represents the approximate boundary of the original pond

Original pond

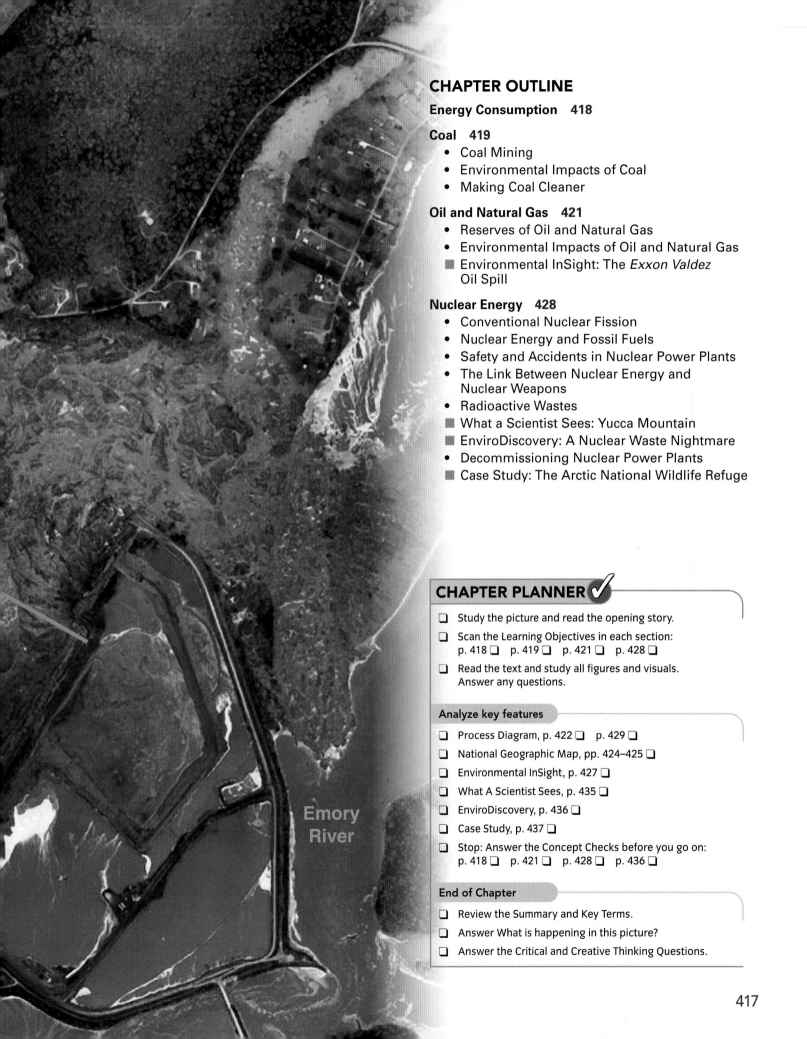

CHAPTER OUTLINE

Emory
River

CHAPTER PLANNER ✔

- ❏ Study the picture and read the opening story.
- ❏ Scan the Learning Objectives in each section:
  p. 418 ❏ p. 419 ❏ p. 421 ❏ p. 428 ❏
- ❏ Read the text and study all figures and visuals. Answer any questions.

Analyze key features

- ❏ Process Diagram, p. 422 ❏ p. 429 ❏
- ❏ National Geographic Map, pp. 424–425 ❏
- ❏ Environmental InSight, p. 427 ❏
- ❏ What A Scientist Sees, p. 435 ❏
- ❏ EnviroDiscovery, p. 436 ❏
- ❏ Case Study, p. 437 ❏
- ❏ Stop: Answer the Concept Checks before you go on:
  p. 418 ❏ p. 421 ❏ p. 428 ❏ p. 436 ❏

End of Chapter

- ❏ Review the Summary and Key Terms.
- ❏ Answer What is happening in this picture?
- ❏ Answer the Critical and Creative Thinking Questions.

Energy Consumption

LEARNING OBJECTIVE

1. **Compare** per person and total energy consumption in highly developed and developing countries.

Human society depends on energy. We use it to warm our homes in winter and cool them in summer; to grow, store, and cook our food; to light our homes; to extract and process natural resources for manufacturing items we use daily; and to power various forms of transportation. Many of the conveniences of modern living depend on a ready supply of energy.

Per person energy consumption in highly developed countries far exceeds that in developing countries (**Figure 17.1**). In the United States, industry uses

Energy consumption • Figure 17.1 _____

a. Annual per person commercial energy consumption in selected countries.

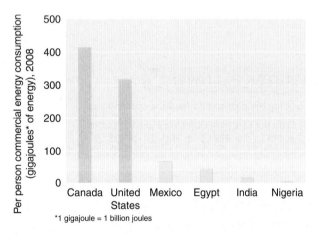

*1 gigajoule = 1 billion joules

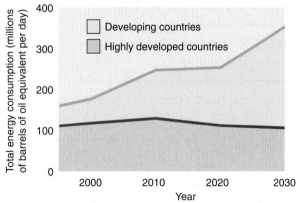

b. Projected total energy consumption, to 2030.

Demand for energy in India • Figure 17.2 ____

Powerlines criss-cross above buildings in Delhi.

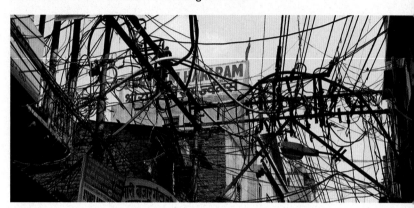

31 percent of the nation's total energy, buildings such as homes and offices consume 41 percent, and transportation uses 28 percent. In developing countries, a much larger fraction is used at the household level.

World energy consumption has increased every year since 1982, with most of the increase occurring in developing countries. From 2007 to 2008, for example, energy consumption increased worldwide by about 2.4 percent, most of it in China and India (**Figure 17.2**). A goal of most developing countries is to improve the standard of living through economic development, a process usually accompanied by a rise in per person energy consumption. Furthermore, the world's energy requirements will continue to increase during the 21st century, as the human population becomes larger, particularly in developing countries.

In contrast, population size in highly developed nations is more stable. Per person energy consumption in such countries as Japan and Germany is decreasing as technological advances increase the energy efficiency of appliances, automobiles, and home insulation (see Chapter 18).

CONCEPT CHECK

1. **How** does per person energy consumption compare in highly developed and developing countries?

Coal

LEARNING OBJECTIVES

1. **Distinguish** between surface mining and subsurface mining.

2. **Summarize** the environmental problems associated with using coal.

3. **Describe** two technologies that can be used to make coal a cleaner fuel.

oal, the most abundant fossil fuel in the world, is found primarily in the Northern Hemisphere (**Figure 17.3**). The largest coal deposits are in the United States, Russia, China, Australia, India, Germany, and South Africa. The United States has 25 percent of the world's coal supply in its massive deposits. According to the World Resources Institute, known world coal reserves could last for more than 200 years at the present rate of consumption. Coal resources currently too expensive to develop have the potential to provide enough coal to last for 1000 or more years at current consumption rates.

Utility companies use coal to produce electricity, and heavy industries use coal for steel production. Coal consumption has surged in recent years, particularly in the rapidly growing economies of India and China (**Figure 17.4**).

Coal Mining

The two basic types of coal mines are surface and subsurface (underground) mines. If the coal bed is within 30 m (100 ft) or so of the surface, **surface mining** is usually done. In **strip mining**, one type of surface

> **surface mining** The extraction of mineral and energy resources near Earth's surface by first removing the soil, subsoil, and overlying rock strata.

Distribution of coal deposits • Figure 17.3 ____

Data are presented as percentages of the 2005 estimated recoverable reserves—that is, of coal known to exist that can be recovered under present economic conditions with existing technologies. (The map is color-coded with the bar graph.)

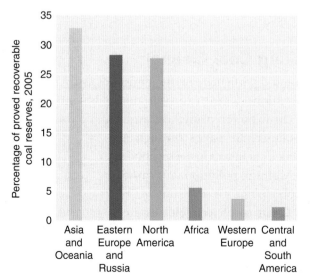

Chinese energy consumption on the rise • Figure 17.4 ____

Workers in Shanxi Province load coal onto trucks. Coal provides 65 percent of China's energy. Consumption of coal in China, the highest in the world, may double within 20 years as its economy grows.

mining, a trench is dug to extract the coal, which is scraped out of the ground and loaded into railroad cars or trucks. Surface mining is used to obtain approximately 60 percent of the coal mined in the United States.

When the coal is deeper in the ground or runs deep into the ground from an outcrop on a hillside, it is mined underground. **Subsurface mining** accounts for approximately 40 percent of the coal mined in the United States.

> **subsurface mining** The extraction of mineral and energy resources from deep underground deposits.

Surface mining has several advantages over subsurface mining: It is usually less expensive and safer for miners, and it generally allows more complete removal of coal from the ground. However, surface mining disrupts the land much more extensively than subsurface mining and has the potential to cause serious environmental problems.

Environmental Impacts of Coal

Coal mining, especially surface mining, has substantial effects on the environment (**Figure 17.5**). Prior to the 1977 **Surface Mining Control and Reclamation Act (SMCRA)**, abandoned surface coal mines were usually left as large open pits or trenches. Acid and toxic mineral drainage from such mines, along with the removal of topsoil, which was buried or washed away by erosion, prevented most plants from naturally recolonizing the land. Streams were polluted with sediment and **acid mine drainage**, which is produced when rainwater seeps

> **acid mine drainage** Pollution caused when sulfuric acid and dangerous dissolved materials, such as lead, arsenic, and cadmium, wash from coal and metal mines into nearby lakes and streams.

through iron sulfide minerals exposed in mine wastes (see Chapter 12). Dangerous landslides occurred on hills that were unstable due to the lack of vegetation.

One of the most land-destructive types of surface mining is **mountaintop removal**. According to Environmental Media Services, mountaintop removal has leveled between 15 and 25 percent of the mountaintops in southern West Virginia. The valleys and streams between the mountains are gone as well, filled with mine tailings and debris. At the current rate, half the peaks in that area will be gone by 2020. Mountaintop removal is also occurring in Kentucky, Pennsylvania, Tennessee, and Virginia.

Coal burning generally contributes more of the common air pollutants than burning either oil or natural gas. In the United States, coal-burning electric power plants currently produce one-third of all airborne mercury emissions. Some coal contains sulfur and nitrogen that, when burned, are released into the atmosphere as sulfur oxides (SO_2 and SO_3) and nitrogen oxides (NO, NO_2, and N_2O), many of which form acids when they react with water. These reactions result in **acid deposition**, which is particularly prevalent downwind from coal-burning electric power plants (**Figure 17.6**). Acid deposition and forest decline are discussed in greater detail in Chapter 9.

Burning any fossil fuel releases carbon dioxide (CO_2) into the atmosphere. CO_2 is a potent greenhouse gas that traps heat, thereby increasing the temperature of the atmosphere. Burning coal causes a more severe CO_2 problem than burning other fossil fuels because coal releases more CO_2 per unit of heat energy produced than does burning either oil or natural gas.

Making Coal Cleaner

Sulfur and particulate matter emissions associated with the combustion of coal can be reduced by using **scrubbers** that clean power plants' exhaust. As polluted air passes through a scrubber, chemicals in the scrubber react with the pollution and cause it to precipitate, or settle out.

"Clean coal" technologies are methods of burning coal that reduce air pollution. In **fluidized-bed combustion**, crushed coal is mixed

> **fluidized-bed combustion** A clean-coal technology in which crushed coal is mixed with limestone to neutralize acidic compounds produced during combustion.

Surface coal mine near Cabin Creek, West Virginia • Figure 17.5

In surface mining, overlying vegetation, soil, and rock are stripped away; coal is extracted out of the ground; and rubble is dumped into the surrounding valleys.

with limestone particles in a strong air current during combustion. This technology produces fewer nitrogen oxides and removes sulfur from the coal. It produces more heat from a given amount of coal, thereby somewhat reducing CO_2 emissions per unit of electricity produced.

In the United States several large power plants are testing fluidized-bed combustion, and a few small plants are already using this technology. **The Clean Air Act Amendments of 1990** provide incentives for utility companies to convert to clean coal technologies. A few coal-fired power plants are currently experimenting with technologies that capture CO_2 before it can be emitted.

Dead trees enveloped in acid fog on Mt. Mitchell, North Carolina • Figure 17.6 _____

Forest decline was first documented in Germany and eastern Europe. More recently, it has been observed in eastern North America, particularly at higher elevations. Acid deposition contributes to forest decline.

CONCEPT CHECK

1. **Which** type of coal mining—surface or subsurface mining—is more land intensive?
2. **What** are the environmental impacts of mining and burning coal?
3. **What** technologies can make coal a cleaner fuel?

Oil and Natural Gas

LEARNING OBJECTIVES

1. **Describe** existing reserves of oil and natural gas.
2. **Discuss** the environmental problems caused by using oil and natural gas.

Oil and natural gas supplied approximately 60.9 percent of the energy used in the United States in 2007. In comparison, other U.S. energy sources include coal (22.5 percent), nuclear power (8.5 percent), and hydropower (3.1 percent). Globally, oil and natural gas provided 60.6 percent of the world's energy (**Figure 17.7**).

Petroleum, or **crude oil**, is a liquid composed of hundreds of hydrocarbon compounds. Oil is also used to produce **petrochemicals**, compounds used to make products such as fertilizers, plastics, paints, pesticides, medicines, and synthetic fibers. During petroleum refining, the compounds are separated into different

World commercial energy sources • Figure 17.7 _____

Note the overwhelming reliance on oil, coal, and natural gas as commercial energy sources. "Alternatives" include geothermal, solar, wind, and wood electric power.

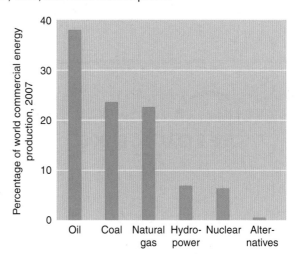

Petroleum refining
• Figure 17.8

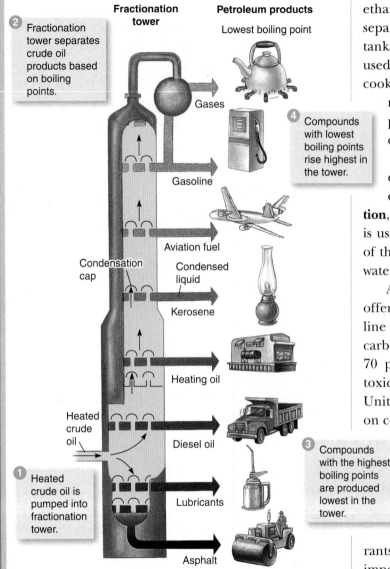

Fractionation tower

Petroleum products

Lowest boiling point

② Fractionation tower separates crude oil products based on boiling points.

Gases

④ Compounds with lowest boiling points rise highest in the tower.

Gasoline

Aviation fuel

Condensation cap

Condensed liquid

Kerosene

Heated crude oil

Heating oil

Diesel oil

③ Compounds with the highest boiling points are produced lowest in the tower.

① Heated crude oil is pumped into fractionation tower.

Lubricants

Asphalt

Highest boiling point

WILEY
RESOURCE KIT

products—such as gases, jet fuel, heating oil, diesel, and asphalt—based on their different boiling points (**Figure 17.8**).

Compared to petroleum, natural gas contains only a few hydrocarbons: methane and smaller amounts of ethane, propane, and butane. Propane and butane are separated from the natural gas, stored in pressurized tanks as a liquid called **liquefied petroleum gas**, and used primarily in rural areas as fuel for heating and cooking. Methane is used to heat residential and commercial buildings, to generate electricity in power plants, and for a variety of purposes in the organic chemistry industry.

Natural gas use is increasing in three main areas: electricity generation, transportation, and commercial cooling. Natural gas is often used in **cogeneration**, a clean and efficient process in which natural gas is used to produce both electricity and steam; the heat of the exhaust gases provides energy to make steam for water and space heating (see Chapter 18).

As a fuel for trucks, buses, and cars, natural gas offers significant environmental advantages over gasoline or diesel: Natural gas vehicles emit 33 percent less carbon dioxide, 80 to 93 percent fewer hydrocarbons, 70 percent less carbon monoxide, 90 percent fewer toxic emissions, and almost no soot. As of 2008, the United States had more than 150,000 vehicles running on compressed natural gas. The city of Los Angeles has the largest fleet of natural gas–powered transit buses in North America.

Natural gas efficiently fuels residential and commercial air-cooling systems. One example is the use of natural gas in a desiccant-based (air-drying) cooling system, which is ideal for restaurants and supermarkets, where humidity control is as important as temperature control.

The main disadvantage of natural gas is that deposits are often located far from where the energy is used. Because it is a gas and less dense than a liquid, natural gas costs four times more to transport through pipelines than crude oil. To transport natural gas over long distances, it must first be compressed to form liquefied natural gas (LNG) and then carried on specially constructed refrigerated ships (**Figure 17.9**). After LNG arrives at its destination, it must be returned to the gaseous state at regasification plants before being piped to where it will be used. Currently, the United States has only four such plants, which severely restricts

Liquified natural gas ship • Figure 17.9

Tanker ships transport liquified natural gas in large, cylindrical tanks.

Double-hulled tanker

Insulated storage tanks

the importation of natural gas from other countries. American energy companies claim that the United States needs at least 40 regasification plants to keep costs down for natural gas and to meet increasing demands.

Reserves of Oil and Natural Gas

Oil and natural gas deposits exist on every continent, but their distribution is uneven. More than half of the world's total estimated reserves are situated in the Persian Gulf region, which includes Iran, Iraq, Kuwait, Oman, Qatar, Saudi Arabia, Syria, the United Arab Emirates, and Yemen. Major oil fields also exist in Venezuela, Mexico, Russia, Kazakhstan, Libya, and the United States (in Alaska and the Gulf of Mexico; **Figure 17.10** on pages 424 and 425). Almost half of the world's proved recoverable reserves of natural gas are located in two countries, Russia and Iran. The United States has more deposits of natural gas than western Europe.

Large oil deposits probably exist under the **continental shelves**, the relatively flat underwater areas that surround continents, and in deep-water areas adjacent to the continental shelves. Despite problems such as storms at sea and the potential for oil spills, many countries engage in offshore drilling. As many as 18 billion barrels (756 billion gals) of oil and natural gas may exist in the deep water of the Gulf of Mexico, just off the continental shelf from Texas to Alabama. Continental shelves off the coasts of western Africa and Brazil are also promising potential sources of oil. Environmentalists and coastal industries such as fishing generally oppose opening the continental shelves for oil and natural gas exploration because of the threat of a major oil spill.

How Long Will Oil and Natural Gas Supplies Last? We cannot predict how many reserves will be discovered, whether technological breakthroughs will allow us to extract more fuel from each deposit, or whether

Oil and Natural Gas **423**

Global energy supply and consumption • Figure 17.10

Oil supplies are distributed unequally around the world, and energy consumption varies dramatically between regions.

▶ **ENERGY CONSUMPTION**
The use and availability of primary energy resources are unequally distributed around the globe. More than 86 percent of energy consumed globally is from nonrenewable fossil fuels—coal, oil, and natural gas. Consumption of these fuels is greatest in industrialized nations, with the United States using up nearly one-quarter. Developing countries, especially those in sub-Saharan Africa, rely on more traditional sources of energy, such as firewood and dung.

HYDROPOWER

NUCLEAR

SOLAR

WIND

GEOTHERMAL

ALTERNATIVE ENERGIES
Hydropower provides nearly 18 percent of the world's electricity, but it is limited to countries with adequate water resources, and it poses threats to local watersheds. **Nuclear energy** makes up 17 percent of Earth's electricity, but few countries have adopted it because of potential environmental risks and waste disposal issues. **Solar** and **wind energy** are inexhaustible and are the focus of new energy technologies and research. **Geothermal energy** is efficient but limited to countries with ready sources of hot groundwater, such as Iceland.

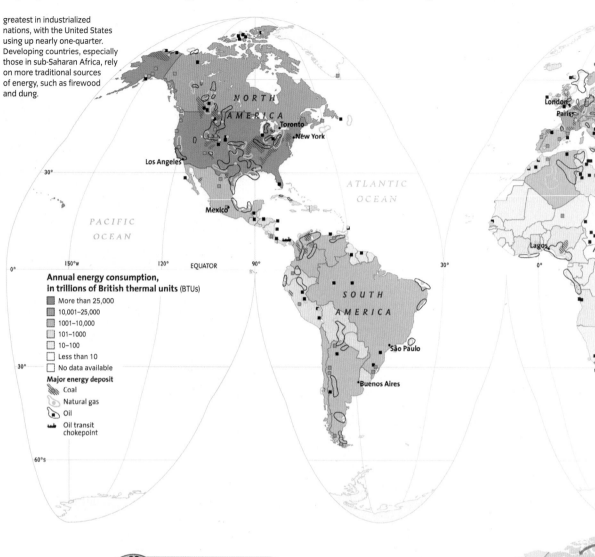

Annual energy consumption, in trillions of British thermal units (BTUs)
- More than 25,000
- 10,001–25,000
- 1001–10,000
- 101–1000
- 10–100
- Less than 10
- No data available

Major energy deposit
- Coal
- Natural gas
- Oil
- Oil transit chokepoint

Is the country you live in a net importer or exporter of oil?

▼ **RENEWABLE ENERGY**
Renewable sources of energy—geothermal, solar, and wind—make up a small percentage of the world's energy supply. They have a significant impact, however, on local and regional energy supplies, especially for electricity, in places such as the United States, Japan, and Germany. These sources of energy can be regenerated or renewed in a relatively short time, whereas fossil fuels form over geologic time

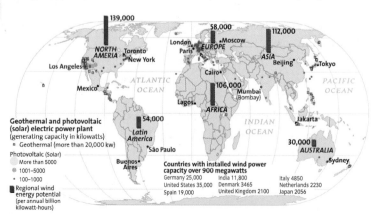

Geothermal and photovoltaic (solar) electric power plant (generating capacity in kilowatts)
- ■ Geothermal (more than 20,000 kw)
- Photovoltaic (Solar)
  - ● More than 5000
  - ● 1001–5000
  - • 100–1000
- ■ Regional wind energy potential (per annual billion kilowatt-hours)

Countries with installed wind power capacity over 900 megawatts
Germany 25,000	India 11,800	Italy 4850
United States 35,000	Denmark 3465	Netherlands 2230
Spain 19,000	United Kingdom 2100	Japan 2056

Oil imports, 2004 (in million metric tons)
- More than 250
- 175–250
- 75–174
- Less than 75

▶ **FLOW OF OIL WORLDWIDE**
Major oil reserves are clustered in a handful of countries, more than half of which are in the Middle East, whereas the greatest demand for oil is in the United States, Europe, Japan, and China. Other major oil exporters include the Russian Federation, Norway, Venezuela, and Mexico.

WILEY RESOURCE KIT

E nergy enables us to cook our food, heat our homes, move about our planet, and run industry. Every day the world uses some 320 billion kilowatt-hours of energy—equivalent to each person burning 22 lightbulbs nonstop—and over the next century, demand may increase threefold. Consumption is not uniform around the globe. People in industrialized countries consume far greater amounts of energy than those in developing countries. The world's energy supply is still fossil fuel based, despite advances in alternative energy sources. To meet demand, many countries must import fuels, making the trade of energy a critical, often volatile global political issue. Instability where most oil is found—the Persian Gulf, Nigeria, and Venezuela—makes this global economic powerline fragile. Insatiable demand where most energy is consumed—the United States, Japan, China, India, and Germany—makes national economies increasingly dependent. Further-more, extraction and use of fossil fuel have serious environmental effects, such as air pollution and climate warming. The challenge for the future? Reducing reliance on fossil fuels, developing alterna-tive energies to meet demand, and mediating the trade-offs between the environment and energy.

GeoBytes

LACK OF ACCESS
More than 2 billion people, mostly in the developing world, do not have access to electricity. Increasingly, small-scale wind and solar projects bring power to poor rural areas.

WINDS OF CHANGE
Worldwide, wind supplies less than 1 percent of electric power, but it is the fastest-growing source, especially in Europe. Denmark gets 20 percent of its electricity from wind.

POWER OF THE SUN
Near Leipzig, Germany, some 33,000 photovoltaic panels produce up to 5 megawatts of power. It is one of the world's largest solar arrays.

GOING NUCLEAR
France gets 78 percent of its electricity from nuclear power. Developing nations, such as China and India, are building new reactors to reduce pollution and meet soaring energy demands.

GROWING PAINS
China is fueling its economic growth with huge quantities of coal, and it suffers from energy-related environmen-tal problems. China is second only to the United States in greenhouse gas emissions that contribute to global warming.

◀ **WORLD OIL SUPPLY**
The world's hunger for oil is insatiable, but the supply is finite and unequally distributed, making it one of the world's most valuable commodities. It is the leading source of energy worldwide, and in industrialized countries it accounts for more than one-third of all energy consumed. Pressure on the world's oil supply continues to mount as both industrialized and developing countries grow more dependent on it to meet increasing energy needs.

Map labels

ASIA, Beijing, Tokyo, Mumbai (Bombay), Jakarta, AUSTRALIA, Sydney, PACIFIC OCEAN, INDIAN OCEAN, 60°N, 30°, 30°, 60°S, 120°, 150°E, 180°, 90°, 60°, scow, ICA

ASIA-PACIFIC, U.S. & MEXICO, ASIA-PACIFIC, U.S., LATIN AMERICA

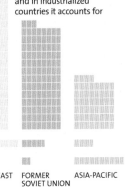

World oil
(1 block=100 metric tons*)

Reserves
Production
Consumption

*100 metric tons=733 barrels of crude oil

CANADA, UNITED STATES, MEXICO, LATIN AMERICA (excluding Mexico), AFRICA, EUROPE, MIDDLE EAST, FORMER SOVIET UNION, ASIA-PACIFIC

NATIONAL GEOGRAPHIC

Rapid expansion of oil exploration in the Gulf of Mexico, 1961–2001 • Figure 17.11 _____

Active oil leases give drilling companies the right to obtain oil in a given location.

world consumption of oil and natural gas will increase, remain the same, or decrease. Even with technological advances, the most optimistic predictions are for global oil production to peak around 2035. Natural gas is more plentiful than oil. Experts estimate that, at current rates of consumption, readily recoverable reserves of natural gas will keep production rising for at least 10 years after conventional supplies of petroleum have begun to decline.

Environmental Impacts of Oil and Natural Gas

Extracting, transporting, and burning oil and natural gas create a variety of environmental problems. As with coal, burning oil and natural gas produces CO_2 that contributes to global climate change. Every gallon of gasoline your car or truck burns releases about 9 kg (20 lb) of CO_2 into the atmosphere. Burning oil also leads to acid deposition and the formation of photochemical smog. Gasoline combustion contributes about half the nitrogen oxides released into the atmosphere by human activities. (Coal combustion contributes the rest.)

Natural gas, on the other hand, is a relatively clean, efficient source of energy that contains almost no sulfur and releases far less CO_2, fewer hydrocarbons, and almost no particulate matter compared to oil and coal.

One risk of oil and natural gas production relates to their transport, often over long distances by pipelines or ocean tankers. Natural gas leaks, while rare, can lead to massive explosions. Oil spills create environmental damage, particularly in aquatic ecosystems, where an oil slick can travel great distances.

The Largest Oil Tanker Spill in the United States As this book goes to press, oil leaking from a failed deep-water platform is spreading along the Louisiana coast. This leak will probably be the largest spill ever in the United States. Such leaks are not a surprise, given the huge increase in oil exploration and extraction in the Gulf of Mexico (**Figure 17.11**). In 1989 the supertanker *Exxon Valdez* hit Bligh Reef and spilled 260,000 barrels (10.9 million gals) of crude oil into Prince William Sound along the coast of Alaska, the largest oil spill from a tanker in U.S. history (**Figure 17.12**). More than 30,000 birds and between 3500 and 5500 sea otters died as a result of the spill. The area's orca and harbor seal populations declined, salmon migration was disrupted, and the fishing season in the area was halted that year.

Although Exxon declared the cleanup "complete" in late 1989, it left behind contaminated shorelines; continued damage to some species of birds, fishes, and mammals; and a reduced commercial salmon catch, among other problems.

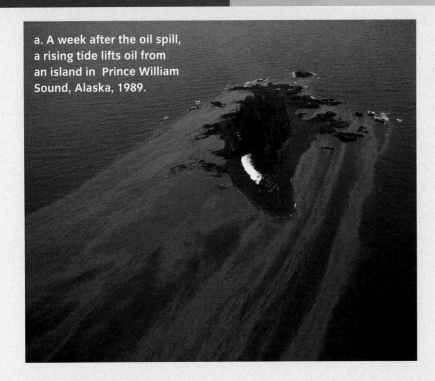

a. A week after the oil spill, a rising tide lifts oil from an island in Prince William Sound, Alaska, 1989.

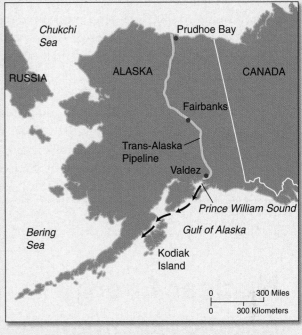

c. The extent of the spill (black arrows). Water currents caused it to spread rapidly for hundreds of kilometers.

b. Workers try to clean the rocky shoreline of Eleanor Island, Alaska, several months after the spill.

d. Nearshore coastal wildlife status, Prince William Sound, Alaska, 2005.

Populations that are recovering	Populations that are not yet recovering
Clams	Common loons
Intertidal communities	Cormorants (3 species)
Orcas	Harbor seals
Marbled murrelets	Harlequin ducks
Mussels	Pacific herring
Sea otters	Pigeon guillemots

NATIONAL GEOGRAPHIC

One positive outcome of the disaster was passage of the **Oil Pollution Act** of 1990. This legislation establishes liability for damages to natural resources resulting from a catastrophic oil spill, including a trust fund that pays to clean up spills when the responsible party cannot, and it requires, by 2015, double hulls on all oil tankers that enter U.S. waters.

Earth's Largest Oil Spill The world's most massive oil spill occurred in 1991 during the Persian Gulf War, when about 6 million barrels (250 million gals) of crude oil—more than 20 times the amount of the *Exxon Valdez* spill—were deliberately dumped into the Persian Gulf.

Many oil wells were set on fire, and lakes of oil spilled into the desert around the burning oil wells. In 2001 Kuwait began a massive remediation project to clean up its oil-contaminated desert. Progress is slow, and it may take a century or more for the area to completely recover.

CONCEPT CHECK

1. **Where** are oil and natural gas reserves most abundant?
2. **What** are three environmental problems associated with using oil and natural gas as energy resources?

Nuclear Energy

LEARNING OBJECTIVES

1. **Define** *nuclear energy* and describe a typical nuclear power reactor.
2. **Discuss** the pros and cons of electric power produced by nuclear energy versus coal.
3. **Describe** safety issues associated with nuclear power plants and risks associated with the storage of radioactive wastes.

All atoms are composed of positively charged protons, negatively charged electrons, and electrically neutral neutrons (**Figure 17.13**). Protons and neutrons, which have approximately the same mass, are clustered in the center of an atom, making up its nucleus. Electrons, which possess little mass compared to protons and neutrons, orbit the nucleus in distinct regions.

As a way to obtain energy, nuclear processes are fundamentally different from the combustion that produces energy from fossil fuels. Combustion is a chemical reaction. In chemical reactions, atoms of one element do not change into atoms of another element, nor does any of their mass (matter) change into energy. The energy released in chemical reactions comes from changes in the chemical bonds that hold together the atoms. Chemical bonds are associations between electrons, and chemical reactions involve the rearrangement of electrons.

Nuclear energy involves changes in the nuclei of atoms; small amounts of matter from the nucleus are converted into large amounts of energy. Nuclear reactions produce 100,000 times more energy per atom than is available from a chemical bond between two atoms.

> **nuclear energy**
> The energy released by nuclear fission or fusion.

Atomic structure • Figure 17.13

An atom contains a nucleus made of protons and neutrons. Circling the nucleus is a "cloud" of electrons.

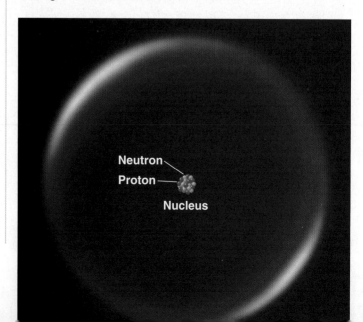

Nuclear fission • Figure 17.14

Starting at the left side of the figure, neutron bombardment ❶ of a uranium 235 (U-235) nucleus causes it to split into two smaller raadioactive fragments ❷ and several free neutrons ❸. The free neutrons bombard nearby U-235 nuclei, causing them to split and release still more free neutrons ❹ in a chain reaction. Many different pairs of radioactive atomic fragments are produced during the fission of U-235.

❷ U-235 nucleus splits into two smaller radioactive fragments and several free neutrons.

❸ Free neutrons bombard nearby U-235 nuclei, causing them to split.

❶ Uranium-235 (U-235) nucleus undergoes nuclear bombardment.

❹ More free neutrons are released in a chain reaction.

Neutron

Atomic fragment

Free neutron

Energy

Uranium-235

Uranium-235

ENERGY

Free neutron

Energy

Atomic fragment

Atomic fragment

Free neutrons

Atomic fragments

Energy

Free neutrons

Atomic fragment

WILEY RESOURCE KIT

---

fission The splitting of an atomic nucleus into two smaller fragments, accompanied by the release of a large amount of energy.

There are two different nuclear reactions that release energy: fission and fusion. In nuclear **fission**, the process nuclear power plants use, energy is released when a single neutron crashes into a large atom of one element, such as uranium, and splits it into two smaller atoms of different elements (**Figure 17.14**). In **fusion**, the process that powers the sun and other stars, two small atoms are combined, forming one larger atom of a different element.

Conventional Nuclear Fission

Uranium ore, the mineral fuel used in conventional nuclear power plants, is a nonrenewable resource present in limited amounts in sedimentary rock in Earth's crust. Approximately 11 percent of the world's uranium deposits are located in the United States.

Uranium ore contains three isotopes: U-238 (which makes up 99.28 percent of uranium), U-235 (0.71 percent), and U-234 (less than 0.01 percent). Because U-235, the isotope used in conventional fission reactions, is such a minor part of uranium ore, uranium ore must be refined after mining to increase the concentration of U-235 to about 3 percent. This refining process, called **enrichment**, requires a great deal of energy.

After enrichment, uranium is processed into small pellets of uranium dioxide; each pellet contains the

enrichment The process by which uranium ore is refined after mining to increase the concentration of fissionable U-235.

Uranium fuel • Figure 17.15

a. Uranium dioxide pellets contain about 3 percent uranium-235, the fission fuel in a nuclear reactor. Each pellet contains the energy equivalent of 1 ton of coal.

b. The uranium pellets are loaded into long fuel rods, which are grouped into square fuel assemblies like the one being inspected here by technicians.

> **nuclear reactor** A device that initiates and maintains a controlled nuclear fission chain reaction to produce energy for electricity.

energy equivalent of 1 ton of coal (**Figure 17.15a**). The pellets are then placed in closed pipes, often as long as 3.7 m (12 ft), called **fuel rods**. The fuel rods are grouped into square **fuel assemblies**, generally made up of 200 rods each (**Figure 17.15b**). A typical nuclear reactor contains 150 to 250 fuel assemblies.

The fission of U-235 releases an enormous amount of heat, which is used to transform water into steam. The steam, in turn, is used to generate electricity. The production of electricity is possible because the fission reaction is controlled. Operators of a nuclear power plant can start or stop and increase or decrease the fission reactions in the reactor to produce the desired amount of heat energy. Recall that nuclear bombs make use of uncontrolled fission reactions. If the control mechanism in a nuclear power plant were to fail, a bomblike nuclear explosion could not take place because nuclear fuel has such a low percentage of U-235 compared to bomb-grade material.

A typical nuclear power plant has four main parts: the reactor core, the steam generator, the turbine, and the condenser (**Figure 17.16**). Nuclear fission occurs in the **reactor core**, and the heat produced by fission is used to produce steam from liquid water in the **steam generator**. The **turbine** uses the steam to generate electricity, and the **condenser** cools the steam, converting it back to a liquid.

Nuclear Energy and Fossil Fuels

Worldwide, nuclear power production is steadily increasing. In part, this is due to concerns about climate change, but it is primarily driven by an increasing demand for energy. In 2009, 31 countries around the world were operating 436 nuclear power plants, producing 373 gigawatts. Another 465 plants were under construction, planned, or proposed, many of them in China and India.

Supporters of nuclear energy argue that we should generate more because nuclear energy affects the environment less than fossil fuels such as coal (**Table 17.1**). The combustion of coal to generate electricity is responsible for more than one-third of the air pollution in the United States and contributes to acid precipitation and climate warming. In comparison, nuclear energy emits few pollutants into the atmosphere. Nuclear energy can also provide power without producing climate-altering CO_2.

Pressurized water reactor • Figure 17.16

Approximately two-thirds of all nuclear power plants in the United States are of this type.

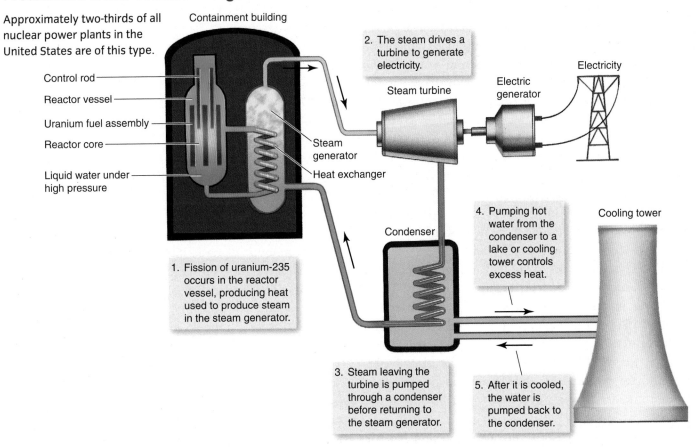

Containment building

Control rod
Reactor vessel
Uranium fuel assembly
Reactor core
Liquid water under high pressure

Steam generator
Heat exchanger

2. The steam drives a turbine to generate electricity.

Steam turbine

Electric generator

Electricity

1. Fission of uranium-235 occurs in the reactor vessel, producing heat used to produce steam in the steam generator.

Condenser

4. Pumping hot water from the condenser to a lake or cooling tower controls excess heat.

Cooling tower

3. Steam leaving the turbine is pumped through a condenser before returning to the steam generator.

5. After it is cooled, the water is pumped back to the condenser.

Comparison of environmental impacts of 1000-megawatt coal and conventional nuclear power plants* • Table 17.1		
Impact	**Coal**	**Nuclear (conventional fission)**
Land use	7000 hectares	800 hectares
Daily fuel requirement	9000 tons (of coal)/day	3 kg (of enriched uranium)/day
Availability of fuel, based on present economics	A few hundred years	100 years, maybe longer
Air pollution	Moderate to severe, depending on pollution controls	Low
Climate change risk (from CO_2 emissions)	Severe	Small**
Radioactive emissions, routine	1 curie	28,000 curies
Water pollution	Often severe	Potentially severe at nuclear waste disposal sites
Risk from catastrophic accidents	Short-term local risk	Long-term risk over large areas
Link to nuclear weapons	No	Yes
Annual occupational deaths	0.5 to 5	0.1 to 1

*Impacts include extraction, processing, transportation, and conversion. Assumes that coal is strip-mined. (A 1000-MWe utility, at a 60 percent load factor, produces enough electricity for a city of 1 million people.)

**Currently, nuclear power requires the use of fossil fuels for mining, construction, processing, transportation, and waste management, and so it cannot be considered carbon free.

spent fuel Used fuel elements that were irradiated in a nuclear reactor

However, nuclear energy generates radioactive waste in the form of **spent fuel**. Nuclear power plants also produce radioactive coolant fluids and gases in the reactor. These radioactive wastes are extremely dangerous, and the hazards of their health and environmental impacts require that special measures be taken to ensure their safe storage and disposal.

International crises, such as the Gulf War of the early 1990s and the Iraq War in the 2000s, occasionally threaten the supply of oil to the United States. Some supporters of nuclear energy assert that our dependence on foreign oil would be reduced if all oil-burning power plants were converted to nuclear plants.

However, oil is responsible for generating only about 3 percent of the electricity in the United States. Replacing electricity generated by oil with electricity generated by nuclear power would do little in the short term to lessen our dependence on foreign oil because we would still need oil for heating buildings and for gasoline. Technological advances could change nuclear power's potential contribution in the future.

As electric heat pumps and hydrogen and electric motor vehicles become more common, however, nuclear power plants could provide a much larger fraction of heating and transportation energy demand, thus decreasing our reliance on foreign oil.

Safety and Accidents in Nuclear Power Plants

Although conventional nuclear power plants cannot explode like atomic bombs, accidents do happen in which dangerous levels of radiation are released into the environment and result in human casualties. At high temperatures, the metal encasing uranium fuel can melt, releasing radiation; this is called a **meltdown**. Also, the water used in a nuclear reactor to transfer heat can boil away during an accident, contaminating the atmosphere with radioactivity.

The nuclear industry considers the probability that a major accident will occur low, but public perception of the risk is high for several reasons. Nuclear power risks are involuntary and potentially catastrophic. In addition, many people are distrustful of the nuclear industry.

The consequences of such accidents are drastic and life threatening, both immediately and long after the accidents have occurred.

Three Mile Island The most serious commercial nuclear reactor accident in the United States occurred in 1979 at the Three Mile Island power plant in Pennsylvania, as a result of human error after the cooling system failed. A partial meltdown of the reactor core took place. Had there been a complete meltdown of the fuel assembly, dangerous radioactivity would have been emitted into the surrounding countryside. Fortunately, the containment building kept almost all the radioactivity released by the core material from escaping. Although a small amount of radiation entered the environment, there were neither substantial environmental damages nor immediate human casualties. Numerous studies have failed to link abnormal health problems (other than increased stress) to the accident.

In the aftermath of the accident, public wariness prompted construction delays and cancellations of several new nuclear power plants across the United States. New safety regulations were put in place, including more frequent safety inspections, new risk assessments, and improved emergency and evacuation plans for nuclear power plants and surrounding communities.

Chernobyl The world's worst nuclear power plant accident took place in 1986 at the Chernobyl plant, located in the former Soviet republic of the Ukraine. One or possibly two explosions ripped apart a nuclear reactor and expelled large quantities of radioactive material into the atmosphere (**Figure 17.17**). The effects of this accident were not confined to the area immediately surrounding the power plant: Significant amounts of radioisotopes quickly spread across large portions of Europe. The Chernobyl accident affected and will continue to affect many nations.

Although cleanup in the immediate vicinity of Chernobyl is finished, the people in Ukraine face many long-term problems. Ultimately, more than 170,000 people permanently abandoned their homes. Much of the farmland and forests are so contaminated that they cannot be used for more than a century. Inhabitants of many areas of Ukraine cannot drink the water or consume locally produced milk, meat, fish, fruits, or vegetables. Mothers do not nurse their babies because their milk is contaminated

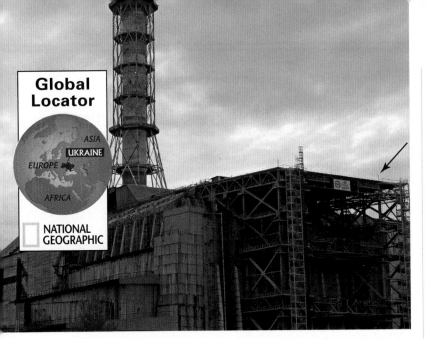

Chernobyl, Ukraine • Figure 17.17

The arrow indicates the site of the explosion. The upper part of the reactor was completely destroyed.

by radioactivity. The frequency of birth defects and mental retardation in newborns has increased in affected areas, and children exposed to the Chernobyl fallout experienced increased incidences of leukemia, thyroid cancer, and abnormalities of the immune system (**Figure 17.18**).

Health consequences of Chernobyl • Figure 17.18

This 14-year-old is recovering from thyroid cancer. A significant (25-fold) increase in thyroid cancer in children and adolescents occurred within a few years after the accident.

The Link Between Nuclear Energy and Nuclear Weapons

Fission is involved in both the production of electricity by nuclear energy and the destructive power of nuclear weapons. Countries that own nuclear power plants have access to the fuel needed for nuclear weapons (by reprocessing spent fuel to make plutonium). Responsible world leaders are concerned about the consequences of terrorist groups and states of concern (such as Iran and North Korea) building nuclear weapons. These concerns have caused many people to shun nuclear energy and to seek alternatives that are not so intimately connected with nuclear weapons.

There are several hundred metric tons of weapons-grade plutonium worldwide. Storing plutonium is a security nightmare because it takes only several kilograms to make a nuclear bomb as powerful as the ones that destroyed the Japanese cities of Nagasaki and Hiroshima in World War II. However, since the 2001 terrorist attacks in the United States, the security of plutonium stockpiles and nuclear power plants has been increased substantially to reduce the chance that terrorist groups could steal plutonium and enriched uranium and use them to make nuclear weapons.

Radioactive Wastes

Radioactive wastes are classified as either "low level" or "high level." **Low-level radioactive wastes** include glassware, tools, paper, clothing, and other items contaminated by radioactivity. They are produced by nuclear power plants, university research labs, nuclear medicine departments in hospitals, and industries. Four sites—located in Washington State, South Carolina, Nevada, and Utah—currently store most of the country's low-level radioactive wastes.

High-level radioactive wastes produced during nuclear fission include the reactor metals (fuel rods and assemblies), coolant fluids, and air or other gases found in the reactor. High-level radioactive wastes are also generated during the reprocessing of spent fuel. Produced by nuclear power

low-level radioactive wastes Solids, liquids, or gases that give off small amounts of ionizing radiation.

high-level radioactive wastes Radioactive solids, liquids, or gases that initially give off large amounts of ionizing radiation.

plants and nuclear weapons facilities, high-level radioactive wastes are among the most dangerous human-made hazardous wastes.

As the radioisotopes in spent fuel decay, they produce considerable heat and are extremely toxic to organisms; they remain radioactive for thousands of years, and their dangerous level of radioactivity requires special handling. Secure storage of these materials must be guaranteed for thousands of years, until the materials decay sufficiently to be safe. The safe disposal of radioactive wastes is one of the main difficulties that must be overcome if we are to realize the potential of nuclear energy in the 21st century.

What are the best sites for the long-term storage of high-level radioactive wastes? Many scientists recommend storing the wastes in stable rock formations deep in the ground. People's reluctance to have radioactive wastes stored near their homes complicates the selection of these sites. Meanwhile, radioactive wastes continue to accumulate. Commercially operated nuclear power plants store their spent fuel in huge indoor pools of water or in storage casks on-site. However, none of these plants was designed for long-term storage of spent fuel (**Figure 17.19**).

Nuclear waste management is the biggest obstacle to increased nuclear power in the United States and abroad. None of the 31 countries that currently have nuclear power plants has decided where or how to dispose of their high-level nuclear waste. Most countries expect in the long term to dispose of nuclear power underground, but as of early 2010, no sites had been selected. It will probably be years, if not decades, before an underground high-level nuclear waste storage facility is established.

Yucca Mountain In 1982 the passage of the **Nuclear Waste Policy Act** put the burden of developing permanent sites for civilian and military radioactive wastes on the federal government and required the first site to be operational by 1998. (The deadline was postponed several times.)

In a 1987 amendment to the Nuclear Waste Policy Act, Congress identified Yucca Mountain in Nevada as the only candidate for a permanent underground storage site for high-level radioactive wastes from commercially operated power plants (see *What a Scientist Sees*). Since 1983 the U.S. Department of Energy has spent billions of dollars conducting feasibility studies on Yucca Mountain's geology. In 2002 Congress approved the choice of Yucca Mountain as the U.S. nuclear-waste repository, despite controversy and opposition from the state of Nevada. In 2009 the Obama administration

Storage casks for spent fuel • Figure 17.19

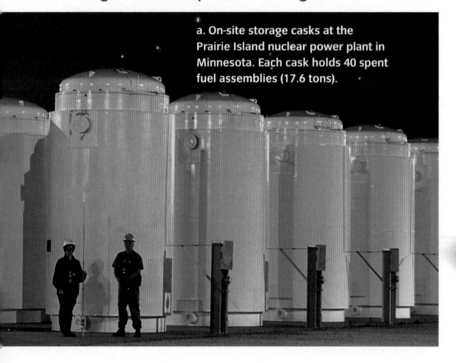

a. On-site storage casks at the Prairie Island nuclear power plant in Minnesota. Each cask holds 40 spent fuel assemblies (17.6 tons).

as length 5.2 m (ft)

- ro tective cover
- id with metallic seals
- torage cas (neutron shield)
- anister of steel
- pent fuel rods

b. Details of a storage cask. Each cask, designed to last at least 40 years, is monitored and will be replaced if leakage occurs.

withdrew support for Yucca Mountain; as of early 2010, a new siting process had not been announced.

Transporting high-level wastes from nuclear reactors and weapons sites is a major concern of opponents of the Yucca Mountain site. A typical shipment would travel an average of 3700 km (2300 miles), and 43 states would have waste passing through them on their way to Yucca Mountain.

Whether or not nuclear waste is eventually stored in Yucca Mountain, the scientific community generally agrees that storage of high-level radioactive waste in deep underground repositories is the best long-term option. Using an underground waste facility is far safer than storing high-level nuclear waste as we do now; storing this waste at many different commercial nuclear reactors poses a risk of terrorist attacks, theft, and, possibly, human health problems.

WHAT A SCIENTIST SEES

Yucca Mountain

a. This tunnel provides access to nuclear waste storage at Yucca Mountain, an arid, sparsely populated area of Nevada.

b. An environmental scientist thinks about the huge complex of interconnected tunnels located in dense volcanic rock 300 m (1000 ft) beneath the mountain crest. Canisters containing high-level radioactive waste can be stored in the tunnels.

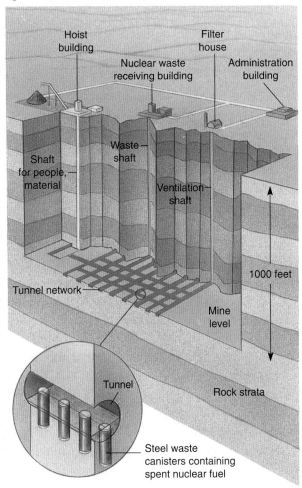

Hoist building

Filter house

Nuclear waste receiving building

Administration building

Shaft for people, material

Waste shaft

Ventilation shaft

1000 feet

Tunnel network

Mine level

Tunnel

Rock strata

Steel waste canisters containing spent nuclear fuel

EnviroDiscovery

A Nuclear Waste Nightmare

Over the past three decades, Soviet (and now Russian) practices for radioactive waste disposal have often violated international standards:

- Billions of gallons of liquid radioactive wastes were pumped directly underground, without being stored in protective containers. Russian officials claim that layers of clay and shale at the sites prevent leakage but admit that more leaks than expected have occurred.

- Highly radioactive wastes were dumped into the ocean in amounts double those of dumped wastes from 12 other nuclear nations combined.

- Both underground injection and underwater dumping of radioactive wastes continue because Russia lacks alternatives for nuclear waste processing. Potential health and environmental hazards associated with these wastes are unknown because so little data exist for these types of long-term storage.

Murmansk nuclear waste site.

Decommissioning Nuclear Power Plants

As nuclear power plants age, certain critical sections, such as the reactor vessel, become brittle or corroded. At the end of their operational usefulness, nuclear power plants are not simply abandoned or demolished because many parts have become contaminated with radioactivity.

When a nuclear power plant is closed, it undergoes **decommissioning**. The International Atomic Energy Agency (IAEA) defines three options for decommissioning: storage, entombment, and immediate dismantling. If an old plant is put into **storage**, the utility company guards it for 50 to 100 years while some of the radioactive materials decay, making it safer to dismantle the plant later. Accidental leaks during the storage period are an ongoing concern.

Most experts do not consider **entombment**, permanently encasing the entire power plant in concrete, a viable option because the tomb would have to remain intact for at least 1000 years. Accidental leaks would probably occur during that time, and we cannot guarantee that future generations would inspect and maintain the site.

The third option for the retirement of a nuclear power plant is to dismantle the plant immediately after it closes. Advances in robotics may make it feasible to tear down sections of old plants that are too "hot" (radioactive) for workers to safely dismantle. As the plant is torn down, small sections of it can be transported to a permanent storage site.

According to the IAEA, 107 nuclear power plants worldwide were permanently retired as of 2004 (23 of them in the United States), and many nuclear power plants are nearing retirement age. During the 21st century, we may find that we are paying more in our utility bills to decommission old plants than we are to construct new ones.

CONCEPT CHECK

1. **How** does a nuclear reactor produce electricity?
2. **What** are the environmental effects of generating electricity with conventional nuclear fission, and how do they compare to the environmental effects associated with burning coal?
3. **Why** is waste disposal an important aspect of nuclear waste management?

RESOURCE KIT

The Arctic National Wildlife Refuge

In 1960 Congress declared a section of northeastern Alaska protected because of its distinctive wildlife. In 1980 Congress expanded this wilderness area to form the Arctic National Wildlife Refuge (ANWR; see **Figure a**). The proposed opening of ANWR to oil exploration has been an ongoing environment-versus-economy conflict since the refuge's inception. On one side are those who seek to protect rare and fragile natural environments; on the other side are those whose higher priority is the development of some of the last major U.S. oil supplies.

The refuge, called "America's Serengeti," is home to many animal species, including polar bears, arctic foxes, peregrine falcons, musk oxen, Dall sheep, wolverines, and snow geese. It is the calving area for a large migrating herd of caribou (see **Figure b**). Although it is biologically rich, the tundra is an extremely fragile ecosystem, in part because of its harsh climate.

In the mid-1990s, prodevelopment interests became more vocal, partly because in 1994, for the first time in its history, the United States imported more than half the oil it used. Although the Department of the Interior concluded that oil drilling in the wildlife refuge would harm the area's ecosystem, both the Senate and the House of Representatives passed measures to allow it. (President Bill Clinton vetoed the bill.) In 2001, President George W. Bush announced his support for opening the refuge to oil drilling, but after a contentious debate in Congress in 2005, the Senate voted against doing so. To date President Barak Obama has not supported drilling in ANWR.

Supporters cite economic considerations as the main reason for drilling in the refuge. Development of domestic oil would improve the balance of trade and make the United States less dependent on foreign countries for its oil. Oil companies are

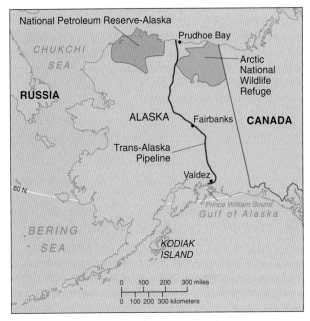

a. Located in the northeastern part of Alaska, the Arctic National Wildlife Refuge is situated close to the Trans-Alaska Pipeline, which begins at Prudhoe Bay and extends south to Valdez. The National Petroleum Reserve–Alaska is also shown.

eager to develop this particular site because it is near Prudhoe Bay, where large oil deposits are already being tapped.

Conservationists think oil exploration poses permanent threats to the delicate balance of nature in the Alaskan wilderness, in exchange for a temporary oil supply. They suggest that the money spent drilling for oil would be better used for research into alternative, renewable energy sources and energy conservation—a more permanent solution to the energy problem.

b. Members of the caribou herd whose calving grounds are on the Arctic National Wildlife Refuge.

Summary

1 Energy Consumption 418

1. The per person energy consumption in highly developed nations is eight times higher than that in developing nations. At present, highly developed countries consume more total energy than do developing countries; experts expect this to reverse within the next decade.

2 Coal 419

1. **Surface mining** is the extraction of mineral and energy resources near Earth's surface by first removing the soil, subsoil, and overlying rock strata. **Subsurface mining** is the extraction of mineral and energy resources from deep underground deposits. Surface mining is less expensive and safer but causes more serious environmental problems than subsurface mining.

2. Coal mining can lead to landslides and can pollute streams with sediment and **acid mine drainage**, when sulfuric acid and dangerous dissolved materials such as lead, arsenic, and cadmium wash from coal and metal mines into nearby lakes and streams. In **mountaintop removal** a huge shovel removes an entire mountaintop to reach coal located below. Burning coal releases more CO_2 and contributes more extensively to global climate warming than burning other fossil fuels. The combustion of coal contributes to **acid deposition**, in which acid falls from the atmosphere to the surface as precipitation or as dry acid particles.

3. Power plants can make coal a cleaner fuel by installing **scrubbers** to clean the power plants' exhaust. **Fluidized-bed combustion** is a clean-coal technology in which crushed coal is mixed with limestone to neutralize the acidic sulfur compounds produced during combustion.

3 Oil and Natural Gas 421

1. More than half of the world's total estimated oil and natural gas reserves are located in the Persian Gulf region.

2. A serious spill along an oil transportation route creates an environmental crisis, particularly in aquatic ecosystems. The burning of oil and natural gas produces CO_2 that can contribute to global climate change. Burning oil also leads to acid deposition by producing nitrogen oxides. Natural gas contains almost no sulfur and produces less CO_2 and other pollutants compared to oil and coal.

4 Nuclear Energy 428

1. **Nuclear energy** is the energy released by nuclear **fission** or fusion. A **nuclear reactor** is a device that initiates and maintains a controlled nuclear fission chain reaction to produce energy for electricity. A typical reactor contains a **reactor core**, where nuclear fission occurs; a **steam generator**; a **turbine**; and a **condenser**.

2. Generating electric power through nuclear energy emits few pollutants (such as CO_2) into the atmosphere compared to the combustion of coal but generates highly dangerous radioactive waste, such as **spent fuel**, the used fuel elements that were irradiated in a nuclear reactor.

3. Accidents at nuclear power plants can release dangerous levels of radiation into the environment and result in human casualties. The safe storage of radioactive wastes is another concern associated with nuclear energy. **Low-level radioactive wastes** are radioactive solids, liquids, or gases that give off small amounts of ionizing radiation. **High-level radioactive wastes** are radioactive solids, liquids, or gases that initially give off large amounts of ionizing radiation. Radioactive wastes must be isolated securely for thousands of years. One option for the retirement of an aging nuclear power plant is to decommission it by dismantling it after it closes.

Key Terms

- acid mine drainage 420
- enrichment 429
- fission 429
- fluidized-bed combustion 420

- high-level radioactive wastes 433
- low-level radioactive wastes 433
- nuclear energy 428
- nuclear reactor 430

- spent fuel 432
- subsurface mining 420
- surface mining 419

What is happening in this picture?

In April, 2010, a deep-water oil drilling rig owned by British Petroleum broke off in the Gulf of Mexico. This image from NASA shows the location of the spill (white arrow), as well as the spreading oil, as of early June 2010.

- Research this oil spill. How far has the oil travelled?

- How does this oil spill compare to the Exxon Valdez accident (see pages 426–428)?

Critical and Creative Thinking Questions

1. Distinguish among coal, oil, natural gas, and nuclear energy, and compare the environmental impacts of each.

2. What economic priorities and environmental concerns might be shared by coal-mining regions in West Virginia and in China?

3. Some environmental analysts think that the latest war in Iraq was related in part to gaining control over the supply of Iraqi oil. Do you think this is plausible? Explain why or why not.

4. Do you think oil drilling should be permitted in the Arctic National Wildlife Refuge? Why or why not?

5. Which major consumer of oil is most vulnerable to disruption in the event of another energy crisis: electric power generation, motor vehicles, heating and air conditioning, or industry? Why?

6. Which of the following produces the most CO_2 per unit of energy: oil, natural gas, or coal?

7. India, China, and the United States all have large coal reserves. Do you think any or all of them should build more coal-fired power plants over the next 20 years? If so, what technologies will help them minimize environmental impacts?

8. Examine the bar graph, which shows the six countries with the greatest natural gas deposits. Is most of the world's natural gas located in North and South America, or in Europe and the Persian Gulf countries?

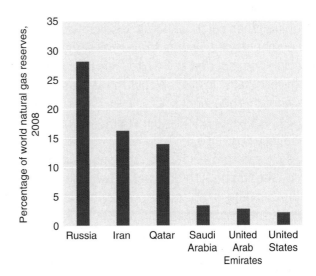

9. Some scholars think the Industrial Revolution may have been concentrated in Europe and North America because coal is located there. Explain the connection between coal and the Industrial Revolution.

Renewable Energy Resources

REDUCING HEAT LOSS

Two strategies can decrease our reliance on fossil fuels: finding alternative energy sources and reducing energy use. While energy reduction can sometimes require infrastructure investments or lifestyle changes, it can also be relatively simple and inexpensive. Older technologies and buildings often use much more energy than do newer models, and upgrading or replacing them can save energy.

The picture to the right is a thermal image of a house in New Haven, Connecticut. In contrast to typical photographs, thermal images capture the heat escaping from an object. The dark blue and green represent cold, while red and white are hottest. Thermal images allow homeowners to identify where their homes are leaking heat and reduce those losses by insulating, caulking cracks, and replacing older windows.

Most of the windows in this house appear to be very cool (blue/green color), although the first-floor window on the right may be older or broken, as suggested by its red color. The roof (red and white) may be the biggest energy loser; attic insulation could save considerable energy for this household.

Cars, appliances, and other energy-intensive technologies can also be evaluated with thermal imaging. Heat from a lightbulb represents wasted energy (see inset). Of three types of lightbulbs, an LED (bottom left) is the coolest, while a traditional incandescent (top) emits much more heat. A compact fluorescent (bottom right) is intermediate—more efficient than the incandescent but less efficient than the LED.

NATIONAL GEOGRAPHIC

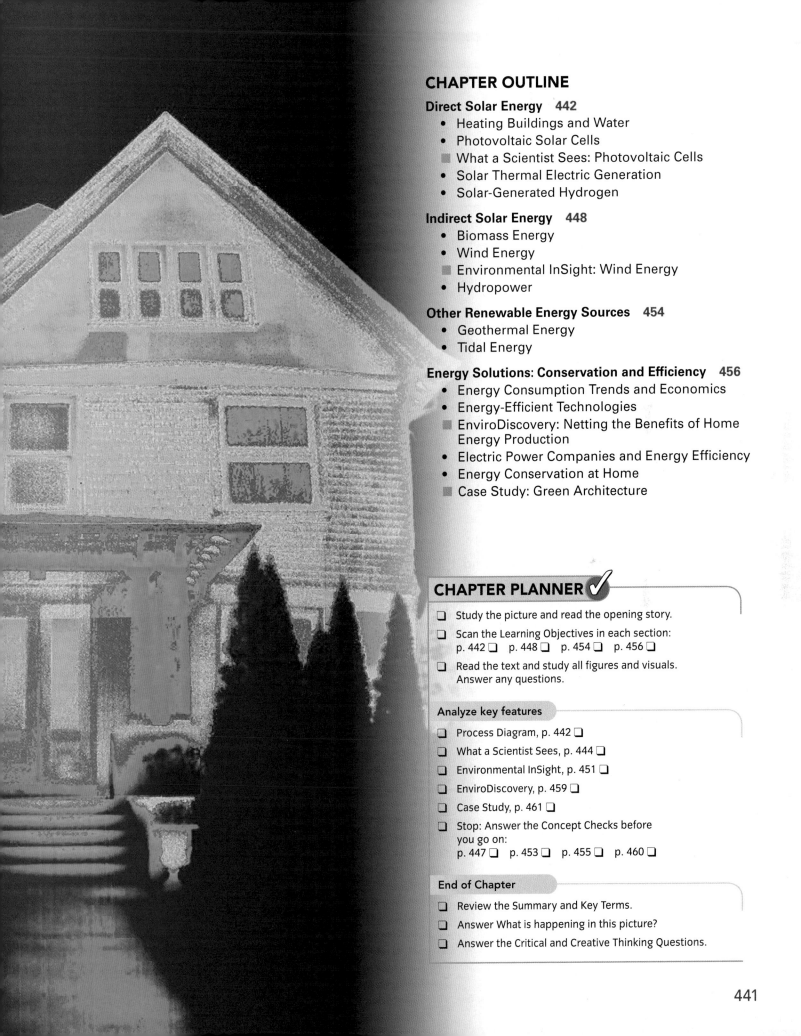

CHAPTER OUTLINE

CHAPTER PLANNER ✓

- ❏ Study the picture and read the opening story.
- ❏ Scan the Learning Objectives in each section:
  p. 442 ❏   p. 448 ❏   p. 454 ❏   p. 456 ❏
- ❏ Read the text and study all figures and visuals. Answer any questions.

Analyze key features

- ❏ Process Diagram, p. 442 ❏
- ❏ What a Scientist Sees, p. 444 ❏
- ❏ Environmental InSight, p. 451 ❏
- ❏ EnviroDiscovery, p. 459 ❏
- ❏ Case Study, p. 461 ❏
- ❏ Stop: Answer the Concept Checks before you go on:
  p. 447 ❏   p. 453 ❏   p. 455 ❏   p. 460 ❏

End of Chapter

- ❏ Review the Summary and Key Terms.
- ❏ Answer What is happening in this picture?
- ❏ Answer the Critical and Creative Thinking Questions.

Direct Solar Energy

LEARNING OBJECTIVES

1. **Distinguish** between active and passive solar heating and describe how each is used.
2. **Contrast** the advantages and disadvantages of photovoltaic solar cells and solar thermal electric generation in converting solar energy into electricity.
3. **Explain** how fuel cells work.

The sun produces a tremendous amount of energy, most of which dissipates into space. Only a small portion is radiated to Earth. Solar energy differs from fossil and nuclear fuels in that it is always available; we will run out of solar energy only when the sun's nuclear fire burns out. To make solar energy useful, we must collect it.

Heating Buildings and Water

In **active solar heating**, a series of collection devices mounted on a roof or in a field gather solar energy. The most common solar collection device is a panel or plate of black metal, which absorbs the sun's energy (**Figure 18.1**). Active solar heating is used primarily for heating water, either for household use or for swimming pools. Heat absorbed by a solar collector is transferred to a fluid inside the panel, which is then pumped to the heat exchanger, where the heat is transferred to water that will be stored in the hot water tank. Solar domestic water heating can

> **active solar heating** A system of putting the sun's energy to use in which collectors absorb solar energy and pumps or fans distribute the collected heat.

Active solar water heating • Figure 18.1

THE PLANNER

1 Solar panels are mounted on the roof of a building.

2 Sunlight enters the solar panels and warms liquid flowing through pipes to a heat exchanger.

3 Each solar panel is a box with a black metal base and glass covering.

4 In the heat exchanger, the fluid heats water, which then moves into a hot water storage tank.

5 A back-up heater runs on electricity or natural gas and keeps the water hot.

Hot water out

Hot water tank

Cold water in

Solar collector or panel

Pump

Heat exchanger

Pump

Back-up heater

Think Critically Where do you use hot water at home? How is it heated?

WILEY **RESOURCE KIT**

provide a family's hot water needs year-round. Because more than 8 percent of energy consumed in the United States goes toward heating water, active solar heating could potentially supply a significant amount of the nation's energy demand.

Active solar energy is not used for space heating as commonly as it is used for heating water, but it may become more important as natural gas, oil, and electricity prices continue to rise.

In **passive solar heating**, solar energy heats buildings without the need for pumps or fans to distribute the heat. Certain design features are incorporated into a passive solar heating system to warm buildings in winter and help them remain cool in summer (**Figure 18.2**). In the Northern Hemisphere, large south-facing windows receive more total sunlight during the day than windows facing other directions. Sunlight entering through the windows provides heat, which is then stored in floors and walls made of concrete, packed earth, or stone, or in containers of water. This stored

> **passive solar heating** A system of putting the sun's energy to use that does not require mechanical devices to distribute the collected heat.

heat is transmitted throughout the building naturally by **convection**, the circulation that occurs because warm air rises and cooler air sinks.

Buildings with passive solar heating systems must be well insulated so that accumulated heat doesn't escape. Depending on a building's design and location, passive heating can save as much as 50 percent of heating costs. Currently, about 7 percent of new homes built in the United States have passive solar heating features.

Photovoltaic Solar Cells

Photovoltaic (PV) solar cells can convert sunlight directly into electricity (see *What a Scientist Sees* on the next page). They are usually arranged on large panels that absorb sunlight even on cloudy or rainy days.

PV cells generate electricity with no pollution and minimal maintenance. They can be used

> **photovoltaic (PV) solar cell** A wafer or thin film of solid-state materials, such as silicon or gallium arsenide, that is treated with certain metals in such a way that the film generates electricity when solar energy is absorbed.

Passive solar heating • Figure 18.2

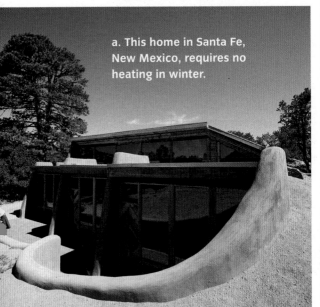

a. This home in Santa Fe, New Mexico, requires no heating in winter.

b. Several passive design features are incorporated into this home.

Summer sun

Winter sun

Overhang blocks summer sun.

Vent allows hot air to escape (in summer).

Attic and north-facing wall are heavily insulated.

South-facing double-paned glass allows winter light to enter directly into the room. Double panes reduce heat loss on cold nights.

Insulated drapes or window shades prevent heat loss at night (in winter).

Warm air escapes (in summer).

Thick adobe or stone walls and floor store heat (in winter).

WHAT A SCIENTIST SEES

Photovoltaic Cells

a. A student seeing the roof of the Intercultural Center of Georgetown University probably knows that it has arrays of photovoltaic (PV) cells to collect solar energy. The PV system supplies about 10 percent of the school's electricity.

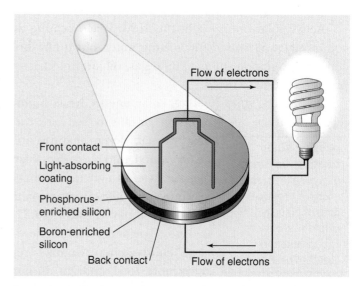

b. A scientist looking at those arrays knows that photovoltaic cells contain silicon and other materials. Sunlight excites electrons, which are ejected from silicon atoms. Useful electricity is generated when the ejected electrons flow out of the PV cells through a wire.

on any scale, from small portable modules attached to camping lanterns to large, multimegawatt power plants, and can power satellites, uncrewed airplanes, highway signals, wristwatches, and calculators. The widespread use of PV cells to generate electricity is currently limited by their low efficiency at converting solar energy to electricity and by the amount of land needed to hold the number of solar panels required for large-scale use. Several thousand acres of today's PV panels would be required to produce the electricity generated by a single large conventional power plant.

In remote areas not served by electric power plants, such as the rural areas of developing countries, it is more economical to use PV cells for electricity than to extend powerlines. Photovoltaics generate energy that can pump water, refrigerate vaccines, grind grain, charge batteries, and supply rural homes with lighting. According to the Institute for Sustainable Power,

more than 1 million households in developing countries have installed rooftop PV solar cells. A PV panel the size of two pizza boxes supplies a rural household with enough electricity for five lights, a radio, and a television.

Utility companies can purchase PV devices in modular units, which can become operational in a short period, allowing them to increase generating capacity in small increments. The PV units can provide the additional energy, for example, to power irrigation pumps on hot, sunny days.

The cost of manufacturing PV modules has steadily declined over the past 35 years, from an average factory price of almost $90 per watt in 1975 to about $4.00 per watt in 2010. The cost of producing electricity from PVs has steadily declined from 1970 to the present. Despite this progress, in 2009 the cost was still about $.15 to $.25 per kilowatt-hour. **Table 18.1** compares the costs

Generating costs of electric power plants, 2009 • Table 18.1

Energy source	Generating costs (cents per kilowatt-hour)*
Hydropower	2–10
Biomass	6–9
Geothermal	3–8
Wind	4–7
Solar thermal	5–13
Photovoltaics	15–25
Natural gas	5–7
Coal	6–8
Nuclear power	6–8

*Electricity production and consumption are measured in kilowatt-hours (kWh). As an example, one 50-watt light bulb that is on for 20 hours uses 1 kWh of electricity (50 × 20 = 1000 watt-hours = 1 kWh).

of generating electricity using different energy sources, including photovoltaics.

Future technological progress may make PVs economically competitive with electricity produced using conventional energy sources. The production of "thin-film" solar cells (**Figure 18.3**), which are much less expensive to manufacture than standard PVs, has decreased costs for PVs across the board. More than 120,000 Japanese homes have installed PV solar-energy roofing in the past few years. Through the Million Solar Roofs initiative, California expects to have solar roofing on 1 million buildings by 2016. Another technological advance that shows promise is dye-sensitized solar cells, which can be produced at about one-fifth the cost of conventional silicon panels.

While operating PVs creates no air or water pollution, the manufacturing process requires industrial chemicals, many of which are toxic. Finding cleaner ways to produce solar panels will be important as more are produced.

Solar shingles • Figure 18.3

These thin-film solar cells look much like conventional roofing materials.

Solar Thermal Electric Generation

Systems that concentrate solar energy to heat fluids have long been used for buildings and industrial processes. In **solar thermal electric generation**, electricity is produced by systems that collect sunlight and concentrate it using mirrors or lenses to heat a fluid to high temperatures.

In one such system, computer-guided trough-shaped mirrors track the sun for optimum efficiency, center sunlight on nearby oil-filled pipes, and heat the oil to 390°C (735°F) (**Figure 18.4**). The hot oil is circulated to a water storage system and used to boil water into super-heated steam, which turns a turbine to generate electricity.

Solar thermal systems often have a backup—usually natural gas—to generate electricity at night and during cloudy days when solar power isn't operating. The world's largest solar thermal system of this type currently operates in the Mojave Desert in southern California.

Solar thermal energy systems are inherently more efficient than other solar technologies because they concentrate the sun's energy. With improved engineering,

> **solar thermal electric generation**
> A means of producing electricity in which the sun's energy is concentrated using mirrors or lenses onto a fluid-filled pipe; the heated fluid is used to generate electricity.

manufacturing, and construction methods, solar thermal energy is becoming cost-competitive with fossil fuels (see Table 18.1). In addition, the environmental benefits of solar thermal plants are significant: These plants don't produce air pollution or contribute to acid rain or global climate change.

Solar-Generated Hydrogen

Increasingly, people think of hydrogen as the fuel of the future, as it is abundant as well as easily produced. Electricity generated by any energy source can split water into the gases oxygen and hydrogen. Consequently, producing hydrogen results in the same environmental and security problems associated with the underlying energy source, including those discussed in Chapter 17.

Hydrogen is a clean fuel; it produces water and heat as it burns, but it produces no sulfur oxides, carbon monoxide, hydrocarbon particulates, or CO_2 emissions. It does produce some nitrogen oxides, though in amounts fairly easy to control. Hydrogen has the potential to provide energy for transportation (in the form of hydrogen-powered automobiles) as well as for heating buildings and producing electricity.

Solar thermal electric generation • Figure 18.4

a. A solar thermal plant in California uses troughs to focus sunlight on a fluid-filled tube, as shown in **b.** The heated oil is pumped to a water tank, where it generates steam used to produce electricity. For simplicity, arrows show sunlight converging on several points; sunlight actually converges on the pipe throughout its length.

Sunlight

Pipe with heated oil

Solar collector

Fuel cells • Figure 18.5

a. These experimental fuel cells combine hydrogen and oxygen to create electricity.

Flow of negatively charged electrons provides electricity

Fuel (hydrogen) →

Air (oxygen)

$H_2 \rightarrow 2H^+ + 2e$

H_2O

Electrolyte solution

Electrodes

b. Cross-section of a fuel cell.

c. Fuel cells can be adapted to many applications.

It may seem wasteful to use electricity generated from solar energy to make hydrogen that will then be used to generate electricity. However, the electricity generated by existing photovoltaic cells must be used immediately, whereas hydrogen offers a convenient way to store solar energy as chemical energy. It can be transported by pipeline, possibly less expensively than electricity is transported by wire.

Production of hydrogen from PV electricity currently has a relatively low efficiency (perhaps 10 percent), which means that very little of the solar energy absorbed by the PV cells is actually converted into the chemical energy of hydrogen fuel. Low efficiency translates into high costs. Scientists are working to improve this efficiency and make solar-generated hydrogen fuel commercially viable.

We face other challenges besides high costs if we are to replace gasoline with hydrogen as a transportation fuel. First, we would need to develop a complex infrastructure (such as hydrogen pipelines) to provide hydrogen to service stations. Because hydrogen is extremely volatile, it must be stored, handled, and transported very carefully. Another challenge is developing **fuel cells** for

> **fuel cell** A device that directly converts chemical energy into electricity. A fuel cell requires hydrogen and oxygen from the air.

motor vehicles that are inexpensive, safe, and allow the vehicle to drive a long distance without the need to refuel. A fuel cell is an electrochemical cell similar to a battery (**Figure 18.5**). Fuel cells represent the most promising way to use hydrogen.

Whereas batteries store a fixed amount of energy, fuel cells produce power as long as they are supplied with fuel. Fuel cells are available to power everything from cell phones to city buses. However, for most applications (including automobiles), they remain very expensive.

CONCEPT CHECK   STOP

1. **What** is active solar energy? passive solar energy?

2. **What** are the advantages of producing electricity by solar thermal energy? using hydrogen and photovoltaic (PV) solar cells?

3. **How** do fuel cells work?

Indirect Solar Energy

LEARNING OBJECTIVES

1. **Define** *biomass* and outline its use as a source of energy.

2. **Compare** the potential of wind energy and hydropower.

Some renewable energy sources indirectly use the sun's energy. Combustion of *biomass* (organic matter) is an example of indirect solar energy because plants use solar energy for photosynthesis and store the energy in biomass. Windmills, or wind turbines, use *wind energy* to generate electricity. The damming of rivers and streams to generate electricity is a type of *hydropower*—the energy of flowing water.

Biomass Energy

Biomass, one of the oldest fuels known to humans, consists of materials such as wood, fast-growing plant and algal crops, crop wastes, sawdust and wood chips, and animal wastes. Biomass contains chemical energy that comes from the sun's radiant energy, which photosynthetic organisms use to form organic molecules. Biomass is a renewable form of energy if managed properly.

> **biomass** Plant and animal material used as fuel.

Biomass fuel, which may be a solid, liquid, or gas, is burned to release its energy. Solid biomass fuels such as wood, charcoal (wood turned into coal by partial burning), animal dung, and peat (partly decayed plant matter found in bogs and swamps) supply a substantial portion of the world's energy. At least half of the human population relies on biomass as their main source of energy. In developing countries, wood is the primary fuel for cooking and heat (**Figure 18.6**).

It is possible to convert biomass, particularly animal wastes, into **biogas**. Biogas, which is usually composed of a mixture of gases (mostly methane), is like natural gas. It is a clean fuel—its combustion produces fewer pollutants than either coal or biomass. In India and China, several million family-sized **biogas digesters** use microbial decomposition of household and agricultural

Biomass • Figure 18.6

Firewood is the major energy source for most of the developing world. Photographed in Garadawa, Niger.

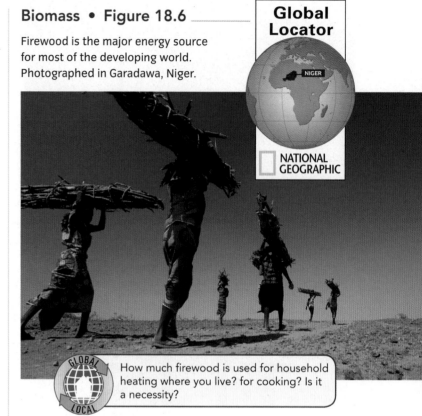

GLOBAL LOCAL How much firewood is used for household heating where you live? for cooking? Is it a necessity?

wastes to produce biogas for cooking and lighting (**Figure 18.7**). When biogas conversion is complete, the solid remains are removed from the digester and used as fertilizer.

Biogas has the potential to power fuel cells to generate electricity. A pilot program at Boston's main sewage treatment plant began producing electricity from biogas in 1997. Sewage sludge in large biogas digesters produces methane, which is then burned in a methane fuel cell to produce enough electricity for 150 homes. Like the hydrogen fuel cells discussed earlier in the chapter, methane fuel cells produce relatively few pollutants.

Biomass can also be converted into liquid fuels, especially **methanol** (methyl alcohol) and **ethanol** (ethyl alcohol), which can replace gasoline in internal combustion engines. In many parts of the world, automotive fuels must contain 10 percent or more ethanol. *Biodiesel*, made from plant or animal oils, is becoming popular as an alternative fuel for diesel engines in trucks, farm equipment, and boats. The oil is often refined from

Biogas digester in India • Figure 18.7

This small-scale biogas digester is being evaluated at a research center. Animal manure placed in the digester decomposes, releasing methane gas that can be used as cooking fuel.

waste oil produced at restaurants (such as the oil used to make french fries); biodiesel burns much cleaner than diesel fuel.

Some U.S. energy companies convert sugarcane, corn, or wood crops to alcohol; others are interested in the commercial conversion of agricultural and municipal wastes into ethanol. Currently, the profitability of ethanol is possible only because of government **subsidies** that reduce ethanol's cost.

Biomass is attractive as a source of energy—and popular with U.S. politicians—because it reduces dependence on fossil fuels. It is popular with consumers because they can easily shift from gasoline and diesel without disruptive lifestyle changes. Also, biomass often makes use of waste products, thereby reducing our waste disposal problem. Although it is not completely free of the pollution problems of fossil fuels, biomass combustion produces levels of sulfur and ash that are lower than those that coal produces.

Some problems associated with obtaining energy from biomass include the use of land and water that might otherwise be dedicated to agriculture. This shift toward energy production might decrease food production, contributing to higher food prices and reducing food supplies even as population is growing.

Also, as mentioned earlier, at least half of the world's population relies on biomass as its main source of energy. Unfortunately, in many areas people burn wood faster than they replant trees. Intensive use of wood for energy has resulted in severe damage to the environment, including soil erosion, deforestation and desertification, air pollution, and degradation of water supplies.

Excessive use of crop biomass can also harm soil quality. Crop residues, such as cornstalks, are increasingly being used for energy. Crop residues left in and on the ground prevent erosion by holding the soil in place; their removal would eventually deplete the soil of minerals and reduce its productivity.

Wind Energy

Wind results from the sun warming the atmosphere. **Wind energy** is an indirect form of solar energy: The radiant energy of the sun is transformed into mechanical energy through the movement of air molecules. Wind is sporadic over much of Earth's surface, varying in direction and magnitude. Like direct solar energy, wind energy is a highly dispersed form of energy. Harnessing wind energy to generate electricity has great potential.

> **wind energy**
> Electric energy obtained from surface air currents caused by the solar warming of air.

New wind turbines are huge—100 m (328 ft) tall—and have long blades designed to harness wind energy efficiently (**Figure 18.8a, b**). As turbines have become larger and more efficient, costs for wind power have declined rapidly—from $.40 per kilowatt-hour in 1980 to a current cost of $.04 to $.07 per kilowatt-hour. Wind power is cost-competitive with many forms of conventional energy. Advances such as turbines that use variable-speed operation may make wind energy an important global source of electricity during the first half of the 21st century.

During the 1990s and early 2000s, wind became the world's fastest-growing source of energy (**Figure 18.8c**). Germany and the United States currently lead the world as the top producers of wind energy. Denmark currently generates 21 percent of its electricity using wind energy. Much of this power is generated offshore because ocean winds are strong. Other leading wind energy countries include Spain and India.

Harnessing wind energy is most profitable in areas with consistent winds, such as islands, coastal areas, mountain passes, and grasslands. The world's largest concentration of wind turbines is currently located in the Tehachapi Pass at the southern end of the Sierra Nevada mountain range in California.

In the continental United States, some of the best locations for large-scale electricity generation from wind energy are on the Great Plains. The 10 states with the greatest wind energy potential, according to the American Wind Energy Association, are North Dakota, Texas, Kansas, South Dakota, Montana, Nebraska, Wyoming, Oklahoma, Minnesota, and Iowa. In fact, if we developed the wind energy in North Dakota, Texas, and Kansas to their full potential, we could supply enough electricity to meet the current needs of the entire United States! Wind power projects are under way in these and many other states.

Currently, wind energy is captured and placed into regional electricity grids. Deploying wind energy on a national scale—for example, wind energy produced in Texas and used in New York City—requires the development of new technologies for storing and distributing energy.

Wind produces no waste and is a clean source of energy. It produces no emissions of sulfur dioxide, carbon dioxide, or nitrogen oxides. Every kilowatt-hour of electricity generated by wind power rather than fossil fuels prevents as much as 1 kg (2.2 lb) of the greenhouse gas CO_2 from entering the atmosphere.

Although the use of wind power doesn't cause major environmental problems, one concern is the deaths of birds and bats. The California Energy Commission estimated that several hundred birds turned up dead in the vicinity of the 7000 turbines at Altamont Pass in California during a 2-year study; most had collided with the turbines. Studies later determined that Altamont Pass is a major bird migration pathway. Wind power sites have implemented technical "fixes" such as painted blades and anti-perching devices, or have shut down operations during peak bird migration periods. Developers of future wind farm sites currently conduct voluntary wildlife studies and try to locate sites away from bird and bat routes.

Not all people welcome wind power projects with open arms. The Maple Ridge Wind Farm in upstate New York has almost 200 windmills. Many local residents appreciate the extra money that wind-farm leases provide to the local economy. However, others think the wind turbines ruin their view of the Adirondack Mountains. A similar dispute is occurring over the proposed Nantucket Sound wind farm off the coast of Massachusetts. This 130-turbine wind farm, if built, will be the first offshore wind project in the United States. Massachusetts has some of the highest electricity costs in the nation, and proponents of the wind farm point out that wind energy will help lower energy costs. However, many people are concerned that the wind farm will adversely affect the local economy because tourists will find it offensive.

Rotor blade

Rotor diameter

Gear box   Generator

Wind direction for an upwind rotor

Drive shaft

Tower

a. Kansas Wind Farm. Wind energy represents a promising alternative that could diminish our dependence on fossil fuels.

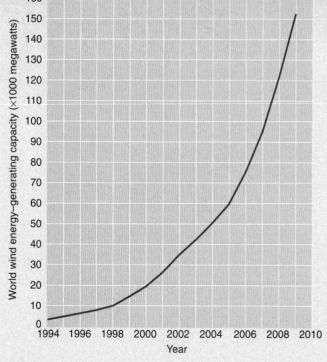

b. Wind Turbine Design. This basic wind turbine design has a horizontal axis (horizontal refers to the orientation of the drive shaft). Airflow causes the turbine's blades to turn 15 to 60 revolutions per minute (rpm). As the blades turn, gears within the turbine spin the drive shaft. This spinning powers the generator, which sends electricity through underground cables to a nearby utility. (The tower isn't drawn to scale and is much taller than depicted.)

Interpreting Data
What was the world wind energy–generating capacity in 1998? 2008? If the trend in this figure continues, what will it be in 2018?

NATIONAL GEOGRAPHIC

WILEY RESOURCE KIT

c. Trends in wind energy use. Global wind energy has shown record growth in recent years.

Hydropower

The sun's energy drives the hydrologic cycle, which includes precipitation, evaporation, transpiration, and drainage and runoff (see Figure 5.9). As water flows from higher elevations back to sea level through rivers and streams, dams can harness and make use of its energy. The potential energy of water held back by a dam is converted to kinetic energy as the water turns turbines to generate electricity (**Figure 18.9**). **Hydropower**, which is more concentrated than solar energy, is more efficient than any other energy source for producing electricity; about 90 percent of available hydropower energy is converted into consumable electricity.

> **hydropower**
> A form of renewable energy that relies on flowing or falling water to generate electricity.

Hydropower generates approximately 19 percent of the world's electricity, making it the form of solar energy in greatest use. The 10 countries with the greatest hydroelectric production are, in decreasing order, Canada, the United States, Brazil, China, Russia, Norway, Japan, India, Sweden, and France. In the United States, approximately 2200 hydropower plants produce about 7 percent of the country's electricity, the most of any renewable energy source. Highly developed countries have already built dams at most of their potential sites. In many developing nations—particularly in undeveloped, unexploited parts of Africa and South America—hydropower represents a great potential source of electricity.

Hydroelectric power • Figure 18.9

a. Water generates electricity as it moves through the Dalles Dam, which spans the Columbia River between Oregon and Washington.

b. A controlled flow of water released down the penstock turns a turbine, which generates electricity.

Traditional hydropower technology is only suited for large dams with rapidly flowing water, such as those found on the Columbia River and its tributaries in the Pacific Northwest and British Columbia. Therefore, only about 3 percent of existing U.S. dams generate electricity. New designs allow modern turbines to harness electricity from large, slow-moving rivers or from streams with small flow capacities. These new technologies have the potential to increase the amount of hydroelectric power generated by existing dams.

Building a dam changes the natural flow of a river: Water backs up, flooding large areas of land and forming a reservoir, which destroys plant and animal habitats. Native fishes are particularly being harmed by dams because the original river ecosystem is so altered. The migration of spawning fish is also altered (see Chapter 10). Below the dam, the once-powerful river is reduced to a relative trickle. The natural beauty of the countryside is affected, and certain forms of wilderness recreation are made impossible or less enjoyable, although the dams permit water sports in the reservoir.

With at least 200 large dams around the world, earthquakes may occur during and after the filling of the reservoir behind the dam. The larger the reservoir and the faster it is filled, the greater the intensity of seismic activity. An area need not be seismically active to have earthquakes induced by reservoirs.

If a dam breaks, people and property downstream may be affected. In addition, waterborne diseases may spread through the population. **Schistosomiasis** is a tropical disease caused by a parasitic worm. As much as half the population of Egypt suffers from this disease, largely as a result of the Aswan Dam, built on the Nile River in 1902 to control flooding and used since 1960 to provide electric power. The large reservoir behind the dam provides habitat for the worm, which spends part of its life cycle in the water. Humans are infected by the worm through bathing, swimming, walking barefoot along water banks, and drinking infected water.

In arid regions, the creation of a reservoir results in greater evaporation because it has a larger surface area in contact with the air than did the stream or river.

As a result, serious water loss and increased salinity of the remaining water may occur.

When an area behind a dam is flooded, the trees and other vegetation die and are decomposed, releasing the large quantities of carbon that were tied up in organic molecules in the plant bodies as carbon dioxide and methane. These gases absorb infrared radiation and are therefore associated with global climate change.

The construction of large dams involuntarily displaces people from lands flooded by reservoirs. The *Three Gorges Project*, the largest dam ever built, was recently completed on the Yangtze River in China (see Chapter 10). The reservoir behind this dam is 632 km (412 mi) long. The tops of as many as 100 mountains have become small islands, fragmenting habitat and threatening 57 endangered species. The reservoir has displaced almost 2 million people, the largest number for any dam project.

The environmental and social impacts of a dam may not be acceptable to the people living in a particular area. Laws prevent or restrict the building of dams in certain locations. In the United States, the Wild and Scenic Rivers Act prevents the hydroelectric development of certain rivers, although this law protects less than 1 percent of the nation's total river system. Other countries, such as Norway and Sweden, have similar laws.

Dams cost a great deal to build but are relatively inexpensive to operate. A dam has a limited life span, usually 50 to 100 years, because over time the reservoir fills in with silt until it cannot hold enough water to generate electricity. This trapped silt, which is rich in nutrients, is prevented from enriching agricultural lands downstream. For example, Egypt must rely on heavy applications of chemical fertilizer downstream from the Aswan Dam to maintain the fertility of the Nile River Valley.

CONCEPT CHECK 🛑 STOP

1. **What** is biomass, and how is it used?

2. **What** are the advantages and disadvantages of using wind to produce electricity? of using hydropower to produce electricity?

Other Renewable Energy Sources

LEARNING OBJECTIVE

1. **Describe** geothermal energy and tidal energy.

Geothermal energy and tidal energy are renewable energy sources that are not derived from solar energy. *Geothermal energy* is the naturally occurring heat within Earth. This heat is used for space heating and to generate electricity. *Tidal energy*, caused by the changes in water level between high and low tides, is exploited to generate electricity on a limited scale.

Geothermal Energy

Geothermal energy, the natural heat within Earth, arises from Earth's core, from friction along continental plate boundaries, and from the decay of radioactive elements. The amount of geothermal energy is enormous. Scien-

> **geothermal energy** Energy from Earth's interior, used for space heating or generation of electricity.

tists estimate that just 1 percent of the heat contained in the uppermost 10 km (6 mi) of Earth's crust is equivalent to 500 times the energy contained in all of Earth's oil and natural gas resources.

Geothermal energy is typically associated with volcanism. Large underground reservoirs of heat exist in areas of geologically recent volcanism. As groundwater in these areas travels downward and is heated, it becomes buoyant and then rises until it is trapped by an impermeable layer in Earth's crust, forming a **hydrothermal reservoir**. Hydrothermal reservoirs contain hot water and possibly steam, depending on the temperature and pressure of the fluid. Some of the hot water or steam may escape to the surface, creating hot springs or geysers. Hot springs have been used for thousands of years for bathing, cooking, and heating buildings. Drilling a well brings the hot fluid from a hydrothermal reservoir to the surface, where a power station may use it to supply heat directly to consumers or to generate electricity (**Figure 18.10a**). The electricity these power stations generate is inexpensive and reliable.

The United States is the world's largest producer of geothermal electricity. Electric power is currently produced at 17 different geothermal fields in California, Nevada, Utah, and Hawaii. The world's largest geothermal power plant—The Geysers in northern California—provides electricity for 1.7 million homes. Other important producers of geothermal energy include the Philippines, Italy, Japan, Mexico, Indonesia, and Iceland (**Figure 18.10b**).

Iceland, a country with minimal oil and natural gas resources, is located on the mid-Atlantic ridge, a boundary between two continental plates. Iceland is therefore an island of intense volcanic activity with considerable geothermal resources. Iceland uses geothermal energy to generate electricity to heat two-thirds of its homes. In addition, most of the fruits and vegetables required by the people of Iceland are grown in geothermally heated greenhouses.

Is geothermal energy renewable? As a source of heat for geothermal energy, the planet is inexhaustible on a human time scale. However, the water used to transfer the heat to the surface isn't inexhaustible. Some geothermal applications recirculate all the water back into the underground reservoir, ensuring many decades of heat extraction from a given reservoir.

Geothermal energy is considered environmentally benign because it emits only a fraction of the air pollutants released by conventional fossil fuel–based energy technologies. The most common environmental hazard is the emission of hydrogen sulfide (H_2S) gas, which comes from the very low levels of dissolved minerals and salts found in the steam or hot water. Hydrogen sulfide smells like rotten eggs and is toxic to humans in high concentrations. A lesser concern is that the surrounding land may subside, or sink, as the water from hot springs and their connecting underground reservoirs is removed.

Scientists are studying how to economically extract some of the vast amount of geothermal energy stored in hot, dry rock. Such a technology could greatly expand the extent and use of geothermal resources.

Heating and Cooling Buildings with Geothermal Energy Geothermal heat pumps (GHPs) take advantage of the difference in temperature between Earth's surface and subsurface (at depths from 1 m to about 100 m or about 3 to 300 ft). In an underground arrangement of pipes containing circulating fluids, GHPs extract natural heat in winter, when Earth acts as a heat source, and transfer excess heat underground in summer, when

Geothermal power plant • Figure 18.10

a. Steam separated from hot water pumped from underground turns a turbine and generates electricity.

b. A geothermal power plant in Iceland.

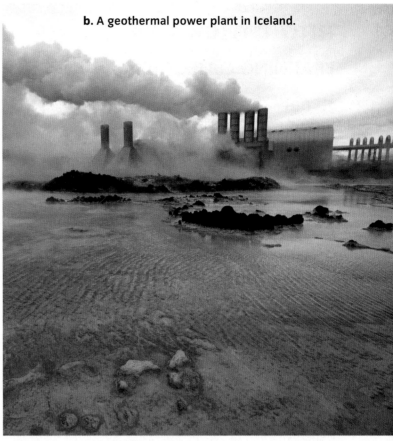

Earth acts as a heat sink. Geothermal heating systems can be modified to provide supplemental hot water.

Though GHPs have been available for many years, they aren't widely used because their installation is expensive. However, with the growth of *green architecture* (see the *Case Study* at the end of this chapter) and rising fuel costs, commercial and residential use of GHPs is on the rise. The system's benefits include low operating costs—which may be half those of conventional systems—and high efficiency.

Tidal Energy

Tides, or the rise and fall of the surface waters of the ocean and seas that occur twice each day, are the result of the gravitational pull of the moon and the sun. A dam built across a bay can harness the energy of large tides to generate electricity. As the tide falls, water flowing back to the ocean over the dam's spillway turns a turbine and generates electricity through **tidal energy**.

Power plants using tidal power are in operation in France, Russia, China, and Canada. Tidal energy can't become a significant resource worldwide because few areas experience large enough differences in water level between high and low tides to make power generation feasible. The most promising locations for tidal power in North America include the Bay of Fundy in Nova Scotia, Passamaquoddy Bay in Maine, Puget Sound in Washington, and Cook Inlet in Alaska.

Other problems associated with tidal energy include the high cost of building a tidal power station and potential environmental problems associated with tidal energy in **estuaries**, coastal areas where river currents meet ocean tides. Fishes and countless invertebrates migrate to estuaries to spawn. Building a dam across the mouth of an estuary would prevent these animals from reaching their breeding habitats.

CONCEPT CHECK **STOP**

1. **What** are the pros and cons of using geothermal energy to produce electricity and what are the pros and cons of using tidal power to produce electricity?

Energy Solutions: Conservation and Efficiency

LEARNING OBJECTIVES

1. **Distinguish** between energy conservation and energy efficiency and give examples of each.
2. **Summarize** options to conserve energy at home.

Human requirements for energy will continue to increase, if only because the human population is growing. In addition, energy consumption continues to increase as developing countries raise their standards of living. We must therefore place a high priority not only on developing alternative sources of energy but on **energy conservation** and **energy efficiency**.

To illustrate the difference between energy conservation and energy efficiency, let's consider automobile gasoline consumption. Energy conservation measures to reduce gasoline consumption would include carpooling and lowering driving speeds, whereas energy efficiency measures would include designing and manufacturing automobiles that travel farther on a gallon of fuel. Conservation and efficiency accomplish the same goal—saving energy.

Many energy experts consider energy conservation and energy efficiency the most promising energy "sources" available because they save energy for future use and buy us time to explore new energy alternatives. Developing technologies for energy conservation and efficiency costs less than developing new sources or supplies of energy; the technologies also improve the economy's productivity. The adoption of energy-efficient technologies generates new business opportunities, including the research, development, manufacture, and marketing of those technologies.

> **energy conservation**
> Using less energy—by reducing energy use and waste, for example.
>
> **energy efficiency**
> Using less energy to accomplish a given task—by using new technology, for example.

Energy-efficient technologies and greater efforts at conservation would also provide important environmental benefits by reducing air pollution. Carbon dioxide emissions that contribute to global climate change, acid precipitation, and other environmental problems are related to large quantities of energy production and consumption.

Energy Consumption Trends and Economics

Although the U.S. economy has become more energy efficient, total energy consumption has increased in recent years, partly due to an increase in population. The per person energy consumption in developing nations is substantially less than that in industrialized countries (see Figure 17.1a), but the greatest per person increase in energy consumption today is occurring in these developing nations, particularly in China and India. The rising energy demand in developing nations is caused by increases in economic development and population, as well as the use of older, less expensive, and less energy-efficient technologies.

Developing countries are forced to balance developing their economies with controlling environmental degradation. At first glance, these two goals appear mutually exclusive. However, both can be realized by adopting the new technologies now being developed in industrialized nations to achieve greater energy efficiency.

Energy-Efficient Technologies

The development of more efficient appliances, automobiles, buildings, and industrial processes has helped reduce energy consumption in highly developed countries. Compact fluorescent lightbulbs produce light of comparable quality to that of incandescent lightbulbs but require only 25 percent of the energy and last up to 15 times longer. Although relatively expensive, the energy-efficient bulbs more than pay for themselves in energy savings. New condensing furnaces require approximately 30 percent less fuel than conventional gas furnaces. "Superinsulated" buildings use 70 to 90 percent less heat energy than buildings insulated using standard methods (**Figure 18.11**).

The **National Appliance Energy Conservation Act (NAECA)** sets national appliance efficiency standards for refrigerators, freezers, washing machines, clothes dryers, dishwashers, room air conditioners, and ranges and ovens (including microwaves). For example, refrigerators built today consume 75 percent less energy than comparable models built in the mid-1970s.

Energy costs often account for 30 percent of a company's operating budget. It makes good economic sense for businesses in older buildings to invest in energy improvements, which often pay for themselves in a few years. Implementing energy improvements may be as simple as fine-tuning existing heating, ventilation, and air conditioning systems or as major as replacing all the windows and lights. Both the environment and a company's bottom line can benefit from any energy improvements.

Automobile efficiency has improved dramatically since the mid-1970s as a result of the use of lighter materials and designs that reduce air drag. The average fuel efficiency of new passenger cars doubled between the mid-1970s and the mid-1980s. It declined after that, as larger vehicles became more popular; light trucks and sport utility vehicles (SUVs) are now a substantial fraction of the passenger vehicles in the United States. Further, significant gains could easily be made using current technology. Automobiles with fuel efficiencies of 60 to 65 mpg could be routinely manufactured

Superinsulated buildings • Figure 18.11

a. A superinsulated home is so well insulated and airtight that it doesn't require a furnace in winter. Heat from the inhabitants, lightbulbs, and appliances provides most or all the necessary heat.

b. A superinsulated office building in Toronto, Canada, has south-facing windows with insulating glass. The building is so well insulated it uses no furnace.

Excellent insulation throughout walls, floors, and roof; almost airtight

Few or no windows on north, east, and west sides

Small, south-facing windows with insulating glass

No large furnace

Concrete foundation

Air-to-air heat exchanger provides ventilation to improve air quality.

The Daihatsu prototype UFE (ultra fuel economy) III hybrid can achieve 169 miles per gallon.

within the next decade or so, and manufacturers are developing even more efficient models for the future (**Figure 18.12**).

Cogeneration One energy technology that has a bright future is **cogeneration**, or **combined heat and power (CHP)**. Cogeneration involves the production of two useful forms of energy from the same fuel. A CHP system generates electricity, and then the steam produced during this process is used rather than wasted. The system's overall conversion efficiency (that is, the ratio

> **cogeneration (CHP)** An energy technology that involves recycling "waste" heat.

of useful energy produced to fuel energy used) is high because some of what is usually waste heat is incorporated into the process.

Cogeneration can be cost-effective on both small and large scales. Modular CHP systems enable hospitals, hotels, restaurants, factories, and other businesses to harness steam that would otherwise be wasted to heat buildings, cook food, or operate machinery before it cools and gets pumped back into the boiler as water (**Figure 18.13**). Larger CHP systems can produce electricity for local utilities.

Cogeneration • Figure 18.13

In this example of a cogeneration system, fuel combustion generates electricity in a generator. The electricity produced is used in-house or sold to a local utility. The waste heat (leftover hot gases or steam) is recovered for useful purposes, such as industrial processes, heating of buildings, hot water heating, and generation of additional electricity.

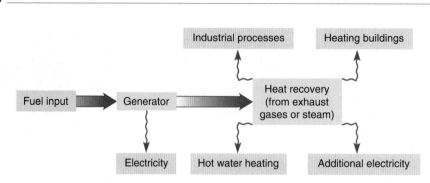

EnviroDiscovery

Netting the Benefits of Home Energy Production

Environmentally aware homeowners have often paid a large economic price when installing renewable energy sources such as solar panels or wind-driven generators. Such systems can be expensive to purchase and install, and they provide electricity only intermittently, when weather conditions permit. During unfavorable conditions, residents must purchase electricity from their local utility.

In a growing trend, utility companies are permitting homeowners who produce their own energy to "net meter" their electricity. Any excess energy homeowners generate is supplied to the utility's power grid, and the homeowners' electric meters run backward. Net metering offsets energy costs over a billing period by essentially providing individuals the full retail price for the electricity they generate. By itself, net metering doesn't turn residential energy generation into an inexpensive prospect, but it certainly makes it more affordable, encouraging the additional development of renewable energy for homes. Some utilities and governments have made it easier for consumers to finance installations, for example by providing tax incentives or promoting long-term loans.

Electric meters can run backward for homeowners participating in net metering.

Electric Power Companies and Energy Efficiency

Changes in the regulations that govern electric utilities allow these companies to make more money by generating less electricity. Such programs provide incentives for energy conservation and thereby reduce power plant emissions that contribute to environmental problems.

Electric utilities can often avoid the massive expenses of building new power plants or purchasing additional power by helping electricity consumers save energy. Some utilities support energy conservation and efficiency by offering cash awards to consumers who install energy-efficient technologies. Other utilities give customers energy-efficient compact fluorescent lightbulbs, air conditioners, or other appliances. They then charge slightly higher rates or a small leasing fee, but the greater efficiency results in savings for both the utility company and the consumer.

The utility company makes more money from selling less electricity because it does not have to invest in additional power generation to meet increased demand. The consumer saves because the efficient lightbulbs or appliances use less energy, which more than offsets the higher rates.

According to the American Council for an Energy-Efficient Economy, U.S. electric power plants are themselves an important target for improved energy efficiency. Much heat is lost during the generation of electricity. If all that wasted energy were harnessed—by cogeneration, for example—it could be used productively, thereby conserving energy.

Another way to increase energy efficiency would be to improve our electric grids because about 10 percent of electricity is lost during transmission. To accomplish this, some energy experts envision that future electricity will be generated far from population centers, converted to supercooled hydrogen, and transported through underground superconducting pipelines. The technology to build such conduits has not yet been developed.

Some energy-saving measures for the home • Figure 18.14

Close doors and partially close vents in unused rooms.

Set winter thermostat at 68° or lower and summer thermostat at 78°F or higher.

Use compact fluorescent bulbs. Turn off lights when leaving a room.

Install programmable thermostat.

Close damper when fireplace is not in use. Install tempered glass fireplace doors.

Wash only full loads in washer, dryer, and dishwasher. Wash clothes in cold water.

Unplug televisions and other appliances when away for an extended period.

Do not block vents.

Clean or replace air filters in heating/cooling systems. Check ducts for leakage.

Set water heater at 140°F with dishwasher or 120°F without. Wrap older water heaters and pipes with insulation.

Keep coils clean on refrigerator. Also, keep refrigerator full.

Use a microwave rather than stove to heat/cook small portions of food.

Install water-saving faucets and shower heads. Take short showers.

Caulk and weather-strip windows and doors.

Energy Conservation at Home

The average U.S. household spends several thousand dollars each year on utility bills. This cost could be reduced considerably with investment in energy-efficient technologies (**Figure 18.14**). Although a more energy-efficient house might cost more up front, depending on the technologies employed, the improvements usually pay for themselves in 2 or 3 years. Energy efficiency has become an essential element of design codes nationwide.

Some energy-saving improvements, such as thicker wall insulation, are easiest to install while a home is being built. Other improvements can be made to older homes to reduce heating and cooling costs and enhance energy efficiency. Examples include installing thicker attic insulation, installing storm windows and doors, caulking cracks around windows and doors, replacing inefficient furnaces and refrigerators, and adding heat pumps. Most local utility companies will also perform an energy audit on a home for little or no charge.

CONCEPT CHECK STOP

1. **What** is the difference between energy conservation and energy efficiency?

2. **How** can you conserve energy at home?

CASE STUDY

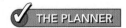

Green Architecture

A recent addition to the Midtown Manhattan skyline represents a new standard in green architecture in New York City (**see photograph**). Completed in 2006, the Hearst Tower, home of Hearst Publishing, was designed to achieve a degree of energy efficiency 26 percent higher than that of standard office buildings. The building's innovations earned it the city's first Gold LEED (Leadership in Energy and Environmental Design) certification from the U.S. Green Building Council (USGBC).

The Hearst Tower's striking "diagrid" frame, which incorporates a system of steel and glass triangles, floods the interior with natural light while using approximately 2000 tons less steel than a conventional frame. Most of the steel (90 percent) used in the 46-story, 80,000 m^2 (856,000-square-ft) building is recycled.

Other energy-saving design features in the Hearst Tower boost the efficiency of the building's heating and cooling systems. Windows are coated to reduce solar radiation, and heating and air conditioning equipment cools and ventilates using only outside air for three-quarters of the year. In addition, embedded polyethylene within the atrium's limestone floor circulates water for both cooling and heating. Unique to the building's design, a 10-story waterfall called the "Icefall" chills the atrium. The Icefall's water supply comes from rainwater collected at the roof and drawn into a 53,000-L (14,000-gal) basement reclamation tank. This water is also used to irrigate the building's plants.

Natural light is enhanced by the placement of few internal walls and only low partitions. Sensors turn off lights in empty rooms and control the amount of artificial light provided based on the natural light being received.

Finally, building health is protected by several interior features. Low-vapor paints and low-toxicity sealants coat surfaces, while furniture and carpet make use of low-toxicity recycled content and materials obtained from sustainable forests.

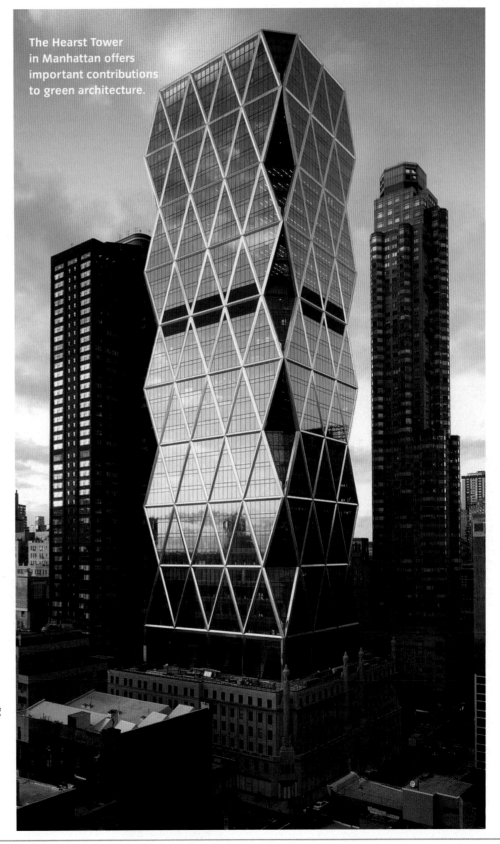

The Hearst Tower in Manhattan offers important contributions to green architecture.

Summary

1 Direct Solar Energy 442

1. An **active solar heating** system collects solar energy and then relies on pumps or fans to distribute heat. **Passive solar heating** systems distribute heat using mechanical devices.

2. A **photovoltaic (PV) solar cell** is a wafer or thin film of solid-state materials, such as silicon or gallium arsenide, that is treated with certain metals in such a way that it generates electricity when solar energy is absorbed. Manufacturing PVs requires toxic industrial chemicals, but PVs generate electricity with no pollution and minimal maintenance. PVs are limited by their low efficiency and by the amount of land needed for their large-scale use. **Solar thermal electric generation** is a means of producing electricity in which the sun's energy is concentrated by mirrors or lenses onto a fluid-filled pipe; the heated fluid is used to generate electricity. Solar thermal energy systems are efficient and provide significant environmental benefits; but they are only now becoming more cost-competitive with fossil fuels.

3. **Fuel cells** convert chemical energy into electricity; a fuel cell requires hydrogen fuel and oxygen (from the air).

2 Indirect Solar Energy 448

1. **Biomass** is plant and animal material used as fuel. Biomass fuel—materials such as wood, fast-growing plant and algal crops, crop wastes, sawdust and wood chips, and animal wastes—is burned to release energy.

2. **Wind energy** is electric energy obtained from surface air currents caused by the solar warming of air. Restricted primarily to areas with consistent winds, wind power is a clean and cost-effective source of energy. Wind turbines can kill birds and bats, and some people find their appearance unpleasant. **Hydropower** is a form of renewable energy that relies on flowing or falling water to generate electricity. Hydropower is highly efficient, but the dams built in traditional hydropower projects can greatly disrupt the natural environment and displace local residents.

3 Other Renewable Energy Sources 454

1. **Geothermal energy** is the use of energy from Earth's interior for either space heating or generation of electricity. **Tidal energy** is a form of renewable energy that relies on the ebb and flow of the tides to generate electricity.

4 Energy Solutions: Conservation and Efficiency 456

1. **Energy conservation** is using less energy—for example, by reducing energy use and waste. **Energy efficiency** is using less energy to accomplish a given task—for example, with new technology. Examples of energy conservation measures that reduce gasoline consumption include carpooling and lowering driving speeds; energy efficiency measures include designing and manufacturing more fuel-efficient automobiles.

2. Households can conserve energy in many ways, including by keeping vents and refrigerator coils clean, replacing inefficient lightbulbs and appliances, and adjusting thermostats.

Key Terms

- active solar heating 442
- biomass 448
- cogeneration (CHP) 458
- energy conservation 456
- energy efficiency 456
- fuel cell 447
- geothermal energy 454
- hydropower 452
- passive solar heating 443
- photovoltaic (PV) solar cell 443
- solar thermal electric generation 446
- wind energy 450

What is happening in this picture?

These hoses suck methane, which is used as fuel, from decomposing trash in California.

- Is this a renewable energy source? a form of conservation? Explain.

- What would be some advantages and disadvantages of this type of energy production?

Critical and Creative Thinking Questions

1. Do you think that a vehicle fuel efficiency of 60 miles per gallon could be reasonably achieved at present? Explain your answer.

2. Japan wants to make use of solar power, but it does not have extensive tracts of land for building large solar power plants. Which solar technology do you think is best suited to Japan's needs? Why?

3. Explain the following statement: Unlike fossil fuels, solar energy is not resource limited but is technology limited.

4. One advantage of the various forms of renewable energy, such as solar thermal and wind energy, is that they cause no net increase in atmospheric carbon dioxide. Is this true for biomass? Why or why not?

5. Give an example of how one or more of the renewable energy sources discussed in this chapter could have a negative effect on each of these aspects of ecosystems: soil preservation, natural water flow, foods used by wild plant and animal populations, and preservation of the diversity of organisms found in an area.

6. Explain how energy conservation and efficiency are major "sources" of energy.

7. Evaluate which forms of energy, other than fossil fuels and nuclear power, have the greatest potential where you live.

8. List energy conservation measures you could adopt in each of the following aspects of your life: washing laundry, lighting, bathing, cooking, buying a car, and driving a car.

The map below shows the average daily total of solar energy (on an annual basis) received on a solar collector that tilts to compensate for latitude.

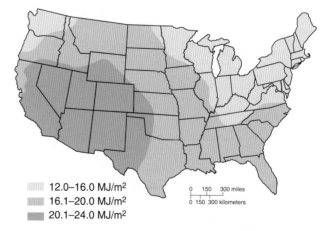

12.0–16.0 MJ/m^2
16.1–20.0 MJ/m^2
20.1–24.0 MJ/m^2

0   150   300 miles
0   150   300 kilometers

9. Based on the map, is Nevada or Maine better suited for producing electricity from solar energy?

10. Which of the following do you think would be best for reducing transportation energy use in the United States: more efficient cars, fuel cell cars, or public transportation? Explain.

✓ THE PLANNER

Glossary

acid deposition A type of air pollution that includes sulfuric and nitric acids in precipitation, as well as dry acid particles that settle out of the air.

acid mine drainage Pollution caused when sulfuric acid and dangerous dissolved materials, such as lead, arsenic, and cadmium, wash from coal and metal mines into nearby lakes and streams.

acute toxicity Adverse effects that occur within a short period after high-level exposure to a toxicant.

age structure The number and proportion of people at each age in a population.

air pollution Various chemicals (gases, liquids, or solids) present in the atmosphere in high enough levels to harm humans, other organisms, or materials.

aquaculture The growing of aquatic organisms (fishes, shellfish, and seaweeds) for human consumption.

aquifer depletion The removal of groundwater faster than it can be recharged by precipitation or melting snow.

artificial eutrophication Overnourishment of an aquatic ecosystem by nutrients such as nitrates and phosphates due to human activities such as agriculture and discharge from sewage treatment plants.

atmosphere The gaseous envelope surrounding Earth.

benthic environment The ocean floor, which extends from the intertidal zone to the deep-ocean trenches.

biocentric preservationist A person who believes in protecting nature from human interference because all forms of life deserve respect and consideration.

biochemical oxygen demand (BOD) The amount of oxygen that microorganisms need to decompose biological wastes into carbon dioxide, water, and minerals.

biodiversity hotspots Relatively small areas of land that contain an exceptional number of endemic species and are at high risk from human activities.

biological diversity The number and variety of Earth's organisms; consists of three components: genetic diversity, species richness, and ecosystem diversity.

biological magnification The increase in toxicant concentrations as a toxicant passes through successive levels of the food chain.

biomass Plant and animal material used as fuel.

biome A large, relatively distinct terrestrial region with similar climate, soil, plants, and animals, regardless of where it occurs in the world.

biosphere The layer of Earth that contains all living organisms.

biotic potential The maximum rate at which a population could increase under ideal conditions.

boreal forest A region of coniferous forest (such as pine, spruce, and fir) in the Northern Hemisphere; located just south of the tundra. Also called *taiga*.

broad-spectrum pesticide A pesticide that kills a variety of organisms, including beneficial organisms, in addition to the target pest.

bycatch The fishes, marine mammals, sea turtles, seabirds, and other animals caught unintentionally in a commercial fishing catch.

carbon management Ways to separate and capture the CO_2 produced during the combustion of fossil fuels and then sequester (store) it.

carcinogen Any substance (for example, chemical, radiation, virus) that causes cancer.

carrying capacity (*K*) The largest population a particular environment can support sustainably (long term), if there are no changes in that environment.

chaparral A biome with mild, moist winters and hot, dry summers; vegetation is typically small-leaved evergreen shrubs and small trees.

chlorofluorocarbons (CFCs) Human-made organic compounds that contain chlorine and fluorine; now banned because they attack the stratospheric ozone layer.

chronic toxicity Adverse effects that occur after a long period of low-level exposure to a toxicant.

clear-cutting A logging practice in which all the trees in a stand of forest are cut, leaving just the stumps.

climate The typical weather patterns that occur in a place over a period of years.

cogeneration (CHP) An energy technology that involves recycling "waste" heat.

command and control regulation Pollution control laws that work by setting limits on levels of pollution.

community A natural association that consists of all the populations of different species that live and interact together within an area at the same time.

compact development Design of cities in which tall, multiple-unit residential buildings are close to shopping and jobs, and all are connected by public transportation.

competition The interaction among organisms that vie for the same resources in an ecosystem (such as food or living space).

conservation biology The scientific study of how humans affect organisms and of the development of ways to protect biological diversity.

conservation easement A legal agreement that protects privately owned forest, rangeland, or other property from development for a specified number of years.

conservation tillage A method of cultivation in which residues from previous crops are left in the soil, partially covering it and helping to hold it in place until the newly planted seeds are established.

consumption overpopulation A situation in which each individual in a population consumes too large a share of resources.

contour plowing Plowing that matches the natural contour of the land.

Coriolis effect The tendency of moving air or water to be deflected from its path and swerve to the right in the Northern Hemisphere and to the left in the Southern Hemisphere.

cost–benefit diagram A diagram that helps policymakers make decisions about costs of a particular action and benefits that would occur if that action were implemented.

crop rotation The planting of a series of different crops in the same field over a period of years.

deep ecology worldview A worldview based on harmony with nature, a spiritual respect for life, and the belief that humans and all other species have an equal worth.

deforestation The temporary or permanent clearance of large expanses of forest for agriculture or other uses.

degradation (of land) Natural or human-induced reduction in the potential ability of the land to support crops or livestock.

demographic transition The process whereby a country moves from relatively high birth and death rates to relatively low birth and death rates.

demographics The applied branch of sociology that deals with population statistics.

desert A biome in which the lack of precipitation limits plant growth; deserts are found in both temperate and tropical regions.

desertification Degradation of once-fertile rangeland or tropical dry forest into nonproductive desert.

dose–response curve In toxicology, a graph that shows the effects of different doses on a population of test organisms.

dust dome A dome of heated air that surrounds an urban area and contains a lot of air pollution.

ecological niche The totality of an organism's adaptations, its use of resources, and the lifestyle to which it is fitted.

ecological succession The process of community development over time, which involves species in one stage being replaced by different species.

ecology The study of the interactions among organisms and between organisms and their abiotic environment.

economic development An expansion in a government's economy, viewed by many as the best way to raise the standard of living.

ecosystem A community and its physical environment.

ecosystem services Important environmental benefits, such as clean air to breathe, clean water to drink, and fertile soil in which to grow crops, that the natural environment provides.

El Niño–Southern Oscillation (ENSO) A periodic, large-scale warming of surface waters of the tropical eastern Pacific Ocean that temporarily alters both ocean and atmospheric circulation patterns.

endangered species A species that faces threats that may cause it to become extinct within a short period.

endemic species Organisms that are native to or confined to a particular region.

energy conservation Using less energy—by reducing energy use and waste, for example.

energy efficiency Using less energy to accomplish a given task—by using new technology, for example.

energy flow The passage of energy in a one-way direction through an ecosystem.

enhanced greenhouse effect Additional atmospheric warming produced as human activities increase atmospheric concentrations of greenhouse gases.

enrichment The process by which uranium ore is refined after mining to increase the concentration of fissionable U-235.

environmental ethics A field of applied ethics that considers the moral basis of environmental responsibility.

environmental justice The right of every citizen to adequate protection from environmental hazards.

environmental science The interdisciplinary study of humanity's relationship with other organisms and the physical environment.

environmental sustainability The ability to meet humanity's current needs without compromising the ability of future generations to meet their needs.

environmental worldview A worldview based on how the environment works, our place in the environment, and right and wrong environmental behaviors.

epidemiology The study of the effects of chemical, biological, and physical agents on the health of human populations.

estuary A coastal body of water, partly surrounded by land, with access to the open ocean and a large supply of fresh water from a river.

evolution The cumulative genetic changes in populations that occur during successive generations.

exponential population growth The accelerating population growth that occurs when optimal conditions allow a constant reproductive rate.

external cost A harmful environmental or social cost that is borne by people not directly involved in selling or buying a product.

extinction The elimination of a species from Earth.

first law of thermodynamics A physical law which states that energy cannot be created or destroyed, although it can change from one form to another.

fission The splitting of an atomic nucleus into two smaller fragments, accompanied by the release of a large amount of energy.

flowing-water ecosystem A freshwater ecosystem such as a river or stream in which water flows in a current.

fluidized-bed combustion A clean-coal technology in which crushed coal is mixed with limestone to neutralize acidic compounds produced during combustion.

food insecurity The condition in which people live with chronic hunger and malnutrition.

forest decline A gradual deterioration and eventual death of many trees in a forest.

freshwater wetlands Lands that shallow fresh water covers for at least part of the year; wetlands have a characteristic soil and water—tolerant vegetation.

fuel cell A device that directly converts chemical energy into electricity. A fuel cell requires hydrogen and oxygen from the air.

full cost accounting The process of evaluating and presenting to decision makers the relative benefits and costs of various alternatives.

genetic engineering The manipulation of genes (for example, taking a specific gene from one species and placing it into an unrelated species) to produce a particular trait.

genetic resistance An inherited characteristic that decreases the effect of a given agent (such as a pesticide) on an organism (such as a pest).

geothermal energy Energy from Earth's interior, used for space heating or generation of electricity.

germplasm Any plant or animal material that may be used in breeding.

green chemistry A subdiscipline of chemistry in which commercially important chemical processes are redesigned to significantly reduce environmental harm.

greenhouse gases Gases—including water vapor, carbon dioxide, methane, and certain other gases—that absorb infrared radiation.

groundwater The supply of fresh water under Earth's surface that is stored in underground aquifers.

growth rate (r) The rate of change (increase or decrease) of a population's size, expressed in percentage per year.

gyres Large, circular ocean current systems that often encompass an entire ocean basin.

habitat fragmentation The breakup of large areas of habitat into small, isolated patches.

hazardous waste A discarded chemical that threatens human health or the environment.

high-level radioactive wastes Radioactive solids, liquids, or gases that initially give off large amounts of ionizing radiation.

highly developed countries Countries with complex industrialized bases, low rates of population growth, and high per person incomes.

hydropower A form of renewable energy that relies on flowing or falling water to generate electricity.

incentive-based regulation Pollution control laws that work by establishing emission targets and providing industries with incentives to reduce emissions.

industrialized agriculture Modern agricultural methods that require large capital inputs and less land and labor than traditional methods.

infant mortality rate The number of deaths of infants under age 1 per 1000 live births.

integrated waste management A combination of the best waste management techniques into a consolidated program to deal effectively with solid waste.

intertidal zone The area of shoreline between low and high tides.

invasive species Foreign species that spread rapidly in a new area if free of predators, parasites, or resource limitations that may have controlled their population in their native habitat.

landscape A region that includes several interacting ecosystems.

less developed countries Countries with low levels of industrialization, very high rates of population growth, very high infant mortality rates, and very low per person incomes relative to highly developed countries.

low-level radioactive wastes Solids, liquids, or gases that give off small amounts of ionizing radiation.

marginal cost of pollution abatement The added cost of reducing one unit of a given type of pollution.

marginal cost of pollution The added cost of an additional unit of pollution.

mass burn incinerator A large furnace that burns all solid waste except for unburnable items such as refrigerators.

microirrigation A type of irrigation that conserves water by piping it to crops through sealed systems.

minerals Elements or compounds of elements that occur naturally in Earth's crust.

moderately developed countries Countries with medium levels of industrialization and per person incomes lower than those of highly developed countries.

monoculture Ecological simplification in which only one type of plant is cultivated over a large area.

municipal solid waste Solid materials discarded by homes, offices, stores, restaurants, schools, hospitals, prisons, libraries, and other commercial and institutional facilities.

national income accounts Measures of the total income of a nation's goods and services for a given year.

natural capital Earth's resources and processes that sustain living organisms, including humans; includes minerals, forests, soils, water, clean air, wildlife, and fisheries.

natural selection The tendency of better-adapted individuals—those with a combination of genetic traits best suited to environmental conditions—to survive and reproduce, increasing their proportion in the population.

neritic province The part of the pelagic environment that overlies the ocean floor from the shoreline to a depth of 200 m (650 ft).

nonmunicipal solid waste Solid waste generated by industry, agriculture, and mining.

nonpoint source pollution Pollution that enter bodies of water over large areas rather than being concentrated at a single point of entry.

nonrenewable resources Natural resources that are present in limited supplies and are depleted as they are used.

nuclear energy The energy released by nuclear fission or fusion.

nuclear reactor A device that initiates and maintains a controlled nuclear fission chain reaction to produce energy for electricity.

nutrient cycling The pathway of various nutrient minerals or elements from the environment through organisms and back to the environment.

oceanic province The part of the pelagic environment that overlies the ocean floor at depths greater than 200 m (650 ft).

optimum amount of pollution The amount of pollution that is economically most desirable.

overburden Soil and rock overlying a useful mineral deposit.

overgrazing A situation that occurs when too many grazing animals consume the plants in a particular area, leaving the vegetation destroyed and unable to recover.

overnutrition A type of malnutrition in which there is an overconsumption of calories that leaves the body susceptible to disease.

ozone thinning The removal of ozone from the stratosphere by human-produced chemicals or natural processes.

passive solar heating A system of putting the sun's energy to use that does not require mechanical devices to distribute the collected heat.

pathogen An agent (usually a microorganism) that causes disease.

people overpopulation A situation in which there are too many people in a given geographic area.

persistent organic pollutants (POPs) Persistent toxicants that bioaccumulate in organisms and travel through air and water to contaminate sites far from their source.

pesticide A toxic chemical used to kill pests.

photochemical smog A brownish-orange haze formed by chemical reactions involving sunlight, nitrogen oxides, and hydrocarbons.

photosynthesis The biological process that captures light energy and transforms it into the chemical energy of organic molecules, which are manufactured from carbon dioxide and water.

photovoltaic (PV) solar cell A wafer or thin film of solid-state materials, such as silicon or gallium arsenide, that is treated with certain metals in such a way that the film generates electricity when solar energy is absorbed.

plate tectonics The study of the processes by which the lithospheric plates move over the asthenosphere.

point source pollution Water pollution that can be traced to a specific spot.

population A group of organisms of the same species that live together in the same area at the same time.

population ecology The branch of biology that deals with the number of individuals of a particular species found in an area and why those numbers increase or decrease over time.

poverty A condition in which people are unable to meet their basic needs for food, clothing, shelter, education, or health.

precautionary principle The idea that new technologies, practices, or materials should not be adopted until there is strong evidence that they will not adversely affect human or environmental health.

predation The consumption of one species (the prey) by another (the predator).

primary air pollutants Harmful chemicals that enter directly into the atmosphere due to either human activities or natural processes.

primary treatment Treatment of wastewater that involves removing suspended and floating particles through mechanical processes.

rangeland Land that is not intensively managed and is used for grazing livestock.

renewable resources Resources that are replaced by natural processes and that can be used forever, provided they are not overexploited in the short term.

replacement-level fertility The number of children a couple must produce to "replace" themselves.

restoration ecology The study of the historical condition of a human-damaged ecosystem, with the goal of returning it as closely as possible to its former state.

risk The probability of harm (such as injury, disease, death, or environmental damage) occurring under certain circumstances.

risk assessment The quantitative and qualitative characterization of risks so that they can be compared, contrasted and managed.

runoff The movement of fresh water from precipitation and snowmelt to rivers, lakes, wetlands, and the ocean.

salinization The gradual accumulation of salt in soil, often as a result of improper irrigation methods.

saltwater intrusion The movement of seawater into a freshwater aquifer near the coast.

sanitary landfill The most common disposal site for solid waste, where waste is compacted and buried under a shallow layer of soil.

savanna A tropical grassland with widely scattered trees or clumps of trees.

scientific method The way a scientist approaches a problem, by formulating a hypothesis and then testing it.

second law of thermodynamics A physical law which states that when energy is converted from one form to another, some of it is degraded into heat, a less usable form that disperses into the environment.

secondary air pollutants Harmful chemicals that form in the atmosphere when primary air pollutants react chemically with one another or with natural components of the atmosphere.

secondary treatment Biological treatment of wastewater to decompose suspended organic material; secondary treatment reduces the water's biochemical oxygen demand.

sewage Wastewater from drains or sewers (from toilets, washing machines, and showers); includes human wastes, soaps, and detergents.

shelterbelt A row of trees planted as a windbreak to reduce soil erosion of agricultural land.

sick building syndrome Eye irritations, nausea, headaches, respiratory infections, depression, and fatigue caused by indoor air pollution.

smelting The process in which ore is melted at high temperatures to separate impurities from the molten metal.

soil The uppermost layer of Earth's crust, which supports terrestrial plants, animals, and microorganisms.

soil erosion The wearing away or removal of soil from the land.

soil horizons Horizontal layers into which many soils are organized, from the surface to the underlying parent material.

solar thermal electric generation A means of producing electricity in which the sun's energy is concentrated using mirrors or lenses onto a fluid-filled pipe; the heated fluid is used to generate electricity.

source reduction An aspect of waste management in which products are designed and manufactured in ways that decrease the amount of solid and hazardous waste in the solid waste stream.

species richness The number of different species in a community.

spent fuel Used fuel elements that were irradiated in a nuclear reactor

spoil bank A hill of loose rock created when the overburden from a new trench is put into the already excavated trench during strip mining.

standing-water ecosystem A body of fresh water surrounded by land and whose water does not flow; a lake or a pond.

subsistence agriculture Traditional agricultural methods that are dependent on labor and a large amount of land to produce enough food to feed oneself and one's family.

subsurface mining The extraction of mineral and energy resources from deep underground deposits.

surface mining The extraction of mineral and energy resources near Earth's surface by first removing the soil, subsoil, and overlying rock strata.

surface water Precipitation that remains on the surface of the land and does not seep down through the soil.

sustainable agriculture Agricultural methods that maintain soil productivity and a healthy ecological balance while having minimal long-term impacts.

sustainable consumption The use of goods and services that satisfy basic human needs and improve the quality of life but that also minimize resource use.

sustainable development Economic growth that meets the needs of the present without compromising the ability of future generations to meet their needs.

sustainable forestry The use and management of forest ecosystems in an environmentally balanced and enduring way.

sustainable soil use The wise use of soil resources, without a reduction in the amount or fertility of soil, so it is productive for future generations.

sustainable water use The wise use of water resources, without harming the essential functioning of the hydrologic cycle or the ecosystems on which present and future humans depend.

symbiosis An intimate relationship or association between members of two or more species; includes mutualism, commensalism, and parasitism.

temperate deciduous forest A forest biome that occurs in temperate areas where annual precipitation ranges from about 75 cm to 126 cm (30 to 50 in).

temperate grassland A grassland with hot summers, cold winters, and less rainfall than is found in the temperate deciduous forest biome.

temperate rain forest A coniferous biome with cool weather, dense fog, and high precipitation.

temperature inversion A layer of cold air temporarily trapped near the ground by a warmer upper layer.

tertiary treatment Advanced wastewater treatment methods that are sometimes employed after primary and secondary treatments.

threatened species A species whose population has declined to the point that it may be at risk of extinction.

total fertility rate (TFR) The average number of children born to each woman.

toxicology The study of toxicants, chemicals with adverse effects on health.

tropical rain forest A lush, species-rich forest biome that occurs where the climate is warm and moist throughout the year.

tundra The treeless biome in the far north that consists of boggy plains covered by lichens and mosses; it has harsh, cold winters and extremely short summers.

ultraviolet (UV) radiation Radiation from the part of the electromagnetic spectrum with wavelengths just shorter than visible light; can be lethal to organisms at high levels of exposure.

undernutrition A type of malnutrition in which there is an underconsumption of calories or nutrients that leaves the body weakened and susceptible to disease.

urban heat island Local heat buildup in an area of high population.

urbanization A process whereby people move from rural areas to densely populated cities.

utilitarian conservationist A person who values natural resources because of their usefulness to humans but uses them sensibly and carefully.

water pollution A physical or chemical change in water that adversely affects the health of humans and other organisms.

Western worldview A worldview based on human superiority over nature, the unrestricted use of natural resources, and economic growth to manage an expanding industrial base.

wilderness A protected area of land in which no human development is permitted.

wildlife corridor A protected zone that connects isolated unlogged or undeveloped areas.

wind energy Electric energy obtained from surface air currents caused by the solar warming of air.

zero population growth The state in which the population remains the same size because the birth rate equals the death rate.

Text, Table & Line Art Credits

Chapter 1

Figure 1.1b: Data from Population Reference Bureau; **Figure 1-4:** Data from *Living Planet Report 2008*; **Graph in Creative and Critical Thinking Questions:** Data from *Science* Vol. 315 (February 16, 2007), page 913.

Chapter 2

First section on sustainable development, page 28: *Our Common Future.* World Commission on Environment and Development. Cary, NC: Oxford University Press (1987), p. 22; **Eight principles of deep ecology, page 33:** Naess, A. and D. Rothenberg. *Ecology, Community and Lifestyle.* Cambridge, UK: Cambridge University Press (2001). Reprinted with permission of Cambridge University Press; **Five recommendations in section on overall plan for sustainable living, p. 36:** Brown, L. *Plan B 2.0: Rescuing a Planet under Stress and a Civilization in Trouble.* New York: W.W. Norton (2006); **Graphs in Creative and Critical Thinking Questions:** Data from U.S. National Climate Assessment.

Chapter 3

Page 54, excerpt from the "Wilderness Essay": a letter written by Wallace Stegner to David Pesonen of the U. of California's Wildland Research Center; **National Income Accounts, pages 62–64:** Main source: Levin, J. "The Economy and the Environment: Revising the National Accounts." *IMF Survey* (June 4, 1990); **An Economist's View of Pollution, pages 65–66:** Main source. Goodstein, E. *Economics and the Environment,* 5th edition (2008).

Chapter 4

Chapter introduction inset, Guilette, E.A., M.M. Meza, M.G. Aguilar, A.D. Soto, and I.E. Garcia. "An Anthropological Approach to the Evaluation of Preschool Children Exposed to Pesticides in Mexico." *Environmental Health Perspectives* (May 1998); **Figure 4.1:** Data from National Safety Council (2008); art design from Health feature in *National Geographic* (August 2006); **Figure 4.2:** Adapted from *Science and Judgment in Risk Assessment.* Washington, D.C.: National Academy Press (1994). Data in graph from "CDC's Second National Report on Human Exposure to Environmental Chemicals" (2003); **Table 4.2:** Adapted from U.S. Centers for Disease Control and Prevention, National Center for Chronic Disease Prevention and Health Promotion, Division of Adolescent and School Health, Youth Risk Behavior Survey; **Figure 4.8b:** Data from Grier, J.W. "Ban of DDT and Subsequent Recovery of Reproduction in Bald Eagles" Copyright 1982, American Association for the Advancement of Science; **Figure 4.8d:** Data from Woodwell, G.M., C.F. Worster, Jr. and P.A. Isaacson. "DDT Residues in an East Coast Estuary: A Case of Biological Concentration of a Persistent Insecticide. *Science* Vol 156 (May 12, 1967); **Table 4.4:** Josten, M.D. and J.L. Wood. *World of Chemistry,* 2nd edition. Philadelphia: Saunders College Publishing, (1996); **Case Study, page 92:** Fujisaki; I.; Rice, K; Woodward, A; and Percival H.F. "Possible Generational Effect of Habitat Degradation on Alligator Reproduction" *The Journal of Wildlife Management* (Sept. 2006).

Chapter 5

Figures 5.8, 5.9, 5.10, 5.12, and 5.13: Values are from Schlesinger, W.H. *Biogeochemistry: An Analysis of Global Change,* 2nd edition. Academic Press, San Diego (1997) and based on several sources; **What a Scientist Sees: Resource Partitioning:** Adapted from MacArthur, R.H. "Population Ecology of Some Warblers of Northeastern Coniferous Forests." *Ecology,* Vol. 39 (1958).

Chapter 6

Figure 6.1: Based on data from the World Wildlife Fund; **Figure 6.2:** Based on Holdridge, L. *Life Zone Ecology.* Tropical Science Center, San Jose, Costa Rica (1967); **Data for all climate graphs (in Figures 6.4–6.12)**

from www.worldclimate.com; **Figure 6.18:** Adapted from Figure 14.11 on p. 428 in B.W. Murck, B.J. Skinner, and D. Mackenzie. *Visualizing Geology,* Hoboken, NJ: John Wiley & Sons, Inc. (2008); **Figures 6.19b and c:** Adapted from Figure 15.13 on p. 244 and Figure 15.19 on p. 247, respectively, in S.A. Alters and B. Alters *Biology: Understanding Life,* Hoboken, NJ: John Wiley & Sons, Inc. (2006); **Case Study, page 154:** Wildfire data from the California Dept. of Forestry and Fire Protection.

Chapter 7

Figure 7.5b: After V.C. Scheffer. "The Rise and Fall of a Reindeer Herd." *Sci. Month.,* Vol. 73 (1951); **Figures 7.6, 7.12, 7.13, 7.14, 7.22, Table 7.1, and graph for Critical and Creative Thinking Question 12:** Data from Population Reference Bureau; **Figure 7.8:** Data from *World Population Prospects, The 2004 Revision,* United Nations Population Division; **Figure 7.17b:** Adapted from Data Table 12 in *World Resources 2002–2004: Decisions for the Earth: Balance, Voice, and Power,* Washington, D.C., World Resources Institute (2003). **What a Scientist Sees: Education and Fertility:** Data from U.S. Census Bureau; **Figure 7.23:** "Urban Agglomerations 2005," U.N. Department of Economic and Social Affairs, Population Division; **Case Study Figure, Page 186,** Courtesy of Karl Gude.

Chapter 8

Quote by Ulf Merbold: from Ulf Merbold of the German Aerospace Liaison; **Figure 8.15:** Adapted from Figure 6.11 in A.F. Arbogast *Discovering Physical Geography,* Hoboken, NJ: John Wiley and Sons, Inc. (2007). **What a Scientist Sees (graph, part b):** temperature data from the National Aeronautics and Space Administration; **Figure 8.14:** Air Quality Planning and Standards, Office of Air and Radiation, EPA.

Chapter 9

Figure 9.5: Global Land-Ocean Temperature Index, Goddard Institute of Space Studies, NASA; **Figure 9.6:** Dave Keeling and Tim Whorf, Scripps Institution of Oceanography, La Jolla, California; **Figure 9.11d:** Richard Feely and Dana Greeley, NOAA Pacific Marine Environmental Laboratory, as shown on page 111 in *National Geographic* (November 2007); **Figure 9.12:** After design from Vattenfall Corporation, Stockholm, Sweden; **Figure 9.13d:** Data from New Zealand National Institute of Water and Atmospheric Research; **Graph in Case Study:** U.S. Department of Energy Carbon Dioxide Information Analysis Center; **Map in Creative and Critical Thinking Questions:** U.S. Global Change Research Program.

Chapter 10

Figure 10.2: Adapted from Figure 2.8 on p. 28 in Strahler, A. and A. Strahler. *Physical Geography: Science and Systems of the Human Environment.* Hoboken, NJ: John Wiley & Sons, Inc. (2002); **Figure 10.5:** *Control of Water Pollution from Urban Runoff.* Paris: Organization for Economic Cooperation and Development (1986); **Figure 10.8:** U.S. Geological Survey; **Figure 10.16:** Adapted from Joesten, M.D., and J.L. Wood. *World of Chemistry,* 2nd edition. Philadelphia: Saunders College Publishing (1996).

Chapter 11

Figure 11.1b: Adapted from Figure 12.5A on p. 347 in Murck, B.W., B.J. Skinner, and D. Mackenzie. *Visualizing Geology,* Hoboken, NJ: John Wiley & Sons, Inc. (2008); **Figure 11.2:** Adapted from Figure 6.31 on p. 148 in A.F. Arbogast. *Discovering Physical Geography.* Hoboken, NJ: John Wiley & Sons, Inc. (2007); **Figure 11.5:** Adapted from Figure 14-1 in Karleskint, G. *Introduction to Marine Biology.* Philadelphia: Harcourt College Publishers (1998); **Figure 11.6a:** Adapted from Marangos, J.E., M.P. Crosby, and J.W. McManus. "Coral Reefs and Biodiversity: A Critical and

Threatened Relationship." *Oceanography*, Vol. 9 (1996); **Figure 11.11a:** After Halpern, B.S. et al. "A Global Map of Human Impact on Marine Ecosystems." *Science*, Vol. 319, No. 5865, pp. 948–952 (February 15, 2008); **What a Scientist Sees: Ocean Warming and Coral Bleaching, Figure b:** Adapted from National Assessment Synthesis Team, *Climate Change Impacts on the United States: The Potential Consequences of Climate Variability and Change* (Report for the U.S. Global Change Research Program). Cambridge, UK: Cambridge University Press (2001) and based on data from the National Oceanic and Atmospheric Administration; **Figure 11.15b:** From www.noaa.gov; **graph in Critical and Creative Thinking Questions:** Adapted from *Vital Signs 2006–2007,* page 27, Worldwatch Institute (2006), and based on data from the FAO.

Chapter 12

Figure 12.7: Adapted from Joesten, M.D., and J.L. Wood. *World of Chemistry*, Second Edition. Philadelphia: Saunders College Publishing (1996); **Figure 12.17:** *One Planet, Many People: Atlas of Our Changing Environment.* United Nations Environment Program (2005)—Part a based on graph on page 30 (FAO 2000), Part b based on table on page 29 (*World Atlas of Desertification*). **Enviro-Discovery art, page 305:** Larmer, B. "The Price of Gold." *National Geographic* (January 2009); **Graph in Critical and Creative Thinking Questions:** Williams, J.R. and D.L. Tanaka. "Economic Evolution of Topsoil Loss in Spring Wheat Production in the Northern Great Plains, U.S." *Soil and Tillage Research*, Vol. 37.

Chapter 13

Table 13.1: U.S. Dept. of Interior, U.S. Dept. of Agriculture, and U.S. Dept. of Defense; **Figure 13.10b:** National Park Service; **Table 13.2:** From Box 1.1 in Noss, R.F., M.A. O'Connell, and D.D. Murphy. *The Science of Conservation Planning: Habitat Conservation Under the Endangered Species Act*, Island Press: World Wildlife Fund (1997). Reprinted with permission; **Figure 13.1:** Adapted from the map, "Federal Lands in the Fifty States," produced by the Cartographic Division of the National Geographic Society (October 1996); **Figure 13.3:** Values are from Schlesinger, W.H. *Biogeochemistry: An Analysis of Global Change*, 2nd edition. Academic Press, San Diego (1997) and based on several sources; **Figure 13.6a:** Based on Goode Base Map (S.R. Eyre 1968); **Figure 13.7:** Data obtained from Marsh, W.M. and J.M. Grossa, Jr., *Environmental Geography*, 2nd edition, John Wiley & Sons, Inc. (2002); **Graph in Critical and Creative Thinking Questions:** World Resources Institute and U.N. Food and Agricultural Organization.

Chapter 14

Figure 14.1d: Based on data from U.N. Food and Agricultural Organization and UNICEF; **Figure 14.2:** Adapted from FAO data, as reported in *Vital Signs 2009*. Washington, DC: Worldwatch Institute (2009); **Figure 14.3:** Adapted from G.H. Heichel, "Agricultural Production and Energy Resources." *American Scientist*, Vol. 64 (January/February 1976); **Figure 14.7:** Data from USDA; **Figure 14.9a:** Adapted from D. Livermore, Health Protection Agency's Resistance Monitoring and Reference Laboratory, United Kingdom; **Figure 14.12:** Data from the International Service for the Acquisition of Agri-Biotech Applications, 2007; **Figure 14.4:** Data from Mark Whalon, Michigan State University, as reported in Brown, L.R. et al. *Vital Signs 1994*. New York: W.W. Norton & Co. (1994); **Figure a in Case Study on page 366:** Adapted from Gardner, G., "IPM and the War on Pests." *World Watch*, Vol. 9, No. 2 (March/April 1996). Redrawn by permission of Worldwatch Institute; **What a Scientist**

Sees: Pesticide Use and New Pest Species (graph): Adapted from Debach, P. *Biological Control by Natural Enemies.* New York: Cambridge University Press (1974); **graph in Critical and Creative Thinking Questions 3 and 4:** Data from UNESCO/FAO; **graphics in Critical and Creative Thinking Question #11:** Data from the International Service for the Acquisition of Agri-Biotech Applications, 2007.

Chapter 15

Table 15.1: Adapted from page 527 of *Climate Change Impacts on the United States*, A report of the National Assessment Synthesis Team, U.S. Global Change Research Program, Cambridge University Press (2001); **Table 15.2:** U.S. Fish and Wildlife Service; **Data in first section on number of species:** Data from Reaka-Kudla, M.L., D.E. Wilson, and E.O. Wilson. *Biodiversity II.* Washington, D.C.: Joseph Henry Press (1997); **Figure 15.1b:** After M.L. Cody and J.M. Diamond, eds., *Ecology and Evolution of Communities.* Harvard University, Cambridge, (1975); **What a Scientist Sees: Where Is Declining Biological Diversity the Greatest Problem:** Map by Conservation International; **Case Study:** Map adapted by EPA, original source University of Wisconsin Sea Grant Institute; **quote in Critical and Creative Thinking Questions:** Aldo Leopold, *A Sand County Almanac.* Oxford University Press, Inc. (1991); **graph in Critical and Creative Thinking Questions:** Data from The Nature Conservancy.

Chapter 16

Figures 16.2 and 16.3: Data from EPA; **Figure 16.8a:** Adapted from T. Zeller, Jr., "Recycling: The Big Picture," *National Geographic* (January 2008); **Figure 16.8d:** Data from Argonne National Laboratory; **Figure 16.14:** Adapted from Rocky Mountain Arsenal Remediation Venture Office; **graph in Critical and Creative Thinking Questions:** EPA.

Chapter 17

Unless noted otherwise, all energy facts cited in this chapter were obtained from the Energy Information Administration (EIA), the statistical agency of the U.S. Department of Energy (DOE). **Figure 17.12d:** Integral Consulting. "2005 Assessment of Lingering Oil and Resource Injuries from the *Exxon Valdez* Oil Spill (Restoration Project 040776)"; **Table 17.1:** Adapted from two sources: Hinrichs, R.A. and M. Kleinbach. *Energy: Its Use and the Environment*, 3rd edition. Philadelphia: Harcourt College Publishers (2002) and *Science for Democratic Action*, Vol. 8, No. 3 (May 2000); **Figure 17.1b:** Adapted from Figure 13-3a in Harris, J.M. *Environmental and Natural Resource Economics: A Contemporary Approach*, second edition. Houghton Mifflin (2006) and based on data from the World Bank.

Chapter 18

Unless noted otherwise, all energy facts cited in this chapter were obtained from the Energy Information Administration (EIA), the statistical agency of the U.S. Department of Energy (DOE). **Table 18.1:** Data from Kammen, D.M. "The Rise of Renewable Energy." *Scientific American* (September 2006) with updates from EIA (DOE); **Figure 18.5b:** Adapted from Hinrichs, R.A. *Energy: Its Use and the Environment*, 2nd ed. Philadelphia: Saunders College Publishing (1996); **Figure 18.8c:** From *Vital Signs 2005*, Worldwatch Institute, W.W. Norton (2005) and based on data from BTM Consult, American Wind Energy Association, and European Wind Energy Association; **map in Critical and Creative Thinking Questions:** DOE.

Photo Credits

Chapter 1

Page 2: NASA/NG Image Collection; page 5 (top): Earth Imaging/Stone/Getty Images; page 5 (bottom): ALEXANDRA AVAKIAN/NG Image Collection; page 5 (inset): NG Maps; page 6 (left): Peter Menzel; page 6 (right): Peter Menzel; page 7: Peter Menzel; page 8: DIANE COOK AND LEN JENSHEL/NG Image Collection; page 10: Macduff Everton/Iconica/Getty Images; page 13 (top): NORBERT WU/MINDEN PICTURES/NG Image Collection; page 13 (center): Visuals Unlimited/©Corbis Images; page 13 (bottom): Tom Bonaventure/Photographer's Choice/Getty Images, Inc.; page 14 (top left): Gene Carl Feldman/Sea WIFs/NASA; page 14 (top left): Lara Hansen, Adam Markham/WWF/NG Maps; page 14 (top right): Photodisc/Getty Images; page 14 (top right): Global Forest Watch/WRI/NG Maps; page 14 (bottom left): Caroline Rogers/USGS; page 14 (bottom left): NOAA/NMFS/UNEP-WCMC/NG Maps; page 14 (bottom right): Bruce Dale/NG Image Collection; page 14 (bottom right): USDA Global Desertification Map/National Geographic Maps; page 15 (top): National/Naval Ice Center/NG Maps; page 15 (center): NG Maps; page 15 (bottom): NASA Earth Observatory; page 16: PHOTOTAKE Inc./Alamy; page 17: Jim Richardson/NG Image Collection; page 17 (inset): Guy Croft SciTech/Alamy; page 19 (left): Nicole Duplaix/NG Image Collection; page 19 (bottom): Courtesy Mathew Price; page 20 (bottom): Ben Osborne/Stone/Getty Images; page 22: Raymond Gehman/NG Image Collection; page 23: UPI Photo/IKONOS/NewsCom; page 25: David H. Wells/CORBIS

Chapter 2

Page 26: Gavin Newman/Alamy; page 26 (inset): ©Vic Cox, photographer; page 29: Dan Lamont/©Corbis Images; page 29 (right): DICK DURRANCE II/NG Image Collection; page 29: NG Maps; page 30: Courtesy PhillyCarShare; page 31: Richard Nowitz/NG Image Collection; page 32 (left): Minnesota Historical Image Collection/CORBIS; Page 32 (right): Mark Moffett/Minden Pictures/NG Image Collection; page 32 (top right): NG Maps; Page 33: Cristina Redinger-Libolt/Botanica/Getty Images, Inc.; page 34: Priit Vesilind/NG Image Collection; page 35: Jenny Hager/The Image Works; page 37 (top left): Peter Turnley/©Corbis; page 37 (top center): Tim Laman/NG Image Collection; page 37 (top right): Ragnar Th Sigurdsson/Alamy; page 37 (center right): Rafael Macia/Photo Researchers, Inc.; page 37 (bottom right): Iain Masterton/Alamy; page 38: Justin Guariglia/NG Image Collection; page 38 (inset): NG Maps; page 40: Michael Nichols/ NG Image Collection; page 41: Charlotte Thege/Alamy; page 43: Peter Scoones/Photo Researchers, Inc.; page 43: Peter Treanor/Alamy; page 44: Jim Richardson/NG Image Collection; page 45: Michael Nichols/NG Image Collection; page 46: ©AP/Wide World Photos

Chapter 3

Page 48: GREG DALE/NG Image Collection; page 49 (inset): MICHAEL MELFORD/NG Image Collection; page 50 (left): USDA/NG Image Collection; page 50 (right): John Schwieder/Alamy; page 51: Courtesy Library of Congress; page 52: Bettman/Corbis Images; page 53 (top): Courtesy National Archives; page 53 (bottom): Eugene Fisher/NG Image

Collection; page 54: Photo by Tom Coleman/Courtesy Aldo Leopold Foundation, Baraboo, WI.; Page 55: Erich Hartmann/Magnum Photos, Inc.; page 56: Todd Gipstein/NG Image Collection; Page 57 (bottom left): NewsCom; page 57 (center): David McNew/Staff/Getty Images, Inc.; page 57 (bottom right): Peter Macdiarmid/Staff/Getty Images, Inc.; page 58 (top): Bob Sacha/NG Image Collection; page 58 (bottom): Michael Nichols/NG Image Collection; page 59: Michael Melford/NG Image Collection; page 60: Greg Dale/NG Image Collection; page 61: Rick Poley/Visuals Unlimited; page 62 (bottom right): NG Maps; page 63 (bottom left): Creatas/SUPERSTOCK; page 63 (bottom right): age fotostock/SUPERSTOCK; Page 63 (bottom): Randy Olson/NG Image Collection; page 64: Tyrone Turner/NG Image Collection; page 65: Tim Fitzharris/NG Image Collection; page 68 (left): Skip Brown/NG Image Collection; page 68 (right): Ted Spiegel/NG Image Collection; page 69 (left): USDA/NG Image Collection; page 69 (right): Todd Gipstein/NG Image Collection; page 70 (top): Greg Dale/NG Image Collection; page 70 (bottom): Reuters Newmedia Inc./Corbis Images

Chapter 4

Page 73: David Alan Harvey/NG Image Collection; page 76: Sisse Brimberg/NG Image Collection; page 77: Scott Camazine/Alamy; page 79: SPL/Photo Researchers, Inc.; a,b Page 80 (top): Courtesy Millipore Corp; page 80 (bottom): NG Maps; page 80 (bottom): Bruce Dale/NG Image Collection; page 80 (top): Courtesy Millipore Corp; page 81: Justin Guariglia/NG Image Collection; page 82: Lou Linwei/Alamy; page 83 (bottom left): Klaus Nigge/NG Image Collection; page 83 (top left): FLIP DE NOOYER/FOTO NATURA/MINDEN PICTURES/NG Image Collection; page 85: KAREN KASMAUSKI/NG Image Collection; page 86: Wolfgang Flamish/zefa/©Corbis; page 88: Courtesy Centers for Disease Control ; page 89 (bottom): Karen Kasmauski/NG Image Collection; page 90: Daniel LeClair/Reuters/Landov; page 91: Andy Levin/Photo Researchers; page 92: Raymond Gehman/NG Image Collection; page 93: Stuart Bauer; page 94 (bottom left): Lou Linwei/Alamy; page 94 (top left): Sisse Brimberg/NG Image Collection; page 94 (center right): Klaus Nigge/NG Image Collection; page 95 (top): Raymond Gehman/NG Image Collection; page 95 (bottom): James L. Amos/NG Image Collection

Chapter 5

Page 96 (left): NSP-RF/Alamy; page 96 (insert right): WILDLIFE GmbH/Alamy; page 97: Amy Mikler/Alamy; page 100 (top): George Grall/NG Image Collection; page 100 (bottom): George Grall/NG Image Collection; page 101 (top right): ROY TOFT/NG Image Collection; page 101 (top left): Taylor S. Kennedy/NG Image Collection; page 102 (left): Romeo Gacad/AFP/Getty Images; page 102 (right): Max Rossi/Reuters/Corbis; page 103: Raymond Gehman/NG Image Collection; page 104 (top left): John Eastcott and Yva Momatiuk/NG Image Collection; page 104 (bottom left): Roy Toft/NG Image Collection; page 104 (top right): TIM LAMAN/NG Image Collection; page 104 (bottom right): Norbert Rosing/NG Image Collection; page 111 (left): Dr Jeremy Burgess/Photo Researchers; page 111 (right): STEPHEN SHARNOFF/NG Image Collection; page 115 (left): SuperStock; page 115 (right): Alan Briere/SuperStock; page 116: Rob and Ann Simpson/Visuals Unlimited; page 117: Photo Resource Hawaii/Danita Delimont; page 118 (top right): Minden

Pictures/Getty Images; page 118 (bottom): Tim Laman/NG Image Collection; page 118 (center left): Darlyne A. Murawski/ NG Image Collection; page 119: CNImaging/NewsCom; page 120 (left): Chris Johns/NG Image Collection; page 120 (right): Rich Reid/NG Image Collection; page 121 (top left): MATTIAS KLUM/NG Image Collection; page 121 (right): NG Maps; page 121 (bottom): NORBERT WU/MINDEN PICTURES/NG Image Collection; page 122 (left): Joel Sartore/NG Image Collection; page 122 (right): PHIL SCHERMEISTER/ NG Image Collection; page 123: ©AP/Wide World Photos; page 124 (left): George Grall/NG Image Collection; page 124 (bottom right): SuperStock; page 125: Darlyne A. Murawski/NG Image Collection; page 126 (bottom left): Geroge F. Mobley/NG Image Collection

Chapter 6

Page 126: Raymond Gehman/NG Image Collection; page 129 (top right): JOEL SARTORE/NG Image Collection; page 129 (bottom right): RICHARD NOWITZ/NG Image Collection; page 129 (top left): Michael Melford/NG Image Collection; page 130: Paul Davis/The Global Land Cover Facility , University of Maryland Institute for Advanced Computer Studies/NG Maps; page 130 (top left): Brand X/SUPERSTOCK; page 130 (center left): Hemis. Fr/SUPERSTOCK; page 130 (bottom left): Photodisc/SUPERSTOCK; page 130 (bottom second from left): age fotostock/SUPERSTOCK; page 130 (bottom center left): Polka Dot Images/SUPERSTOCK; page 130 (bottom center right): Michael McCoy/Photo Researchers, Inc.; page 130 (bottom second from right): Corbis/SUPERSTOCK; page 130 (bottom right): age fotostock/SUPERSTOCK; page 131 (bottom left): age fotostock/SuperStock, Inc.; page 131 (bottom second from left): James Steinberg/Photo Researchers, Inc.; page 131 (bottom center left): PAUL CHESLEY/NG Image Collection; page 131 (bottom center right): Imagemore/SUPERSTOCK; page 131 (bottom second from right): MICHAEL S. YAMASHITA/NG Image Collection; page 131 (bottom right): Ernest Manewal/SUPERSTOCK; page 132: Paul Nicklen/NG Image Collection; page 133: RAYMOND GEHMAN/NG Image Collection; page 134: Sarah Leen/NG Image Collection; page 135: Don Johnston/Alamy; page 136: FRANS LANTING/NG Image Collection; page 137: Ken Lucas/Visuals Unlimited; page 138: Steve McCurry/NG Image Collection; page 139: TIM FITZHARRIS/MINDEN PICTURES/NG Image Collection; page 140: YVA MOMATIUK & JOHN EASTCOTT/MINDEN PICTURES/NG Image Collection; page 141: Danita Delimont/Alamy; Page 143: Kathleen Revis/NG Image Collection; page 144 (right): Frans Lanting/NG Image Collection; page 145: John Eastcott and Momatiuk/NG Image Collection; page 146 (left): RAYMOND GEHMAN/NG Image Collection; page 146 (right): TIM LAMAN/NG Image Collection; Page 146 (right): NG Maps; page 147: JAMES L. STANFIELD/NG Image Collection; page 148: BEVERLY JOUBERT/NG Image Collection; page 149 (center): MARK MOFFETT/MINDEN PICTURES/ NG Image Collection; page 149 (top left): age fotostock/ SuperStock; page 149 (center left): TIM LAMAN/NG Image Collection; page 149 (center right): HIROYA MINAKUCHI/ MINDEN PICTURES/NG Image Collection; page 149 (bottom right): TUI DE ROY/MINDEN PICTURES/NG Image Collection; page 149 (bottom left): Tierbild Okapia/Photo Researchers, Inc.; page 149 (bottom center): Eric Hosking/Photo

Researchers, Inc.; page 151: O. Louis Mazzatenta/NG Image Collection; page 152 (top): BRUCE DALE/NG Image Collection; page 152 (bottom left): Wolfgang Kaehler/Wolfgang Kaehler; page 152 (bottom center): Glenn N. Oliver/ Visuals Unlimited; page 152 (bottom right): Mira/Alamy; page 154 (center): ©AP/Wide World Photos; page 154 (top right): ©AP/ Wide World Photos; page 155: TIM FITZHARRIS/MINDEN PICTURES/NG Image Collection; page 156 (bottom): Jodi Cobb/NG Image Collection; page 157 (left): Joseph T. Collins/ Photo Researchers

Chapter 7

Page 158: Reuters/Pawan Kumar/©Corbis Images; page 158: Reuters/Pawan Kumar/©Corbis Images; page 160 (bottom left): TIM FITZHARRIS/MINDEN PICTURES/NG Image Collection; page 160 (bottom right): MEDFORD TAYLOR/NG Image Collection; page 162: CNRI/Science Photo Library/Photo Researchers; page 163: Michael Abbey/Photo Researchers; page 164: Yvona Momatiuk and John Eastcott/Photo Researchers; page 166: Karen Kasmauski/NG Image Collection; page 167: James P. Blair/NG Image Collection; page 167: NG Maps; page 168 (left): Jodi Cobb/NG Image Collection; page 168: NGS Maps; page 168 (bottom): NG Maps; page 168: NG Maps; page 169 (bottom right): NG Maps; page 171 (left): Annie Griffiths Belt/NG Image Collection; page 171 (right): Karen Kasmauski/NG Image Collection; page 171 (right): Karen Kasmauski/NG Image Collection; page 175 (left): Pablo Corral Vega/NG Image Collection; page 175 (right): James P. Blair/NG Image Collection; page 176: Jodi Cobb/NG Image Collection; Page 177 (bottom right): INTERFOTO/ Alamy; page 177 (left): NG Maps; page 178: Karen Kasmauski/ NG Image Collection; page 179: Karen Kasmauski/NG Image Collection; page 180: RAFIQUR RAHMAN/Reuters/Landov LLC; page 182: mediacolor s/Alamy; page 183: URA Resource Center; page 185 (bottom): Stuart Franklin/NG Image Collection; page 188: Lynn Johnson/NG Image Collection

Chapter 8

Page 191: Norbert Rosing/NG Image Collection; page 193 (top right): Roger Harris/Photo Researchers, Inc.; page 194 (top right): Kenneth Garrett/NG Image Collection; page 199: Sam Abell/NG Image Collection; page 200 (top): USGS; page 200 (center right): Emory Kristof/NG Image Collection; page 200 (bottom left): AFP PHOTO/Halldor Kolbeins/NewsCom; page 201 (top right): Mario Tama/Getty Images; page 204: Frank Lukasseck/©Corbis; Page 205: James P. Blair/NG Image Collection; page 207 (bottom center): John D. Cunningham/ Visuals Unlimited; page 207 (bottom right): John D. Cunningham/Visuals Unlimited; page 209: George Steinmetz/NG Image Collection; page 209 (right): NG Maps; page 211 (top left): Joerg Boethling/Peter Arnold, Inc.; page 213: Raymond Gehman/NG Image Collection; page 214 (right): Emory Kristof/NG Image Collection; page 215: Ted Spiegal/NG Image Collection; page 215 (bottom right): NGS Maps

Chapter 9

Page 217: National Snow and Ice Data Center, W. O. Field, B. F. Molnia; page 219: DIANE COOK AND LEN JENSHEL/ NG Image Collection; page 219: DIANE COOK AND LEN

JENSHEL/NG Image Collection; page 222 (right): SuperStock, Inc.; page 226 (top): Stock Trek/PhotoDisc Green/Getty; page 226 (bottom): Jamest L. Stanfield/NG Image Collection; page 226 (top right): NG Maps; page 227: Michael Nichols/NG Image Collection; page 227 (top right): NG Maps; page 228 (top left): FLPA/Alamy; page 228 (top right): Thomas R. Fletcher/Alamy; page 228 (top right): Peter Scoones/Photo Researchers, Inc.; page 228 (bottom): NG Maps; page 228 (top right): Peter Scoones/Photo Researchers, Inc.; page 232 (bottom left): NASA Media Services; page 234 (bottom left): Guillermo Aldana Espinosa/NG Image Collection; page 234 (bottom right): Peter Essick/NG Image Collection; page 236 (bottom left): Alfred Pasieka/Photo Researchers; page 236 (bottom right): Nigel Cattlin/Visuals Unlimited/Getty Images, Inc.; page 236 (center right): Al Petteway/NG Image Collection; page 237: Mike Dobel/Alamy; page 240 (top right): MARIA STENZEL/NG Image Collection

Chapter 10

Page 241: AMY TOENSING/NG Image Collection; page 243 (bottom left): Jodi Cobb/NG Image Collection; page 246 (top right): DAVID TULIS/Reuters/Landov LLC; page 247: Courtesy NASA; Page 248: NG Maps; page 249: Annie Griffiths Belt/NG Image Collection; page 250: NASA/USGS; Page 250: NG Maps; page 251 (top left): NG Maps; page 251 (bottom left): NG Maps; page 251 (top right): NG Maps; page 252 (bottom right): NG Maps; page 252 (bottom right): NG Maps; Page 253 (left): NG Maps; page 254 (bottom left): Courtesy U. S. Dept. of Energy; page 254 (bottom right): Philip James Corwin/©Corbis; page 256 (top left): RICHARD NOWITZ/NG Image Collection; page 256 (top right): Xinhua/Landov LLC; page 259 (top left): age fotostock/SuperStock; page 259 (top right): W.A. Banaszewski/Visuals Unlimited; page 260: Peter Arnold, Inc./Alamy; page 265: Carrie Garcia/Alamy; page 267 (left): Steve McCurry/NG Image Collection; page 267 (right): ©AP/Wide World Photos; page 268 (bottom): China Images/Alamy; page 268 (bottom): China Photos/Getty Images, Inc.

Chapter 11

Page 271: Brian J. Skerry/NG Image Collection; page 271 (inset): Jose Cort/Courtesy NOAA; page 274 (bottom): Courtesy NASA; page 278 (top left): Bill Curtsinger/NG Image Collection; page 278 (center left): Alex Kerstitch/Visuals Unlimited; page 278 (top right): Stuart Westmorland/Photo Researchers; page 278 (center right): NORBERT WU/MINDEN PICTURES/NG Image Collection; page 279 (top): TIM LAMAN/NG Image Collection; page 280 (left): Raul Touzon/NG Image Collection; page 280 (right): TIM LAMAN/NG Image Collection; page 281 (right): Hugh Rose/Danita Delimont; page 281 (bottom): Joe Stancampiano/NG Image Collection; page 282 (top): Dante Fenolio/Photo Researchers, Inc.; page 282 (center): E. Widder/HBOI/Alamy; page 283 (top left): Courtesy B. Halpern, K. Selkoe, C. Kappel, F. Micheli; National Center for Ecological Analysis and Synthesis; page 284: Courtesy B. Halpern, K. Selkoe, C. Kappel, F. Micheli; National Center for Ecological Analysis and Synthesis; page 285 (top left): ALASKA STOCK IMAGES/NG Image Collection; Page 286 (bottom): Brian J. Skerry/NG Image Collection; page 286 (bottom left): NGS Maps ; page 287 (top left): © 2009 Scripps Institution of Oceanography/UC San Diego; page 288 (bottom left): Reinhard Dirscherl/Alamy; page 289 (bottom): Science VU/Visuals Unlimited; page 291: FLIP NICKLIN/NG Image

Collection; page 292 (bottom): Time & Life Pictures/Getty Images, Inc.; page 294 (right): SANDRA CRITELLI/NG Image Collection

Chapter 12

Page 294: Ben Faulkner/Glenn Springs Holdings; page 295 (inset): EMORY KRISTOF/NG Image Collection; page 300 (center left): Phil Degginger/ Alamy; page 300 (top): DAVID EDWARDS/NG Image Collection; page 300 (center right): RAUL TOUZON/NG Image Collection; page 300 (bottom): TAYLOR S. KENNEDY/NG Image Collection; page 302 (top left): Mikhail Pozhenko/iStockphoto; page 302 top third: Matthew Ragen/iStockphoto; page 302 (top right): iStockphoto; Page 302 (left): Mike Clarke/iStockphoto; page 302 top second: Christoph Ermel/iStockphoto; Page 302 top third: Andrew Johnson/iStockphoto; page 302 (top right): Dave White/iStockphoto; Page 302 center left: Jack Cobben/iStockphoto; page 302 center second: Matt Meadows/Alamy; page 302 center third: Vladimir Melnik/iStockphoto; page 302 (center right): Wesley VanDinter/iStockphoto; page 302 (bottom left): Predrag Novakovic/iStockphoto; page 302 bottom second: Nikki Lowry/iStockphoto; page 302 bottom third: Don Wilkie/iStockphoto; page 302 (bottom right): Jozsef Szasz-Fabian/iStockphoto; page 303 top second: iStockphoto; page 303 (top left): JAMES L. AMOS/NG Image Collection; page 305 (right): Inga Spence/Visuals Unlimited; page 306 (bottom right): R. Ashley/Visuals Unlimited; page 307 (bottom): NewsCom; Page 308 (right): JIM RICHARDSON/NG Image Collection; page 309 (left): Photo by Ray Weil, courtesy Martin Rabenhorst; page 309 (bottom right): Courtesy U.S. Dept. of Agriculture; page 312 (left): Richard R. Hansen/Photo Researchers; page 312 (bottom right): JIM RICHARDSON/NG Image Collection; page 314 (bottom left): Courtesy U. S. Department of Agriculture; page 314 (bottom right): PAUL CHESLEY/NG Image Collection; page 315 (left): Kevin Flaming/Corbis Images; Page 315 (center left): NGS Maps; page 317 (center left): Phil Degginger/Alamy; page 318 (center right): RANDY OLSON/NG Image Collection

Chapter 13

Page 319: Cosmo Condina North America/Alamy; page 320 (inset) (top): Jane Leaman/Alamy; page 322 (bottom left): Sam Abell/NG Image Collection; page 324: Kirtley-Perkins/Visuals Unlimited; page 326 (top): Stephen Sharnoff/NG Image Collection; page 327 (center): NewsCom; page 328 (bottom right): PETER ESSICK/NG Image Collection; page 330 (center left): NRSC/Photo Researchers, Inc.; page 330 (bottom): MATTIAS KLUM/NG Image Collection; page 330 (center right): FRANS LANTING/NG Image Collection; page 332 (bottom): Steve Smith/SUPERSTOCK; page 333 (bottom): Steve McCurry/NG Image Collection; Page 333 (bottom right): NGS Maps ; page 336 (top left): Radius Images/Getty Images, Inc.; page 336 (top right): BWAC Images/Alamy; page 336 (bottom right): Jim Parkin/Alamy; page 337 (top left): Tom Murphy/NG Image Collection; page 338 (top): Rich Reid/NG Image Collection; page 338 (center left): Kevin Schafer/Alamy; page 338 (center right): James Schwabel/Alamy; page 339: JAMES P. BLAIR/NG Image Collection; page 341 (top left): Pacific Stock/SUPERSTOCK; page 341 (top left): Corbis/SUPERSTOCK; page 341 (top left): Prisma/SUPERSTOCK; page 341

Department of Energy/USGS Mineral Resources Program; page 424 (bottom left): The Geothermal Energy Association/ Barry Soloman/Michigan Technological University/Sandia National Laboratories/Windpower Monthly News Magazine; page 425 (bottom right): NG Image Collection/BP Statistical View of World Energy; page 427 (top left): NATALIE B. FOBES/ NG Image Col- lection; page 427 (bottom): Bob Hallinen/ Anchorage Daily News/MCT/NewsCom; page 430 (left): JACK FLETCHER/NG Image Collection; page 430 (right): Jayson Mellom Photography; page 433 (top): GENIA SAVILOV/ AFP/Getty Images/Getty Images, Inc.; page 433: NGS Maps; page 433 (bottom): Caroline Penn/Corbis Images; 15 Page 434 (left): Peter Essick/Aurora Photos; page 435 (bottom left): PETER ESSICK/NG Image Collection; page 436: Fedoseyev Lev/ITAR-TASS/Landov LLC; page 437 (bottom): Steven J. Kazlowski/Alamy; page 438: MELISSA FARLOW/NG Image Collection; page 439 (left): Courtesy NASA; page 439 (right): Win McNamee//Getty Images, Inc.; page 441: Tyrone Turner/ NG Image Collection; page 441: Tyrone Turner/NG Image Collection

Chapter 18

Page 444 (bottom): James P.Blair/NG Image Collection; page 445 (inset) (top): Courtesy Solar Design Associates; page 446: Courtesy of Uni-Solar. Energy Conservation Devices, Discover Magazine; inset, United Solar Systems, Discover Magazine.; page 447: Hank Morgan/Getty Images; page 448: Pasquale Sorrentino/ Photo Researchers, Inc.; page 448 (bottom): © KIMIMASA MAYAMA/©Corbis; page 449: Jim Richardson/NG Image Collection; page 449 (right): NG Maps; page 450 (top): Prof. David Hall/Photo Researchers, Inc.; Page 450 (left): NGS Maps; page 451: Mark Thiessen/NG Image Collection; Page 453 (left): Corbis/SUPERSTOCK; page 456 (left): CYRIL RUOSO/JH EDI-TORIAL/MINDEN PICTURES/NG Image Collection; page 458 (right): Courtesy Ontario Hydro; page 458: Courtesy Ontario Hydro; page 459: Iain Materton/Alamy; page 460: Greg Dale/ NG Image Collection; page 462: Nick Wood/Alamy; page 462: Emmanuel Dunand/Getty Images; Page 464 (top): Sarah Leen/ NG Image Collection; page 464: NG Image Collection; page 465: Michael Poliza/NG Image Collection;

Index

Note: f indicates figures; t indicates tables

Clear-cutting, 324, 325f, 342
Climate. *See also* Global climate change
    atmospheric changes and, 232
    precipitation and, 220–221
    shaping terrestrial biomes and, 129
    solar intensity and latitude, 220, 220f
    solar radiation and, 219–220, 219f
Clinton, Bill, 437
Closed shrublands, 130f
Coal
    acid deposition and, 420
    "clean coal" technologies and, 420
    distribution of, 419, 419f
    environmental impact of, 420, 421
    fluidized-bed combustion and,
        420–421
    making cleaner, 420–421
    mining of, 419–420
    mountaintop removal of, 420
    strip mining and, 419–420
    using scrubbers and, 420
Coastal development, oceans and,
    283f, 286
Coastal upwelling, 274, 275f
Coasts, ecosystem services and, 374t
Cobalt, 301f
Cobia, 285f
Coca Cola, 254
Coevolution, 116, 116f
Coffee plantations, bird populations
    and, 378
Cogeneration, natural gas and, 422
Cogeneration systems, 458, 458f
Colony collapse disorder, 118
Colorado
    Bear Lake, Rocky Mountain National
        Park, 143
    Great Sand Dunes National Park, 318
    Yankee Boy Basin, 160f
Colorado potato beetle, 357f
Colorado River Basin, 248, 248f
Colorado River Compact, 248
Columbia River, 253, 253f
Combined heat and power (CHP),
    458, 458f
Combustion, 107
Command and control regulation, 67
Commensalism, 116–118, 117f
Commercial harvests, 383, 384f
Communities, organization of, 98, 99f
Comoros Islands, 379
Compact development, 183
Comparative anatomy, evolution and, 150f
Competition, 120–121
Composting, 401–402, 401f
Comprehensive Environmental Response,
    Compensation, and Liability Act
    (CERCLA), 410
Comprehensive Nutrient Management
    Plans, 260
Computer equipment, obsolete, 413
Computer waste, 413
Condensers, nuclear fission and, 430

Congaree National Park, South
    Carolina, 318
Conservation biology
    captive-breeding programs, 386, 387f
    ex situ conservation, 384
    habitat protection and, 384–385, 385f
    habitat restoration and, 385–386, 386f
    land resources and, 339
    reintroducing endangered species, 386
    seed banks and, 386–388, 387f
    in situ conservation, 384
    species conservation, 386–388, 387f
    wildlife refuges, 385
Conservation easements, 333, 353
Conservation Reserve Program
    (CRP), 314
Conservation tillage, 41
Consumers, 102–103, 103f
Consumption, sustainable, 28–30
Consumption levels, 4
Consumption overpopulation, 9, 9f, 28–29
Continental shelves, oil deposits and, 423
Contour plowing, 312
Contraceptive use, fertility rates
    and, 178–180, 178f
Control groups, 18
Convection, 443
Convention on Biological Diversity, 390
Convention on International Trade in
    Endangered Species of Wild Flora
    and Fauna (CITES), 390
Copper Basin, Tennessee, 294
Copper ore tailings, 304, 395f
Coqui tree frog, 227f
Coral bleaching, 42f, 287, 287f
Coral reefs, 278, 278f
Corals, ocean warming and, 227f
Coriolis effect, 195, 195f, 272
Cost-benefit diagrams, 66–67, 67f
Costa Rica, tropical rain forest, 129
Council on Environmental Quality, 59
Country demographics
    age structure and, 172–174, 173f
    demographic transition and,
        170–172, 171f
    fertility changes and, 172–174, 172f
    gap between rich and poor, 4–6
    infant mortality rate, 170, 171f
    most populous countries, 170t
    prereproductive and elderly popula-
        tions and, 174, 174f
    replacement-level fertility and, 170
Cowles, Henry, 152
Crabs, 103f
Crater Lake, Oregon, 258
Crop biomass, 449
Crop yields, 351, 354–355, 354f
Croplands, 131f
Crude oil, 421
Cryptosporidiosis, 78t
Cryptosporidium, 77
Cultivation methods, pest control
    and, 365t

Cultural diversity, biological diversity and,
    39–40, 40f
Culture, fertility and, 175–177, 175f
Culver City, California, 137f
Curitiba, Brazil, 186
Cuyahoga River, Cleveland, Ohio, 61
Cyanide heap leaching, 305
Cyanobacteria, 110, *110f*

D

Daihatsu prototype UFE III hybrid, 458f
Dalles Dam, Columbia River, 452f
Dams, 253, 253f
Darwin, Charles, 147, 147f
Darwinism. *See* Natural selection
Darwin's finches, 149f
Data collection, environmental science
    and, 17, 17f
DDT (dichlorodiphenyltrichloroethane),
    81, 81f, 84t
DDT spraying, red scale infestation
    and, 364
Dead zone in Gulf of Mexico, 291
Death, probability by selected causes,
    74, 74f
Deciduous broad-leaved forests, 130f
Deciduous needle-leaved forests, 130f
Decommissioning, nuclear power plants
    and, 436
Decomposers, 102–103, 103f
Deep ecology worldview, 32–34, 34f
Deep-well injection, 412
Defenders of Wildlife, 339
Deforestation
    agricultural causes and, 329f
    boreal forests and, 327, 327f
    fuelwood and, 329f
    rangelands and, 331–332, 332f
    results of, 327
    slash-and-burn agriculture, 328
    tropical forests and, 328–329, 329f, 377
    world map and, 14f
Degraded environments, poor neighbor-
    hoods and, 35, 35f
Deltas, 144f
Dematerialization, 403
Demographic transition, 170–172, 171f
Demographics. *See* Country demographics
Denitrification, 110
Denmark
    Kalundborg, 315
    wind energy and, 450
Density, seawater, 272
Derelict land restoration, 306
Desert, 141, 141f
Desertification, 14f
Detritus, 103
Devall, Bill, 32
Developing countries. *See also* Less
        developed countries
    air pollution and, 208–209, 208f
    child labor and, 176, 176f

Reproductive controls, pest control and, 365t
Reproductive success, differential, 148, 149f
Reservoirs, 253, 253f
Resource Conservation and Recovery Act (RCRA), 410
Resources
    conservation and preservation of, 50, 50f
    degradation and pollution, 62–63
    national income accounting and, 65
    nonrenewable, 7
    partitioning and, 114–115, 115f
    population size and consumption of, 9–11
    renewable, 7
    types of, 7
Respiratory disease in children, 89, 89f
Restoration ecology, 385–386, 386f
Rhine River Basin, 252, 252f
Rhinoceros populations, 383
Rhizobium, 110
Rice, high-yield varieties, 355f
Richter scale, 298
Ring of Fire, 297
Risk analysis, environmental problems and, 21, 21f
Risk assessment, 75–76, 75f
Risk characterization, 75f
Risk management, 74–75
Rivers, features of, 142–144, 144f
Roadless Area Conservation Rule, 342
Rock cycle, 298–299, 299f
Rocky shores, zonation along, 277f
Rodenticides, 362
Roosevelt, Franklin, 38, 54
Roosevelt, Theodore, 52, 52f, 385
Roraima State, Brazil, 40f
Rosy periwinkle, 375f
Runoff, 108, 307
Russia
    Kamchatka, 341f
    radioactive waste disposal and, 436

S

Safe Drinking Water Act, 92, 264
Salinization
    impact and, 356
    irrigated soil and, 248–249
    plant poisoning and, 311, 311f
Sally Lightfoot crabs, 103f
Salmon River, Idaho, 331f
Salt marshes, 144f, 146–147
Saltwater intrusion, 229, 246
San Simeon, California, 311f
Sand County Almanac (Leopold), 34
Sand dunes, 152f
Sanitary landfills, 398–399
Sarek National Park, Sweden, 341f
Slash-and-burn agriculture, 328
Saskatchewan, Canada, 133f

Savannas, 140, 140f
Schistosomiasis, 78t, 453
Science as a process, 17–20
Scientific assessment, environmental problems and, 21, 21f
Scientific method, 18, 19f
Sea grasses, 278–279, 279f
Sea-level rise, 225, 288
Sea stars, 99f
Seasons
    progression of, 202f
    temperature changes and, 220, 220f
Seattle, Washington, 29f
Second green revolution, 359
Second law of thermodynamics, 102
Secondary consumers, 103
Secondary municipal sewage treatment, 263
Secondary sludge, 264
Secondary succession, 153, 153f
Sediment pollution, 257t
Sedimentary rocks, 299, 299f
Sedona, Arizona, 141f
Seed banks, 386–388, 387f
Seed tree cutting, 324, 325f
Seismic waves, 297–298
Selective cutting, 324, 325f
Semiarid lands, 245
Sessions, George, 32
Sewage, 256, 256f
Sewage-contaminated water, 77
Sewage sludge, biomass energy and, 448
Shade-intolerant plants, 154
Shelterbelts, 314, 314f
Shelterwood cutting, 324, 325f
Shifting cultivation, 352
Shipping, ocean dumping and, 286, 286f
Short-grass prairies, 139
Shrewsbury, England, 20f
Sick building syndrome, 209–210
Sierra Club, 53, 318
Silent Spring (Carson), 54–55
Silicon, 301f
Silver, 301f
Simon, Julian, 55
Sitka spruce, 235f
Slag, 303
Slash-and-burn agriculture, 352
Slope mines, 303, 303f
Sludge, secondary, 264
Smelting, 303
Smith, Robert Angus, 233
Smithsonian Migratory Bird Center (SMBC), 378
Smog, 201
Smoking, risks and, 87–88
Smoky Mountains National Park, North Carolina, 65f
Snail species, 373f
Social behavior, animal prey and, 119–120
Socolow, Robert, 122
Soil. *See also* Agriculture
    conservation, 312–314, 313f
    conservation tillage and, 312, 313f

contour plowing and, 312
crop rotation and, 312, 313f
degradation and, 312, 313f
erosion and, 311, 311f
formation and composition of, 307–309
humus and, 308–309, 308f
leaching and, 309
organisms and, 309–310, 309f
pollution and, 311, 311f
profile of, 308
reclamation of, 312–314
salinization and, 311, 311f
soil horizons, 309
strip cropping and, 312
terracing and, 312, 312f
Solar energy. *See* Direct solar energy; Indirect solar energy
Solar-generated hydrogen, 446–447, 447f
Solar intensity and latitude, 220, 220f
Solar panels, 37f
Solar radiation, climate and, 219–220, 219f
Solar shingles, 445, 445f
Solar thermal electric generation, 446, 446f
Solid waste
    composting and, 401–402, 401f
    dematerialization and, 403
    disposal of, 397–402, 397f
    incineration and, 398–401, 401f
    integrated waste management and, 406, 406f
    municipal solid waste, 396–397, 397f
    open dumps and, 397
    plastics and, 398
    recycling materials and, 403–406, 405f
    reducing, 402–406
    reusing products and, 403
    sanitary landfills and, 398–399
    source reduction and, 403
    types of, 396–397
Songbird populations, 378
Source reduction, waste management and, 403, 411
South Asia, chronic hunger and malnutrition in, 348–350, 349f
South Bronx, New York City, 90f
South Carolina, Congaree National Park, 318
Species, endangered. *See* Endangered species
Species conservation, 386–388, 387f
Species richness, 372–374, 372f
Spent fuel, 432
Spiritual value, organisms and, 375
Spoil bank, 302
Squatter settlements, 43, 43f
Stabilization wedges, 122
Standing-water ecosystems, 142
Starfish, 99f
Stationary sources, air pollution and, 200, 200f
Steam generator, nuclear fission and, 430
Stockholm Convention on Persistent Organic Pollutants, 84